Lectures in Applied Mathematics

Proceedings of the Summer Seminar, Boulder, Colorado, 1960

VOLUME 1 LECTURES IN STATISTICAL MECHANICS
G. E. Uhlenbeck and G. W. Ford with E. W. Montroll

VOLUME 2 MATHEMATICAL PROBLEMS OF RELATIVISTIC PHYSICS
I. E. Segal with G. W. Mackey

VOLUME 3 PERTURBATION OF SPECTRA IN HILBERT SPACE
K. O. Friedrichs

VOLUME 4 QUANTUM MECHANICS
R. Jost

Proceedings of the Summer Seminar, Ithaca, New York, 1963

VOLUME 5 SPACE MATHEMATICS, PART 1
J. Barkley Rosser, Editor

VOLUME 6 SPACE MATHEMATICS, PART 2
J. Barkley Rosser, Editor

VOLUME 7 SPACE MATHEMATICS, PART 3
J. Barkley Rosser, Editor

Proceedings of the Summer Seminar, Ithaca, New York, 1965

VOLUME 8 RELATIVITY THEORY AND ASTROPHYSICS
1. RELATIVITY AND COSMOLOGY
Jürgen Ehlers, Editor

VOLUME 9 RELATIVITY THEORY AND ASTROPHYSICS
2. GALACTIC STRUCTURE
Jürgen Ehlers, Editor

VOLUME 10 RELATIVITY THEORY AND ASTROPHYSICS
3. STELLAR STRUCTURE
Jürgen Ehlers, Editor

Proceedings of the Summer Seminar, Stanford, California, 1967

VOLUME 11 MATHEMATICS OF THE DECISION SCIENCES, PART 1
George B. Dantzig and Arthur F. Veinott, Jr., Editors

VOLUME 12 MATHEMATICS OF THE DECISION SCIENCES, PART 2
George B. Dantzig and Arthur F. Veinott, Jr., Editors

Proceedings of the Summer Seminar, Troy, New York, 1970

VOLUME 13 MATHEMATICAL PROBLEMS IN THE GEOPHYSICAL SCIENCES
1. GEOPHYSICAL FLUID DYNAMICS
William H. Reid, Editor

VOLUME 14 MATHEMATICAL PROBLEMS IN THE GEOPHYSICAL SCIENCES
2. INVERSE PROBLEMS, DYNAMO THEORY, AND TIDES
William H. Reid, Editor

Mathematical Problems in the Geophysical Sciences

2. Inverse Problems, Dynamo Theory, and Tides

William H. Reid, Editor

The University of Chicago

LECTURES IN APPLIED MATHEMATICS, VOLUME 14

AMERICAN MATHEMATICAL SOCIETY, PROVIDENCE, RHODE ISLAND, 1971

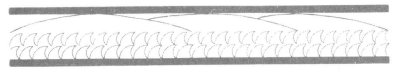

Prepared by the American Mathematical Society under Contract N00014-69-C-0381 with the Office of Naval Research, Grant SSF(70)-16 of the New York State Science and Technology Foundation, and Grant GZ-1509 of the National Science Foundation

International Standard Book Number 0-8218-1114-2
Library of Congress Catalog Number 62-21481
AMS 1970 Primary Subject Classification 86-02

Printed in the United States of America

Contents

Preface

The Sixth Summer Seminar on Applied Mathematics, sponsored jointly by the American Mathematical Society and the Society for Industrial and Applied Mathematics, was held at the Rensselaer Polytechnic Institute from July 6 to 31, 1970.

The seminar program was intended to provide an opportunity for graduate students and recent recipients of the Ph.D. degree to become familiar with current developments in a number of areas of the geophysical sciences in which applied mathematics plays a central role. In the geophysical sciences at the present time, there is an important interplay between the observational and experimental work on the one hand and the mathematical developments on the other, and this aspect of the subject is clearly evident in many of the papers which appear in these proceedings.

The program for the seminar was organized by a committee which included Hirsh G. Cohen (I.B.M., T. J. Watson Research Center), Richard C. DiPrima (Rensselaer Polytechnic Institute), Dave Fultz (The University of Chicago), C. C. Lin (Massachusetts Institute of Technology), and William H. Reid.

Thanks are due to the Office of Naval Research, the National Science Foundation, and the New York State Science and Technology Foundation for their financial support of the seminar. I am also grateful to Dr. Gordon L. Walker, executive director of the American Mathematical Society, and Mrs. Lillian R. Casey for their generous help in the planning and administration of the seminar; to Professor Lester A. Rubenfeld for his efficient handling of the local arrangements; and to Mrs. Mary A. Coccoli for her devoted work in the day-to-day operation of the conference office.

I also wish to thank the speakers for preparing their lecture notes in advance so that they could be distributed at the time of the seminar and thus permit the rapid publication of these proceedings.

WILLIAM H. REID, EDITOR
Departments of Mathematics and
the Geophysical Sciences
The University of Chicago

Inference from Inadequate and Inaccurate Data

George Backus

I. Introduction

A. **A qualitative variety of scientific inference.** Many quantitative problems in scientific inference can be described as follows: we have measured D numerical properties of a physical object E which requires many more than D parameters for its complete specification. To make these measurements we have performed a number (not necessarily D) of experiments, some of which give us values for more than one of the numerical properties of E. The outcome of these experiments is D real numbers $\gamma_1, \ldots, \gamma_D$, together with an estimate of the statistical distribution of the D-tuple of experimental errors $(\delta\gamma_1, \ldots, \delta\gamma_D)$. From these data we want to predict the values $\tilde{\gamma}_1, \ldots, \tilde{\gamma}_P$ of P more numerical properties of E and to estimate the errors of our predictions.

We will suppose that, as far as the experiments are concerned which measure the D data and the P predictions, the physical object E is adequately described by one unknown member m_E of a known manifold $\mathfrak{M}$ of possible models of E. The construction of $\mathfrak{M}$ is, of course, the creative part of the scientific inference, and we do not pretend to discuss the very difficult questions involved in inventing $\mathfrak{M}$. Our discussion begins with $\mathfrak{M}$ given and the D data measured, along with the joint distribution of their experimental errors. If dim $\mathfrak{M}$, the dimension of $\mathfrak{M}$, is D or less, then making the predictions $\tilde{\gamma}_1, \ldots, \tilde{\gamma}_P$ is an exercise in conventional statistical inference (treated perhaps by least squares or maximum likelihood techniques), which presents only computational difficulties. If dim $\mathfrak{M} > D$ there is a conceptual difficulty as well.

AMS 1970 *subject classifications*. Primary 62–02, 62M20, 46C05, 46C10, 60B05, 46E20, 86–02, 86A30; Secondary 62F15, 86A15.

The example which stimulated the present work is a geophysical inverse problem [Backus and Gilbert, 1968 and 1970]. The object E is the earth. The measured data are the total mass and moment of the earth, the travel times of seismic waves from certain sources to certain receivers, and, for certain seismograms of finite length, those Fourier components which are above the noise level. Such a seismogram yields the periods of a finite number of the earth's elastic-gravitational normal modes of oscillation [Benioff et al., 1961], [Ness et al., 1961], [Alsop et al., 1961] and the amplitudes and phases to which these modes were excited by the source. If the amplitudes and phases are ignored, then the manifold $\mathfrak{M}$ of relevant models m consists of all ordered triples $m = (\rho, \kappa, \mu)$ of real valued functions defined and square-integrable inside the earth. Here $\rho(r)$, $\kappa(r)$, and $\mu(r)$ are the density, bulk modulus, and shear modulus at position r in the earth. The properties $\tilde{\gamma}_1, \ldots, \tilde{\gamma}_P$ which we want to predict from the data might be the periods of P not-yet-observed normal modes, or the value of the density or seismic velocity at P different locations in the earth. For example, detection of a low-velocity zone requires values of the velocity at three depths. The manifold just described reflects our expectation that other properties of E, such as anisotropy, will influence neither our data $\gamma_1, \ldots, \gamma_D$ nor our predictions $\tilde{\gamma}_1, \ldots, \tilde{\gamma}_P$. If we find that no member of $\mathfrak{M}$ will explain our data, we may have to admit anisotropy [Morris et al., 1969] in our models, which will then include not two but 21 elastic coefficients as functions of position in the earth.

In the geophysical example, $\mathfrak{M}$ is infinite dimensional while *all* the experimental information about the earth at any one epoch consists of a finite number of data. A similar infinite insufficiency of data occurs, for example, in the interpretation of electron scattering by a nucleus and of spectral emission by the top of a star's atmosphere.

The unknown model in $\mathfrak{M}$ which represents E is m_E. We suppose that if we knew m_E we could correctly calculate the outcomes of the $D + P$ measurements in which we are interested. Our methods of calculation are $D + P$ rules which assign real numbers $g_1(m), \ldots, g_D(m)$, $\tilde{g}_1(m), \ldots, \tilde{g}_P(m)$ to every model m in some open subset of $\mathfrak{M}$. These rules are functionals (real valued functions) on $\mathfrak{M}$. Our data are summarized by the statement that to a certain accuracy m_E satisfies the D equations $g_i(m_E) = \gamma_i, i = 1, \ldots, D$. From the measured data γ_i and the known functionals $g_i : \mathfrak{M} \to R$ (R is the real line) we want to learn enough about the unknown m_E to estimate $\tilde{g}_k(m_E)$, $k = 1, \ldots, P$.

B. **Problems of nonlinearity.** To date I have not studied in any generality manifolds $\mathfrak{M}$ which are not linear spaces, so throughout these lectures I shall assume that $\mathfrak{M}$ is a real linear space. In the problems I have encountered so far, either $\mathfrak{M}$ was a Hilbert space in an obvious norm or there was a not-so-obvious way of introducing an inner product on some subspace of $\mathfrak{M}$ which contained m_E. This subspace could then be completed to a Hilbert space $\mathfrak{H}$. Later we will discuss some examples of this "Hilbertizing" process, so until further notice we will assume that the space of models is a Hilbert space, which we call $\mathfrak{H}$ instead of $\mathfrak{M}$. We will denote by h_E instead of m_E the model which correctly represents the real object E.

If all the data functionals $g_i : \mathfrak{H} \to R$, $i = 1, \ldots, D$, and all the prediction functionals $\tilde{g}_k : \mathfrak{H} \to R$, $k = 1, \ldots, P$, are bounded linear functionals on $\mathfrak{H}$, we will be able to propose a complete solution to the problem of estimating $\tilde{g}_1(h_E), \ldots, \tilde{g}_P(h_E)$ from $g_1(h_E), \ldots, g_D(h_E)$. This solution is the subject of §II. Even if the g_i and $\tilde{g}_k$ are unbounded, we can usually alter the inner product on $\mathfrak{H}$ so as to make them bounded. §II.D contains examples of such "quelling" of unbounded linear functionals.

If one of $g_1, \ldots, g_D, \tilde{g}_1, \ldots, \tilde{g}_P$ is nonlinear, there are questions of existence and uniqueness of h_E which are at present unanswered. Backus and Gilbert [1967] have used a numerical version of Newton's method in Hilbert space to construct one solution h_0 of the D nonlinear equations $g_i(h) = \gamma_i$, $i = 1, \ldots, D$, for some seismological problems in which $\mathfrak{H}$ is infinite dimensional. Their method, described in §III, was designed to overcome the unsurprising fact that under suitable smoothness conditions on g_i the set $\mathfrak{A}$ of all solutions of these D equations (i.e. the set of all models h which satisfy the data and might be h_E) is either empty or an infinite-dimensional manifold. No general method is known for discovering the shape and extent of the manifold $\mathfrak{A}$ of acceptable models.

If we assume that all the g_i and $\tilde{g}_k$ are Fréchet differentiable, then the linear techniques of §II can be used to study that part of $\mathfrak{A}$ which lies close to h_0. In some particular problems, a few of which are described in §IV, certain infinite collections of data have been shown to admit only a single model, which must then be h_E. In one seismic problem [Gerver, 1970] in which $\mathfrak{H}$ is a function space, an infinite collection of data has been shown to determine h_E if h_E satisfies certain smoothness conditions, and furthermore it has been shown that there exist finite subcollections of data which restrict that part of $\mathfrak{A}$ consisting of suitably smooth functions to lie in an arbitrarily small sphere around h_E. Gerver's estimates seem to be very unsharp, and, to my knowledge, so far no one has succeeded in computing rigorous numerically useful bounds on $\tilde{g}_k(h_E)$ from $g_i(h_E)$ in a nonlinear geophysical inverse problem.

II. Linear Inference

A. Mathematical preliminaries.

1. *Notation for pre-Hilbert and Hilbert spaces.* In most of §II we will assume that the space of models is a real Hilbert space and that the data

functionals and prediction functionals are bounded and linear. By a pre-Hilbert space we mean simply any real vector space endowed with a real, bilinear, positive-definite inner product. A Hilbert space is complete, but a pre-Hilbert space need not be.

We will suppose that the reader is familiar with the Schwarz and triangle inequalities, the norm of a bounded linear transformation, the closed graph theorem, orthogonal complementation and projection, the fact that every bounded linear functional on a Hilbert space is obtainable by taking inner products with a fixed vector in that space, and the construction of adjoints of bounded linear operators on Hilbert spaces. Most of these matters are discussed, for example, in Halmos [1951], Riesz and Nagy [1955], or Lorch [1962]. Our point of view will be closest to that of Lorch.

If f is a function, or mapping, which assigns to every element s of some set S an element $f(s)$ in a set T, then we write $f: S \to T$, which we read as "f maps S into T." We will also often read "$f:S \to T$" as a substantive, "the function f, which maps S into T." If $f:S \to T$ and $R \subseteqq S$, then the function $\tilde{f}:R \to T$, defined by requiring $\tilde{f}(r) = f(r)$ for every r in R, is called the restriction of f to R and is written $f|R$. The set whose members are $s_1, \ldots, s_n$ will be written $\{s_1, \ldots, s_n\}$. If $P(s)$ is any statement about an arbitrary object s, then $\{s: s \text{ in } S \text{ and } P(s)\}$ will denote the set of all elements s of the set S for which $P(s)$ is true. Sometimes we will write for this set simply $\{s:P(s)\}$ if S is clear from the context.

Old German capitals will stand for closed subspaces of real Hilbert spaces, a capital script letter will denote the orthogonal projection operator from the whole Hilbert space onto the subspace named by the same letter in old German, lower case italics will denote vectors in real Hilbert spaces, and real numbers. The inner product of two vectors f and g will be written both as $\langle f, g \rangle$ and as $f \cdot g$, and $\|f\|$ will stand for $\langle f, f \rangle^{1/2}$. If $V: \mathfrak{H} \to \mathfrak{H}$ is a bounded linear operator sending Hilbert space $\mathfrak{H}$ into itself, and h is in $\mathfrak{H}$, the vector $V(h)$ will be written $V \cdot h$. The norm of V will be written $\|V\|$; thus $\|V\| = \sup\{\|V \cdot h\| : h \text{ in } \mathfrak{H} \text{ and } \|h\| = 1\}$.

If $V: \mathfrak{H} \to \mathfrak{H}$ is a bounded linear operator, we can define a mapping $\tilde{V}: \mathfrak{H} \times \mathfrak{H} \to R$ which assigns to any ordered pair (f, g) of vectors in $\mathfrak{H}$ a real number, namely $\tilde{V}(f, g) = \langle f, V \cdot g \rangle = f \cdot (V \cdot g)$. The number $\tilde{V}(f, g)$ depends linearly on f for each fixed g and linearly on g for each fixed f, whence $\tilde{V}$ is a bilinear functional on $\mathfrak{H} \times \mathfrak{H}$. In addition, $\tilde{V}$ is bounded, in the sense that there is a real M such that for each f and g in $\mathfrak{H}$ we have $|\tilde{V}(f, g)| \leqq M\|f\| \|g\|$. One M which works (in fact the smallest) is $\|V\|$. Any bounded bilinear functional on $\mathfrak{H} \times \mathfrak{H}$ is called a second-order tensor over $\mathfrak{H}$. Every bounded linear operator on $\mathfrak{H}$

determines a unique second-order tensor over $\mathfrak{H}$ and vice-versa. We will sometimes think of linear operators as second-order tensors.

If V^* is the adjoint of the bounded linear operator V, then $\langle f, V \cdot g\rangle = \langle V^* \cdot f, g\rangle = \langle g, V^* \cdot f\rangle$, so $f \cdot (V \cdot g) = g \cdot (V^* \cdot f)$, an equation whose validity for all f and g in $\mathfrak{H}$ determines V^*. We will sometimes write $V^* \cdot f$ as $f \cdot V$. Then we have $f \cdot (V \cdot g) = (V^* \cdot f) \cdot g = (f \cdot V) \cdot g$, so following Dirac and Gibbs, we can write $f \cdot V \cdot g$ without ambiguity. Clearly $f \cdot V \cdot g = g \cdot V^* \cdot f$.

If h and k are fixed vectors in $\mathfrak{H}$, the linear operator which assigns to any vector g in $\mathfrak{H}$ the vector $h(k \cdot g)$ will be denoted by hk. Thus $(hk) \cdot g = h(k \cdot g)$. The adjoint of hk is evidently kh. The operator hk and its tensor are "dyadics."

If $V:\mathfrak{H} \to \mathfrak{H}$ and $W:\mathfrak{H} \to \mathfrak{H}$ are bounded linear operators, then $V \cdot W$ will denote the composition of V with W, so that $(V \cdot W) \cdot h = V \cdot (W \cdot h)$ and $\|V \cdot W\| \leqq \|V\| \, \|W\|$. If S is any subset of $\mathfrak{H}$, then $V \cdot S$ will denote the set of all vectors in $\mathfrak{H}$ which have the form $V \cdot s$ for some s in S.

If $\mathfrak{R}$ and $\mathfrak{S}$ are closed subspaces of Hilbert space $\mathfrak{H}$ and $\mathfrak{S} \subseteqq \mathfrak{R}$ then $\mathfrak{R} \ominus \mathfrak{S}$ will denote the set of all vectors in $\mathfrak{R}$ which are orthogonal to every vector in $\mathfrak{S}$. The space $\mathfrak{H} \ominus \mathfrak{R}$ will be written $\mathfrak{R}^\perp$, and $\mathscr{R}:\mathfrak{H} \to \mathfrak{R}$ and $\mathscr{R}^\perp:\mathfrak{H} \to \mathfrak{R}^\perp$ will denote the orthogonal projection operators from $\mathfrak{H}$ onto $\mathfrak{R}$ and $\mathfrak{R}^\perp$. The identity operator on $\mathfrak{H}$ is written $\mathscr{H}$.

DEFINITION 1. Let $V:\mathfrak{H} \to \mathfrak{H}$ be a bounded linear operator on Hilbert space $\mathfrak{H}$, and let $\mathfrak{R}$ be a closed subspace of $\mathfrak{H}$. Suppose that $V \cdot \mathfrak{R} \subseteqq \mathfrak{R}$, and $V \cdot \mathfrak{R}^\perp = \{0\}$, and $(V|\mathfrak{R})^{-1}:\mathfrak{R} \to \mathfrak{R}$ exists and is bounded. Then we will say that V lives on $\mathfrak{R}$, and we will define $V^{\text{inv}}:\mathfrak{H} \to \mathfrak{H}$ as $V^{\text{inv}} = (V|\mathfrak{R})^{-1} \cdot \mathscr{R}$.

Note that V^{inv} is the only linear mapping $U:\mathfrak{H} \to \mathfrak{H}$ such that $U \cdot V = V \cdot U = \mathscr{R}$ and $U \cdot \mathscr{R}^\perp = 0$. Clearly V^{inv} has these properties, and if U does then $U \cdot V \cdot V^{\text{inv}} = \mathscr{R} \cdot V^{\text{inv}}$, or $U \cdot \mathscr{R} = V^{\text{inv}}$. Hence $U = U \cdot (\mathscr{R} + \mathscr{R}^\perp) = V^{\text{inv}}$. The operator V^{inv} is the generalized inverse of V in the sense of Moore [1920] and Penrose [1955].

If V lives on $\mathfrak{R}$, then so does V^{inv}, and $(V^{\text{inv}})^{\text{inv}} = V$. If V and W live on $\mathfrak{R}$ then so does $V \cdot W$, and $(V \cdot W)^{\text{inv}} = W^{\text{inv}} \cdot V^{\text{inv}}$. Finally, if V lives on $\mathfrak{R}$ then so does V^*, and $(V^*)^{\text{inv}} = (V^{\text{inv}})^*$. To see this, note that the requirements $V \cdot \mathfrak{R} \subseteqq \mathfrak{R}$ and $V \cdot \mathfrak{R}^\perp = \{0\}$ are equivalent to the equations $\mathscr{R} \cdot V = V$ and $V \cdot \mathscr{R}^\perp = 0$, which in turn are equivalent to the equations $V = \mathscr{R} \cdot V = V \cdot \mathscr{R}$. By taking the adjoints of these equations we see that if V satisfies them so does V^*. It remains to show that $(V^*|\mathfrak{R})^{-1}$ exists and is bounded. For any vectors r_1 and r_2 in $\mathfrak{R}$ we have $\langle (V|\mathfrak{R})^* \cdot r_1, r_2\rangle = \langle r_1, (V|\mathfrak{R}) \cdot r_2\rangle = \langle r_1, V \cdot r_2\rangle = \langle V^* \cdot r_1, r_2\rangle = \langle (V^*|\mathfrak{R}) \cdot r_1, r_2\rangle$. Since $(V^*|\mathfrak{R}):\mathfrak{R} \to \mathfrak{R}$, it follows that $V^*|\mathfrak{R} = (V|\mathfrak{R})^*$, where the second adjoint is taken over the

Hilbert space $\mathfrak{R}$. But $(V|\mathfrak{R})^{-1}$ exists and is bounded, so[1] $[(V|\mathfrak{R})^*]^{-1}$ exists, is bounded, and equals $[(V|\mathfrak{R})^{-1}]^*$. Hence V^* lives on $\mathfrak{R}$ and $(V^*)^{\text{inv}} = (V^{\text{inv}})^*$.

2. Oblique subspaces.

(a) *The angle between two subspaces.* If $\mathfrak{S}$ and $\mathfrak{T}$ are two subspaces of $\mathfrak{H}$, $\mathfrak{S} + \mathfrak{T}$ will denote the subspace consisting of all vectors $s + t$ with s in $\mathfrak{S}$ and t in $\mathfrak{T}$. If $\mathfrak{S}$ and $\mathfrak{T}$ are mutually orthogonal, $\mathfrak{S} + \mathfrak{T}$ will be written $\mathfrak{S} \oplus \mathfrak{T}$. The angle between $\mathfrak{S}$ and $\mathfrak{T}$, $\angle(\mathfrak{S}, \mathfrak{T})$, will be defined as the infimum (greatest lower bound) of the positive angles between pairs of vectors (s, t) with s in $\mathfrak{S}$ and t in $\mathfrak{T}$. The angle between 0 and any vector will be taken as $\pi/2$. Then $\angle(\mathfrak{S}, \mathfrak{T})$ is that angle θ between 0 and $\pi/2$ such that $\cos\theta = \sup\{s \cdot t : s \text{ in } \mathfrak{S}, t \text{ in } \mathfrak{T}, \text{ and } \|s\| = \|t\| = 1\}$. Clearly $\angle(\mathfrak{S}, \mathfrak{T}) = \angle(\mathfrak{T}, \mathfrak{S})$. If $\mathfrak{S} \cap \mathfrak{T} \neq \{0\}$ then $\angle(\mathfrak{S}, \mathfrak{T}) = 0$, but the converse is false [Lorch, 1939]. If $\mathfrak{S}$ and $\mathfrak{T}$ are orthogonal, $\angle(\mathfrak{S}, \mathfrak{T}) = \pi/2$ and conversely.

(b) *Parallel projections.* Suppose that $\mathfrak{S}$ and $\mathfrak{T}$ are two subspaces of Hilbert $\mathfrak{H}$ such that $\mathfrak{S} \cap \mathfrak{T} = \{0\}$. Then we can define two linear mappings, $\mathscr{S}_{\mathscr{T}} : \mathfrak{S} + \mathfrak{T} \to \mathfrak{S}$ and $\mathscr{T}_{\mathscr{S}} : \mathfrak{S} + \mathfrak{T} \to \mathfrak{T}$, by requiring that for any s in $\mathfrak{S}$ and any t in $\mathfrak{T}$ we have $\mathscr{S}_{\mathscr{T}} \cdot (s + t) = s$ and $\mathscr{T}_{\mathscr{S}} \cdot (s + t) = t$. We will call $\mathscr{S}_{\mathscr{T}}$ the projection of $\mathfrak{S} + \mathfrak{T}$ onto $\mathfrak{S}$ parallel to $\mathfrak{T}$. Evidently $\mathscr{S}_{\mathscr{T}} + \mathscr{T}_{\mathscr{S}}$ is the identity operator on $\mathfrak{S} + \mathfrak{T}$.

Lorch [1939] has proved the following lemma:

LEMMA 2. *If $\mathfrak{S}$ and $\mathfrak{T}$ are closed subspaces of Hilbert space $\mathfrak{H}$ and $\mathfrak{S} \cap \mathfrak{T} = \{0\}$, then $\mathscr{S}_{\mathscr{T}} : \mathfrak{S} + \mathfrak{T} \to \mathfrak{S}$ is a closed operator.*

PROOF. Let h_n be a sequence of vectors in $\mathfrak{S} + \mathfrak{T}$ such that as $n \to \infty$, $h_n \to h$ and $\mathscr{S}_{\mathscr{T}} \cdot h_n \to s$ for some vectors h and s in $\mathfrak{H}$. We want to show that h is in $\mathfrak{S} + \mathfrak{T}$ and that $\mathscr{S}_{\mathscr{T}} \cdot h = s$. We can write $h_n = s_n + t_n$ with s_n in $\mathfrak{S}$ and t_n in $\mathfrak{T}$. Then $s_n = \mathscr{S}_{\mathscr{T}} \cdot h_n$ so $s_n \to s$. Since $\mathfrak{S}$ is closed, s is in $\mathfrak{S}$. Moreover, $t_n = h_n - s_n \to h - s = t$, and since $\mathfrak{T}$ is also closed, t is in $\mathfrak{T}$. Then $h = s + t$ is in $\mathfrak{S} + \mathfrak{T}$ and $\mathscr{S}_{\mathscr{T}} \cdot h = s$, as claimed.

Using the closed graph theorem and Lemma 2 we can now prove a theorem most of which is due to Lorch [1939].

THEOREM 3. *If $\mathfrak{S}$ and $\mathfrak{T}$ are closed subspaces of Hilbert space $\mathfrak{H}$ such that $\mathfrak{S} \cap \mathfrak{T} = \{0\}$, then the following statements are equivalent:*

(i) *$\mathfrak{S} + \mathfrak{T}$ is closed;*

(ii) *$\mathscr{S}_{\mathscr{T}}$ is bounded;*

(iii) $\angle(\mathfrak{S}, \mathfrak{T}) > 0$.

---

[1] Here we are using a simple fact. If $W : \mathfrak{R} \to \mathfrak{R}$ has a bounded inverse, so does W^*, and $(W^*)^{-1} = (W^{-1})^*$. To see this let $U = W^{-1}$. Then $U \cdot W = W \cdot U = \mathscr{R}$, so, taking adjoints, $W^* \cdot U^* = U^* \cdot W^* = \mathscr{R}$. Hence W^* has an inverse, and it is U^*. But $\|U^*\| = \|U\|$, so U^* is bounded.

Moreover, if these statements are true, $\|\mathscr{S}_{\mathscr{T}}\| = \operatorname{cosec} \angle(\mathfrak{S}, \mathfrak{T})$. *Finally, these statements are true if either* $\mathfrak{S}$ *or* $\mathfrak{T}$ *is finite dimensional.*

PROOF. (i) ⇒ (ii). If $\mathfrak{S} + \mathfrak{T}$ is closed then it is a Hilbert space, and the closed operator $\mathscr{S}_{\mathscr{T}}$ is defined and linear on all of it. Therefore, according to the closed graph theorem, $\mathscr{S}_{\mathscr{T}}$ is bounded.

(ii) ⇒ (i). Let $\{s_n\}$ and $\{t_n\}$ be sequences of vectors from $\mathfrak{S}$ and $\mathfrak{T}$ such that as $n \to \infty$, $s_n + t_n \to h$. We want to prove h in $\mathfrak{S} + \mathfrak{T}$. We have $s_n = \mathscr{S}_{\mathscr{T}} \cdot (s_n + t_n)$, and since $\{s_n + t_n\}$ is a Cauchy sequence and $\mathscr{S}_{\mathscr{T}}$ is bounded, $\{s_n\}$ is also a Cauchy sequence. Therefore $s_n \to s$ for some s in $\mathfrak{H}$, and since $\mathfrak{S}$ is closed, s is in $\mathfrak{S}$. Then $t_n \to t = h - s$, and since $\mathfrak{T}$ is closed, t is in $\mathfrak{T}$. Thus $h = s + t$ is in $\mathfrak{S} + \mathfrak{T}$, so $\mathfrak{S} + \mathfrak{T}$ is closed.

(ii) ⇒ (iii). The boundedness of $\mathscr{S}_{\mathscr{T}}$ implies that for any s in $\mathfrak{S}$ and t in $\mathfrak{T}$, $\|s\| \leqq \|\mathscr{S}_{\mathscr{T}}\| \, \|s - t\|$. If we square this inequality we have $\|s\|^2 \leqq \|\mathscr{S}_{\mathscr{T}}\|^2[\|s\|^2 - 2(s, t) + \|t\|^2]$. If $\|s\| = \|t\| = 1$, this implies $(s, t) \leqq 1 - \frac{1}{2}\|\mathscr{S}_{\mathscr{T}}\|^{-2} < 1$. Thus $\cos \angle(\mathfrak{S}, \mathfrak{T}) < 1$ and $\angle(\mathfrak{S}, \mathfrak{T}) > 0$.

(iii) ⇒ (ii). For any s and t in $\mathfrak{S}$ and $\mathfrak{T}$, $\|s - t\|^2 = \|s\|^2 - 2(s, t) + \|t\|^2 \geqq \|s\|^2 - 2\|s\| \, \|t\| \cos \theta + \|t\|^2$ where $\theta = \angle(\mathfrak{S}, \mathfrak{T})$. But the minimum of $\|s\|^2 - 2\|s\| x \cos \theta + x^2$ over all real x is $\|s\|^2 \sin^2\theta$. Thus $\|s - t\|^2 \geqq \|s\|^2 \sin^2\theta$, or $\|\mathscr{S}_{\mathscr{T}} \cdot (s - t)\| \leqq \|s - t\| \operatorname{cosec} \angle(\mathfrak{S}, \mathfrak{T})$. This proves not only the boundedness of $\mathscr{S}_{\mathscr{T}}$ but also that $\|\mathscr{S}_{\mathscr{T}}\| \leqq \operatorname{cosec} \angle(\mathfrak{S}, \mathfrak{T})$. To prove that this inequality is equality, let $\{s_n\}$ be a sequence of vectors from $\mathfrak{S}$ and $\{t_n\}$ a sequence of vectors from $\mathfrak{T}$ such that $\|s_n\| = 1$, $\|t_n\| = \cos \theta = \cos \angle(\mathfrak{S}, \mathfrak{T})$, and $(s_n, t_n) \to \cos^2 \theta$ as $n \to \infty$. Then $\|\mathscr{S}_{\mathscr{T}} \cdot (s_n - t_n)\| = \|s_n\| = 1$, while $\|s_n - t_n\|^2 = 1 - 2(s_n, t_n) + \cos^2 \theta \to \sin^2 \theta$. Thus $\|\mathscr{S}_{\mathscr{T}} \cdot (s_n - t_n)\| / \|s_n - t_n\| \to \operatorname{cosec} \theta$, showing that $\|\mathscr{S}_{\mathscr{T}}\| \geqq \operatorname{cosec} \angle(\mathfrak{S}, \mathfrak{T})$. To complete the proof of Theorem 3, suppose that $\mathfrak{S}$ is finite dimensional. (If it is $\mathfrak{T}$ which is finite dimensional, we observe that $\mathscr{S}_{\mathscr{T}} + \mathscr{T}_{\mathscr{S}}$ is the identity on $\mathfrak{S} + \mathfrak{T}$, so $\mathscr{T}_{\mathscr{S}}$ is bounded if and only if $\mathscr{S}_{\mathscr{T}}$ is.) If $\mathscr{S}_{\mathscr{T}}$ were unbounded, there would be a sequence $\{s_n\}$ in $\mathfrak{S}$ and a sequence $\{t_n\}$ in $\mathfrak{T}$ such that $\|s_n\| = 1$ and $\|s_n - t_n\| \to 0$. The unit sphere in a finite-dimensional space is compact, so $\{s_n\}$ has a limit point s with $\|s\| = 1$. We can delete enough members of $\{s_n\}$ to assume that $s_n \to s$. Then $t_n \to s$. Since both $\mathfrak{S}$ and $\mathfrak{T}$ are closed, s is in $\mathfrak{S} \cap \mathfrak{T}$. Hence, by hypothesis, $s = 0$, contradicting $\|s\| = 1$. Therefore $\mathscr{S}_{\mathscr{T}}$ must be bounded.

An immediate corollary of Theorem 3 is that $\|\mathscr{S}_{\mathscr{T}}\| = \|\mathscr{T}_{\mathscr{S}}\|$.

DEFINITION 4. If $\mathfrak{S}$ and $\mathfrak{T}$ are closed subspaces of Hilbert space $\mathfrak{H}$ such that $\mathfrak{S} \cap \mathfrak{T} = \{0\}$ and $\mathfrak{H} = \mathfrak{S} + \mathfrak{T}$, we will write $\mathfrak{H} = \mathfrak{S} \dotplus \mathfrak{T}$.

When $\mathfrak{H} = \mathfrak{S} \dotplus \mathfrak{T}$, Theorem 3 shows that $\mathscr{S}_{\mathscr{T}} : \mathfrak{H} \to \mathfrak{S}$ and $\mathscr{T}_{\mathscr{S}} : \mathfrak{H} \to \mathfrak{T}$ are bounded and that $\angle(\mathfrak{S}, \mathfrak{T}) > 0$. Moreover, we have $\mathscr{S}_{\mathscr{T}} + \mathscr{T}_{\mathscr{S}} = \mathscr{H}$, $\mathscr{S}_{\mathscr{T}}^2 = \mathscr{S}_{\mathscr{T}}$, $\mathscr{T}_{\mathscr{S}}^2 = \mathscr{T}_{\mathscr{S}}$, and $\mathscr{S}_{\mathscr{T}} \cdot \mathscr{T}_{\mathscr{S}} = \mathscr{T}_{\mathscr{S}} \cdot \mathscr{S}_{\mathscr{T}} = 0$.

(c) *Adjoints of parallel projections.* Lorch [1962] notes

THEOREM 5. *If* $S:\mathfrak{H} \to \mathfrak{H}$ *is any bounded linear operator on Hilbert space* $\mathfrak{H}$ *such that* $S^2 = S$, *then there are closed subspaces* $\mathfrak{S}$ *and* $\mathfrak{T}$ *of* $\mathfrak{H}$ *such that* $\mathfrak{H} = \mathfrak{S} \dotplus \mathfrak{T}$, $S = \mathscr{S}_{\mathscr{T}}$, and $\mathscr{H} - S = \mathscr{T}_{\mathscr{S}}$.

PROOF. Let $T = \mathscr{H} - S$, let $\mathfrak{S} = S \cdot \mathfrak{H}$, and let $\mathfrak{T} = T \cdot \mathfrak{H}$. Observe that $T^2 = T$ and $S \cdot T = T \cdot S = 0$. From the definition of T, $\mathfrak{H} = \mathfrak{S} + \mathfrak{T}$. If s is in $\mathfrak{S}$ then there is an h in $\mathfrak{H}$ such that $s = S \cdot h$. Then $S \cdot s = S^2 \cdot h = S \cdot h = s$. Similarly, if t is in $\mathfrak{T}$, $T \cdot t = t$. Therefore if h is in $\mathfrak{S} \cap \mathfrak{T}$, $h = S \cdot h = S \cdot T \cdot h = 0 \cdot h = 0$. Hence, $\mathfrak{S} \cap \mathfrak{T} = \{0\}$. Next, suppose that $\{s_n\}$ is a sequence in $\mathfrak{S}$ such that $s_n \to s$ for some s in $\mathfrak{H}$. Then $s_n = S \cdot s_n$, and since S is bounded we can take limits as $n \to \infty$, obtaining $s = S \cdot s$. Thus s is in $\mathfrak{S}$, so $\mathfrak{S}$ is closed. Similarly $\mathfrak{T}$ is closed, so $\mathfrak{H} = \mathfrak{S} \dotplus \mathfrak{T}$. Therefore $\mathscr{S}_{\mathscr{T}}$ and $\mathscr{T}_{\mathscr{S}}$ are well-defined, bounded linear operators on $\mathfrak{H}$. Both $\mathscr{S}_{\mathscr{T}}$ and S are the identity on $\mathfrak{S}$ when restricted to $\mathfrak{S}$ and the zero operator on $\mathfrak{T}$ when restricted to $\mathfrak{T}$, so $\mathscr{S}_{\mathscr{T}} = S$. Hence $\mathscr{T}_{\mathscr{S}} = T$.

An immediate corollary of Theorem 5 is that if $\mathfrak{H} = \mathfrak{S} \dotplus \mathfrak{T}$, then $\mathfrak{H} = \mathfrak{S}^{\perp} \dotplus \mathfrak{T}^{\perp}$ and

$$\mathscr{T}^{\perp}_{\mathscr{S}^{\perp}} = \mathscr{S}^{*}_{\mathscr{T}}. \tag{1}$$

To prove these results we note that since $\mathscr{S}^{2}_{\mathscr{T}} = \mathscr{S}_{\mathscr{T}}$, taking adjoints gives $(\mathscr{S}^{*}_{\mathscr{T}})^2 = \mathscr{S}^{*}_{\mathscr{T}}$. Moreover, $\mathscr{T}^{*}_{\mathscr{S}} = \mathscr{H} - \mathscr{S}^{*}_{\mathscr{T}}$. Therefore Theorem 5 gives $\mathfrak{H} = \mathfrak{S}^{\perp} \dotplus \mathfrak{T}^{\perp}$ and (1) if we can show that $\mathfrak{S}^{\perp} = \mathscr{T}^{*}_{\mathscr{S}} \cdot \mathfrak{H}$ and $\mathfrak{T}^{\perp} = \mathscr{S}^{*}_{\mathscr{T}} \cdot \mathfrak{H}$. The two demonstrations are identical, so we give the second. If t is in $\mathfrak{T}$ and h is in $\mathfrak{H}$, then $\langle t, \mathscr{S}^{*}_{\mathscr{T}} \cdot h\rangle = \langle \mathscr{S}_{\mathscr{T}} \cdot t, h\rangle = \langle 0, h\rangle = 0$, so $\mathscr{S}^{*}_{\mathscr{T}} \cdot \mathfrak{H} \subseteq \mathfrak{T}^{\perp}$. Then $\mathfrak{T}^{\perp} = \mathscr{S}^{*}_{\mathscr{T}} \cdot \mathfrak{H} \oplus [\mathfrak{T}^{\perp} \ominus \mathscr{S}^{*}_{\mathscr{T}} \cdot \mathfrak{H}]$, since $\mathfrak{T}^{\perp}$ is complete. If $t^{\perp}$ is in $\mathfrak{T}^{\perp} \ominus \mathscr{S}^{*}_{\mathscr{T}} \cdot \mathfrak{H}$, then for any h in $\mathfrak{H}$, $\langle t^{\perp}, \mathscr{S}^{*}_{\mathscr{T}} \cdot h\rangle = 0$, so $\langle \mathscr{S}_{\mathscr{T}} \cdot t^{\perp}, h\rangle = 0$. Then $\mathscr{S}_{\mathscr{T}} \cdot t^{\perp} = 0$, so $t^{\perp}$ is in $\mathfrak{T}$ and hence vanishes. Thus $\mathfrak{T}^{\perp} \ominus \mathscr{S}^{*}_{\mathscr{T}} \cdot \mathfrak{H} = \{0\}$ and $\mathfrak{T}^{\perp} = \mathscr{S}^{*}_{\mathscr{T}} \cdot \mathfrak{H}$.

We will make considerable use of the relations between parallel projections, orthogonal projections, and their adjoints which we next record. Suppose that $\mathfrak{H} = \mathfrak{S} \dotplus \mathfrak{T}$; then

$$\mathscr{S}_{\mathscr{T}} \cdot \mathscr{S} = \mathscr{S}, \qquad \mathscr{S} \cdot \mathscr{S}_{\mathscr{T}} = \mathscr{S}_{\mathscr{T}}, \tag{2a, b}$$

$$\mathscr{S}_{\mathscr{T}} \cdot \mathscr{T} = 0, \qquad \mathscr{T} \cdot \mathscr{S}_{\mathscr{T}} = \mathscr{T} - \mathscr{T}_{\mathscr{S}}, \tag{3a, b}$$

$$\mathscr{S}_{\mathscr{T}} \cdot \mathscr{S}^{\perp} = \mathscr{S}_{\mathscr{T}} - \mathscr{S}, \qquad \mathscr{S}^{\perp} \cdot \mathscr{S}_{\mathscr{T}} = 0, \tag{4a, b}$$

$$\mathscr{S}_{\mathscr{T}} \cdot \mathscr{T}^{\perp} = \mathscr{S}_{\mathscr{T}}, \qquad \mathscr{T}^{\perp} \cdot \mathscr{S}_{\mathscr{T}} = \mathscr{T}^{\perp}. \tag{5a, b}$$

Equations (2a), (2b), (3a) and (4b) are immediate consequences of the definitions. To prove (3b), write $\mathscr{T} \cdot \mathscr{S}_{\mathscr{T}} = \mathscr{T} \cdot (\mathscr{H} - \mathscr{T}_{\mathscr{S}})$ and use (2b). To prove (4a) write $\mathscr{S}_{\mathscr{T}} \cdot \mathscr{S}^{\perp} = \mathscr{S}_{\mathscr{T}} \cdot (\mathscr{H} - \mathscr{S})$ and use (2a). To prove (5a) write $\mathscr{S}_{\mathscr{T}} \cdot \mathscr{T}^{\perp} = \mathscr{S}_{\mathscr{T}} \cdot (\mathscr{H} - \mathscr{T})$ and use (3a). To prove (5b) write $\mathscr{T}^{\perp} \cdot \mathscr{S}_{\mathscr{T}} = \mathscr{T}^{\perp} \cdot (\mathscr{H} - \mathscr{T}_{\mathscr{S}})$ and use (4b).

The adjoint relations are

(6a, b) $\mathscr{S}^*_{\mathscr{T}} \cdot \mathscr{S} = \mathscr{S}^*_{\mathscr{T}}$, $\qquad \mathscr{S} \cdot \mathscr{S}^*_{\mathscr{T}} = \mathscr{S}$,

(7a, b) $\mathscr{S}^*_{\mathscr{T}} \cdot \mathscr{T} = \mathscr{T} - \mathscr{T}_{\mathscr{S}}$, $\qquad \mathscr{T} \cdot \mathscr{S}^*_{\mathscr{T}} = 0$,

(8a, b) $\mathscr{S}^*_{\mathscr{T}} \cdot \mathscr{S}^{\perp} = 0$, $\qquad \mathscr{S}^{\perp} \cdot \mathscr{S}^*_{\mathscr{T}} = \mathscr{S}^*_{\mathscr{T}} - \mathscr{S}$,

(9a, b) $\mathscr{S}^*_{\mathscr{T}} \cdot \mathscr{T}^{\perp} = \mathscr{T}^{\perp}$, $\qquad \mathscr{T}^{\perp} \cdot \mathscr{S}^*_{\mathscr{T}} = \mathscr{S}^*_{\mathscr{T}}$.

These relations are obtained by taking the adjoints of (2), (3), (4) and (5) and interchanging a and b. Recall that orthogonal projections are self-adjoint.

One other set of relations obviously belongs with (2)–(5) and (6)–(9). We shall not use these relations but give them only for completeness. They are

(10a, b) $\mathscr{S}_{\mathscr{T}} \cdot \mathscr{S}_{\mathscr{T}} = \mathscr{S}_{\mathscr{T}}$, $\qquad \mathscr{S}_{\mathscr{T}} \cdot \mathscr{S}_{\mathscr{T}} = \mathscr{S}_{\mathscr{T}}$,

(11a, b) $\mathscr{S}_{\mathscr{T}} \cdot \mathscr{T}_{\mathscr{S}} = 0$, $\qquad \mathscr{T}_{\mathscr{S}} \cdot \mathscr{S}_{\mathscr{T}} = 0$,

(12a, b) $\mathscr{S}_{\mathscr{T}} \cdot \mathscr{S}^*_{\mathscr{T}} \geqq 0$, $\qquad \mathscr{S}^*_{\mathscr{T}} \cdot \mathscr{S}_{\mathscr{T}} \geqq 0$,

(13a, b) $\mathscr{S}_{\mathscr{T}} \cdot \mathscr{T}^*_{\mathscr{S}} = \mathscr{S}_{\mathscr{T}} - \mathscr{S}_{\mathscr{T}} \cdot \mathscr{S}^*_{\mathscr{T}}$, $\qquad \mathscr{T}^*_{\mathscr{S}} \cdot \mathscr{S}_{\mathscr{T}} = \mathscr{S}^*_{\mathscr{T}} - \mathscr{S}^*_{\mathscr{T}} \cdot \mathscr{S}_{\mathscr{T}}$.

Equations (10a, b) and (11a, b) have already been noted. The inequalities (12a, b) are obvious, and I know of no useful equalities with which to replace them. Equation (13a) is obtained by writing $\mathscr{S}_{\mathscr{T}} \cdot \mathscr{T}^*_{\mathscr{S}} = \mathscr{S}_{\mathscr{T}} \cdot (\mathscr{H} - \mathscr{S}^*_{\mathscr{T}})$, and (13b) is obtained similarly.

(d) *The geometry of* $\mathfrak{S} \dotplus \mathfrak{T}$ *when* $\dim \mathfrak{T} < \infty$. If $\mathfrak{H} = \mathfrak{S} \dotplus \mathfrak{T}$ and $\mathfrak{T}$ is N-dimensional, with $N < \infty$, the geometrical relation between $\mathfrak{S}$ and $\mathfrak{T}$ is easy to obtain and is likely to be useful in numerical computation of $\mathscr{S}_{\mathscr{T}}$ and $\mathscr{T}_{\mathscr{S}}$. Since all the applications we envision satisfy $\dim \mathfrak{T} < \infty$, we shall examine this geometry in some detail.

Let $\mathfrak{R}_0 = \mathfrak{S} \ominus \mathscr{S} \cdot \mathfrak{T}$. Then $\mathfrak{S} = \mathfrak{R}_0 \oplus \mathscr{S} \cdot \mathfrak{T}$, and we claim that $\mathfrak{R}_0 \subseteqq \mathfrak{T}^{\perp}$. Indeed, if r is in $\mathfrak{R}_0$ and t is in $\mathfrak{T}$, we have $\langle r, t \rangle = \langle \mathscr{S} \cdot r, t \rangle = \langle r, \mathscr{S} \cdot t \rangle = 0$, where we have used the facts that r is in $\mathfrak{S}$, that $\mathscr{S}^* = \mathscr{S}$, and that r is orthogonal to $\mathscr{S} \cdot \mathfrak{T}$. Clearly $\mathfrak{R}_0 \subseteqq (\mathscr{S} \cdot \mathfrak{T})^{\perp}$, so $\mathfrak{R}_0 \subseteqq (\mathfrak{T} + \mathscr{S} \cdot \mathfrak{T})^{\perp}$. But $\mathfrak{H} = \mathfrak{S} + \mathfrak{T} = \mathfrak{R}_0 + \mathscr{S} \cdot \mathfrak{T} + \mathfrak{T}$, so

(14) $$\mathfrak{H} = \mathfrak{R}_0 \oplus (\mathfrak{T} + \mathscr{S} \cdot \mathfrak{T}).$$

Since $\mathfrak{R}_0 \subseteqq \mathfrak{S}$, $\mathscr{S}_{\mathscr{T}} | \mathfrak{R}_0$ is the identity map on $\mathfrak{R}_0$ and $\mathscr{T}_{\mathscr{S}} | \mathfrak{R}_0$ is the zero operator on $\mathfrak{R}_0$. Moreover, both $\mathscr{S}_{\mathscr{T}}$ and $\mathscr{T}_{\mathscr{S}}$ map $\mathfrak{T} + \mathscr{S} \cdot \mathfrak{T}$ into itself. Therefore, to understand $\mathscr{S}_{\mathscr{T}}$ and $\mathscr{T}_{\mathscr{S}}$ it suffices to understand their action on the finite-dimensional space $\mathfrak{T} + \mathscr{S} \cdot \mathfrak{T}$.

The operator $\mathscr{T} \cdot \mathscr{S} \cdot \mathscr{T}$ maps $\mathfrak{T}$ into itself and $\mathfrak{T}^{\perp}$ into $\{0\}$, so it is essentially finite dimensional; all its nonzero structure lies in $\mathfrak{T}$. It is selfadjoint and positive semi-definite because $\mathscr{T}\cdot\mathscr{S}\cdot\mathscr{T}=(\mathscr{S}\cdot\mathscr{T})^*\cdot(\mathscr{S}\cdot\mathscr{T})$. Therefore $\mathfrak{T}$ has an orthonormal basis $t_1,\ldots,t_N$ consisting of eigenvectors of $\mathscr{T}\cdot\mathscr{S}\cdot\mathscr{T}$, and all the eigenvalues are nonnegative. Since $\|\mathscr{T}\cdot\mathscr{S}\cdot\mathscr{T}\|\leqq\|\mathscr{S}\|\,\|\mathscr{T}\|\,\|\mathscr{S}\|=1$, no eigenvalue exceeds 1. Moreover, 1 is not an eigenvalue, for if it were there would be a nonzero vector t in $\mathfrak{T}$ with $t=\mathscr{T}\cdot\mathscr{S}\cdot\mathscr{T}\cdot t=\mathscr{T}\cdot\mathscr{S}\cdot t$. Then $\|t\|\leqq\|\mathscr{S}\cdot t\|\leqq\|t\|$, so $\|t\|=\|\mathscr{S}\cdot t\|$. But $t=\mathscr{S}\cdot t+\mathscr{S}^{\perp}\cdot t$, so $\|t\|^2=\|\mathscr{S}\cdot t\|^2+\|\mathscr{S}^{\perp}\cdot t\|^2$. Hence $\mathscr{S}^{\perp}\cdot t=0$, so $t=\mathscr{S}\cdot t$, so t is in $\mathfrak{S}$, contradicting the hypothesis that $\mathfrak{S}\cap\mathfrak{T}=\{0\}$.

The foregoing shows that we can write the eigenvalues of $\mathscr{T}\cdot\mathscr{S}\cdot\mathscr{T}$ corresponding to the eigenvectors $t_1,\ldots,t_N$ as $\cos^2\theta_1,\ldots,\cos^2\theta_N$ and that, by reordering if necessary, we can assume that $0<\theta_1\leqq\theta_2\leqq\cdots\leqq\theta_N\leqq\pi/2$.

Now let $s_\nu=\mathscr{S}\cdot t_\nu$, $\nu=1,\ldots,N$. Then clearly $s_1,\ldots,s_N$, $t_1,\ldots,t_N$ span $\mathfrak{T}+\mathscr{S}\cdot\mathfrak{T}$. Let $\mathfrak{R}_\nu$ be the space spanned by s_ν and t_ν, $\nu=1,\ldots,N$. Then $\mathfrak{T}+\mathscr{S}\cdot\mathfrak{T}=\mathfrak{R}_1+R_2+\cdots+\mathfrak{R}_N$. We claim that $\mathfrak{R}_\nu$ is orthogonal to $\mathfrak{R}_\mu$ when $\mu\neq\nu$, and that $\dim\mathfrak{R}_\nu=1$ or 2 according as $\theta_\nu=\pi/2$ or $<\pi/2$. To substantiate these claims, observe that $\langle s_\mu,s_\nu\rangle=\langle\mathscr{S}\cdot t_\mu,\mathscr{S}\cdot t_\nu\rangle=\langle t_\mu,\mathscr{S}\cdot t_\nu\rangle=\langle t_\mu,s_\nu\rangle=\langle\mathscr{T}\cdot t_\mu,\mathscr{S}\cdot\mathscr{T}\cdot t_\nu\rangle=\langle t_\mu,\mathscr{T}\cdot\mathscr{S}\cdot\mathscr{T}\cdot t_\nu\rangle=\cos^2\theta_\nu\langle t_\mu,t_\nu\rangle$. Thus $\langle s_\mu,s_\nu\rangle=\langle t_\mu,s_\nu\rangle=0$ if $\mu\neq\nu$, which shows that $\mathfrak{R}_\mu$ is orthogonal to $\mathfrak{R}_\nu$ if $\mu\neq\nu$. Hence we can write

$$\mathfrak{T}+\mathscr{S}\cdot\mathfrak{T}=\mathfrak{R}_2\oplus\mathfrak{R}_1\oplus\cdots\oplus\mathfrak{R}_N. \tag{15}$$

Also, $\|s_\nu\|=\|t_\nu\|\cos\theta_\nu$, which is 0 iff $\theta_\nu=\pi/2$. Thus $\dim\mathfrak{R}_\nu=1$ if $\theta_\nu=\pi/2$, and in that case t_ν is orthogonal to $\mathfrak{S}$. If $\theta_\nu<\pi/2$, then the cosine of the angle between s_ν and t_ν is $s_\nu\cdot t_\nu\|s_\nu\|^{-1}\|t_\nu\|^{-1}=\cos\theta_\nu$, so θ_ν is the angle between s_ν and t_ν. Since $\theta_\nu>0$, s_ν and t_ν are linearly independent and $\dim\mathfrak{R}_\nu=2$.

Now $\mathscr{S}_{\mathscr{T}}\cdot\mathfrak{R}_\nu\subseteqq\mathfrak{R}_\nu$, $\mathscr{T}_{\mathscr{S}}\cdot\mathfrak{R}_\nu\subseteqq\mathfrak{R}_\nu$ and on $\mathfrak{R}_\nu$ the operator $\mathscr{S}_{\mathscr{T}}|\mathfrak{R}_\nu$ is simply projection parallel to t_ν onto the space spanned by s_ν. Let s^ν, t^ν be the basis dual to s_ν,t_ν; this basis is defined by the requirements that it lie in $\mathfrak{R}_\nu$ and that $s_\nu\cdot s^\nu=t_\nu\cdot t^\nu=1$ and $s_\nu\cdot t^\nu=t_\nu\cdot s^\nu=0$. Because $\|t_\nu\|=1$, the four vectors are related as shown in Figure 1. For any vector r in $\mathfrak{R}_\nu$, $r=s_\nu(s^\nu\cdot r)+t_\nu(t^\nu\cdot r)$, so, if $\theta_\nu<\pi/2$, $\mathscr{S}_{\mathscr{T}}|\mathfrak{R}_\nu=s_\nu s^\nu$ and $\mathscr{T}_{\mathscr{S}}|\mathfrak{R}_\nu=t_\nu t^\nu$. If we agree that $s_\nu=0$ and $t^\nu=t_\nu$ when $\theta_\nu=\pi/2$, then $\mathscr{S}_{\mathscr{T}}|\mathfrak{R}_\nu=s_\nu s^\nu$ and $\mathscr{T}_{\mathscr{S}}|\mathfrak{R}_\nu=t_\nu t^\nu$ even in that case. Therefore we have

$$\mathscr{S}_{\mathscr{T}}=\mathscr{R}_0+\sum_{\nu=1}^{N}s_\nu s^\nu,\qquad \mathscr{T}_{\mathscr{S}}=\sum_{\nu=1}^{N}t_\nu t^\nu. \tag{16}$$

The geometry we have just discussed enables us to show that if $\dim\mathfrak{T}<\infty$, $\angle(\mathfrak{S},\mathfrak{T})=\theta_1$. To see this, recall that $\angle(\mathfrak{S},\mathfrak{T})$ is the angle θ

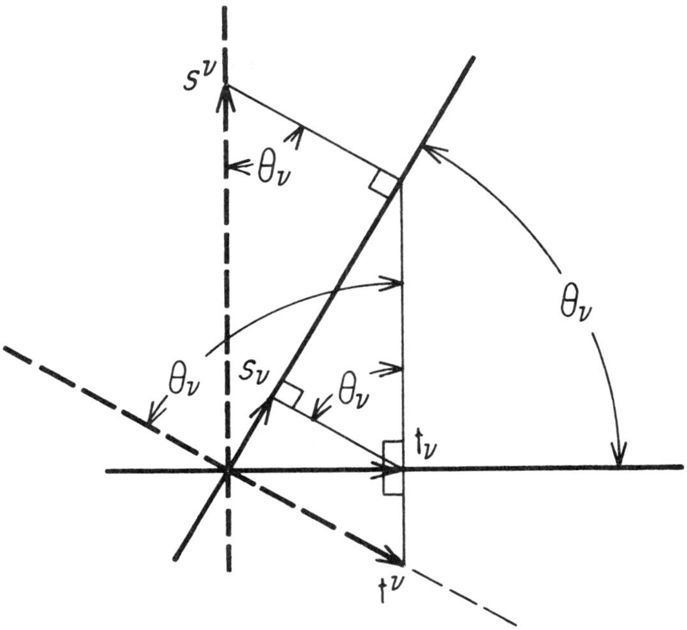

FIGURE 1

between 0 and $\pi/2$ such that

$$\cos\theta = \sup\{s \cdot t/\|s\|\,\|t\| : s \text{ in } \mathfrak{S},\ t \text{ in } \mathfrak{T},\ s \neq 0,\ t \neq 0\}.$$

Since $s_1 \cdot t_1/\|s_1\|\,\|t_1\| = \cos\theta_1$, it is clear that $\cos\theta \geqq \cos\theta_1$, so $\theta \leqq \theta_1$. If s and t are any vectors in $\mathfrak{S}$ and $\mathfrak{T}$, we can write $s = r_0 + \sum_{\nu=1}^N a_\nu s_\nu$, $t = \sum_{\nu=1}^N b_\nu t_\nu$, with r_0 in $\mathfrak{R}_0$. Then $\|s\|^2 = \|r_0\|^2 + \sum_{\nu=1}^N a_\nu^2\|s_\nu\|^2 = \|r_0\|^2 + \sum_{\nu=1}^N a_\nu^2\cos^2\theta_\nu$, and $\|t\|^2 = \sum_{\nu=1}^N b_\nu^2$ while $s \cdot t = \sum_{\nu=1}^N a_\nu b_\nu s_\nu \cdot t_\nu = \sum_{\nu=1}^N a_\nu b_\nu \cos^2\theta_\nu$. By Schwarz's inequality, $(s \cdot t)^2 \leqq [\sum_{\nu=1}^N a_\nu^2\cos^2\theta_\nu] \cdot [\sum_{\nu=1}^N b_\nu^2\cos^2\theta_\nu]$, and $\sum_{\nu=1}^N b_\nu^2\cos^2\theta_\nu \leqq \cos^2\theta_1\sum_{\nu=1}^N b_\nu^2$. Thus $(s\cdot t)^2 \leqq \cos^2\theta_1\|s\|^2\|t\|^2$, and so $\cos\theta \leqq \cos\theta_1$. Hence $\cos\theta = \cos\theta_1$, $\theta = \theta_1$ as asserted.

(e) *A variational characterization of* $\mathscr{S}_{\mathscr{T}}$. For any h in $\mathfrak{H}$ it is well known that $\mathscr{S} \cdot h$ is that vector s in $\mathfrak{S}$ which minimizes $\|s - h\|$. Here we give the corresponding characterization of $\mathscr{S}_{\mathscr{T}} \cdot h$.

Suppose that $\mathfrak{H}$ is a Hilbert space and $\mathfrak{H} = \mathfrak{S} \dotplus \mathfrak{T}$. Suppose that $\mathfrak{U}$ is a closed subspace of $\mathfrak{H}$ such that $\mathfrak{U} = (\mathfrak{S} \cap \mathfrak{U}) \oplus \mathfrak{T}$. Then for any h in $\mathfrak{H}$, $\mathscr{S}_{\mathscr{T}} \cdot h$ is that vector s in $\mathfrak{S} \cap (h + \mathfrak{U})$ which minimizes $\|s - h\|$. To see this, note first that $\mathscr{S}_{\mathscr{T}} \cdot h = h - \mathscr{T}_{\mathscr{S}} \cdot h$, so $\mathscr{S}_{\mathscr{T}} \cdot h$ is indeed in

$\mathfrak{S} \cap (h + \mathfrak{U})$. If s is any vector in $\mathfrak{S} \cap (h + \mathfrak{U})$ then $s - \mathscr{S}_{\mathscr{T}} \cdot h$ is in $\mathfrak{S} \cap \mathfrak{U}$ and hence orthogonal to $\mathfrak{T}$. Therefore $\|s - h\|^2 = \|s - \mathscr{S}_{\mathscr{T}} \cdot h - \mathscr{T}_{\mathscr{S}} \cdot h\|^2 = \|s - \mathscr{S}_{\mathscr{T}} \cdot h\|^2 + \|\mathscr{S}_{\mathscr{T}} \cdot h - h\|^2$. Thus $\|s - h\| > \|\mathscr{S}_{\mathscr{T}} \cdot h - h\|$ unless $s = \mathscr{S}_{\mathscr{T}} \cdot h$. Figure 2 illustrates the content of this remark.

(f) *Generalized inverses on oblique subspaces.* We will find very useful an analogue in Hilbert space of the expression for the inverse of a 2×2 triangular matrix. Suppose that $\mathfrak{H}$ is a Hilbert space, that $\mathfrak{H} = \mathfrak{S} \dotplus \mathfrak{T}$, that $A: \mathfrak{H} \to \mathfrak{S}$ is a bounded linear operator living on $\mathfrak{S}$, that $B: \mathfrak{H} \to \mathfrak{T}$ is a bounded linear operator living on $\mathfrak{T}$, and that $Q: \mathfrak{H} \to \mathfrak{H}$ is any bounded linear operator. Then $A + \mathscr{S} \cdot Q \cdot \mathscr{T} + B$ has a bounded inverse on $\mathfrak{H}$, and

$$(17)\quad \begin{aligned}&(A + \mathscr{S} \cdot Q \cdot \mathscr{T} + B)^{-1} \\ &\quad = \mathscr{S}^*_{\mathscr{T}} \cdot A^{\text{inv}} \cdot \mathscr{S}_{\mathscr{T}} - \mathscr{S}^*_{\mathscr{T}} \cdot A^{\text{inv}} \cdot Q \cdot B^{\text{inv}} \cdot \mathscr{T}_{\mathscr{S}} + \mathscr{T}^*_{\mathscr{S}} \cdot B^{\text{inv}} \cdot \mathscr{T}_{\mathscr{S}}.\end{aligned}$$

In proof, let $U = A + \mathscr{S} \cdot Q \cdot \mathscr{T} + B$ and let V be the right-hand side of (17). We want to show that $U \cdot V = V \cdot U = \mathscr{H}$. We have

$$\begin{aligned} V \cdot A &= \mathscr{S}^*_{\mathscr{T}} \cdot A^{\text{inv}} \cdot \mathscr{S}_{\mathscr{T}} \cdot A && (\text{because } \mathscr{T}_{\mathscr{S}} \cdot A = 0) \\ &= \mathscr{S}^*_{\mathscr{T}} \cdot A^{\text{inv}} \cdot A && (\text{because } \mathscr{S}_{\mathscr{T}} \cdot A = A) \\ &= \mathscr{S}^*_{\mathscr{T}} \cdot \mathscr{S} && (\text{because } A^{\text{inv}} \cdot A = \mathscr{S}) \\ &= \mathscr{S}^*_{\mathscr{T}} && (\text{see (6a)}). \end{aligned}$$

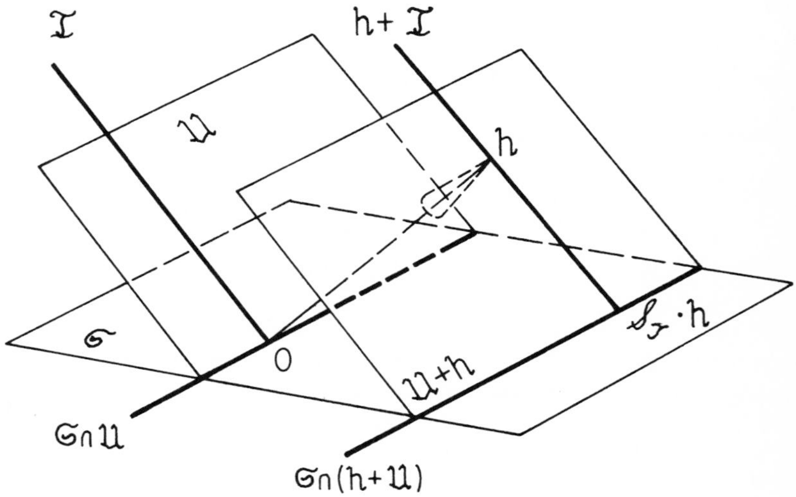

FIGURE 2

We also have

$$\begin{aligned} V \cdot \mathscr{S} \cdot Q \cdot \mathscr{T} &= \mathscr{S}^*_{\mathscr{T}} \cdot A^{\mathrm{inv}} \cdot \mathscr{S}_{\mathscr{T}} \cdot \mathscr{S} \cdot Q \cdot \mathscr{T} && \text{(because } \mathscr{T}_{\mathscr{S}} \cdot \mathscr{S} = 0\text{)} \\ &= \mathscr{S}^*_{\mathscr{T}} \cdot A^{\mathrm{inv}} \cdot \mathscr{S} \cdot Q \cdot \mathscr{T} && \text{(because } \mathscr{S}_{\mathscr{T}} \cdot \mathscr{S} = \mathscr{S}\text{)} \\ &= \mathscr{S}^*_{\mathscr{T}} \cdot A^{\mathrm{inv}} \cdot Q \cdot \mathscr{T} && \text{(because } A^{\mathrm{inv}} \cdot \mathscr{S} = A^{\mathrm{inv}}\text{).} \end{aligned}$$

And we have

$$\begin{aligned} V \cdot B &= -\mathscr{S}^*_{\mathscr{T}} \cdot A^{\mathrm{inv}} \cdot Q \cdot B^{\mathrm{inv}} \cdot \mathscr{T}_{\mathscr{S}} \cdot B + \mathscr{T}^*_{\mathscr{S}} \cdot B^{\mathrm{inv}} \cdot \mathscr{T}_{\mathscr{S}} \cdot B \\ & \qquad\qquad \text{(because } \mathscr{S}_{\mathscr{T}} \cdot B = 0\text{)} \\ &= -\mathscr{S}^*_{\mathscr{T}} A^{\mathrm{inv}} \cdot Q \cdot \mathscr{T} + \mathscr{T}^*_{\mathscr{S}} \cdot \mathscr{T} \\ & \qquad\qquad \text{(because } \mathscr{T}_{\mathscr{S}} \cdot B = B \text{ and } B^{\mathrm{inv}} \cdot B = \mathscr{T}\text{)} \\ &= -\mathscr{S}^*_{\mathscr{T}} A^{\mathrm{inv}} \cdot Q \cdot \mathscr{T} + \mathscr{T}^*_{\mathscr{S}} \qquad \text{(see (6a)).} \end{aligned}$$

If we add the three foregoing expressions we have $V \cdot U = \mathscr{S}^*_{\mathscr{T}} + \mathscr{T}^*_{\mathscr{S}} = \mathscr{H}$. In order to discuss $U \cdot V$, we note that $A \cdot \mathscr{S}^*_{\mathscr{T}} \cdot A^{\mathrm{inv}} = A \cdot \mathscr{S} \cdot \mathscr{S}^*_{\mathscr{T}} \cdot A^{\mathrm{inv}} = A \cdot \mathscr{S} \cdot A^{\mathrm{inv}} = A \cdot A^{\mathrm{inv}} = \mathscr{S}$. Similarly $B \cdot \mathscr{T}^*_{\mathscr{S}} \cdot B^{\mathrm{inv}} = \mathscr{T}$. Now

$$\begin{aligned} A \cdot V &= (A \cdot \mathscr{S}^*_{\mathscr{T}} \cdot A^{\mathrm{inv}}) \cdot (\mathscr{S}_{\mathscr{T}} - Q \cdot B^{\mathrm{inv}} \cdot \mathscr{T}_{\mathscr{S}}) \\ & \qquad\qquad \text{(because, from (1), } A \cdot \mathscr{T}^*_{\mathscr{S}} = 0\text{)} \\ &= \mathscr{S} \cdot (\mathscr{S}_{\mathscr{T}} - Q \cdot B^{\mathrm{inv}} \cdot \mathscr{T}_{\mathscr{S}}) \\ &= \mathscr{S}_{\mathscr{T}} - \mathscr{S} \cdot Q \cdot B^{\mathrm{inv}} \cdot \mathscr{T}_{\mathscr{S}}. \end{aligned}$$

Also,

$$\begin{aligned} (\mathscr{S} \cdot Q \cdot \mathscr{T}) \cdot V &= \mathscr{S} \cdot Q \cdot \mathscr{T} \cdot \mathscr{T}^*_{\mathscr{S}} \cdot B^{\mathrm{inv}} \cdot \mathscr{T}_{\mathscr{S}} && \text{(because } \mathscr{T} \cdot \mathscr{S}^*_{\mathscr{T}} = 0\text{)} \\ &= \mathscr{S} \cdot Q \cdot \mathscr{T} \cdot B^{\mathrm{inv}} \cdot \mathscr{T}_{\mathscr{S}} && \text{(because } \mathscr{T} \cdot \mathscr{T}^*_{\mathscr{S}} = \mathscr{T}\text{)} \\ &= \mathscr{S} \cdot Q \cdot B^{\mathrm{inv}} \cdot \mathscr{T}_{\mathscr{S}} && \text{(because } \mathscr{T} \cdot B^{\mathrm{inv}} = B^{\mathrm{inv}}\text{).} \end{aligned}$$

Finally,

$$\begin{aligned} B \cdot V &= B \cdot \mathscr{T}^*_{\mathscr{S}} \cdot B^{\mathrm{inv}} \cdot \mathscr{T}_{\mathscr{S}} \quad \text{(because, from (1), } B \cdot \mathscr{S}_{\mathscr{T}} = 0\text{)} \\ &= \mathscr{T} \cdot \mathscr{T}_{\mathscr{S}} \\ &= \mathscr{T}_{\mathscr{S}}. \end{aligned}$$

If we add the three foregoing expressions we obtain $U \cdot V = \mathscr{S}_{\mathscr{T}} + \mathscr{T}_{\mathscr{S}} = \mathscr{H}$. This completes the proof of (17).

3. *Cylinder measures.*

(a) *Rings, σ-rings, and probability measures.* To discuss the effects of errors in our measurements, we need probability distributions on Hilbert spaces. We cannot hope to find probability *measures* in the classical sense of Kolmogoroff [1956]. It is known [Gross [1964] describing a result of Loewner [1939]] that any countably additive, positive, bounded set function on an infinite-dimensional real Hilbert space $\mathfrak{H}$ which is invariant under all rigid rotations of $\mathfrak{H}$ (all unitary transformations of $\mathfrak{H}$) assigns measure 0 to the unit ball, the set $\{h : h \text{ in } \mathfrak{H} \text{ and } \|h\| \leqq 1\}$.

The analogue of probability measure which seems best to suit our purposes is that of cylinder measure. §II.A.3 is a brief survey of the elementary part of this theory as it now exists in the literature. In reporting it we shall lean heavily on Gross [1962, 1964, 1967] and Getoor [1957]. The recent activity in the field seems to have been stimulated by a fundamental and now classical paper of I.E. Segal [1956].

Before examining cylinder measures, we include a brief summary of the ideas from classical probability theory which we will use. [Kolmogoroff, 1956] and [Halmos, 1950] contain the proofs of the result we will quote.

If S is a set and A is a subset of S then the complement of A in S, written $S - A$, is the subset of S consisting of all those members of S which are not members of A. Two subsets A and B of S are "disjoint" if the set $A \cap B$ consisting of their common members is the empty set $\varnothing$. Clearly A and $S - A$ are disjoint.

For any set S, a ring of subsets of S is a collection Ω of subsets of S with the following properties:

(i) S is a member of Ω;
(ii) if A is a member of Ω, so is $S - A$;
(iii) if A and B are members of Ω, so is $A \cap B$.

If A and B are members of Ω, then so is $A \cup B = S - [(S - A) \cap (S - B)]$.

A σ-ring of subsets of a set S is a ring Ω of subsets of S such that if $\{A_1, A_2, \ldots\}$ is any denumerable sequence of members of Ω then $A_1 \cap A_2 \cap \cdots$ is also a member of Ω. Clearly, if Ω is a σ-ring and $\{A_1, A_2, \ldots\}$ are members of Ω then $A_1 \cup A_2 \cup \cdots$ is also a member of Ω.

A premeasurable space is an ordered pair (S, Ω), S being a set and Ω a ring of subsets of S. A measureable space is a premeasurable space (S, Ω) in which Ω is a σ-ring of subsets of S.

If $\boldsymbol{C}$ is a nonempty collection of σ-rings Ω_c of subsets of a fixed set S, let $\bigcap_{\boldsymbol{C}} \Omega_c$ denote the collection of subsets of S consisting of those subsets of S which are members of every σ-ring Ω_c in $\boldsymbol{C}$. Then $\bigcap_{\boldsymbol{C}} \Omega_c$ is itself a σ-ring of subsets of S, and $(S, \bigcap_{\boldsymbol{C}} \Omega_c)$ is a measurable space. If S is $\mathfrak{T}$, a finite-dimensional Hilbert space, and $\boldsymbol{C}$ is the collection of all σ-rings of subsets of $\mathfrak{T}$ which include every open subset of $\mathfrak{T}$ as a member, then $\boldsymbol{C}$ is nonempty because it includes the set of all subsets of $\mathfrak{T}$. The σ-ring $B(\mathfrak{T}) = \bigcap_{\boldsymbol{C}} \Omega_c$ in this case is the smallest σ-ring of subsets of $\mathfrak{T}$ which includes as members all open subsets of $\mathfrak{T}$. The members of $B(\mathfrak{T})$ are called

Borel subsets of $\mathfrak{T}$. If $f:\mathfrak{S} \to \mathfrak{T}$ is a linear transformation from finite-dimensional Hilbert space $\mathfrak{S}$ into finite-dimensional Hilbert space $\mathfrak{T}$ and B is a Borel subset of $\mathfrak{S}$ then $f(B)$ is a Borel subset of $\mathfrak{T}$. If $f:\mathfrak{S} \to \mathfrak{T}$ is one-to-one into then a subset $B \subseteqq \mathfrak{S}$ is a Borel subset of $\mathfrak{S}$ iff $f(B)$ is a Borel subset of $\mathfrak{T}$. If $f:\mathfrak{S} \to \mathfrak{T}$ is onto then subset $C \subseteqq \mathfrak{T}$ is a Borel subset of $\mathfrak{T}$ iff $f^{-1}(C)$ is a Borel subset of $\mathfrak{S}$.

A real-valued set function μ on a premeasurable space (S, Ω) is a function $\mu:\Omega \to R$ (R is the real line) which assigns a real number $\mu(A)$ to every subset A of S which is a member of Ω. Such a set function is "additive" if for any two disjoint members A and B of Ω we have $\mu(A \cup B) = \mu(A) + \mu(B)$. If μ is additive and $A_1, \ldots, A_N$ are any pairwise disjoint members of Ω then by induction $\mu(A_1 \cup A_2 \cup \cdots \cup A_N) = \mu(A_1) + \cdots + \mu(A_N)$. Moreover, if A and B are members of Ω and $A \subseteqq B$ then $\mu(B - A) = \mu(B) - \mu(A)$. Therefore $\mu(\varnothing) = 0$ if μ is additive.

A real-valued set function μ on a premeasurable space (S, Ω) is countably additive if $\mu(A_1 \cup A_2 \cup \cdots) = \mu(A_1) + \mu(A_2) + \cdots$ whenever $\{A_1, A_2, \cdots\}$ is a denumerable sequence of pairwise disjoint members of Ω and $A_1 \cup A_2 \cup \cdots$ is also a member of Ω.

A measure space is a triple (S, Ω, μ) consisting of a measurable space (S, Ω) and a countably additive set function μ on (S, Ω). If $\mu(S) = 1$ and $\mu(A) \geqq 0$ for every member of Ω, then (S, Ω, μ) is a probability measure space and μ is a probability measure.

If $f:A \to B$ is any function mapping set A into set B, and if $S \subseteqq B$, then the inverse image of S under f, written $f^{-1}(S)$, is the set of all members a of A such that $f(a)$ is a member of S. Clearly $f^{-1}(S) \subseteqq A$. Note that if $f:A \to B$, $g:B \to C$, and S is a subset of C then

$$(g \circ f)^{-1}(S) = f^{-1}[g^{-1}(S)]. \tag{18}$$

Here $g \circ f: A \to C$ is the mapping defined by $(g \circ f)(a) = g[f(a)]$; it is the composition of f and g.

Suppose (S, Ω) is a measurable space, $\mathfrak{T}$ is a finite-dimensional Hilbert space, and $f:S \to \mathfrak{T}$ is a function. Then f is called "Borel measurable" on (S, Ω) iff $f^{-1}(B)$ is a member of Ω whenever B is a Borel subset of $\mathfrak{T}$. For example, if $f:R^m \to R^n$ is piecewise continuous, it is Borel measurable on $(R^m, B(R^m))$. Here R^n is the Hilbert space of ordered n-tuples of real numbers, endowed with the Pythagorean inner product.

Let $(S, \Omega\, \mu)$ be a probability measure space and let $f:S \to R$ be a Borel-measurable function from S to the real line R. Let A be a subset of S which is a member of Ω. The function f is called "integrable over A with respect to μ" if the series

$$q(\lambda) = \sum_{n=-\infty}^{\infty} n\lambda\mu\{s:s \text{ in } A \text{ and } n\lambda \leqq f(s) < (n+1)\lambda\}$$

converges absolutely for every real λ. If f is integrable over A with respect to μ, then $\lim_{\lambda \to 0} q(\lambda)$ exists [Kolmogoroff, 1956, p. 37] (see also [Halmos, 1950, Chapter V]) and is called the integral of f over A with respect to μ. This integral is written $\int_A f(s)\mu(ds)$. For fixed f and A it is linear in μ, for fixed A and μ it is linear in f, and for fixed f and μ it is a countably additive set function on the measurable space (A, Ω_A), where Ω_A consists of those members of Ω which are subsets of A. Moreover, f is integrable iff $|f|$ is. If $|f| \leqq g$ and g is integrable, so is f. And if f and g are integrable and $f \leqq g$ then $\int_A f(s)\mu(ds) \leqq \int_A g(s)\mu(ds)$.

Suppose that (S, Ω, μ) is a probability measure space, that $\mathfrak{T}$ is a finite-dimensional Hilbert space, and that $f: S \to \mathfrak{T}$ is measurable. Suppose that A is a member of Ω and that for each fixed t in $\mathfrak{T}$, $t \cdot f: S \to R$ is integrable over A with respect to μ. Then we say that f is integrable over A with respect to μ, and we define $\int_A f(s)\mu(ds)$ as that unique vector in $\mathfrak{T}$ such that for every t in $\mathfrak{T}$

$$t \cdot \int_A f(s)\mu(ds) = \int_A t \cdot f(s)\, \mu(ds). \tag{19}$$

If F is a function which assigns to every member s of S a linear operator $F(s): \mathfrak{T} \to \mathfrak{T}$, and if for every t_1 and t_2 in $\mathfrak{T}$ the function $t_1 \cdot F \cdot t_2 : S \to R$ is integrable over A with respect to μ then we say that F is integrable over A with respect to μ, and we define $\int_A F(s)\mu(ds)$ as that unique linear operator on $\mathfrak{T}$ such that for every t_1 and t_2 in $\mathfrak{T}$,

$$t_1 \cdot \left[\int_A F(s)\mu(ds)\right] \cdot t_2 = \int_A (t_1 \cdot F(s) \cdot t_2)\mu(ds). \tag{20}$$

If $\mathfrak{T}$ is a finite-dimensional Hilbert space and $(\mathfrak{T}, B(\mathfrak{T}), \mu)$ is a probability measure space, then the function $\phi: \mathfrak{T} \to C$ (C is the complex plane) defined by

$$\phi(h) = \int_{\mathfrak{T}} \exp(ih \cdot t)\mu(dt)$$

is called the characteristic function of the probability measure μ. For every h in $\mathfrak{H}$ we obviously have $|\phi(h)| \leqq 1$. It is a well-known result from probability theory [Cramer, 1946, p. 101] that different probability measures on $\mathfrak{T}$ have different characteristic functions. It is also true from the bounded convergence theorem [Halmos, 1950] that ϕ depends continously on h. Moreover, ϕ is a positive-definite function; that is, for any integer N and any vectors $h_1, \ldots, h_N$ in $\mathfrak{T}$, the $N \times N$ matrix $\Phi_{mn} = \phi(h_m - h_n)$ is Hermitian and positive definite. To see this, let $c_1, \ldots, c_N$

be any complex numbers. Then

$$\sum_{m,n} c_m c_n^* \Phi_{mn} = \sum_{m,n} c_m c_n^* \int_{\mathfrak{T}} \exp(ih_m \cdot t)(\exp(ih_n \cdot t))^* \mu(dt)$$

$$= \int_{\mathfrak{T}} f(t) f^*(t) \mu(dt)$$

where $f(t) = \sum_m c_m \exp(ih_m \cdot t)$. Thus $\sum_{m,n} c_m c_n^* \Phi_{mn} \geqq 0$ as asserted.

Finally, Bochner [1955] has shown that if $\phi : \mathfrak{T} \to C$ is any continuous, positive-definite function on $\mathfrak{T}$ with $|\phi| \leqq 1$ then ϕ is the characteristic function of exactly one probability measure on $\mathfrak{T}$. It follows from Bochner's theorem that the product of two continuous, positive-definite functions ϕ and ψ on $\mathfrak{T}$ is continuous and positive definite. The reason is that $\phi\psi$ is the characteristic function of the probability measure obtained by convoluting the probability measures whose characteristic functions are ϕ and ψ.

If (S, Ω, μ) and (S', Ω', μ') are two probability measure spaces, let $S'' = S \times S'$. Then S'' is the set of all ordered pairs (s, s') with s a member of S and s' a member of S'. Let Ω'' be the smallest σ-ring which includes as members all the sets $B \times B'$ where B is a member of Ω and B' is a member of Ω'. Then [Halmos, 1950] there is a unique probability measure μ'' on the σ-ring Ω'' such that for any member B of Ω and any member B' of Ω', $\mu''(B \times B') = \mu(B)\mu'(B')$. The measure μ'' is called the product measure of μ and μ', and the probability measure space (S'', Ω'', μ'') is called the product of the probability measure spaces (S, Ω, μ) and (S', Ω', μ'). Fubini's Theorem [Halmos, 1950, p. 148] says that if $f'' : S \times S' \to R$ is any integrable function on $S \times S'$ then the functions $f(s) = \int_{S'} f''(s, s')\mu'(ds')$ and $f'(s') = \int_S f''(s, s')\mu(ds)$ are integrable on S and S' respectively and $\int_{S''} f''(s'')\mu''(ds'') = \int_S f(s)\mu(ds) = \int_{S'} f'(s')\mu'(ds')$. Conversely, if $f(s, s')$ is integrable in s' over S' for almost every fixed s (i.e., except for a subset $A \subseteqq S$ with $\mu(A) = 0$) and the integral $f(s)$ defined above is integrable over S, then $f(s, s')$ is integrable over $S \times S'$.

(b) *Cylinder sets and cylinder functions.* Let $\mathfrak{T}$ be a finite-dimensional subspace of Hilbert space $\mathfrak{H}$. Let B be a Borel subset of $\mathfrak{T}$. Then $\mathscr{T}^{-1}(B)$ is called a cylinder set with base B and generator $\mathfrak{T}^{\perp}$. It consists of all vectors h in $\mathfrak{H}$ whose orthogonal projections $\mathscr{T} \cdot h$ onto $\mathfrak{T}$ are members of B. If $\mathfrak{H}$ is infinite dimensional, the cylinder subsets of $\mathfrak{H}$ are all unbounded. If $\mathfrak{H}$ is finite dimensional, its cylinder sets are precisely its Borel sets. Obviously every cylinder subset of $\mathfrak{H}$ is a Borel set when $\dim \mathfrak{H} < \infty$, and if B is a Borel subset of $\mathfrak{H}$ then it is a cylinder set with base B and generator $\mathfrak{H}^{\perp} = \{0\}$.

The base and generator of a cylinder set are not unique. To see this, denote by $c_{\mathscr{H}}(\mathfrak{T})$ the collection of all cylinder subsets of Hilbert space $\mathfrak{H}$ which have as bases Borel subsets of the finite-dimensional subspace $\mathfrak{T} \subseteq \mathfrak{H}$ and have generator $\mathfrak{T}^{\perp}$. We have

REMARK 6. *Suppose $\mathfrak{S}$ and $\mathfrak{T}$ are finite-dimensional subspaces of Hilbert space $\mathfrak{H}$ and $\mathfrak{S} \subseteq \mathfrak{T}$. Then $c_{\mathscr{H}}(\mathfrak{S}) \subseteq c_{\mathscr{H}}(\mathfrak{T})$.*

To prove this remark, let H be any member of $c_{\mathscr{H}}(\mathfrak{S})$. Then there is a Borel subset S of $\mathfrak{S}$ such that $H = \mathscr{S}^{-1}(S)$. Define $T = \mathfrak{T} \cap H$. Then $T = (\mathscr{S}|\mathfrak{T})^{-1}(S)$, so T is a Borel subset of $\mathfrak{T}$. Further,

$$\mathscr{T}^{-1}(T) = [(\mathscr{S}|\mathfrak{T}) \circ \mathscr{T}]^{-1}(S) = \mathscr{S}^{-1}(S) = H$$

so H is a member of $c_{\mathscr{H}}(\mathfrak{T})$.

Another characterization of cylinder sets is given in

REMARK 7. *A subset H of Hilbert space $\mathfrak{H}$ is a cylinder set iff there are finitely many vectors $t_1, \ldots, t_N$ in $\mathfrak{H}$ and a Borel set $\tilde{T}$ in R^N such that H consists of all those h in $\mathfrak{H}$ for which the N-tuple $(t_1 \cdot h, \ldots, t_N \cdot h)$ is in $\tilde{T}$. If $\mathfrak{T} = \mathrm{sp}\{t_1, \ldots, t_N\}$ is the space spanned by $t_1, \ldots, t_N$, and $T = H \cap \mathfrak{T}$, then T is a Borel subset of $\mathfrak{T}$ and $H = \mathscr{T}^{-1}(T)$.*

To prove Remark 7, let $t_1, \ldots, t_N$ be any set of vectors which spans the finite-dimensional subspace $\mathfrak{T}$ of $\mathfrak{H}$. Define a linear mapping $f: \mathfrak{H} \to R^N$ by requiring, for any h in $\mathfrak{H}$, that $f(h) = (t_1 \cdot h, \ldots, t_N \cdot h)$. A subset T of $\mathfrak{T}$ is a Borel subset of $\mathfrak{T}$ iff there is a Borel subset $\tilde{T}$ of R^N such that $f(T) = \tilde{T}$. For any h in $\mathfrak{H}$, $f(h) = f(\mathscr{T} \cdot h)$, so $\mathscr{T} \cdot h$ is in T iff $f(h)$ is in $\tilde{T}$. In other words, $\mathscr{T}^{-1}(T) = f^{-1}(\tilde{T})$, which is Remark 7.

An immediate consequence is

REMARK 8. *Suppose that $\mathfrak{S}$ and $\mathfrak{T}$ are finite-dimensional subspaces of Hilbert space $\mathfrak{H}$, that S Borel subset of $\mathfrak{S}$ and T is a Borel subset of $\mathfrak{T}$. Then $\mathscr{S}^{-1}(S) \cap \mathscr{T}^{-1}(T)$ is a member of $c_{\mathscr{H}}(\mathfrak{S} + \mathfrak{T})$.*

PROOF. Let $s_1, \ldots, s_M$ be a finite collection of vectors which spans $\mathfrak{S}$ and $t_1, \ldots, t_N$ a finite collection of vectors which spans $\mathfrak{T}$. Then there are Borel subsets $\tilde{S}$ of R^M and $\tilde{T}$ of R^N such that $\mathscr{S}^{-1}(S) = \{h : h \text{ in } \mathfrak{H} \text{ and } (s_1 \cdot h, \ldots, s_M \cdot h) \text{ is in } \tilde{S}\}$, and $\mathscr{T}^{-1}(T) = \{h : h \text{ in } \mathfrak{H} \text{ and } (t_1 \cdot h, \ldots, t_N \cdot h) \text{ is in } \tilde{T}\}$. Clearly, $\mathscr{S}^{-1}(S) \cap \mathscr{T}^{-1}(T) = \{h : h \text{ in } \mathfrak{H} \text{ and } (s_1 \cdot h, \ldots, s_M \cdot h, t_1 \cdot h, \ldots, t_N \cdot h) \text{ is in } \tilde{S} \times \tilde{T}\}$, where $\tilde{S} \times \tilde{T}$ is the Borel subset of R^{M+N} consisting of ordered pairs $(\tilde{s}, \tilde{t})$ where $\tilde{s}$ is an M-tuple from $\tilde{S}$ and $\tilde{t}$ is an N-tuple from $\tilde{T}$. Then, according to Remark 7, $\mathscr{S}^{-1}(S) \cap \mathscr{T}^{-1}(T)$ is a cylinder set based on a subset of $\mathrm{sp}\{s_1, \ldots, s_M, t_1, \ldots, t_N\}$. This latter space is simply $\mathfrak{S} + \mathfrak{T}$.

Let $C(\mathfrak{H})$ denote the collection of all cylinder subsets of Hilbert space $\mathfrak{H}$. If $\mathfrak{T}$ is any finite-dimensional subspace of $\mathfrak{H}$, evidently $(\mathfrak{H}, c_{\mathscr{H}}(\mathfrak{T}))$ is a measurable space (i.e., $c_{\mathscr{H}}(\mathfrak{T})$ is a σ-ring). However, we have

REMARK 9. *If $\mathfrak{H}$ is an infinite-dimensional Hilbert space, then $(\mathfrak{H}, C(\mathfrak{H}))$ is a premeasurable but not a measurable space (i.e. $C(\mathfrak{H})$ is a ring but not a σ-ring).*

PROOF. First we show that $C(\mathfrak{H})$ is a ring of subsets of $\mathfrak{H}$. The set $\mathfrak{H}$ is a member of $C(\mathfrak{H})$, being a cylinder set with base $\{0\}$ and generator $\{0\}^{\perp} = \mathfrak{H}$. If H is a member of $C(\mathfrak{H})$, we can write $H = \mathscr{T}^{-1}(T)$ for some Borel subset T of some finite-dimensional space $\mathfrak{T} \subseteqq \mathfrak{H}$. Then $\mathfrak{H} - H = \mathscr{T}^{-1}(\mathfrak{T} - T)$, so $\mathfrak{H} - H$ is in $C(\mathfrak{H})$. Finally, if H and K are in $C(\mathfrak{H})$, Remark 8 shows that $H \cap K$ is in $C(\mathfrak{H})$. Thus $(\mathfrak{H}, C(\mathfrak{H}))$ is indeed a premeasurable space. If $\mathfrak{H}$ is infinite dimensional, it has an infinite-dimensional separable closed subspace $\mathfrak{H}'$. The cylinder subsets of $\mathfrak{H}'$ are the intersections of $\mathfrak{H}'$ with the cylinder subsets of $\mathfrak{H}$, on account of Remark 7. Therefore it suffices to show that $(\mathfrak{H}', C(\mathfrak{H}'))$ is not a measurable space. Let $\{h_1, h_2, \ldots\}$ be a countable dense subset of vectors in $\mathfrak{H}$. Let $S_n = \{h : h \text{ in } \mathfrak{H}' \text{ and } |h_n \cdot h| \leqq \|h_n\|\}$. Then S_n is a cylinder subset of $\mathfrak{H}'$, but $S_1 \cap S_2 \cap \cdots$ is the unit ball in $\mathfrak{H}'$, i.e. $\{h : h \text{ is in } \mathfrak{H}' \text{ and } \|h\| \leqq 1\}$. Since the unit ball is bounded, it cannot be a cylinder set.

The relation between the cylinder subsets of $\mathfrak{H}$ and those of the subspaces of $\mathfrak{H}$ is given by

REMARK 10. *Suppose $\mathfrak{K}$ is a closed subspace of Hilbert space $\mathfrak{H}$. If $\mathfrak{T}$ is a finite-dimensional subspace of $\mathfrak{K}$ and K is a member of $c_{\mathscr{K}}(\mathfrak{T})$, then $\mathscr{K}^{-1}(K)$ is a member of $c_{\mathscr{H}}(\mathfrak{T})$. If $\mathfrak{T}$ is a finite-dimensional subspace of $\mathfrak{H}$ and H is a member of $c_{\mathscr{H}}(\mathfrak{T})$, then $H \cap \mathfrak{K}$ is a member of $c_{\mathscr{K}}(\mathscr{K} \cdot \mathfrak{T})$ and $\mathscr{K} \cdot H$ is a member of $c_{\mathscr{K}}(\mathfrak{K} \cap \mathfrak{T})$.*

PROOF. If $\mathfrak{T}$ is a finite-dimensional subspace of $\mathfrak{K}$ and K is in $c_{\mathscr{K}}(\mathfrak{T})$, there is a Borel subset T of $\mathfrak{T}$ such that $K = (\mathscr{T}|\mathfrak{K})^{-1}(T)$. Then $\mathscr{K}^{-1}(K) = [(\mathscr{T}|\mathfrak{K}) \circ \mathscr{K}]^{-1}(T) = \mathscr{T}^{-1}(T)$, so $\mathscr{K}^{-1}(K)$ is a member of $c_{\mathscr{H}}(\mathfrak{T})$. If $\mathfrak{T}$ is a finite-dimensional subspace of $\mathfrak{H}$, and H is a member of $c_{\mathscr{H}}(\mathfrak{T})$, then there are vectors $t_1, \ldots, t_N$ spanning $\mathfrak{T}$, and a Borel subset $\tilde{T}$ of R^N, such that $H = \{h : h \text{ is in } \mathfrak{H} \text{ and } (h \cdot t_1, \ldots, h \cdot t_N) \text{ is in } \tilde{T}\}$. Then $H \cap \mathfrak{K} = \{k : k \text{ is in } \mathfrak{K} \text{ and } (k \cdot \mathscr{K} \cdot t_1, \ldots, k \cdot \mathscr{K} \cdot t_N) \text{ is in } \tilde{T}\}$. From Remark 7 it follows that $H \cap \mathfrak{K}$ is in $c_{\mathscr{K}}(\mathscr{K} \cdot \mathfrak{T})$. Finally, if H is a member of $c_{\mathscr{H}}(\mathfrak{T})$, there is a Borel subset T of $\mathfrak{T}$ such that $H = \mathscr{T}^{-1}(T)$. Let $\mathscr{K} \wedge \mathscr{T}$ denote the orthogonal projection operator from $\mathfrak{H}$ onto $\mathfrak{K} \cap \mathfrak{T}$. We claim that

$$\mathscr{K} \cdot \mathscr{T}^{-1}(T) = (\mathscr{K} \wedge \mathscr{T}|\mathfrak{K})^{-1}[(\mathscr{K} \wedge \mathscr{T}) \cdot T]. \tag{21}$$

Since $(\mathscr{K} \wedge \mathscr{T}) \cdot T$ is a Borel subset of $\mathfrak{K} \cap \mathfrak{T}$, a finite-dimensional subspace of $\mathfrak{K}$, if we can prove (21) we will have shown that $\mathscr{K} \cdot H$ is a member of $c_{\mathscr{K}}(\mathfrak{K} \cap \mathfrak{T})$. First we prove $\subseteq$ in (21). If h is in $H = \mathscr{T}^{-1}(T)$ then there is a t in T such that $\mathscr{T} \cdot h = t$. Then $(\mathscr{K} \wedge \mathscr{T}|\mathfrak{K}) \cdot \mathscr{K} \cdot h = \mathscr{K} \wedge \mathscr{T} \cdot h = \mathscr{K} \wedge \mathscr{T} \cdot \mathscr{T} \cdot h = \mathscr{K} \wedge \mathscr{T} \cdot t$, so $\mathscr{K} \cdot h$ is in $(\mathscr{K} \wedge \mathscr{T}|\mathfrak{K})^{-1}[\mathscr{K} \wedge \mathscr{T} \cdot T]$. To prove $\supseteq$ in (21), suppose that k is in $(\mathscr{K} \wedge \mathscr{T}|\mathfrak{K})^{-1}[(\mathscr{K} \wedge \mathscr{T}) \cdot T]$. Then there is a t in T such that $\mathscr{K} \wedge \mathscr{T} \cdot k = \mathscr{K} \wedge \mathscr{T} \cdot t$. Let $\mathfrak{S} = \mathfrak{T} \ominus (\mathfrak{T} \cap \mathfrak{K})$. Then $\mathfrak{T} + \mathfrak{K} = \mathfrak{S} \dotplus \mathfrak{K}$, and on the space $\mathfrak{S} \dotplus \mathfrak{K}$ the parallel projection operators $\mathscr{S}_{\mathscr{K}}$ and $\mathscr{K}_{\mathscr{S}}$ and their adjoints are defined. Let

$$h = \mathscr{K} \wedge \mathscr{T} \cdot t + \mathscr{K}_{\mathscr{S}}^{*} \cdot (k - \mathscr{K} \wedge \mathscr{T} \cdot k) + \mathscr{S}_{\mathscr{K}}^{*} \cdot (t - \mathscr{K} \wedge \mathscr{T} \cdot t).$$

Then $\mathscr{K} \cdot h = \mathscr{K} \wedge \mathscr{T} \cdot t + k - \mathscr{K} \wedge \mathscr{T} \cdot k = k$, and $\mathscr{T} \cdot h = (\mathscr{S} + \mathscr{K} \wedge \mathscr{T}) \cdot h = \mathscr{K} \wedge \mathscr{T} \cdot t + \mathscr{S} \cdot (t - \mathscr{K} \wedge \mathscr{T} \cdot t) + \mathscr{K} \wedge \mathscr{T} \cdot (k - \mathscr{K} \wedge \mathscr{T} \cdot k) = (\mathscr{K} \wedge \mathscr{T} + \mathscr{S}) \cdot t = \mathscr{T} \cdot t = t$. (Here we have used the adjoints of the obvious equations $\mathscr{K}_{\mathscr{S}} \cdot \mathscr{K} \wedge \mathscr{T} = \mathscr{K} \wedge \mathscr{T}, \mathscr{S}_{\mathscr{K}} \cdot \mathscr{K} \wedge \mathscr{T} = 0$.) Thus h is in $\mathscr{T}^{-1}(T)$, so k is in $\mathscr{K} \cdot \mathscr{T}^{-1}(T)$. This completes the proof of Equation (21) and Remark 10.

A function $f: \mathfrak{H} \to S$ which assigns to every vector h in Hilbert space $\mathfrak{H}$ a member $f(h)$ of the set S is called a cylinder function based on the finite-dimensional subspace $\mathfrak{T} \subseteq \mathfrak{H}$ if for every h in $\mathfrak{H}$ we have $f(h) = f(\mathscr{T} \cdot h)$. There follows

REMARK 11. *Let $\mathfrak{T} \subseteq \mathfrak{K}$ be finite-dimensional subspaces of Hilbert space $\mathfrak{H}$. If $f: \mathfrak{H} \to S$ is a cylinder function based on $\mathfrak{T}$, then f is a cylinder function based on $\mathfrak{K}$.*

The proof is immediate. If f is a cylinder function based on $\mathfrak{T}$, then for any h in $\mathfrak{H}$, $f(\mathscr{R} \cdot h) = f(\mathscr{T} \cdot \mathscr{R} \cdot h) = f(\mathscr{T} \cdot h) = f(h)$.

From Remark 11 follows immediately

REMARK 12. *Suppose that $\mathfrak{S}$ and $\mathfrak{T}$ are finite-dimensional subspaces of Hilbert space $\mathfrak{H}$, that $\mathfrak{K}$ is another Hilbert space, and that $f: \mathfrak{H} \to \mathfrak{K}$ and $g: \mathfrak{H} \to \mathfrak{K}$ are cylinder functions based on $\mathfrak{S}$ and $\mathfrak{T}$ respectively. Then $(f + g): \mathfrak{H} \to \mathfrak{K}$ and $fg: \mathfrak{H} \to L(\mathfrak{K}, \mathfrak{K})$ are cylinder functions based on $\mathfrak{S} + \mathfrak{T}$.*

(Here $L(\mathfrak{K}, \mathfrak{K})$ is the space of second order tensors, or linear operators, over $\mathfrak{K}$, and $f(h)g(h)$ denotes the dyadic or tensor product of $f(h)$ and $g(h)$.)

Now suppose that $\mathfrak{T}$ is a finite-dimensional subspace of Hilbert space $\mathfrak{H}$, that $\mathfrak{R}$ is a finite-dimensional Hilbert space, and that $f: \mathfrak{H} \to \mathfrak{R}$ is a cylinder function based on $\mathfrak{T}$. We will say that f is measurable iff $(f|\mathfrak{T}): \mathfrak{T} \to \mathfrak{R}$ is measurable.

(c) *Definition of cylinder measures and integrals.* Let $\mathfrak{H}$ be a Hilbert space. A real-valued set function $\mu: \boldsymbol{C}(\mathfrak{H}) \to R$ is called a cylinder measure on $\mathfrak{H}$ iff, for every finite-dimensional subspace $\mathfrak{T} \subseteq \mathfrak{H}$, $\mu|c_{\mathscr{H}}(\mathfrak{T})$ is a probability measure on the σ-ring $c_{\mathscr{H}}(\mathfrak{T})$ and hence on the measurable space $(\mathfrak{H}, c_{\mathscr{H}}(\mathfrak{T}))$. A cylinder measure μ is additive on the premeasurable space $(\mathfrak{H}, \boldsymbol{C}(\mathfrak{H}))$ because, if $\mathfrak{S}$ and $\mathfrak{T}$ are finite-dimensional subspaces of $\mathfrak{H}$ and H and K are cylinder subsets of $\mathfrak{H}$ in $c_{\mathscr{H}}(\mathfrak{S})$ and $c_{\mathscr{H}}(\mathfrak{T})$ respectively then, according to Remark 7, both H and K are in $c_{\mathscr{H}}(\mathfrak{S} + \mathfrak{T})$. If $H \cap K = \varnothing$, it follows that $\mu(H \cup K) = \mu(H) + \mu(K)$ because μ is additive (in fact countably additive) on $c_{\mathscr{H}}(\mathfrak{S} + \mathfrak{T})$.

If $\mathfrak{H}$ is infinite dimensional, cylinder measures on $\mathfrak{H}$ need not be countably additive [Gross, 1964].

Marginal distributions of cylinder measures can be defined and have most of the same interrelations as marginal distributions of measures on finite-dimensional spaces [Parzen, 1960]. We have first

REMARK 13. *Suppose μ is a cylinder measure on Hilbert space $\mathfrak{H}$ and $\mathfrak{K}$ is a closed subspace of $\mathfrak{H}$. Define a set function $\mu_{\mathscr{K}}$ on the premeasurable space $(\mathfrak{K}, \boldsymbol{C}(\mathfrak{K}))$ by requiring for any cylinder set K in $\boldsymbol{C}(\mathfrak{K})$ that*

$$\mu_{\mathscr{K}}(K) = \mu[\mathscr{K}^{-1}(K)]. \tag{22}$$

Then $\mu_{\mathscr{K}}$ is a cylinder measure on $\mathfrak{K}$.

PROOF. First, by Remark 10, $\mu_{\mathscr{K}}(K)$ is a well-defined, nonnegative number for every K in $\boldsymbol{C}(\mathfrak{K})$, and $\mu_{\mathscr{K}}(\mathfrak{K}) = 1$. Next, suppose that $\mathfrak{T}$ is a finite-dimensional subspace of $\mathfrak{K}$ and that $\{K_1, K_2, \ldots\}$ is a denumerable sequence of pairwise disjoint sets in $c_{\mathscr{K}}(\mathfrak{T})$. Because $\mathscr{K}^{-1}(K_i) \cap \mathscr{K}^{-1}(K_j) = \mathscr{K}^{-1}(K_i \cap K_j)$, it follows that the denumerable sequence $\{\mathscr{K}^{-1}(K_1), \mathscr{K}^{-1}(K_2), \ldots\}$ is pairwise disjoint. According to Remark 10 all the sets in this latter sequence are members of $c_{\mathscr{H}}(\mathfrak{T})$, and $\mu | c_{\mathscr{H}}(\mathfrak{T})$ is countably additive. Therefore

$$\mu[\mathscr{K}^{-1}(K_1) \cup \mathscr{K}^{-1}(K_2) \cup \cdots] = \sum_n \mu[\mathscr{K}^{-1}(K_n)] = \sum_n \mu_{\mathscr{K}}(K_n).$$

But

$$\mathscr{K}^{-1}(K_1 \cup K_2 \cup \cdots) = \mathscr{K}^{-1}(K_1) \cup \mathscr{K}^{-1}(K_2) \cup \cdots,$$

so

$$\begin{aligned} \mu[\mathscr{K}^{-1}(K_1) \cup \mathscr{K}^{-1}(K_2) \cup \cdots] &= \mu[\mathscr{K}^{-1}(K_1 \cup K_2 \cup \cdots)] \\ &= \mu_{\mathscr{K}}[K_1 \cup K_2 \cup \cdots]. \end{aligned}$$

Hence $\mu_{\mathscr{K}} | c_{\mathscr{K}}(\mathfrak{T})$ is countably additive. This completes the proof of Remark 13.

The cylinder measure $\mu_{\mathscr{K}}$ defined on $(\mathfrak{K}, \boldsymbol{C}(\mathfrak{K}))$ by Equation (22) we will call the marginal distribution of μ on $\mathfrak{K}$. If $\mathfrak{K}$ is finite dimensional, then

$\mu_{\mathscr{K}}$ must be a probability measure on the measurable space $(\mathfrak{K}, B(\mathfrak{K}))$, since when $\mathfrak{K}$ is finite dimensional, $C(\mathfrak{K}) = B(\mathfrak{K})$.

Now suppose that μ is a cylinder measure on Hilbert space $\mathfrak{H}$ and that $\mathfrak{L} \subseteq \mathfrak{K}$ are closed subspaces of $\mathfrak{H}$. Then μ has marginal distributions $\mu_{\mathscr{K}}$ on $\mathfrak{K}$ and $\mu_{\mathscr{L}}$ on $\mathfrak{L}$, and $\mu_{\mathscr{K}}$ has a marginal distribution $\mu_{\mathscr{K}\mathscr{L}}$ on $\mathfrak{L}$. We claim

REMARK 14. $\mu_{\mathscr{K}\mathscr{L}} = \mu_{\mathscr{L}}$.

The proof is a computation from the definition (22). If L is any member of $C(\mathfrak{L})$ then

$$\begin{aligned} \mu_{\mathscr{K}\mathscr{L}}(L) &= \mu_{\mathscr{K}}[(\mathscr{L}|\mathfrak{K})^{-1}(L)] \\ &= \mu\{\mathscr{K}^{-1}[(\mathscr{L}|\mathfrak{K})^{-1}(L)]\} \\ &= \mu\{[(\mathscr{L}|\mathfrak{K}) \circ \mathscr{K}]^{-1}(L)\} \\ &= \mu[\mathscr{L}^{-1}(L)] \\ &= \mu_{\mathscr{L}}(L). \end{aligned}$$

A collection $\mathscr{C}$ of finite-dimensional subspaces of a Hilbert space $\mathfrak{H}$ is "closed under intersection" if whenever $\mathfrak{S}$ and $\mathfrak{T}$ are members of $\mathscr{C}$ so is $\mathfrak{S} \cap \mathfrak{T}$. For example, if $\mathfrak{D}$ is a fixed finite-dimensional subspace of $\mathfrak{H}$, then the collection $\mathscr{C}(\mathfrak{D}, \mathfrak{H})$ consisting of all finite-dimensional subspaces $\mathfrak{S}$ such that $\mathfrak{D} \subseteq \mathfrak{S} \subseteq \mathfrak{H}$ is closed under intersection.

If $\mathfrak{S}$ and $\mathfrak{T}$ are finite-dimensional subspaces of Hilbert space $\mathfrak{H}$, and $\mu_{\mathscr{S}}$ and $\mu_{\mathscr{T}}$ are probability measures on $(\mathfrak{S}, B(\mathfrak{S}))$ and $(\mathfrak{T}, B(\mathfrak{T}))$, then $\mu_{\mathscr{S}}$ and $\mu_{\mathscr{T}}$ are "consistent" if they have the same marginal distribution on $\mathfrak{S} \cap \mathfrak{T}$.

Suppose that $\mathscr{C}$ is a collection of finite-dimensional subspaces of $\mathfrak{H}$ which is closed under intersection, and that each subspace $\mathfrak{R}$ in $\mathscr{C}$ has a probability measure $\mu_{\mathscr{R}}$ on $(\mathfrak{R}, B(\mathfrak{R}))$. Suppose also that whenever $\mathfrak{R}$ and $\mathfrak{S}$ are in $\mathscr{C}$ and $\mathfrak{R} \subseteq \mathfrak{S}$ then $\mu_{\mathscr{R}}$ and $\mu_{\mathscr{S}}$ are consistent. From this it follows that whenever $\mathfrak{S}$ and $\mathfrak{T}$ are in $\mathscr{C}$, $\mu_{\mathscr{S}}$ and $\mu_{\mathscr{T}}$ are consistent, for $\mathfrak{R} = \mathfrak{S} \cap \mathfrak{T}$ is in $\mathscr{C}$ and $\mu_{\mathscr{R}}$ is the marginal distribution on $\mathfrak{R}$ of both $\mu_{\mathscr{S}}$ and $\mu_{\mathscr{T}}$.

These general comments lead to

REMARK 15. *Suppose that $\mathfrak{D}$ is a finite-dimensional subspace of Hilbert space $\mathfrak{H}$, and $\mathscr{C}(\mathfrak{D}, \mathfrak{H})$ is the collection of all finite-dimensional subspaces of $\mathfrak{H}$ which contain $\mathfrak{D}$. Suppose that for every $\mathfrak{S}$ in $\mathscr{C}(\mathfrak{D}, \mathfrak{H})$ we are given a probability measure $\mu_{\mathscr{S}}$ on $(\mathfrak{S}, B(\mathfrak{S}))$, and whenever $\mathfrak{S} \subseteq \mathfrak{T}$ and both are in $\mathscr{C}(\mathfrak{D}, \mathfrak{H})$ then $\mu_{\mathscr{S}}$ is the marginal distribution of $\mu_{\mathscr{T}}$ on $\mathfrak{S}$. Then there is exactly one cylinder measure μ on $\mathfrak{H}$ such that, for every $\mathfrak{S}$ in $\mathscr{C}(\mathfrak{D}, \mathfrak{H})$, $\mu_{\mathscr{S}}$ is the marginal distribution of μ on $\mathfrak{S}$.*

PROOF. Suppose that H is any cylinder subset of $\mathfrak{H}$. Then there is a finite-dimensional subspace $\mathfrak{P}$ of $\mathfrak{H}$ such that H is a member of $c_{\mathscr{H}}(\mathfrak{P})$. If $\mathfrak{S} = \mathfrak{P} + \mathfrak{D}$, then according to Remark 6, H is a member of $c_{\mathscr{H}}(\mathfrak{S})$. Thus $H = \mathscr{S}^{-1}(S)$ for some Borel subset S of $\mathfrak{S}$. Since $\mathfrak{S}$ is a member of $\mathscr{C}(\mathfrak{D}, \mathfrak{H})$, a probability measure $\mu_{\mathscr{S}}$ is given on $(\mathfrak{S}, B(\mathfrak{S}))$. We define $\mu(H) = \mu_{\mathscr{S}}(S)$. We claim that this definition is unambiguous. Indeed, if $\mathfrak{S}$ and $\mathfrak{T}$ are both members of $\mathscr{C}(\mathfrak{D}, \mathfrak{H})$ and S and T are Borel subsets of $\mathfrak{S}$ and $\mathfrak{T}$ such that $H = \mathscr{S}^{-1}(S) = \mathscr{T}^{-1}(T)$, then according to Remark 6 there is a Borel subset R of $\mathfrak{R} = \mathfrak{S} + \mathfrak{T}$ such that $H = \mathscr{R}^{-1}(R)$. Now h is in H iff $\mathscr{R} \cdot h$ is in R, and iff $\mathscr{S} \cdot h$ is in S. But $\mathscr{S} \cdot h = \mathscr{S} \cdot \mathscr{R} \cdot h$, so a vector r in $\mathfrak{R}$ is in R iff $\mathscr{S} \cdot r$ is in S. Thus $R = (\mathscr{S}|\mathfrak{R})^{-1}(S)$. By hypothesis, $\mu_{\mathscr{S}}(S) = \mu_{\mathscr{R}}[(\mathscr{S}|\mathfrak{R})^{-1}(S)]$, so $\mu_{\mathscr{S}}(S) = \mu_{\mathscr{R}}(R)$. Similarly $\mu_{\mathscr{T}}(T) = \mu_{\mathscr{R}}(R)$, so $\mu_{\mathscr{S}}(S) = \mu_{\mathscr{T}}(T)$ and our definition of $\mu(H)$ is unambiguous. The rest of Remark 15 is obvious from this definition and Remark 14.

From Remark 14 follows

THEOREM 16. *Suppose that μ is a cylinder measure on Hilbert space $\mathfrak{H}$, that $\mathfrak{S}$ and $\mathfrak{T}$ are finite-dimensional subspaces of $\mathfrak{H}$, and that f is a real-valued, vector-valued or tensor-valued cylinder function on $\mathfrak{H}$ which is based on $\mathfrak{S}$ and also on $\mathfrak{T}$. If $f|\mathfrak{S}$ is integrable with respect to $\mu_{\mathscr{S}}$, then $f|\mathfrak{T}$ is integrable with respect to $\mu_{\mathscr{T}}$, and the two integrals are equal.*

PROOF. Let $\mathfrak{R} = \mathfrak{S} + \mathfrak{T}$. It suffices to show that $f|\mathfrak{T}$ is integrable with respect to $\mu_{\mathscr{S}}$ iff $f|\mathfrak{R}$ is integrable with respect to $\mu_{\mathscr{R}}$, and that

$$\int_{\mathfrak{S}} f(s)\mu_{\mathscr{S}}(ds) = \int_{\mathfrak{R}} f(r)\mu_{\mathscr{R}}(dr). \tag{23}$$

Let $\mathscr{S}' = \mathscr{S}|\mathfrak{R}$, and let $f' = f|\mathfrak{S}$. Then the mapping $\mathscr{S}' : \mathfrak{R} \to \mathfrak{S}$ is a measurable transformation (inverse images of Borel sets are Borel sets), and the classical formula for changing variables in an integral [Halmos, 1950, p. 163] is

$$\int_{\mathfrak{S}} f'(s)\mu_{\mathscr{R}}(\mathscr{S}^{-1}\, ds) = \int_{\mathfrak{R}} f'(\mathscr{S} \cdot r)\mu_{\mathscr{R}}(dr),$$

either integral's existence implying that of the other. But $\mu_{\mathscr{R}}(\mathscr{S}^{-1}\, ds) = \mu_{\mathscr{R}\mathscr{S}}(ds) = \mu_{\mathscr{S}}(ds)$, which proves Equation (23).

Theorem 16 permits us to introduce

DEFINITION 17. Suppose that μ is a cylinder measure on Hilbert space $\mathfrak{H}$, that $\mathfrak{T}$ is a finite-dimensional subspace of $\mathfrak{H}$, and that f is a complex-valued, vector-valued or tensor-valued measurable cylinder function based on $\mathfrak{T}$. If $f|\mathfrak{T}$ is integrable over $\mathfrak{T}$ with respect to $\mu_{\mathscr{T}}$, we will say that

f is integrable over $\mathfrak{H}$ with respect to μ, and we will call $\int_{\mathfrak{T}} f(t)\mu_{\mathcal{T}}(dt)$ the integral of f over $\mathfrak{H}$ with respect to μ, written $\int_{\mathfrak{H}} f(h)\mu(dh)$.

Theorem 16 also implies

REMARK 18. *Suppose μ is a cylinder measure on Hilbert space $\mathfrak{H}$, and a and b are complex constants, and $f:\mathfrak{H} \to C$ and $g:\mathfrak{H} \to C$ (C is the complex plane) are measurable cylinder functions on $\mathfrak{H}$. Then*

(i) *f is integrable over $\mathfrak{H}$ with respect to μ iff $|f|$ is so,*

(ii) *if f and g are integrable over $\mathfrak{H}$ with respect to μ and $f \leqq g$ then $\int_{\mathfrak{H}} f(h)\mu(dh) \leqq \int_{\mathfrak{H}} g(h)\mu(dh)$,*

(iii) *if $f \geqq 0$ in $\mathfrak{H}$ and $\int_{\mathfrak{H}} f(h)\,(dh) = 0$, then $f(h) = 0$ except on a cylinder set of cylinder measure 0,*

(iv) *if f and g are integrable over $\mathfrak{H}$ with respect to μ, so is $af + bg$, and $\int_{\mathfrak{H}} [af(h) + bg(h)]\mu(dh) = a\int_{\mathfrak{H}} f(h)\mu(dh) + b\int_{\mathfrak{H}} g(h)\mu(dh)$.*

To prove Remark 18, let $\mathfrak{S}$ and $\mathfrak{T}$ be finite-dimensional subspaces of $\mathfrak{H}$ on which f and g respectively are based. Then both f and g are cylinder functions based on $\mathfrak{S} + \mathfrak{T}$, and parts (i) through (iv) of Remark 18 become, through Theorem 16, simply the usual results about probability integrals over $\mathfrak{S} + \mathfrak{T}$ with respect to the marginal distribution of μ on $\mathfrak{S} + \mathfrak{T}$.

(d) *Mean, variance, and moments of a cylinder measure.* If S is a subset of real Hilbert space $\mathfrak{H}$, denote by sp S the subspace of $\mathfrak{H}$ consisting of all real linear combinations of a finite number of vectors in S. Let μ be a cylinder measure on $\mathfrak{H}$, and let n be a positive integer. We will say that μ has bounded nth moments if

$$M_n = \sup\left\{\int_{\mathrm{sp}\{\hat{h}\}} |\hat{h}\cdot x|^n \mu_{\mathrm{sp}\{\hat{h}\}}(dx):\ \hat{h} \text{ in } \mathfrak{H} \text{ and } \|\hat{h}\| \leqq 1\right\} \tag{24}$$

is a finite number rather than $+\infty$. If m is another positive integer and $m < n$, then for any real positive x, $x^m \leqq 1 + x^n$. Thus $M_m \leqq 1 + M_n$, and if μ has bounded nth moments it has bounded moments of any lower order.

If μ has bounded first moments on $\mathfrak{H}$, then the integral $f(h) = \int_{\mathfrak{H}} h\cdot x\mu(dx)$ exists for every h in $\mathfrak{H}$ and, according to (iv) of Remark 17, the integral depends linearly on h. From (i) and (ii) of Remark 17, together with the definition (24), $|f(h)| \leqq \|h\|M_1$ for all h in $\mathfrak{H}$. Thus f is a bounded linear functional on $\mathfrak{H}$, and there is a unique vector m in $\mathfrak{H}$ such that for any h in $\mathfrak{H}$

$$h\cdot m = \int_{\mathfrak{H}} h\cdot x\mu(dx). \tag{25}$$

We will call m the mean, or center of gravity, of cylinder measure μ. It will be written heuristically as

$$m = \int_{\mathfrak{H}} x\mu(dx). \tag{26}$$

For any h, k and x in $\mathfrak{H}$, we have $h \cdot xx \cdot k = \frac{1}{4}(h + k) \cdot xx \cdot (h + k) - \frac{1}{4}(h - k) \cdot xx \cdot (h - k)$. From definition (25), it follows that if μ has bounded second moments and $\|\hat{h}\| = \|\hat{k}\| = 1$, then $\hat{h} \cdot xx \cdot \hat{k}$ is an integrable function of x over $\mathfrak{H}$ with respect to μ, and that the real-valued function

$$w(h, k) = \int_{\mathfrak{H}} h \cdot xx \cdot k\mu(dx)$$

is well defined for every h and k in $\mathfrak{H}$. Obviously $w(h, k) = w(k, h)$ and $w(h, h) \geqq 0$. From (iv) of Remark 18, w is bilinear, and from definition (24) it follows that $|w(h, h)| \leqq M_2\|h\|^2$ for any h in $\mathfrak{H}$. If $\hat{h}$ and $\hat{k}$ are unit vectors in $\mathfrak{H}$,

$$w(\hat{h}, \hat{k}) = \tfrac{1}{4}w(\hat{h} + \hat{k}, \hat{h} + \hat{k}) - \tfrac{1}{4}w(\hat{h} - \hat{k}, \hat{h} - \hat{k}),$$

so $|w(\hat{h}, \hat{k})| \leqq 2M_2$. Thus, for any h and k in $\mathfrak{H}$, $|w(h, k)| \leqq 2M_2\|h\| \|k\|$. Therefore there is a unique selfadjoint positive semidefinite bounded linear operator $W: \mathfrak{H} \to \mathfrak{H}$ such that for any h and k in $\mathfrak{H}$, $w(h, k) = h \cdot W \cdot k$. That is,

$$h \cdot W \cdot k = \int_{\mathfrak{H}} (h \cdot xx \cdot k)\mu(dx). \tag{27}$$

We will call W the second moment tensor of μ, and we will write it heuristically as

$$W = \int_{\mathfrak{H}} xx\mu(dx).$$

If μ has bounded second moments then it also has bounded first moments as well as a mean m and a second moment tensor W. For any h, k and x in $\mathfrak{H}$, we have

$$h \cdot (x - m)(x - m) \cdot k = h \cdot xx \cdot k - h \cdot mx \cdot k - h \cdot xm \cdot k + h \cdot mm \cdot k \tag{28}$$

so

$$|h \cdot (x - m)(x - m) \cdot k| \leqq |h \cdot xx \cdot k| + |h \cdot m||x \cdot k| + |h \cdot x||m \cdot k| + |h \cdot m||m \cdot k|.$$

Therefore the real-valued function

$$v(h, k) = \int_{\mathfrak{H}} [h \cdot (x - m)(x - m) \cdot k] \mu(dx)$$

is well defined for every h and k in $\mathfrak{H}$. From (28), (27) and (25) it is clear that

$$v(h, k) = h \cdot W \cdot k - h \cdot mm \cdot k.$$

Thus if we define a linear operator $V: \mathfrak{H} \to \mathfrak{H}$ as

$$V = W - mm \tag{29}$$

then for any h and k in $\mathfrak{H}$ we have

$$h \cdot V \cdot k = \int_{\mathfrak{H}} h \cdot (x - m)(x - m) \cdot k \mu(dx). \tag{30}$$

The operator V is selfadjoint, positive semidefinite, and bounded. We will call it the variance tensor of the cylinder measure μ and will write it heuristically as

$$V = \int_{\mathfrak{H}} (x - m)(x - m) \mu(dx). \tag{31}$$

If $\mathfrak{H}$ is infinite dimensional, Equations (26) and (31) are only heuristic reminders of (25) and (30). If, however, $\mathfrak{H}$ is finite dimensional and μ has bounded second moments, then (31) and (26) exist as ordinary integrals with respect to the probability measure μ and give the mean and variance tensor of μ.

Let μ be a cylinder measure on Hilbert space $\mathfrak{H}$ and let $\mathfrak{T}$ be a finite-dimensional subspace of $\mathfrak{H}$. Then the marginal distribution of μ on $\mathfrak{T}$, which we write as $\mu_{\mathscr{T}}$, is a probability measure. If μ has bounded second moments, so does $\mu_{\mathscr{T}}$, and the mean $m_{\mathscr{T}}$ and variance tensor $V_{\mathscr{T}}$ of $\mu_{\mathscr{T}}$ are given by

$$m_{\mathscr{T}} = \int_{\mathfrak{T}} t \mu_{\mathscr{T}}(dt),$$

$$V_{\mathscr{T}} = \int_{\mathfrak{T}} (t - m_{\mathscr{T}})(t - m_{\mathscr{T}}) \mu_{\mathscr{T}}(dt).$$

For any h in $\mathfrak{H}$ we have

$$\begin{aligned} h \cdot m_{\mathscr{T}} &= \int_{\mathfrak{T}} (h \cdot t) \mu_{\mathscr{T}}(dt) \\ &= \int_{\mathfrak{T}} (h \cdot \mathscr{T} \cdot t) \mu_{\mathscr{T}}(dt) \\ &= \int_{\mathfrak{H}} (h \cdot \mathscr{T} \cdot x) \mu(dx) \end{aligned}$$

because $h \cdot \mathscr{T} \cdot x$ is a cylinder function of x based on $\mathfrak{T}$. But from (25) it follows that $h \cdot m_{\mathscr{T}} = h \cdot \mathscr{T} \cdot m$ for all h in $\mathfrak{H}$, so

(32) $$m_{\mathscr{T}} = \mathscr{T} \cdot m.$$

That is, the mean of $\mu_{\mathscr{T}}$ is the orthogonal projection of the mean of μ onto $\mathfrak{T}$.

For any h and k in $\mathfrak{H}$, we have

$$h \cdot V \cdot k = \int_{\mathfrak{T}} h \cdot (t - m_{\mathscr{T}})(t - m_{\mathscr{T}}) \cdot k \mu_{\mathscr{T}}(dt).$$

On account of (32) we can write this as

$$h \cdot V_{\mathscr{T}} \cdot k = \int_{\mathfrak{T}} h \cdot \mathscr{T} \cdot (t - m)(t - m) \cdot \mathscr{T} \cdot k \mu_{\mathscr{T}}(dt).$$

But $h \cdot \mathscr{T} \cdot (x - m)(x - m) \cdot \mathscr{T} \cdot k$ is a cylinder function of x based on $\mathfrak{T}$, so

$$h \cdot V_{\mathscr{T}} \cdot k = \int_{\mathfrak{H}} h \cdot \mathscr{T} \cdot (x - m)(x - m) \cdot \mathscr{T} \cdot k \mu(dx).$$

Comparison with (30) shows that $h \cdot V_{\mathscr{T}} \cdot k = h \cdot \mathscr{T} \cdot V \cdot \mathscr{T} \cdot k$ for every h and k in $\mathfrak{H}$, whence

(33) $$V_{\mathscr{T}} = \mathscr{T} \cdot V \cdot \mathscr{T}.$$

The variance tensor of $\mu_{\mathscr{T}}$ is the orthogonal projection onto $\mathfrak{T}$ of the variance tensor of μ.

(e) *The characteristic function of a cylinder measure.* Let μ be a cylinder measure on Hilbert space $\mathfrak{H}$. For any h in $\mathfrak{H}$, $\exp(ih \cdot x)$ defines a complex-valued cylinder function of x based on $\mathrm{sp}\{h\}$, so

(34) $$\phi(h) = \int_{\mathfrak{H}} \exp(ih \cdot x) \mu(dx)$$

is a well-defined complex-valued function on $\mathfrak{H}$. For any finite-dimensional subspace $\mathfrak{T}$ of $\mathfrak{H}$, $\phi|\mathfrak{T}$ is the characteristic function of $\mu_{\mathscr{T}}$, the marginal distribution of μ on $\mathfrak{T}$. We will call ϕ the characteristic function of the cylinder measure μ; ϕ is also called the Fourier transform of μ. Since $\phi|\mathfrak{T}$ uniquely determines $\mu_{\mathscr{T}}$, ϕ uniquely determines μ.

Getoor [1957] has shown that a function $\phi : \mathfrak{H} \to C$ is the characteristic function of a cylinder measure μ on $\mathfrak{H}$ iff $\phi(0) = 1$ and $\phi|\mathfrak{T}$ is continuous and positive definite on every finite-dimensional subspace $\mathfrak{T}$ of $\mathfrak{H}$. Gross [1963] has described conditions on μ which make ϕ continuous on $\mathfrak{H}$.

(f) *Product cylinder measures.*

DEFINITION 19. Suppose that $\mathfrak{H}$ is a Hilbert space with closed subspaces $\mathfrak{K}$ and $\mathfrak{L}$ such that $\mathfrak{H} = \mathfrak{K} \dotplus \mathfrak{L}$. Suppose that $\mu_{\mathscr{H}}$, $\mu_{\mathscr{K}}$ and $\mu_{\mathscr{L}}$ are cylinder measures on $\mathfrak{H}$, $\mathfrak{K}$ and $\mathfrak{L}$. We will say that $\mu_{\mathscr{H}}$ is a product of $\mu_{\mathscr{K}}$ and $\mu_{\mathscr{L}}$ if for any cylinder subset K of $\mathfrak{K}$ and any cylinder subset L of $\mathfrak{L}$ we have

$$\mu_{\mathscr{H}}[\mathscr{K}^{-1}(K) \cap \mathscr{L}^{-1}(L)] = \mu_{\mathscr{K}}(K)\mu_{\mathscr{L}}(L). \tag{35}$$

Note that if we set $L = \mathfrak{L}$ in (35) we get

$$\mu_{\mathscr{H}}[\mathscr{K}^{-1}(K)] = \mu_{\mathscr{K}}(K), \quad \text{or} \quad \mu_{\mathscr{H}\mathscr{K}}(K) = \mu_{\mathscr{K}}(K).$$

In other words, if (35) holds then $\mu_{\mathscr{K}}$ is the marginal distribution of $\mu_{\mathscr{H}}$ on $\mathfrak{K}$ and is uniquely determined by $\mu_{\mathscr{H}}$. Similarly, $\mu_{\mathscr{L}}$ is the marginal distribution of $\mu_{\mathscr{H}}$ on $\mathfrak{L}$. This suggests a probabilistic interpretation of (35). If we regard $\mu_{\mathscr{H}}$ as a probability distribution for the vector h in $\mathfrak{H}$, then $\mu_{\mathscr{K}}(K) = \mu_{\mathscr{H}}[\mathscr{K}^{-1}(K)]$ is the probability that $\mathscr{K} \cdot h$ is in K, $\mu_{\mathscr{L}}(L)$ is the probability that $\mathscr{L} \cdot h$ is in L, and $\mu_{\mathscr{H}}[\mathscr{K}^{-1}(K) \cap \mathscr{L}^{-1}(L)]$ is the probability that $\mathscr{K} \cdot h$ is in K and also $\mathscr{L} \cdot h$ is in L. If $\mathfrak{H}$ were finite dimensional, (35) would be the condition that $\mathscr{K} \cdot h$ and $\mathscr{L} \cdot h$ be independent random variables on $\mathfrak{H}$ [Parzen, 1960]; we will use this language even in the infinite-dimensional case.

When $\mathfrak{H}$ is finite dimensional, the situation described in definition (35) has been thoroughly studied. We have

REMARK 20. *If $\mathfrak{H}$ is a finite-dimensional Hilbert space with subspaces $\mathfrak{K}$ and $\mathfrak{L}$ such that $\mathfrak{H} = \mathfrak{K} \dotplus \mathfrak{L}$, and if $\mu_{\mathscr{K}}$ and $\mu_{\mathscr{L}}$ are probability measures on $(\mathfrak{K}, B(\mathfrak{K}))$ and $(\mathfrak{L}, B(\mathfrak{L}))$, then there is exactly one probability measure $\mu_{\mathscr{H}}$ on $(\mathfrak{H}, B(\mathfrak{H}))$ which is a product of $\mu_{\mathscr{K}}$ and $\mu_{\mathscr{L}}$.*

PROOF. We identify $\mathfrak{H}$ with the product space $\mathfrak{K} \times \mathfrak{L}$ consisting of all ordered pairs (k, l) with k in $\mathfrak{K}$ and l in $\mathfrak{L}$. Let $U: \mathfrak{H} \to \mathfrak{K} \times \mathfrak{L}$ be defined by $U(h) = (\mathscr{K} \cdot h, \mathscr{L} \cdot h)$. Let $U^{-1}: \mathfrak{K} \times \mathfrak{L} \to \mathfrak{H}$ be defined by $U^{-1}(k, l) = \mathscr{K}^*_{\mathscr{L}} \cdot k + \mathscr{L}^*_{\mathscr{K}} \cdot l$. Then $U^{-1}[U(h)] = h$ for any h in $\mathfrak{H}$, and $U[U^{-1}(k, l)] = (k, l)$ for any (k, l) in $\mathfrak{K} \times \mathfrak{L}$. Thus U is a one-to-one mapping of $\mathfrak{H}$ onto $\mathfrak{K} \times \mathfrak{L}$. Clearly U preserves Borel sets. If K and L are Borel subsets of $\mathfrak{K}$ and $\mathfrak{L}$ then $K \times L$ is a Borel subset of $\mathfrak{K} \times \mathfrak{L}$, and $U^{-1}(K \times L) = \mathscr{K}^{-1}(K) \to \mathscr{L}^{-1}(L)$. Therefore, if $\mu_{\mathscr{H}}$ is a product of $\mu_{\mathscr{K}}$ and $\mu_{\mathscr{L}}$, the correspondence $U: \mathfrak{H} \times \mathfrak{K} \times \mathfrak{L}$ sends $\mu_{\mathscr{H}}$ into the product measure of $\mu_{\mathscr{K}}$ and $\mu_{\mathscr{L}}$ on $\mathfrak{K} \times \mathfrak{L}$. It is well known [Halmos, 1950, Chapter VIII] that this product measure exists and is unique.

When $\mathfrak{H}$ is finite dimensional, integrals with respect to a product measure can be evaluated with the help of

REMARK 21. *Suppose that $\mathfrak{H}$ is a finite-dimensional Hilbert space, that $\mathfrak{H} = \mathfrak{K} \dotplus \mathfrak{L}$ for closed subspaces $\mathfrak{K}$ and $\mathfrak{L}$, that $\mu_{\mathscr{H}}$, $\mu_{\mathscr{K}}$ and $\mu_{\mathscr{L}}$ are probability measures on $(\mathfrak{H}, B(\mathfrak{H}))$, $(\mathfrak{K}, B(\mathfrak{K}))$ and $(\mathfrak{L}, B(\mathfrak{L}))$, and that $\mu_{\mathscr{H}}$ is the product of $\mu_{\mathscr{K}}$ and $\mu_{\mathscr{L}}$. Suppose that $f: \mathfrak{K} \times \mathfrak{L} \to C$ is a complex-valued function on $\mathfrak{K} \times \mathfrak{L}$ such that for almost every k in $\mathfrak{K}$, $f(k, l)$ is an* *integrable function of l, and suppose that* $g(k) = \int_{\mathfrak{L}} f(k, l)\mu_{\mathscr{L}}(dl)$ *is an integrable function of k. Then the function* $\tilde{f}: \mathfrak{H} \to C$ *defined by* $\tilde{f}(h) = f(\mathscr{K} \cdot h, \mathscr{L} \cdot h)$ *is an integrable function of h, and*

$$\int_{\mathfrak{H}} f(\mathscr{K} \cdot h, \mathscr{L} \cdot h)\mu_{\mathscr{H}}(dh) = \int_{\mathfrak{K}} \left[\int_{\mathfrak{L}} f(k, l)\mu_{\mathscr{L}}(dl) \right] \mu_{\mathscr{K}}(dk). \tag{36}$$

PROOF. $\tilde{f} = f \circ U$ where U was defined in the proof of Remark 20. With this observation, Equation (36) becomes Fubini's Theorem [Halmos, 1950, Chapter VII].

The connection between product measures in finitely and infinitely many dimensions is given by

LEMMA 22. *Cylinder measure* $\mu_{\mathscr{H}}$ *on* $\mathfrak{H} = \mathfrak{K} \dotplus \mathfrak{L}$ *is a product of cylinder measures* $\mu_{\mathscr{K}}$ *on* $\mathfrak{K}$ *and* $\mu_{\mathscr{L}}$ *on* $\mathfrak{L}$ *iff for any finite-dimensional subspaces* $\mathfrak{S} \subseteqq \mathfrak{K}$ *and* $\mathfrak{T} \subseteqq \mathfrak{L}$, $\mu_{\mathscr{H}\mathscr{R}}$ *is the product measure on* $\mathfrak{R} = \mathfrak{S} \dotplus \mathfrak{T}$ *of the measures* $\mu_{\mathscr{K}\mathscr{S}}$ *on* $\mathfrak{S}$ *and* $\mu_{\mathscr{L}\mathscr{T}}$ *on* $\mathfrak{T}$.

PROOF. The lemma depends on a set-theoretic equality. We claim that if S and T are any Borel subsets of $\mathfrak{S}$ and $\mathfrak{T}$ then

$$\mathscr{S}^{-1}(S) \cap \mathscr{T}^{-1}(T) = \mathscr{R}^{-1}[(\mathscr{S}|\mathfrak{R})^{-1}(S) \cap (\mathscr{T}|\mathfrak{R})^{-1}(T)]. \tag{37}$$

This equation follows from setting $f = \mathscr{R}$ in the general rule $f^{-1}(A) \cap f^{-1}(B) = f^{-1}(A \cap B)$. Now let $K = (\mathscr{S}|\mathfrak{R})^{-1}(S)$ and $L = (\mathscr{T}|\mathfrak{L})^{-1}(T)$. Then K and L are cylinder subsets of $\mathfrak{K}$ and $\mathfrak{L}$, and $\mathscr{K}^{-1}(K) = \mathscr{S}^{-1}(S)$, $\mathscr{L}^{-1}(L) = \mathscr{T}^{-1}(T)$. Therefore Equation (35) is equivalent to the equation

$$\mu_{\mathscr{H}}\{\mathscr{R}^{-1}[(\mathscr{S}|\mathfrak{R})^{-1}(S) \cap (\mathscr{T}|\mathfrak{R})^{-1}(T)]\} = \mu_{\mathscr{K}}[(\mathscr{S}|\mathfrak{K})^{-1}(S)]\mu_{\mathscr{L}}[(\mathscr{T}|\mathfrak{L})^{-1}(T)],$$

which in turn is equivalent to

$$\mu_{\mathscr{H}\mathscr{R}}[(\mathscr{S}|\mathfrak{R})^{-1}(S) \cap (\mathscr{T}|\mathfrak{R})^{-1}(T)] = \mu_{\mathscr{K}\mathscr{S}}(S)\mu_{\mathscr{L}\mathscr{T}}(T).$$

But this last equation, if true for all S in $B(\mathfrak{S})$ and T in $B(\mathfrak{T})$, is simply the assertion that $\mu_{\mathscr{H}\mathscr{R}}$ is the product of $\mu_{\mathscr{K}\mathscr{S}}$ and $\mu_{\mathscr{L}\mathscr{T}}$. Thus Lemma 22 is established.

Now we can extend Remark 20 to the infinite-dimensional case.

THEOREM 23. *Let* $\mathfrak{H}$ *be a Hilbert space with closed subspaces* $\mathfrak{K}$ *and* $\mathfrak{L}$ *such that* $\mathfrak{H} = \mathfrak{K} \dotplus \mathfrak{L}$. *Let* $\mu_{\mathscr{K}}$ *and* $\mu_{\mathscr{L}}$ *be cylinder measures on* $\mathfrak{K}$ *and* $\mathfrak{L}$. *Then there is exactly one cylinder measure on* $\mathfrak{H}$ *which is a product of* $\mu_{\mathscr{K}}$ *and* $\mu_{\mathscr{L}}$. *We will call it the product of* $\mu_{\mathscr{K}}$ *and* $\mu_{\mathscr{L}}$ *and write it as* $\mu_{\mathscr{K} \wedge \mathscr{L}}$.

PROOF. Let $\mathfrak{U}$ be any finite-dimensional subspace of $\mathfrak{H}$. Let $\mathfrak{S} = \mathscr{K}_{\mathscr{L}} \cdot \mathfrak{U}$ and $\mathfrak{T} = \mathscr{L}_{\mathscr{K}} \cdot \mathfrak{U}$ and $\mathfrak{R} = \mathfrak{S} + \mathfrak{T}$. Then $\mathfrak{R} = \mathfrak{S} \dotplus \mathfrak{T}$, $\mathfrak{U} \subseteqq \mathfrak{R}$, and $\mathfrak{S}$ is a finite-dimensional subspace of $\mathfrak{K}$ while $\mathfrak{T}$ is a finite-dimensional subspace of $\mathfrak{L}$. If $\mu_{\mathscr{H}}$ is a product measure of $\mu_{\mathscr{K}}$ and $\mu_{\mathscr{L}}$, then according to Lemma 37 $\mu_{\mathscr{H}\mathscr{R}}$ is determined uniquely by $\mu_{\mathscr{K}\mathscr{S}}$ and $\mu_{\mathscr{L}\mathscr{T}}$. But then $\mu_{\mathscr{H}\mathscr{R}\mathscr{U}} = \mu_{\mathscr{H}\mathscr{U}}$ is determined. That is, the marginal distribution of $\mu_{\mathscr{H}}$ on any finite-dimensional subspace of $\mathfrak{H}$ is uniquely determined by $\mu_{\mathscr{K}}$ and $\mu_{\mathscr{L}}$. If H is in $c_{\mathscr{H}}(\mathfrak{U})$ then $H = \mathscr{U}^{-1}(U)$ for some Borel subset U of $\mathfrak{U}$, and $\mu_{\mathscr{H}}(H) = \mu_{\mathscr{H}\mathscr{U}}(U)$ by the Remark 13. Thus $\mu_{\mathscr{H}}(H)$ is determined for every H in $\mathbf{C}(\mathfrak{H})$; that is, $\mu_{\mathscr{K}}$ and $\mu_{\mathscr{L}}$ have at most one product cylinder measure.

To prove that they have at least one, let $\mathfrak{U}$, $\mathfrak{S}$, $\mathfrak{T}$, $\mathfrak{R}$ be defined as above. Define $\mu_{\mathscr{R}}$ as the product probability measure of $\mu_{\mathscr{K}\mathscr{S}}$ and $\mu_{\mathscr{L}\mathscr{T}}$ on $\mathfrak{R}$. Define $\mu_{\mathscr{U}}$ as $\mu_{\mathscr{R}\mathscr{U}}$. Then for any Borel subset U of $\mathfrak{U}$, define $\mu_{\mathscr{H}}[\mathscr{U}^{-1}(U)] = \mu_{\mathscr{U}}(U)$. This determines $\mu_{\mathscr{H}}(H)$ for all cylinder sets H in $\mathfrak{U}$. Thus $\mu_{\mathscr{H}}$ is a well-defined set function on $C(\mathfrak{H})$, and is countably additive on $(\mathfrak{H}, c_{\mathscr{H}}(\mathfrak{U}))$ for any finite-dimensional subspace $\mathfrak{U} \subseteq \mathfrak{H}$; that is, $\mu_{\mathscr{H}}$ is a cylinder measure on $\mathfrak{H}$. If $\mathfrak{S} \subseteq \mathfrak{K}$ and $\mathfrak{T} \subseteq \mathfrak{L}$ are finite-dimensional subspaces and $\mathfrak{R} = \mathfrak{S} + \mathfrak{T}$, then from the definition of $\mu_{\mathscr{H}}$, $\mu_{\mathscr{H}\mathscr{R}}$ is the product measure of $\mu_{\mathscr{K}\mathscr{S}}$ and $\mu_{\mathscr{L}\mathscr{T}}$ on $\mathfrak{R}$. Thus by Lemma 22, $\mu_{\mathscr{H}}$ is a product of $\mu_{\mathscr{K}}$ and $\mu_{\mathscr{L}}$. This completes the proof of Theorem 23.

Now we can prove the infinite-dimensional extension of Remark 21. We have

THEOREM 24. *Suppose that $\mathfrak{K}$ and $\mathfrak{L}$ are closed subspaces of Hilbert space $\mathfrak{H}$ and that $\mathfrak{H} = \mathfrak{K} \dotplus \mathfrak{L}$. Suppose that $\mu_{\mathscr{K}}$ and $\mu_{\mathscr{L}}$ are cylinder measures on $\mathfrak{K}$ and $\mathfrak{L}$ and that $\mu = \mu_{\mathscr{K}\vee\mathscr{L}}$ is their product. Suppose that $f : \mathfrak{K} \times \mathfrak{L} \to C$ is a function with the following properties:*

(i) *there are finite-dimensional subspaces $\mathfrak{S} \subseteq \mathfrak{K}$ and $\mathfrak{T} \subseteq \mathfrak{L}$ such that for any (k, l) in $\mathfrak{K} \times \mathfrak{L}$, $f(k, l) = f(\mathscr{S} \cdot k, \mathscr{T} \cdot l)$;*

(ii) *for any fixed k in $\mathfrak{K}$ the function $f_k : \mathfrak{L} \to C$ defined by $f_k(l) = f(k, l)$ is integrable on $\mathfrak{L}$ with respect to $\mu_{\mathscr{L}}$;*

(iii) *the function $g : \mathfrak{K} \to C$ defined by $g(k) = \int_{\mathfrak{L}} f(k, l)\mu_{\mathscr{L}}(dl)$ is integrable on $\mathfrak{K}$ with respect to $\mu_{\mathscr{K}}$.*

Then the function $\tilde{f} : \mathfrak{H} \to C$ defined by $\tilde{f}(h) = f(\mathscr{K} \cdot h, \mathscr{L} \cdot h)$ is a cylinder function on $\mathfrak{H}$, integrable with respect to $\mu = \mu_{\mathscr{K}\vee\mathscr{L}}$, and

$$\int_{\mathfrak{H}} f(\mathscr{K} \cdot h, \mathscr{L} \cdot h)\mu_{\mathscr{K}\vee\mathscr{L}}(dh) = \int_{\mathfrak{K}} \left[\int_{\mathfrak{L}} f(k, l)\mu_{\mathscr{L}}(dl) \right] \mu_{\mathscr{K}}(dk). \tag{38}$$

PROOF. Let $\mathfrak{R} = \mathfrak{S} + \mathfrak{T}$. Then

$$\begin{aligned} \tilde{f}(\mathscr{R} \cdot h) &= f(\mathscr{K} \cdot \mathscr{R} \cdot h, \mathscr{L} \cdot \mathscr{R} \cdot h) = f(\mathscr{S} \cdot h, \mathscr{T} \cdot h) \\ &= f(\mathscr{S} \cdot \mathscr{K} \cdot h, \mathscr{T} \cdot \mathscr{L} \cdot h) = f(\mathscr{K} \cdot h, \mathscr{L} \cdot h) = \tilde{f}(h). \end{aligned}$$

Thus $\tilde{f}$ is in $c_{\mathscr{H}}(\mathfrak{R})$. Therefore, by definition, $\int_{\mathfrak{H}} \tilde{f}(h)\mu(dh) = \int_{\mathfrak{R}} \tilde{f}(r)\mu_{\mathscr{R}}(dr)$, if the latter integral exists. But the hypothesis and Remark 21 assure that

$\int_{\Re} \tilde{f}(r)\mu_{\mathscr{R}}(dr)$ does exist and equals

$$\int_{\mathfrak{S}}\left[\int_{\mathfrak{T}} f(s,t)\mu_{\mathscr{L}\mathscr{T}}(dt)\right]\mu_{\mathscr{K}\mathscr{S}}(ds).$$

For each fixed s in $\mathfrak{S}$,

$$\int_{\mathfrak{T}} f(s,t)\mu_{\mathscr{L}\mathscr{T}}(dt) = \int_{\mathfrak{L}} f(s,l)\mu_{\mathscr{L}}(dl),$$

and the function $\int_{\mathfrak{L}} f(k,l)\mu_{\mathscr{L}}(dl)$ as a cylinder function of k is in $c_{\mathscr{K}}(\mathfrak{S})$. Therefore

$$\int_{\mathfrak{S}}\left[\int_{\mathfrak{T}} f(s,t)\mu_{\mathscr{L}\mathscr{T}}(dt)\right]\mu_{\mathscr{K}\mathscr{S}}(ds) = \int_{\mathfrak{K}}\left[\int_{\mathfrak{L}} f(k,l)\mu_{\mathscr{L}}(dl)\right]\mu_{\mathscr{K}}(dk),$$

which proves Theorem 24.

Now suppose that $\mathfrak{H}$ is a Hilbert space with closed subspaces $\mathfrak{L}$ and $\mathfrak{K}$ such that $\mathfrak{H} = \mathfrak{K} \dotplus \mathfrak{L}$, that $\mu_{\mathscr{K}}$ and $\mu_{\mathscr{L}}$ are cylinder measures on $\mathfrak{K}$ and $\mathfrak{L}$, and that $\mu_{\mathscr{K}\vee\mathscr{L}}$ is the product of $\mu_{\mathscr{K}}$ and $\mu_{\mathscr{L}}$ on $\mathfrak{H}$. Suppose that $\mu_{\mathscr{K}}$ and $\mu_{\mathscr{L}}$ have bounded nth moments on $\mathfrak{K}$ and $\mathfrak{L}$ respectively. Then we claim that $\mu_{\mathscr{K}\vee\mathscr{L}}$ has bounded nth moments on $\mathfrak{H}$. To see this, we observe that for any h and x in $\mathfrak{H}$,

(39) $$h \cdot x = h \cdot \mathscr{K}^*_{\mathscr{L}} \cdot \mathscr{K} \cdot x + h \cdot \mathscr{L}^*_{\mathscr{K}} \cdot \mathscr{L} \cdot x$$

so

$$|h \cdot x|^n \leqq \sum_{m=0}^{n} \binom{n}{m} |h \cdot \mathscr{K}^*_{\mathscr{L}} \cdot \mathscr{K} \cdot x|^m |h \cdot \mathscr{L}^*_{\mathscr{K}} \cdot \mathscr{L} \cdot x|^{n-m},$$

where $\binom{n}{m}$ is a binomial coefficient. If $M_m(\mu_{\mathscr{K}})$ and $M_m(\mu_{\mathscr{L}})$ denote the bounds (24) on the mth moments of $\mu_{\mathscr{K}}$ and $\mu_{\mathscr{L}}$ on $\mathfrak{K}$ and $\mathfrak{L}$ respectively, then Theorem 24 applied to the above inequality gives

$$\int_{\mathfrak{H}} |h \cdot x|^n \mu(dx) \leqq \|h\|^n \|\mathscr{K}_{\mathscr{L}}\|^n \sum_{m=0}^{n} \binom{n}{m} M_m(\mu_{\mathscr{K}}) M_{n-m}(\mu_{\mathscr{L}}).$$

Thus μ has bounded nth moments, and

(40) $$M_n(\mu) \leqq \|\mathscr{K}_{\mathscr{L}}\|^n \sum_{m=0}^{n} \binom{n}{m} M_m(\mu_{\mathscr{K}}) M_{n-m}(\mu_{\mathscr{L}}).$$

If $\mu_{\mathscr{K}}$ and $\mu_{\mathscr{L}}$ have bounded second moments on $\mathfrak{K}$ and $\mathfrak{L}$ then they have means $m_{\mathscr{K}}$ and $m_{\mathscr{L}}$ and variance tensors $V_{\mathscr{K}}$ and $V_{\mathscr{L}}$. Moreover $\mu_{\mathscr{K}\vee\mathscr{L}}$ has bounded second moments, and hence has a mean $m_{\mathscr{K}\vee\mathscr{L}}$ and a variance tensor $V_{\mathscr{K}\vee\mathscr{L}}$. We claim that the relations between these means and variances are

(41) $$m_{\mathscr{K}\vee\mathscr{L}} = \mathscr{K}^*_{\mathscr{L}} \cdot m_{\mathscr{K}} + \mathscr{L}^*_{\mathscr{K}} \cdot m_{\mathscr{L}},$$

(42) $$V_{\mathscr{K}\vee\mathscr{L}} = \mathscr{K}^*_{\mathscr{L}} \cdot V_{\mathscr{K}} \cdot \mathscr{K}_{\mathscr{L}} + \mathscr{L}^*_{\mathscr{K}} \cdot V_{\mathscr{L}} \cdot \mathscr{L}_{\mathscr{K}}.$$

In (42) we regard $V_{\mathscr{K}}$ and $V_{\mathscr{L}}$ as operators on $\mathfrak{H}$ by defining $V_{\mathscr{K}} = (V_{\mathscr{K}}|\mathfrak{K}) \cdot \mathscr{K}$ and $V_{\mathscr{L}} = (V_{\mathscr{L}}|\mathfrak{L}) \cdot \mathscr{L}$. If $(V_{\mathscr{K}}|\mathfrak{K})$ and $(V_{\mathscr{L}}|\mathfrak{L})$ have bounded inverses, then comparison of (42) with (17) shows that

$$V_{\mathscr{K} \vee \mathscr{L}}^{-1} = V_{\mathscr{K}}^{\text{inv}} + V_{\mathscr{L}}^{\text{inv}}. \tag{43}$$

To prove (41), we observe that applying Theorem 24 to the function in Equation (39) results in the equation

$$\int_{\mathfrak{H}} h \cdot x \mu_{\mathscr{K} \vee \mathscr{L}}(dx) = \int_{\mathfrak{K}} h \cdot \mathscr{K}_{\mathscr{L}}^{*} \cdot k \mu_{\mathscr{K}}(dk) + \int_{\mathfrak{L}} h \cdot \mathscr{L}_{\mathscr{K}}^{*} \cdot l \mu_{\mathscr{L}}(dl).$$

From the definitions of $m_{\mathscr{K} \vee \mathscr{L}}$, $m_{\mathscr{K}}$ and $m_{\mathscr{L}}$, it follows that

$$h \cdot m_{\mathscr{K} \vee \mathscr{L}} = (h \cdot \mathscr{K}_{\mathscr{L}}^{*}) \cdot m_{\mathscr{K}} + (h \cdot \mathscr{L}_{\mathscr{K}}^{*}) \cdot m_{\mathscr{L}}.$$

The fact that this equation is true for all h in $\mathfrak{H}$ proves (41).

To prove (42), we observe that for any h, k and x in $\mathfrak{H}$, (41) implies

$$\begin{aligned} &h \cdot (x - m_{\mathscr{K} \vee \mathscr{L}})(x - m_{\mathscr{K} \vee \mathscr{L}}) \cdot k \\ &\quad = h \cdot [\mathscr{K}_{\mathscr{L}}^{*} \cdot (\mathscr{K} \cdot x - m_{\mathscr{K}}) + \mathscr{L}_{\mathscr{K}}^{*} \cdot (\mathscr{L} \cdot x - m_{\mathscr{L}})] \\ &\qquad \cdot [(\mathscr{K} \cdot x - m_{\mathscr{K}}) \cdot \mathscr{K}_{\mathscr{L}} + (\mathscr{L} \cdot x - m_{\mathscr{L}}) \cdot \mathscr{L}_{\mathscr{K}}] \cdot k. \end{aligned}$$

We can use Theorem 24 to integrate this equation over $\mathfrak{H}$ with respect to $\mu_{\mathscr{K} \vee \mathscr{L}}$. Using (30) the result can be written $h \cdot V_{\mathscr{K} \vee \mathscr{L}} \cdot k = h \cdot [\mathscr{K}_{\mathscr{L}}^{*} \cdot V_{\mathscr{K}} \cdot \mathscr{K}_{\mathscr{L}} + \mathscr{L}_{\mathscr{K}}^{*} \cdot V_{\mathscr{L}} \cdot \mathscr{L}_{\mathscr{K}}] \cdot k$. The truth of this equation for all h and k in $\mathfrak{H}$ yields (42).

B. **Bounded linear inference on a Hilbert space.**

1. *General description of the problem.*

(a) *Continuous linear data functionals and prediction functionals.* We return now to the problem of inference from inadequate data. We suppose that the physical object E in which we are interested is adequately described by one member h_E of a Hilbert space $\mathfrak{H}$. We suppose that we have measured D numerical properties of E, which we call the "data," and label $\gamma_1, \ldots, \gamma_D$. From these data we want to predict P other numerical properties of E, which we label $\tilde{\gamma}_1, \ldots, \tilde{\gamma}_P$. We suppose that the experiments which produced the data are completely understood, so that for any member h of $\mathfrak{H}$ we can calculate theoretically what the data would have been if h were the model which represented E. In other words, for each datum we have a functional $g_i : \mathfrak{H} \to R$, $i = 1, \ldots, D$. The functionals $\{g_1, \ldots, g_D\}$ will be called the data functionals. All we know experimentally about h_E is that it satisfies the D equations $g_i(h_E) = \gamma_i$, $i = 1, \ldots, D$.

We also suppose that the properties $\tilde{\gamma}_1, \ldots, \tilde{\gamma}_P$ have been carefully defined, so that for any h in $\mathfrak{H}$ we can calculate what $\tilde{\gamma}_1, \ldots, \tilde{\gamma}_P$ would be if h were the model representing E. That is, we have P "prediction functionals," $\tilde{g}_j:\mathfrak{H} \to R$, $j = 1, \ldots, P$. We would like to use our data about h_E to estimate $\tilde{g}_j(h_E)$, $j = 1, \ldots, P$.

In the present §II.B, we will assume that the data functionals $g_i:\mathfrak{H} \to R$ and the prediction functionals $\tilde{g}_j:\mathfrak{H} \to R$ are *linear*, and until further notice we will assume that they are *continuous*. Since $g_i:\mathfrak{H} \to R$ is a continuous linear functional, it is bounded, and there is a unique vector d_i in $\mathfrak{H}$ such that for every h in $\mathfrak{H}$, $g_i(h) = d_i \cdot h$. The vectors $\{d_1, \ldots, d_D\}$ will be called the data vectors. Similarly there is a unique vector p_j in $\mathfrak{H}$ such that for every h in $\mathfrak{H}$, $\tilde{g}_j(h) = p_j \cdot h$. The vectors $\{p_1, \ldots, p_P\}$ will be called the prediction vectors.

All we know about h_E is summarized in the D equations

(44) $$d_i \cdot h_E = \gamma_i, \qquad i = 1, \ldots, D,$$

where the data vectors d_i and the real numbers γ_i are known, but h_E is unknown. What we would like to know about h_E are the values of the P numbers

(45) $$\tilde{\gamma}_j = p_j \cdot h_E, \qquad j = 1, \ldots, P,$$

where the prediction vectors p_j are known but the $\tilde{\gamma}_j$ are not.

Let $\mathfrak{D} = \mathrm{sp}\{d_1, \ldots, d_D\}$ and $\mathfrak{P} = \mathrm{sp}\{p_1, \ldots, p_P\}$. We will call $\mathfrak{D}$ the data space and $\mathfrak{P}$ the prediction space. We assert

REMARK 25. *Given $d_1, \ldots, d_D$, knowledge of the D numbers γ_i in Equation* (44) *is completely equivalent to knowledge of $\mathscr{D} \cdot h_E$; and given $p_1, \ldots, p_P$, knowledge of the P numbers $\tilde{\gamma}_j$ in* (45) *is completely equivalent to knowledge of $\mathscr{P} \cdot h_E$.*

PROOF. Obviously it suffices to consider the assertion about $\mathscr{D}$. First, if we know $\mathscr{D} \cdot h_E$ then we can calculate $d_i \cdot (\mathscr{D} \cdot h_E)$. But $\mathscr{D}$ is selfadjoint, so $d_i \cdot \mathscr{D} = \mathscr{D} \cdot d_i$. Since d_i is in $\mathfrak{D}$, $\mathscr{D} \cdot d_i = d_i$. Thus $d_i \cdot (\mathscr{D} \cdot h_E) = (d_i \cdot \mathscr{D}) \cdot h_E = d_i \cdot h_E = \gamma_i$. Next, suppose we know the numbers $d_i \cdot h_E$, $i = 1, \ldots, D$. If the d_i are linearly dependent then either the numbers γ_i are inconsistent or some of them are redundant. In the former case no model fits the data, and we must choose a different model space $\mathfrak{H}$. In the latter case we may eliminate enough of the d_i that the remainder form a basis for $\mathfrak{D}$. We will assume that this has been done, so that $\{d_1, \ldots, d_D\}$ are a basis for $\mathfrak{D}$. The basis $\{d^1, \ldots, d^D\}$ dual to $\{d_1, \ldots, d_D\}$ is uniquely determined by the requirements that its members be in $\mathfrak{D}$ and that $d^i \cdot d_j = \delta^i_j$, $i, j = 1, \ldots, D$ [Gibbs, 1901]. Here δ^i_j is the Kronecker delta, $= 1$ if $i = j$ and $= 0$ if $i \neq j$. For any h in $\mathfrak{H}$, $\mathscr{D} \cdot h$ is in $\mathfrak{D}$ so there are real numbers $a_1, \ldots, a_D$ and $a^1, \ldots, a^D$ such that

$$\mathscr{D} \cdot h = \sum_{i=1}^{D} a^i d_i = \sum_{i=1}^{D} a_i d^i.$$

Then, using the defining property of the dual basis, we see that $d^i \cdot \mathscr{D} \cdot h = a^i$ and $d_i \cdot \mathscr{D} \cdot h = a_i$, so $a^i = d^i \cdot h$ and $a_i = d_i \cdot h$. Then

$$\mathscr{D} \cdot h = \sum_{i=1}^{D} d_i(d^i \cdot h) = \sum_{i=1}^{D} d^i(d_i \cdot h). \tag{46}$$

From (46) we see that if the γ_i in (44) are known then we can calculate $\mathscr{D} \cdot h$ as

$$\mathscr{D} \cdot h = \sum_{i=1}^{D} \gamma_i d^i. \tag{47}$$

This completes the proof of Remark 25.

Equation (46) leads to an interesting formula for $\mathscr{D}$. We have, for any h in $\mathfrak{H}$, $\mathscr{D} \cdot h = (\sum d_i d^i) \cdot h = (\sum d^i d_i) \cdot h$, so

$$\mathscr{D} = \sum_{i=1}^{D} d_i d^i = \sum_{i=1}^{D} d^i d_i. \tag{48}$$

In the light of Remark 25, another way of stating our problem is this: given $\mathscr{D} \cdot h_E$, what can we say about $\mathscr{P} \cdot h_E$?

(b) *The two attacks.* When $P = \dim \mathfrak{P} = 1$ and there is only one prediction vector p, I know two ways to attack the problem of estimating $p \cdot h_E$ from $\mathscr{D} \cdot h_E$. One way is to study the limits set on h_E by knowledge of $\mathscr{D} \cdot h_E$ and a hypothesis about the boundedness of h_E, and to evaluate $p \cdot h$ for all h within those limits. The second way is to observe that if we know only $\mathscr{D} \cdot h_E$ then the only vectors d for which we can evaluate $d \cdot h_E$ from the data are the members of $\mathfrak{D}$. Therefore we try to choose that member d of $\mathfrak{D}$ which best approximates p, and estimate $p \cdot h_E$ as $d \cdot h_E$.

The first attack can be generalized fairly easily to the case $P > 1$, and the calculations it requires tend to be quite simple. When $P > 1$ and the data contain experimental errors, as they always will in a real problem, the first attack seems to require the use of Bayesian subjective probabilities [Savage, 1962]. This makes the method suspect to some probabilists. I prefer it myself because of its greater conceptual and computational simplicity.

The second method of attack is not open to doubts by the objectivist school of probabilists and is the method used so far in actual computations in the literature [Backus and Gilbert, 1968, 1970]. Therefore I will discuss this method first and in sufficient detail to show that its results can be obtained from the first method by setting $P = 1$.

2. *Approximating a single prediction vector.*

(a) *Error-free data.* If we know only $\mathscr{D} \cdot h_E$ then from the data we can calculate $d \cdot h_E$ only for vectors d in $\mathfrak{D}$. If we estimate $p \cdot h_E$ as $d \cdot h_E$, our error will be $(d - p) \cdot h_E$. Knowing nothing more about h_E, we can use Schwarz's inequality to obtain the error bound

$$|d \cdot h_E - p \cdot h_E| \leqq \|d - p\| \, \|h_E\|. \tag{49}$$

Now the error $d \cdot h_E - p \cdot h_E$ is not a physically dimensionless number. We would expect $p \cdot h_E$ to be of the order of $\|p\| \, \|h_E\|$, and if it is very much smaller than this, that is important information about h_E. Therefore we might reasonably take, as a dimensionless measure of the error in our estimate, the quotient $|d \cdot h_E - p \cdot h_E|/\|p\| \, \|h_E\|$. Then (49) shows that

$$|d \cdot h_E - p \cdot h_E|/\|p\| \, \|h_E\| \leqq \|d - p\|/\|p\|. \tag{50}$$

Therefore it is reasonable to argue that our best available estimate of $p \cdot h_E$ is $d \cdot h_E$ where d is chosen from $\mathfrak{D}$ so as to minimize $\|d - p\|$. The minimizing d is $\mathscr{D} \cdot p$, so we estimate $p \cdot h_E$ as $(\mathscr{D} \cdot p) \cdot h_E = p \cdot (\mathscr{D} \cdot h_E)$, which is what anyone would do instinctively. The dimensionless error in our estimate is, according to (50), at most $\|\mathscr{D}^\perp \cdot p\|/\|p\|$. If this number is much smaller than 1, we can do a decent job of estimating $(p \cdot h_E)/\|p\| \, \|h_E\|$ from the data. If $\|\mathscr{D}^\perp \cdot p\|/\|p\|$ is not much smaller than 1, the data are irrelevant to the prediction.

It is worth noting that we can slightly improve the bound (49) when we know that $d = \mathscr{D} \cdot p$, because then $(\mathscr{D} \cdot p - p) \cdot h_E = -p \cdot \mathscr{D}^\perp \cdot h_E = -p \cdot \mathscr{D}^\perp \cdot \mathscr{D}^\perp \cdot h_E = (\mathscr{D} \cdot p - p) \cdot (\mathscr{D}^\perp \cdot h_E)$. Thus

$$|(\mathscr{D} \cdot p - p) \cdot h_E| \leqq \|\mathscr{D}^\perp \cdot p\| \, \|\mathscr{D}^\perp \cdot h_E\|. \tag{51}$$

A complication is often injected into the foregoing argument in real calculations. Suppose, for example, that $\mathfrak{H}$ is $L_2[0, 1]$, the Hilbert space of square-integrable functions $h(x)$ on the unit interval $0 \leqq x \leqq 1$, with norm $\|h\|$ given by $\|h\|^2 = \int_0^1 h(x)^2 \, dx$. Suppose that $\int_0^1 p(x) \, dx = 1$, so that we may think of $p(x)$ as an averaging kernel and of $p \cdot h = \int_0^1 p(x)h(x) \, dx$ as a weighted average of h, possibly with some negative weights. Suppose also that there are some values x_1 between 0 and 1 for which (12) $\int_0^1 p(x)(x - x_1)^2 \, dx \ll 1$, so that p's weight is concentrated in the region of these values of x_1. Let x_0 be the value of x_1 which minimizes (12) $\int_0^1 p(x)(x - x_1)^2 \, dx$, i.e. $x_0 = \int_0^1 xp(x) \, dx$. Then we can think of $p \cdot h$ as a localized average of $h(x)$, the values of $h(x)$ near x_0 being weighted very heavily.

When we look in $\mathfrak{D}$ for a vector d which approximates p, we may want to demand that d also be an averaging kernel concentrated around x_0, i.e. that d satisfy $\int_0^1 d(x) \, dx = 1$ and $\int_0^1 x \, d(x) \, dx = x_0$. These demands can be written $d \cdot h_1 = p \cdot h_1$ and $d \cdot h_2 = p \cdot h_2$ where $h_1(x) = 1$ and $h_2(x) = x$ for $0 \leqq x \leqq 1$.

This example shows that in the general situation there may be a finite number of models $h_1, \ldots, h_F$ whose treatment by d we feel should be exact. That is, any acceptable approximation d to the prediction vector p should satisfy

$$(52) \qquad d \cdot h_i = p \cdot h_i, \qquad i = 1, \ldots, F.$$

Then we would choose d to minimize $\|d - p\|$ subject to *two* constraints; d should satisfy (52) and d should be in $\mathfrak{D}$ so that $d \cdot h_E$ can be calculated from the data.

To solve this more complicated problem, let $\mathfrak{F}$ be the orthogonal complement of $\mathrm{sp}\{h_1, \ldots, h_F\}$; i.e. $\mathfrak{F} = \mathfrak{H} \ominus \mathrm{sp}\{h_1, \ldots, h_F\}$. Then conditions (52) can be restated as the demand that $d - p$ lie in $\mathfrak{F}$ or that d lie in $p + \mathfrak{F}$. Thus we want to minimize $\|d - p\|$ subject to the constraint that d lie in $\mathfrak{D} \cap (p + \mathfrak{F})$. If there are no constraints (52), then we take $\mathfrak{F} = \mathfrak{H}$, so that the problem reduces to that already considered, because $\mathfrak{D} \cap (p + \mathfrak{H}) = \mathfrak{D} \cap \mathfrak{H} = \mathfrak{D}$.

If the problem is to have a solution, $\mathfrak{D} \cap (p + \mathfrak{F})$ cannot be empty. But $\mathfrak{D} \cap (p + \mathfrak{F})$ is nonempty iff there is a d' in $\mathfrak{D}$ and an f' in $\mathfrak{F}$ such that $d' = p + f'$. This is true iff p is in $\mathfrak{D} + \mathfrak{F}$.

Since $\mathfrak{F}$ is a closed subspace of $\mathfrak{H}$ and $\mathfrak{D}$ is finite dimensional, $\mathfrak{D} + \mathfrak{F}$ is also a closed subspace of $\mathfrak{H}$. All the data vectors $d_1, \ldots, d_D$ lie in $\mathfrak{D} + \mathfrak{F}$, and if our problem is to have a solution the prediction vector p must also lie in $\mathfrak{D} + \mathfrak{F}$. Therefore the data $\gamma_i = d_i \cdot h_E$ and the prediction $\tilde{\gamma} = p \cdot h_E$ are completely determined by the orthogonal projection of h_E on $\mathfrak{D} + \mathfrak{F}$, and without loss of generality we can take $\mathfrak{D} + \mathfrak{F}$ instead of $\mathfrak{H}$ as our space of possible models for E. The space $\mathfrak{H} \ominus (\mathfrak{D} + \mathfrak{F})$ is irrelevant both to the data and to the prediction. Therefore we will assume that

$$(53) \qquad \mathfrak{H} = \mathfrak{D} + \mathfrak{F}.$$

Next we define $\mathfrak{B} = \mathfrak{F} \ominus (\mathfrak{D} \cap \mathfrak{F})$, so that

$$(54) \qquad \mathfrak{F} = \mathfrak{B} \oplus (\mathfrak{F} \cap \mathfrak{D}).$$

Then $\mathfrak{B}$ is a closed subspace of $\mathfrak{H}$, as is $\mathfrak{D}$, and $\mathfrak{H} = \mathfrak{B} + \mathfrak{D}$. Finally, $\mathfrak{B} \cap \mathfrak{D} = \{0\}$, so we can write $\mathfrak{H} = \mathfrak{B} \dotplus \mathfrak{D}$, and we have bounded parallel projection operators $\mathscr{B}_{\mathscr{D}} : \mathfrak{H} \to \mathfrak{B}$ and $\mathscr{D}_{\mathscr{B}} : \mathfrak{H} \to \mathfrak{D}$. We want to minimize $\|d - p\|$ subject to the constraint that d lie in $\mathfrak{D} \cap (p + \mathfrak{F})$. From (54) we see that this is the variational problem solved in §II.A.2.e.

The minimizing d is $\mathscr{D}_{\mathscr{B}} \cdot p$. Then we estimate $p \cdot h_E$ as $(\mathscr{D}_{\mathscr{B}} \cdot p) \cdot h_E = p \cdot (\mathscr{D}_{\mathscr{B}}^* \cdot h_E)$. The error in our estimate is $(\mathscr{D}_{\mathscr{B}} \cdot p - p) \cdot h_E$. Since $\mathscr{F}^{\perp} \cdot (\mathscr{D}_{\mathscr{B}} \cdot p - p) = 0$ and $\mathscr{F}$ is selfadjoint, we can write this error as $(\mathscr{D}_{\mathscr{B}} \cdot p - p) \cdot$

$(\mathscr{F} \cdot h_E)$, and by Schwarz's inequality

(55) $$\frac{|(\mathscr{D}_{\mathscr{B}} \cdot p) \cdot h_E - p \cdot h_E|}{\|p\| \, \|h_E\|} \leqq \frac{\|\mathscr{B}_{\mathscr{D}} \cdot p\|}{\|p\|} \frac{\|\mathscr{F} \cdot h_E\|}{\|h_E\|}.$$

If there are no constraints (52), then $\mathscr{F} = \mathscr{H}$, $\mathscr{B}_{\mathscr{D}} = \mathscr{D}^{\perp}$, and (55) reduces to (50).

The right side of (55) can be larger than 1, so the additional constraints (52) on d can lead to a less accurate estimate of $p \cdot h_E$ than the simple estimate $p \cdot (\mathscr{D} \cdot h_E)$ in which only the data constrain d. The only clear advantage gained from the constraints (52) is that if they are applied we get a bound on our error which depends only on $\|\mathscr{F} \cdot h_E\|$ and not on $\|\mathscr{D}^{\perp} \cdot h_E\|$. In the example with $\mathfrak{H} = L_2[0, 1]$, already discussed, we had $h_1(x) = 1$ and $h_2(x) = x$. Then $\mathscr{F} \cdot h_E = \tilde{h}_E$ where $\tilde{h}_E(x) = h_E(x) - A - B(x - \frac{1}{2})$, and A and B are chosen to make

$$\int_0^1 \tilde{h}_E(x)\, dx = 0 \quad \text{and} \quad \int_0^1 (x - \tfrac{1}{2})\tilde{h}_E(x)\, dx = 0.$$

Thus $A + B(x - \frac{1}{2})$ is the straight line which fits $h_E(x)$ best in the sense of least squares, and $\|\mathscr{F} \cdot h_E\|^2 = \int_0^1 \tilde{h}_E(x)^2\, dx$ is the mean-square deviation of $h_E(x)$ from its best-fitting straight line.

(b) *Data with known error variance.* Suppose that the data $\gamma_1, \ldots, \gamma_D$ have not been measured exactly, but that the D-tuple of errors, $(\delta\gamma_1, \ldots, \delta\gamma_D)$, has a joint probability distribution on R^D with mean $(0, \ldots, 0)$ and $D \times D$ variance matrix $V_{ij} = E[(\delta\gamma_i)(\delta\gamma_j)]$. Here E means expected value. We will assume that the V_{ij} are estimated by the application of classical statistical techniques applied to repeated measurements of $(\gamma_1, \ldots, \gamma_D)$, and henceforth we will ignore the difference between this estimate of V_{ij} and its true value. We will assume that V_{ij} is known.

If $\gamma_1, \ldots, \gamma_D$ are measured exactly, we can discard data vectors until $\{d_1, \ldots, d_D\}$ are linearly independent. When there are experimental errors in $\gamma_1, \ldots, \gamma_D$, such winnowing discards information, so we really ought to treat the case in which $\{d_1, \ldots, d_D\}$ are linearly dependent. This case will be treated in §II.B.3.b, so for simplicity we will assume in the present discussion (§II.B.2.b) that $\{d_1, \ldots, d_D\}$ are linearly independent. If $\{d^1, \ldots, d^D\}$ is the basis for $\mathfrak{D}$ dual to $\{d_1, \ldots, d_D\}$, the data vectors, then

(56) $$h' = \sum_{i=1}^{D} \gamma_i d^i$$

is only an estimate for $\mathscr{D} \cdot h_E$. This estimate is in error by $\sum_{i=1}^{D} \delta\gamma_i d^i$. The probability distribution of h' in $\mathfrak{D}$ has mean $\mathscr{D}' \cdot h_E$ and variance tensor

$$V = E[(h' - \mathscr{D} \cdot h_E)(h' - \mathscr{D} \cdot h_E)] = \sum_{i,j=1}^{D} V_{ij}\, d^i d^j. \tag{57}$$

The tensor V is symmetric (i.e. selfadjoint) and positive semidefinite. It is positive-definite as long as no nontrivial linear combination of $\delta\gamma_1, \ldots, \delta\gamma_D$ vanishes with probability 1. We will exclude this degenerate case and assume that V is positive-definite.

As before, we attempt to estimate $p \cdot h_E$ by $d \cdot h_E$ for some d in $\mathfrak{D} \cap (p + \mathfrak{F})$, so that $d \cdot h_E$ is calculable from the data and d satisfies any constraints (52) that we have imposed. For any d in $\mathfrak{D} \cap (p + \mathfrak{F})$, what we calculate from the data, $\gamma_1, \ldots, \gamma_D$, is not really $d \cdot h_E$ but $d \cdot h'$, where $h' = \sum \gamma_i d^i$. The error in this estimate of $p \cdot h_E$ is $d \cdot h' - p \cdot h_E = (d - p) \cdot h_E + d \cdot (h' - h_E)$. Because we require that d be in $\mathfrak{D} \cap (p + \mathfrak{F})$, we have $d - p = (d - p) \cdot \mathscr{F}$ so we can write

$$d \cdot h' - p \cdot h_E = (d - p) \cdot \mathscr{F} \cdot h_E + d \cdot (h' - h_E).$$

The expected value of the square of this error is

$$E[(d \cdot h' - p \cdot h_E)^2] = |(d - p) \cdot \mathscr{F} \cdot h_E|^2 + d \cdot V \cdot d \tag{58}$$

where V is given by (57).

Now we come to a crucial part of the argument. If our only information about h_E is the data, $\mathscr{D} \cdot h_E$, then we can put no bound on (58). The error in our estimate of $p \cdot h_E$ can be arbitrarily large. In most physical problems, however, our interest will be concentrated on some *bounded* region of $\mathfrak{H}$. For example, most geophysicists will be surprised if the earth's central density turns out to be more than 30 gm/cm^3 or if its root-mean-square density turns out to be more than 10 gm/cm^3. A rigourous, nontrivial deduction about the earth's internal density structure which required the assumption that the root-mean-square density was less than 100 gm/cm^3 would not be absolutely conclusive, but would convince most geophysicists and be interesting to the remainder.

Suppose we are able to choose a positive number M such that we believe the inequality $\|h_E\| \leqq M$ to be very likely. Then we will be able to bound the error (58) in terms of M. Here the advantage of imposing constraints like (52) on d becomes apparent. If d is required to satisfy those constraints, we need assume not $\|h_E\| \leqq M$ but merely $\|\mathscr{F} \cdot h_E\| \leqq M$. This latter might be a more palatable assumption. In the example with $\mathfrak{H} = L_2[0, 1]$, $h_1(x) = 1$ and $h_2(x) = x$, the assumption $\|\mathscr{F} \cdot h_E\| \leqq M$ simply puts a bound on the extent to which $h_E(x)$ deviates from its best-fitting straight line.

If we are willing to assume that $\|\mathscr{F} \cdot h_E\| \leqq M$ then the largest value which $E[(d \cdot h' - p \cdot h_E)^2]$ can have in (58) is

$$\text{(59)} \qquad \varepsilon_M(d)^2 = \|d - p\|^2 M^2 + d \cdot V \cdot d.$$

We want to choose d in $\mathfrak{D} \cap (p + \mathfrak{F})$ so as to minimize not $\|d - p\|$ but the maximum error $\varepsilon_M(d)$.

For any d in $\mathfrak{D} \cap (p + \mathfrak{F})$, $d + \delta d$ is also in $\mathfrak{D} \cap (p + \mathfrak{F})$ iff δd is in $\mathfrak{D} \cap \mathfrak{F}$. Therefore if a d_0 exists in $\mathfrak{D} \cap (p + \mathfrak{F})$ which minimizes $\varepsilon_M(d)$, then $\varepsilon_M(d_0 + \delta d) - \varepsilon_M(d_0)$ must vanish to first order in δd for all δd in $\mathfrak{D} \cap \mathfrak{F}$. That is, for every δd in $\mathfrak{D} \cap \mathfrak{F}$ we must have $\delta d \cdot [M^2(d_0 - p) + V \cdot d_0] = 0$. This equation is linear in δd so there is no need to make δd small. The minimizing d_0 must satisfy

$$\text{(60)} \qquad \tilde{f} \cdot [M^2(d_0 - p) + V \cdot d_0] = 0$$

for every $\tilde{f}$ in $\mathfrak{D} \cap \mathfrak{F}$. We want to solve (60) for d_0. Let $\mathscr{D} \wedge \mathscr{F}$ denote the orthogonal projection operator from $\mathfrak{H}$ onto $\mathfrak{D} \cap \mathfrak{F}$. Then (60) is equivalent to the assertion that

$$\text{(61)} \qquad \mathscr{D} \wedge \mathscr{F} \cdot [(M^2 \mathscr{D} + V) \cdot d_0 - M^2 p] = 0.$$

Define f by the equation

$$\text{(62)} \qquad d_0 = \mathscr{D}_{\mathscr{B}} \cdot p + f.$$

Since d_0 and $\mathscr{D}_{\mathscr{B}} \cdot p$ are in $\mathfrak{D}$, so is f. Now $p - \mathscr{D}_{\mathscr{B}} \cdot p = \mathscr{B}_{\mathscr{D}} \cdot p$, a vector in $\mathfrak{B}$, and hence in $\mathfrak{F}$. Since $d_0 - p$ is also in $\mathfrak{F}$, f must be in $\mathfrak{F}$. Thus f is in $\mathfrak{D} \cap \mathfrak{F}$. If we replace d_0 in (61) by its expression in (62) we obtain

$$\mathscr{D} \wedge \mathscr{F} \cdot (M^2 \mathscr{D} + V) \cdot f + \mathscr{D} \wedge \mathscr{F} \cdot [V \cdot \mathscr{D}_{\mathscr{B}} \cdot p + M^2(\mathscr{D}_{\mathscr{B}} \cdot p - p)] = 0.$$

But $f = \mathscr{D} \wedge \mathscr{F} \cdot f$ and $\mathscr{D}_{\mathscr{B}} \cdot p - p = -\mathscr{B}_{\mathscr{D}} \cdot p$, a vector in $\mathfrak{B}$ and therefore orthogonal to $\mathfrak{D} \cap \mathfrak{F}$. Hence $\mathscr{D} \wedge \mathscr{F} \cdot (\mathscr{D}_{\mathscr{B}} \cdot p - p) = 0$, and the equation for f becomes

$$\text{(63)} \qquad \mathscr{D} \wedge \mathscr{F} \cdot (M^2 \mathscr{D} + V) \cdot \mathscr{D} \wedge \mathscr{F} \cdot f = -\mathscr{D} \wedge \mathscr{F} \cdot V \cdot \mathscr{D}_{\mathscr{B}} \cdot p.$$

The operator $M^2 \mathscr{D} + V$ is positive-definite on $\mathfrak{D}$, so

$$[\mathscr{D} \wedge \mathscr{F} \cdot (M^2 \mathscr{D} + V) \cdot \mathscr{D} \wedge \mathscr{F}] | \mathfrak{D} \cap \mathfrak{F}$$

is positive-definite, and hence has a bounded inverse. Thus the solution of (63) can be written as

$$f = -[\mathscr{D} \wedge \mathscr{F} \cdot (M^2 \mathscr{D} + V) \cdot \mathscr{D} \wedge \mathscr{F}]^{\text{inv}} \cdot V \cdot \mathscr{D}_{\mathscr{B}} \cdot p.$$

Then the d_0 which minimizes $\varepsilon_M(d)^2$ in $\mathfrak{D} \cap (p + \mathfrak{F})$, if it exists, must be

$$\text{(64)} \qquad d_0 = \mathscr{D}_{\mathscr{B}} \cdot p - [\mathscr{D} \wedge \mathscr{F} \cdot (M^2 \mathscr{D} + V) \cdot \mathscr{D} \wedge \mathscr{F}]^{\text{inv}} \cdot V \cdot \mathscr{D}_{\mathscr{B}} \cdot p.$$

So far we have shown that if there is a d_0 in $\mathfrak{D} \cap (p + \mathfrak{F})$ which minimizes $\varepsilon_M(d)^2$, it must be (64). We have not yet shown that this d_0 actually works. To see that it does, let d be any other vector in $\mathfrak{D} \cap (p + \mathfrak{F})$. Then $d = d_0 + f'$, with f' in $\mathfrak{D} \cap \mathfrak{F}$. We have

$$\varepsilon_M(d)^2 = \varepsilon_M(d_0)^2 + 2f' \cdot [M^2(d_0 - \mathrm{p}) + V \cdot d_0] + M^2 \|f'\|^2 + f' \cdot V \cdot f'.$$

Setting $\tilde{f} = f'$ in (60), we see that if d is in $\mathfrak{D} \cap (p + \mathfrak{F})$,

$$\varepsilon_M(d)^2 = \varepsilon_M(d_0)^2 + M^2 \|d - d_0\|^2 + (d - d_0) \cdot V \cdot (d - d_0),$$

so $\varepsilon_M(d) > \varepsilon_M(d_0)$ unless $d = d_0$.

Having chosen d_0 according to (64), we estimate $p \cdot h_E$ as $d_0 \cdot h'$, where h' is given by (56) using the measured $(\gamma_1, \ldots, \gamma_D)$. We can write (64) as

$$d_0 = p \cdot \mathscr{D}_{\mathscr{B}}^* - p \cdot \mathscr{D}_{\mathscr{B}}^* \cdot V \cdot [\mathscr{D} \wedge \mathscr{F} \cdot (M^2 \mathscr{D} + V) \cdot \mathscr{D} \wedge \mathscr{F}]^{\mathrm{inv}}.$$

Then $d_0 \cdot h' = p \cdot \tilde{h}_{\mathscr{B} \vee \mathscr{D}}$ where, by definition,

$$\tilde{h}_{\mathscr{B} \vee \mathscr{D}} = \mathscr{D}_{\mathscr{B}}^* \cdot h' - \mathscr{D}_{\mathscr{B}}^* \cdot V \cdot [\mathscr{D} \wedge \mathscr{F} \cdot (M^2 \mathscr{D} + V) \cdot \mathscr{D} \wedge \mathscr{F}]^{\mathrm{inv}} \cdot h'. \tag{65}$$

When we estimate $p \cdot h_E$ as $d \cdot h'$, the maximum value of the expected square error in our estimate is $\varepsilon_M(d_0)^2$. To calculate this quantity, let $U = VM^{-2}$, $W = \mathscr{D} \wedge \mathscr{F} \cdot (\mathscr{D} + U) \cdot \mathscr{D} \wedge \mathscr{F}$. Then

$$\varepsilon_M(d)^2/M^2 = \|d - p\|^2 + d \cdot U \cdot d,$$

$$d_0 = \mathscr{D}_{\mathscr{B}} \cdot p - W^{\mathrm{inv}} \cdot U \cdot \mathscr{D}_{\mathscr{B}} \cdot p,$$

$$d_0 - p = -\mathscr{B}_{\mathscr{D}} \cdot p - W^{\mathrm{inv}} \cdot U \cdot \mathscr{D}_{\mathscr{B}} \cdot p,$$

so

$$\|d_0 - p\|^2 = p \cdot \mathscr{B}_{\mathscr{D}}^* \cdot \mathscr{B}_{\mathscr{D}} \cdot p + 2p \cdot \mathscr{B}_{\mathscr{D}}^* \cdot W^{\mathrm{inv}} \cdot U \cdot \mathscr{D}_{\mathscr{B}} \cdot p + p \cdot \mathscr{D}_{\mathscr{B}}^* \cdot U \cdot W^{\mathrm{inv}} \cdot W^{\mathrm{inv}} \cdot U \cdot \mathscr{D}_{\mathscr{B}} \cdot p$$

and

$$d_0 \cdot U \cdot d_0 = p \cdot \mathscr{D}_{\mathscr{B}}^* \cdot [U - 2U \cdot W^{\mathrm{inv}} \cdot U + U \cdot W^{\mathrm{inv}} \cdot U \cdot W^{\mathrm{inv}} \cdot U] \cdot \mathscr{D}_{\mathscr{B}} \cdot p.$$

But $\mathscr{D} \wedge \mathscr{F} \cdot \mathscr{B}_{\mathscr{D}} = 0$, so $\mathscr{B}_{\mathscr{D}}^* \cdot \mathscr{D} \wedge \mathscr{F} = 0$, and $\mathscr{B}_{\mathscr{D}}^* \cdot W^{\mathrm{inv}} = 0$. Then we can write

$$\varepsilon_M(d_0)^2 = p \cdot \tilde{V}_{\mathscr{B} \vee \mathscr{D}} \cdot p \tag{66}$$

with

$$M^{-2} \tilde{V}_{\mathscr{B} \vee \mathscr{D}} = \mathscr{B}_{\mathscr{D}}^* \cdot \mathscr{B}_{\mathscr{D}} + \mathscr{D}_{\mathscr{B}}^* \cdot Q \cdot \mathscr{D}_{\mathscr{B}} \tag{67}$$

where

$$
\begin{aligned}
Q &= U \cdot W^{\mathrm{inv}} \cdot \mathscr{D} \cdot W^{\mathrm{inv}} \cdot U + U \cdot W^{\mathrm{inv}} \cdot U \cdot W^{\mathrm{inv}} \cdot U + U - 2U \cdot W^{\mathrm{inv}} \cdot U \\
&= U \cdot W^{\mathrm{inv}} \cdot (\mathscr{D} + U) \cdot W^{\mathrm{inv}} \cdot U + U - 2U \cdot W^{\mathrm{inv}} \cdot U \\
&= U \cdot W^{\mathrm{inv}} \cdot W \cdot W^{\mathrm{inv}} \cdot U + U - 2U \cdot W^{\mathrm{inv}} \cdot U \\
&= U - U \cdot W^{\mathrm{inv}} \cdot U.
\end{aligned}
$$

We claim that Q lives on $\mathfrak{D}$ and that $Q^{\mathrm{inv}} = U^{\mathrm{inv}} + \mathscr{D} \wedge \mathscr{F}$. To see this, we note that $(U^{\mathrm{inv}} + \mathscr{D} \wedge \mathscr{F}) \cdot \mathscr{D}^{\perp} = 0$ and that

$$
\begin{aligned}
&(U^{\mathrm{inv}} + \mathscr{D} \wedge \mathscr{F}) \cdot Q \\
&\quad = \mathscr{D} - W^{\mathrm{inv}} \cdot U + \mathscr{D} \wedge \mathscr{F} \cdot U - \mathscr{D} \wedge \mathscr{F} \cdot U \cdot \mathscr{D} \wedge \mathscr{F} \cdot W^{\mathrm{inv}} \cdot U \\
&\quad = \mathscr{D} + \mathscr{D} \wedge \mathscr{F} \cdot U - W \cdot W^{\mathrm{inv}} \cdot U = \mathscr{D}
\end{aligned}
$$

and

$$
\begin{aligned}
&Q \cdot (U^{\mathrm{inv}} + \mathscr{D} \wedge \mathscr{F}) \\
&\quad = \mathscr{D} + U \cdot \mathscr{D} \wedge \mathscr{F} - U \cdot W^{\mathrm{inv}} - U \cdot W^{\mathrm{inv}} \cdot U \cdot \mathscr{D} \wedge \mathscr{F} \\
&\quad = \mathscr{D} + U \cdot \mathscr{D} \wedge \mathscr{F} - U \cdot W^{\mathrm{inv}} \cdot W = \mathscr{D}.
\end{aligned}
$$

It follows that we can write (67) as

$$M^{-2} \tilde{V}_{\mathscr{B} \vee \mathscr{D}} = \mathscr{B}^*_{\mathscr{D}} \cdot \mathscr{B} \cdot \mathscr{B}_{\mathscr{D}} + \mathscr{D}^*_{\mathscr{B}} \cdot (U^{\mathrm{inv}} + \mathscr{D} \wedge \mathscr{F})^{\mathrm{inv}} \cdot \mathscr{D}_{\mathscr{B}}.$$

Comparison with formula (17) shows that

$$M^2 \tilde{V}^{-1}_{\mathscr{B} \vee \mathscr{D}} = \mathscr{B} + \mathscr{D} \wedge \mathscr{F} + U^{\mathrm{inv}}$$

so, finally,

$$\tilde{V}^{-1}_{\mathscr{B} \vee \mathscr{D}} = M^{-2} \mathscr{F} + V^{\mathrm{inv}}. \tag{68}$$

In the foregoing calculations, d_0 depends on M, so we should write it $d_0(M)$. We chose M to be any number which we thought was very likely larger than $\|\mathscr{F} \cdot h_E\|$. In any real problem we should numerically compute $\varepsilon_M[d_0(M)]^2$ as a function of M. It will decrease as M decreases, and we are interested in the question: are there values of M sufficiently large to make the hypothesis $\|\mathscr{F} \cdot m_E\| \leqq M$ plausible and yet small enough to make $\varepsilon_M[d_0(M)]^2$ so small that $d_0(M) \cdot h'$ is a useful estimate of $p \cdot h_E$? If the answer is no, then we conclude that no useful estimate of $p \cdot h_E$ can be made from the available data. If the answer is yes, then for any such M the best estimate of $p \cdot h_E$ is $d_0(M) \cdot h' = p \cdot \tilde{h}_{\mathscr{B} \vee \mathscr{D}}$, and its expected square error is given by (66) and (68).

(c) *Trade-off curves.* Recently Backus and Gilbert [1970] discussed the question of estimating $p \cdot h_E$ from erroneous data with known error variance in a way which avoids the assumption $\|\mathscr{F} \cdot h_E\| \leqq M$. The argument was as follows: we want to estimate $p \cdot h_E$ from the data, so we must estimate it as $d \cdot h_E$ with d in $\mathfrak{D}$. We want d to behave like p with respect to the constraints (52), so d must be in $\mathfrak{D} \cap (p + \mathfrak{F})$. At first we are tempted to make d as much like p as these constraints permit; that is, to minimize $\|d - p\|$ subject to d in $\mathfrak{D} \cap (p + \mathfrak{F})$, obtaining $d = \mathscr{D}_{\mathscr{B}} \cdot p$. However, the data permit us to calculate not $d \cdot h_E$ but only $d \cdot h' = d \cdot \sum \gamma_i d^i$. There may be considerable cancellation in the sum $\sum \gamma_i(d \cdot d^i)$, and, since the measured γ_i is not exactly $d_i \cdot h_E$, the accumulation of errors may make $d \cdot h'$ a very poor estimate of $d \cdot h_E$.

The function $\|d - p\|$ for d in $\mathfrak{D} \cap (p + \mathfrak{F})$ has a stationary minimum at $d = \mathscr{D}_{\mathscr{B}} \cdot p$, so vectors d in $\mathfrak{D} \cap (p + \mathfrak{F})$ can be quite far from $\mathscr{D}_{\mathscr{B}} \cdot p$ before $\|d - p\|$ becomes appreciably larger than $\|\mathscr{D}_{\mathscr{B}} \cdot p - p\|$. This raises the possibility that we might find in $\mathfrak{D} \cap (p + \mathfrak{F})$ a vector d with $\|d - p\|$ only slightly larger than $\|\mathscr{D}_{\mathscr{B}} \cdot p - p\|$ and yet producing much less cancellation in $\sum_i \gamma_i(d \cdot d^i)$ than is present in $\sum_i \gamma_i(\mathscr{D}_{\mathscr{B}} \cdot p) \cdot d^i$. That is, we might be able to make $d \cdot V \cdot d$ much smaller than $(\mathscr{D}_{\mathscr{B}} \cdot p) \cdot V \cdot (\mathscr{D}_{\mathscr{B}} \cdot p)$ and yet make $\|d - p\|$ only slightly larger than $\|\mathscr{D}_{\mathscr{B}} \cdot p - p\|$, so that d is nearly as good a fit to p as is $\mathscr{D}_{\mathscr{B}} \cdot p$.

Therefore we choose a real number α slightly larger than $\|\mathscr{D}_{\mathscr{B}} \cdot p - p\|$ and seek to minimize $d \cdot V \cdot d$ among all d in $\mathfrak{D} \cap (p + \mathfrak{F})$ which satisfy $\|d - p\| \leqq \alpha$. As we increase α, the minimum achievable $d \cdot V \cdot d$ will decrease. The graph of the resulting relationship between α, the maximum acceptable value of $\|d - p\|$, and the corresponding minimum achievable $d \cdot V \cdot d$ has been called a trade-off curve between error, $d \cdot V \cdot d$, and resolution or goodness of fit, measured by $\|d - p\|$.

For any $\alpha > \|\mathscr{D}_{\mathscr{B}} \cdot p - p\|$ and not too large, Backus and Gilbert [1970] have shown that the minimum value of $d \cdot V \cdot d$ for d in $\mathfrak{D} \cap (p + \mathfrak{F}) \cap \{d : \|d - p\| \leqq \alpha\}$ is achieved at a single point $\tilde{d}_0$ on the surface $\|d - p\| = \alpha$. At this point, $d \cdot V \cdot d$ is stationary to first order in any changes δd which leave $\tilde{d}_0 + \delta d$ in $\mathfrak{D} \cap (p + \mathfrak{F}) \cap \{d : \|d - p\| = \alpha\}$. That is, $\delta d \cdot V \cdot \tilde{d}_0 = 0$ if δd is in $\mathfrak{D} \cap \mathfrak{F}$ and $\delta d \cdot (d_0 - p) = 0$. Now if δd is in $\mathfrak{D} \cap \mathfrak{F}$, then $\delta d \cdot V \cdot \tilde{d}_0 = \delta d \cdot (\mathscr{D} \wedge \mathscr{F} \cdot V \cdot \tilde{d}_0)$. Moreover, $d_0 - p$ is in $\mathfrak{D} \cap \mathfrak{F}$, so $\delta d \cdot (\mathscr{D} \wedge \mathscr{F} \cdot V \cdot \tilde{d}_0) = 0$ as long as δd is in $\mathfrak{D} \cap \mathfrak{F} \ominus \mathscr{D} \wedge \mathscr{F} \cdot \operatorname{sp}\{\tilde{d}_0 - p\}$. In other words, $\mathscr{D} \wedge \mathscr{F} \cdot V \cdot \tilde{d}_0$ is a vector in $\mathfrak{D} \cap \mathfrak{F}$ orthogonal to $\mathfrak{D} \cap \mathfrak{F} \ominus \mathscr{D} \wedge \mathscr{F} \cdot \operatorname{sp}\{\tilde{d}_0 - p\}$. Therefore $\mathscr{D} \wedge \mathscr{F} \cdot V \cdot \tilde{d}_0$ is a member of $\mathscr{D} \wedge \mathscr{F} \cdot \operatorname{sp}\{d_0 - p\}$, and there is a real λ such that $\mathscr{D} \wedge \mathscr{F} \cdot V \cdot \tilde{d}_0 = -\lambda \mathscr{D} \wedge \mathscr{F} \cdot (\tilde{d}_0 + p)$. Therefore

$$\mathscr{D} \wedge \mathscr{F} \cdot [\lambda \tilde{d}_0 + V \cdot \tilde{d}_0 - \lambda p] = 0. \tag{69}$$

If we set $\lambda = M^2$, Equation (69) is identical with (61), so $\tilde{d}_0 = d_0$ is given by (64). In (69), however, λ is merely a Lagrange multiplier (the same λ as in Backus and Gilbert [1970]), and its interpretation as a likely upper bound for $\|\mathscr{F} \cdot h_E\|^2$ is lost. Also lost is any bound on the error $d_0 \cdot h' - p \cdot h_E$. All we have is a bound on the error $d_0 \cdot h' - d_0 \cdot h_E$ as a function of how well d_0 matches p. In short, if we approach the prediction problem by means of trade-off curves, we reproduce the $\tilde{h}_{\mathscr{B} \vee \mathscr{D}}$ of (65) but there is no analogue of the $\tilde{V}_{\mathscr{B} \vee \mathscr{D}}$ defined by (68). To me this argues in favor of investigating $\varepsilon_M[d_0(M)]^2$ as a function of the bound M we are willing to hypothecate for $\|\mathscr{F} \cdot h_E\|$, rather than discussing the trade-off of accuracy against goodness of fit. The calculations are the same in either case, and only their interpretation is in question.

There is one argument in favor of trade-off curves: we may be as happy to know $d \cdot h_E$ as $p \cdot h_E$ if $\|d - p\|$ is small. For example, if $\mathfrak{H}$ was $L_2[0, 1]$, and $p(x)$ was an averaging kernel chosen arbitrarily subject to $\int_0^1 p(x)\,dx = 1$, $\int_0^1 xp(x)\,dx = x_0$, and the requirement that most of p's weight be near x_0, then if we could find a d in $\mathfrak{D} \cap (p + \mathfrak{F})$ ($\mathfrak{F}$ being $\{h_1, h_2\}^\perp$ with $h_1(x) = 1$, $h_2(x) = x$) for which $\|d - p\|$ was small, d would also satisfy $\int_0^1 d(x)\,dx = 1$, $\int_0^1 x\,d(x)\,dx = x_0$, and most of d's weight would be near x_0. If all we wanted was a weighted average of $h_E(x)$ near x_0, then $d \cdot h_E$ would be as acceptable as $p \cdot h_E$ and would have the advantage that $d \cdot h_E$ is calculable directly from the data, with a variance determined by the observed variance of the data, while $p \cdot h_E$ cannot be estimated from the data alone, without some further hypothesis like $\|\mathscr{F} \cdot h_E\| \leqq M$, except in the trivial case that $\mathfrak{P} \subseteqq \mathfrak{D}$.

In most predictions we do not have the option to vary p slightly. If $\mathfrak{H} = L_2[0, 1]$ and $h_E(x)$ is the density of the earth as a function of distance x from the center, measured in units of the earth's radius, and if we want to predict the frequency of oscillation of an as-yet-unmeasured normal mode from the observed mass, moment, seismic travel times and normal mode frequencies, we are not free to change p at all. (The example is nonlinar, but this does not affect the point at issue. A linear example would be predicting the frictional decay rate of one normal mode from the observed decay rates of others, but a detailed discussion of solid friction would take us too far afield here.) When we are not free to vary p, the estimate of $p \cdot h_E$ based on the hypothesis that $\|\mathscr{F} \cdot h_E\| \leqq M$ seems preferable to the use of trade-off curves.

3. *Subjective bounds on* h_E.

(a) *Error-free data.* When $P = \dim \mathfrak{P} > 1$, the separate components of the P-tuple of predictions, $(p_1 \cdot h_E, \ldots, p_P \cdot h_E)$, can be estimated separately by the method of §II.B.2, but evidently this procedure destroys information. The errors in the data are not likely to permit the errors in all the components of $(p_1 \cdot h_E, \ldots, p_P \cdot h_E)$ to be large at once. For example, to detect a low-velocity zone in the mantle requires estimates of seismic velocity at three depths, and the joint distribution of the triple might have a variance ellipsoid sufficiently small to establish the existence of a low-velocity zone, whereas if the velocity estimates were regarded as independent their variances might be so large as to leave the question in doubt.

The discussions in §II.B.2 were all based on attempts to approximate the prediction vector p by a vector d in $\mathfrak{D} \cap (p + \mathfrak{F})$, so that $d \cdot h_E$ could be calculated or estimated from the data. I know of no way to generalize this procedure when $P > 1$. In §II.B.2.b we pointed out that it is often possible to find a closed subspace $\mathfrak{F} \subseteqq \mathfrak{H}$ and a positive number M such that the hypothesis $\|\mathscr{F} \cdot h_E\| \leqq M$ seems extremely plausible. In the present §II.B.3, we will systematically exploit such a hypothesis, or prejudice. When we are willing to accept such a hypothesis, we need no longer try to approximate p with a member of $\mathfrak{D} \cap (p + \mathfrak{F})$. We can estimate $p \cdot h_E$ directly.

To clarify ideas, we discuss first the unrealistic case that the data, $\gamma_1, \ldots, \gamma_D$, are measured without errors, so that we know $h_0 = \mathscr{D} \cdot h_E$ exactly. We suppose that there is a closed subspace $\mathfrak{F} \subseteqq \mathfrak{H}$ and a positive number M such that it seems very likely that

$$\|\mathscr{F} \cdot h_E\| \leqq M. \tag{70}$$

The data and the prejudice (70) permit us to say nothing more about h_E than that it is a member of the set

$$S(\mathfrak{D}, \mathfrak{F}, M) = \{h : h \text{ in } \mathfrak{H}, \mathscr{D} \cdot h = h_0, \text{and } \|\mathscr{F} \cdot h\| \leqq M\}. \tag{71}$$

Therefore, all we can say about $\mathscr{P} \cdot h_E$ from the data and the prejudice is that $\mathscr{P} \cdot h_E$ is a member of $\mathscr{P} \cdot S(\mathfrak{D}, \mathfrak{F}, M)$. The problem of estimating $\mathscr{P} \cdot h_E$ from the data $\mathscr{D} \cdot h_E = h_0$ and the prejudice $\|\mathscr{F} \cdot h_E\| \leqq M$ is simply the problem of finding a simple description of the set $\mathscr{P} \cdot S(\mathfrak{D}, \mathfrak{F}, M)$.

Define $\mathfrak{K} = \mathfrak{D} + \mathfrak{F}$. Then $\mathfrak{K}$ is a closed subspace of $\mathfrak{H}$, and $\mathscr{D} \cdot h_E = \mathscr{D} \cdot (\mathscr{K} \cdot h_E)$ and $\mathscr{F} \cdot h_E = \mathscr{F} \cdot (\mathscr{K} \cdot h_E)$. Therefore neither the data $\mathscr{D} \cdot h_E = h_0$ nor the prejudice $\|\mathscr{F} \cdot h_E\| \leqq M$ places any limits on $\mathscr{K}^\perp \cdot h_E$. If p is any member of $\mathfrak{P}$ which is not a member of $\mathfrak{K}$, then $\mathscr{K}^\perp \cdot p \neq 0$. But $p \cdot h_E = (\mathscr{K} \cdot p) \cdot (\mathscr{K} \cdot h_E) + (\mathscr{K}^\perp \cdot p) \cdot (\mathscr{K}^\perp \cdot h_E)$ and, as far as the data and the prejudice are concerned, $\mathscr{K}^\perp \cdot h_E$ can be an arbitrarily large multiple of $\mathscr{K}^\perp \cdot p$. Therefore $p \cdot h_E$ can have any value whatever, and in the direction of p, $\mathscr{P} \cdot S(\mathfrak{D}, \mathfrak{F}, M)$ is unbounded.

The data and the prejudice can be used to place limits on $p \cdot h_E$ only if p is in $\mathfrak{K} = \mathfrak{D} + \mathfrak{F}$ as well as $\mathfrak{P}$. Therefore we must restrict our attention to prediction vectors in $\mathfrak{P} \cap \mathfrak{K}$; for the others our problem is hopeless from the outset. If we relabel $\mathfrak{P} \cap \mathfrak{K}$ as $\mathfrak{P}$ then we have

$$\mathfrak{P} \subseteqq \mathfrak{D} + \mathfrak{F}. \tag{72}$$

But when (72) holds, then all the data and predictions and the prejudice $\|\mathscr{F} \cdot h_E\| \leqq M$ depend only on $\mathscr{K} \cdot h_E$; $\mathscr{K}^\perp \cdot h_E$ is irrelevant. Therefore we can replace h_E by $\mathscr{K} \cdot h_E$ and think of $\mathfrak{K}$ as our Hilbert space of models. We promptly rename $\mathfrak{K}$ as $\mathfrak{H}$, so that we have

$$\mathfrak{H} = \mathfrak{D} + \mathfrak{F}. \tag{73}$$

Henceforth we will assume (73), from which, of course, (72) follows. Equation (73) is always true if $\mathfrak{F} = \mathfrak{H}$, i.e. if we believe it likely that $\|h_E\| \leqq M$.

We define $\mathfrak{B} = \mathfrak{F} \ominus (\mathfrak{D} \cap \mathfrak{F})$, so that $\mathfrak{H} = \mathfrak{B} \dot{+} \mathfrak{D}$ and we have parallel projection operators $\mathscr{B}_{\mathscr{D}}: \mathfrak{H} \to \mathfrak{B}$ and $\mathscr{D}_{\mathscr{B}}: \mathfrak{H} \to \mathfrak{D}$. If $\mathscr{D} \wedge \mathscr{F}$ denotes the orthogonal projection operator from $\mathfrak{H}$ onto $\mathfrak{D} \cap \mathfrak{F}$ then $\mathscr{F} \cdot h_E = \mathscr{B} \cdot h_E + \mathscr{D} \wedge \mathscr{F} \cdot h_E$. Since $\mathscr{B} \cdot h_E$ and $\mathscr{D} \wedge \mathscr{F} \cdot h_E$ are mutually orthogonal, $\|\mathscr{F} \cdot h_E\|^2 = \|\mathscr{B} \cdot h_E\|^2 + \|\mathscr{D} \wedge \mathscr{F} \cdot h_E\|^2$. But $\mathscr{D} \wedge \mathscr{F} \cdot h_E = \mathscr{D} \wedge \mathscr{F} \cdot \mathscr{D} \cdot h_E$, which is calculable from the data. Therefore, if we define

$$\tilde{M}^2 = M^2 - \|\mathscr{D} \wedge \mathscr{F} \cdot h_E\|^2 \tag{74}$$

we can write

$$\mathscr{P} \cdot S(\mathfrak{D}, \mathfrak{F}, M) = \mathscr{P} \cdot \{h : h \text{ in } \mathfrak{H}, \mathscr{D} \cdot h = h_0, \text{and } \|\mathscr{B} \cdot h\| \leqq \tilde{M}\}. \tag{75}$$

Our goal in §II.B.3.a is to find an expression for $S(\mathfrak{D}, \mathfrak{F}, M)$ more intelligible than (75).

The vector $\mathscr{D}_{\mathscr{B}}^* \cdot h_E = \mathscr{D}_{\mathscr{B}}^* \cdot \mathscr{D} \cdot h_E$ is calculable from the data. Moreover, it satisfies $\mathscr{D} \cdot \mathscr{D}_{\mathscr{B}}^* \cdot h_E = \mathscr{D} \cdot h_E$, so $h_E = \mathscr{D}_{\mathscr{B}}^* \cdot h_E + d^\perp$ where $d^\perp$ is in $\mathfrak{D}^\perp$. Finally, $\mathscr{B} \cdot \mathscr{D}_{\mathscr{B}}^* \cdot h_E = 0$, so $\mathscr{B} \cdot h_E = \mathscr{B} \cdot d^\perp$, and thus

$$\mathscr{P} \cdot S(\mathfrak{D}, \mathfrak{F}, M) = \mathscr{P} \cdot \mathscr{D}_{\mathscr{B}}^* \cdot h_E + \mathscr{P} \cdot \{d^\perp : d^\perp \text{ in } \mathfrak{D}^\perp \text{ and } \|\mathscr{B} \cdot d^\perp\| \leqq \tilde{M}\}.$$

Now define

$$\mathfrak{N} = \mathfrak{P} \ominus (\mathfrak{D} \cap \mathfrak{P}). \tag{76}$$

Then if $d^\perp$ is in $\mathfrak{D}^\perp$, $(\mathscr{D} \wedge \mathscr{P}) \cdot d^\perp = 0$, so $\mathscr{P} \cdot d^\perp = (\mathscr{N} + \mathscr{D} \wedge \mathscr{P}) \cdot d^\perp = \mathscr{N} \cdot d^\perp$. Furthermore, $d^\perp = \mathscr{B}_{\mathscr{D}}^* \cdot \mathscr{B} \cdot d^\perp + \mathscr{D}_{\mathscr{B}}^* \cdot \mathscr{D} \cdot d^\perp = \mathscr{B}_{\mathscr{D}}^* \cdot \mathscr{B} \cdot d^\perp$ for any $d^\perp$ in $\mathfrak{D}^\perp$, and if b is in $\mathfrak{B}$ then $\mathscr{B}_{\mathscr{D}}^* \cdot b = \mathscr{D}_{\mathscr{B}^\perp}^\perp \cdot b$ is in $\mathfrak{D}^\perp$. Therefore

$$\mathscr{P} \cdot S(\mathfrak{D}, \mathfrak{F}, M) = \mathscr{P} \cdot \mathscr{D}_{\mathscr{B}}^* \cdot h_E + \mathscr{N} \cdot \mathscr{B}_{\mathscr{D}}^* \cdot \{b : b \text{ in } \mathfrak{B} \text{ and } \|b\| \leqq \tilde{M}\}. \tag{77}$$

Now we need two remarks:

REMARK 26. *The mapping* $\mathscr{N} \cdot \mathscr{B}_{\mathscr{D}}^* \cdot \mathscr{B}_{\mathscr{D}} \cdot \mathscr{N} : \mathfrak{N} \to \mathfrak{N}$ *is one-to-one onto.*

Since $\mathfrak{N}$ is finite dimensional, it suffices to prove that this mapping is one-to-one. Since the mapping is linear, it suffices to show that $\mathscr{N} \cdot \mathscr{B}_{\mathscr{D}}^* \cdot \mathscr{B}_{\mathscr{D}} \cdot \mathscr{N} \cdot n = 0$ and n in $\mathfrak{N}$ imply $n = 0$. But if $\mathscr{N} \cdot \mathscr{B}_{\mathscr{D}}^* \cdot \mathscr{B}_{\mathscr{D}} \cdot \mathscr{N} \cdot n = 0$, then $0 = n \cdot \mathscr{B}_{\mathscr{D}}^* \cdot \mathscr{B}_{\mathscr{D}} \cdot n = \|\mathscr{B}_{\mathscr{D}} \cdot n\|^2$, so $\mathscr{B}_{\mathscr{D}} \cdot n = 0$. Then $n = \mathscr{D}_{\mathscr{B}} \cdot n$, so n is in $\mathfrak{D}$. But n is in $\mathfrak{N}$ and hence $\mathfrak{P}$, so n is in $\mathfrak{D} \cap \mathfrak{P}$ as well as $\mathfrak{N}$. Therefore n is orthogonal to itself and must vanish.

REMARK 27. *The null space of* $\mathscr{N}\cdot\mathscr{B}_{\mathscr{D}}^{*}:\mathfrak{B}\to\mathfrak{N}$ *is* $\mathfrak{B}\ominus\mathscr{B}_{\mathscr{D}}\cdot\mathfrak{N}$.

To see this, note that if b is in $\mathfrak{B}$ then for any n in $\mathfrak{N}$ we have $n\cdot(\mathscr{N}\cdot\mathscr{B}_{\mathscr{D}}^{*}\cdot b)=(\mathscr{B}_{\mathscr{D}}\cdot n)\cdot b$. Thus $\mathscr{N}\cdot\mathscr{B}_{\mathscr{D}}^{*}\cdot b=0$ iff b is orthogonal to $\mathscr{B}_{\mathscr{D}}\cdot\mathfrak{N}$.

Now if b is in $\mathfrak{B}$ we can write $b=b'+c$ where b' is in $\mathscr{B}_{\mathscr{D}}\cdot\mathfrak{N}$ and c is in $\mathfrak{B}\ominus\mathscr{B}_{\mathscr{D}}\cdot\mathfrak{N}$. Then $\|b\|^2=\|b'\|^2+\|c\|^2$, and, by Remark 27, $\mathscr{N}\cdot\mathscr{B}_{\mathscr{D}}^{*}\cdot c=0$, so $\mathscr{N}\cdot\mathscr{B}_{\mathscr{D}}^{*}\cdot b=\mathscr{N}\cdot\mathscr{B}_{\mathscr{D}}^{*}\cdot b'$. Therefore, from Equation (77),

$$\mathscr{P}\cdot S(\mathfrak{D},\mathfrak{F},M)=\mathscr{P}\cdot\mathscr{D}_{\mathscr{B}}^{*}\cdot h_E+\mathscr{N}\cdot\mathscr{B}_{\mathscr{D}}^{*}\cdot\{b:b\text{ in }\mathscr{B}_{\mathscr{D}}\cdot\mathfrak{N}\text{ and }\|b\|\leqq\tilde{M}\}$$

$$=\mathscr{P}\cdot\mathscr{D}_{\mathscr{B}}^{*}\cdot h_E+\mathscr{N}\cdot\mathscr{B}_{\mathscr{D}}^{*}\cdot\mathscr{B}_{\mathscr{D}}\cdot\mathscr{N}\cdot\{n':n'\text{ in }\mathfrak{N}\text{ and }\|\mathscr{B}_{\mathscr{D}}\cdot n'\|^2\leqq\tilde{M}^2\}.$$

Given any n' in $\mathfrak{N}$, let $n=\mathscr{N}\cdot\mathscr{B}_{\mathscr{D}}^{*}\cdot\mathscr{B}_{\mathscr{D}}\cdot\mathscr{N}\cdot n'$. Then according to Remark 26, $n'=(\mathscr{N}\cdot\mathscr{B}_{\mathscr{D}}^{*}\cdot\mathscr{B}_{\mathscr{D}}\cdot\mathscr{N})^{\text{inv}}\cdot n$, and $\|\mathscr{B}_{\mathscr{D}}\cdot n'\|^2=n'\cdot\mathscr{N}\cdot\mathscr{B}_{\mathscr{D}}^{*}\cdot\mathscr{B}_{\mathscr{D}}\cdot\mathscr{N}\cdot n'=n\cdot(\mathscr{N}\cdot\mathscr{B}_{\mathscr{D}}^{*}\cdot\mathscr{B}_{\mathscr{D}}\cdot\mathscr{N})^{\text{inv}}\cdot n$. Therefore, finally,

(78) $$\mathscr{P}\cdot S(\mathfrak{D},\mathfrak{F},M)=\mathscr{P}\cdot\mathscr{D}_{\mathscr{B}}^{*}\cdot h_E+\{n:n\text{ in }\mathfrak{N}\text{ and }n\cdot(\mathscr{N}\cdot\mathscr{B}_{\mathscr{D}}^{*}\cdot\mathscr{B}_{\mathscr{D}}\cdot\mathscr{N})^{\text{inv}}\cdot n\leqq\tilde{M}^2\}.$$

It follows from Equation (78) and Remark 26 that $\mathscr{P}\cdot S(\mathfrak{D},\mathfrak{F},M)$ is a nondegenerate solid ellipsoid in $\mathscr{P}\cdot\mathscr{D}_{\mathscr{B}}^{*}\cdot h_E+\mathfrak{N}$ whose center is at $\mathscr{P}\cdot\mathscr{D}_{\mathscr{B}}^{*}\cdot h_E$ and whose principal semiaxes are $\hat{n}_\nu\lambda_\nu$, $\nu=1,\ldots,N$, where $\{\hat{n}_1,\ldots,\hat{n}_N\}$ is an orthonormal basis for $\mathfrak{N}$ consisting of eigenvectors of the selfadjoint positive-definite operator $\mathscr{N}\cdot\mathscr{B}_{\mathscr{D}}^{*}\cdot\mathscr{B}_{\mathscr{D}}\cdot\mathscr{N}|\mathfrak{N}$, and $\{\lambda_1^2,\ldots,\lambda_N^2\}$ are the corresponding eigenvalues of that operator.

The limits on $(p_1\cdot h_E,\ldots,p_P\cdot h_E)$ are found with the help of the mapping $f:\mathfrak{P}\to R^P$ which assigns to any vector p in $\mathfrak{P}$ the P-tuple $f(p)=(p_1\cdot p,\ldots,p_P\cdot p)$. The data $\mathscr{D}\cdot h_E=h_0$ and the prejudice $\|\mathscr{F}\cdot h_E\|\leqq M$ confine $(p_1\cdot h_E,\ldots,p_P\cdot h_E)$ to the ellipsoid $f[\mathscr{P}\cdot S(\mathfrak{D},\mathfrak{F},M)]$, which lies in the hyperplane $f[\mathscr{P}\cdot\mathscr{D}_{\mathscr{B}}^{*}\cdot h_E]+f(\mathfrak{N})$, and is nondegenerate there.

If $P=1$, there are two cases. When $\mathfrak{D}\cap\mathfrak{P}\neq\varnothing$, then p is a member of $\mathfrak{D}$, and the prediction problem is trivial. Otherwise $\mathfrak{N}=\mathfrak{P}=\text{sp}\{p\}$, and

$$p\cdot h_E=p\cdot\mathscr{D}_{\mathscr{B}}^{*}\cdot\mathscr{D}\cdot h_E+p\cdot n$$

where $n=\alpha p$ for some real α and otherwise all we know about n is that $n\cdot(\mathscr{P}\cdot\mathscr{B}_{\mathscr{D}}^{*}\cdot\mathscr{B}_{\mathscr{D}}\cdot\mathscr{P})^{\text{inv}}\cdot n\leqq\tilde{M}^2$. But if $\hat{p}=p/\|p\|$ then

$$n\cdot(\mathscr{P}\cdot\mathscr{B}_{\mathscr{D}}^{*}\cdot\mathscr{B}_{\mathscr{D}}\cdot\mathscr{P})^{\text{inv}}\cdot n=\|n\|^2(\hat{p}\cdot\mathscr{B}_{\mathscr{D}}^{*}\cdot\mathscr{B}_{\mathscr{D}}\cdot\hat{p})^{-1}$$

$$=\|n\|^2\|\mathscr{B}_{\mathscr{D}}\cdot\hat{p}\|^{-2}=\|n\|^2\|\mathscr{B}_{\mathscr{D}}\cdot p\|^{-2}\|p\|^2.$$

Thus $\|n\|\,\|p\|\leqq\tilde{M}\|\mathscr{B}_{\mathscr{D}}\cdot p\|$, so when $\dim\mathfrak{P}=1$ and p is not in $\mathfrak{D}$, we conclude that

(79) $$|p\cdot h_E-p\cdot\mathscr{D}_{\mathscr{B}}^{*}\cdot\mathscr{D}\cdot h_E|\leqq\tilde{M}\|\mathscr{B}_{\mathscr{D}}\cdot p\|.$$

When $\mathfrak{F}=\mathfrak{H}$, then $\mathscr{D}_{\mathscr{B}}^{*}=\mathscr{D}$, and $\mathscr{B}_{\mathscr{D}}=\mathscr{D}^{\perp}$, so the result becomes simply

(80) $$|p\cdot h_E-p\cdot\mathscr{D}\cdot h_E|\leqq\tilde{M}\|\mathscr{D}^{\perp}\cdot p\|,$$

where now $\tilde{M}^2=M^2-\|\mathscr{D}\cdot h_E\|^2$.

(b) *Data with known error variance.* 1. *Subjective probability and Bayes's theorem.* I know of no way to find the joint probability distribution of more than one prediction, when there are errors in the data, except to invoke a generalization of Bayesian subjective probability. The following brief discussion of that subject is modelled after Parzen [1960] and Savage [1962].

Suppose we want to weigh the Hope diamond. We have a pan balance whose error statistics we know. That is, for any possible value t of the true weight of the Hope diamond, we know the probability density $p_t(w)$ of the reading w produced by the balance. For any Borel subset B of R, the real line, $\int_B p_t(w)\,dw$ is the probability that the balance will give a weight w in B when we weigh an object whose true weight is t. According to the objectivist school of probabilists, to find the true weight t of the Hope diamond we should weigh it N times and use the observed weights $w^{(1)}, \ldots, w^{(N)}$ to estimate the parameter t in the density function $p_t(w)$. The method of estimation and the smallest N we can use are determined by the form of p_t. As an example, if p_t is unbiased, i.e. $\int_{-\infty}^{\infty} wp_t(w)\,dw = t$ for every t, we can estimate t as the sample mean $w^{(0)}$:

$$w^{(0)} = \sum_{\nu=1}^{N} w^{(\nu)}/N. \tag{81}$$

This quantity is a random variable with mean t and variance

$$N^{-1} \int_{-\infty}^{\infty} (w - t)^2 p_t(w)\,dw.$$

Its probability density is calculable from p_t and is $p_{\text{meas}}(w^{(0)}|t) = Nf_N(Nw^{(0)})$ where $f_1(w) = p_t(w)$, $f_{\nu+1}(w) = (f_\nu * p_t)(w)$, and $*$ means convolution [Cramer, 1946]. We assume, of course, that the weighings are independent.

Objectivist theory specifically forbids us to choose a constant $C(w^{(0)})$ such that $C(w^{(0)}) \int_{-\infty}^{\infty} p_{\text{meas}}(w^{(0)}|t)\,dt = 1$ and to interpret $C(w^{(0)})p_{\text{meas}}(w^{(0)}|t)$ as the probability density that the true weight of the Hope diamond is t, given that the mean of our N weighings was $w^{(0)}$. To an objectivist, the "probability distribution of the true weight of the Hope diamond" is not a useful mathematical concept. He holds that nothing in the real world is modelled by such an idea, since there is only one Hope diamond.

The subjectivists, or Bayesian school of probabilists, is more optimistic. They argue that even before any of the weighings were performed, I was willing to bet even odds that the Hope diamond weighed less than a ton, but would refuse a bet at even odds that it weighed less than a kilogram. The Bayesian subjectivists argue that before the weighings I behaved as if I ascribed to the true weight t a broad, diffuse probability distribution, which they call my *a priori* subjective probability distribution for t. They claim to be able to estimate the probability density, $p_{\text{prej}}(t)$, of this distribution by observing my reactions to various sporting propositions about the Hope diamond, all proposed before the weighings. After the weighings, I seem to bet much more "knowledgeably," apparently basing my bets on a subjective *a posteriori* probability density $p_{\text{post}}(t)$ which has a much smaller variance than $p_{\text{prej}}(t)$. The subjectivists see the statistical problem of measurement as the question of discussing rationally how to obtain p_{post} from p_{prej} and the observed outcomes of the N weighings.

The subjectivist or Bayesian approach depends heavily on the idea of conditional probability. Let (S, Ω, P) be a probability measure space. The measure on the σ-ring Ω of subsets of the set S is interpreted as a probability. If $B \subseteqq S$ is in Ω, then $P(B)$ is the probability that s is a member of B if s is picked at random from among the members of S according to the probability law P. For any subsets $B \subseteqq S$ and $C \subseteqq S$ which are members of Ω, if $P(B) > 0$ we define

$$P(C|B) = P(C \cap B)/P(B) \tag{82}$$

and we interpret $P(C|B)$ as the conditional probability that an element s from S lies in C, given that it lies in B. For a fixed set B in Ω with $P(B) > 0$, let Ω_B be the σ-ring of all subsets of B which are members of Ω, and let $P_B(C) = P(C|B)$ for any C in Ω_B. Then (B, Ω_B, P_B) is a probability measure space. It is supposed to describe the following sampling process: we pick a subset $C \subseteqq B$ which is a member of Ω_B, and we select so many elements at random from S according to probability law P that a large number of those selected are members of C. Then $P_B(C)$ is an estimate of the ratio of the number of selected elements which are in C to the number which are in B.

If B and C are in Ω and $P(B) > 0$ and $P(C) > 0$ then $P(B \cap C) = P(B|C)P(C) = P(C|B)P(B)$, so

$$P(C|B) = P(B|C)P(C)/P(B). \tag{83}$$

Suppose $\{C_1, C_2, \ldots\}$ is a finite or denumerable sequence of pairwise disjoint sets in Ω such that $B \subseteqq \bigcup_j C_j$. Then $B = \bigcup_j (B \cap C_j)$ so

$$P(B) = \sum_j P(B \cap C_j) = \sum_j P(B|C_j)P(C_j). \tag{84}$$

If we substitute this in (83) and set $C = C_i$ there, we obtain the simplest form of Bayes's theorem:

$$P(C_i|B) = P(B|C_i)P(C_i)/\sum_j P(B|C_j)P(C_j). \tag{85}$$

This noncontroversial formula is a straightforward deduction from the definitions, and is accepted by objectivists and subjectivists alike. The controversy concerns the way in which the subjectivists apply the formula.

Bayesians argue as follows: suppose $C_1, C_2, \ldots$ represent mutually exclusive hypotheses, and suppose B is a set for which we know the conditional probabilities $P(B|C_i)$. Suppose that by introspection or betting I find that to hypothesis C_i I assign an a priori subjective probability $p^{(i)}_{\text{prej}}$. Now someone selects a member of S (performs an experiment) and reports to me only that the result lies in B. In the light of this fact, Bayesians say I should adopt new, a posteriori, subjective probabilities $p^{(i)}_{\text{post}}$, for the hypotheses C_i, as follows: I set $P(C_i) = p^{(i)}_{\text{prej}}$, use (85) to calculate $P(C_i|B)$, and set $p^{(i)}_{\text{post}} = P(C_i|B)$. In short, $p^{(i)}_{\text{post}}$ is the probability of C_i given that B has occurred.

The Bayesian discussion of how to weigh the Hope diamond involves a continuous analogue of (85). Let W and T both be the real line, let $S = W \times T$, and let Ω be the collection of all Borel subsets of S. In the measurable space (S, Ω) a point (w, t) is regarded as the event that the true weight of the Hope diamond is t and that we measure it to be w (for example, by averaging N weighings). For any t and any Borel subset $\tilde{W} \subseteqq W$ we are supposed to know $\mu^t_{\text{meas}}(\tilde{W})$, the probability that we will measure the weight to lie in $\tilde{W}$ when it is really t. We assume that μ^t_{meas} has a density function $p_{\text{meas}}(w|t)$, so that

$$\mu^t_{\text{meas}}(\tilde{W}) = \int_{\tilde{W}} p_{\text{meas}}(w|t)\lambda(dw) \tag{86}$$

where λ is Lebesgue measure on W. Bayesians claim that we also have an a priori subjective probability distribution for t, a probability measure μ_{prej} on $(T, B(T))$. They claim that even before we make any measurements we can establish a subjective probability distribution P on (S, Ω). They argue as follows:

(i) The marginal distribution on T of the probability measure P ought to be μ_{prej}. If $\tilde{T}$ is any Borel subset of T, then $P(W \times \tilde{T})$ is the probability that t is in $\tilde{T}$ if all information about the value of w is ignored.

(ii) If B is any Borel subset of $W \times T$, for any t in T let $B_t = \{w : (w, t) \text{ is in } B\}$. See Figure 3.

If we subdivide T into intervals $T_i = \{t : t_i \leqq t < t_{i+1}\}$ and let $C_i = W \times T_i$ then (84) ought to apply, so

$$P(B) = \sum_i P(B|C_i)P(C_i) = \sum_i P(B|C_i)\mu_{\text{prej}}(T_i).$$

Here $P(B|C_i)$ is the probability that (w, t) is in B, given that it is in C_i. Arguing heuristically for a moment, we observe that if T_i is very short then $P(B|C_i)$ is nearly the probability that w is in B_{t_i} given that t is in C_i.

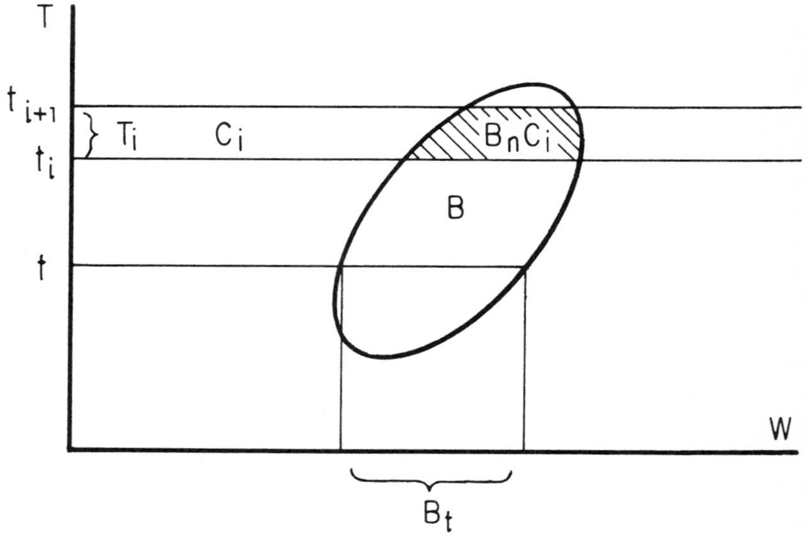

FIGURE 3

That is, $P(B|C_i)$ is approximately $\int_{B_{t_i}} p_{\text{meas}}(w|t_i)\lambda(dw)$. If we let the lengths of the intervals T_i approach zero we obtain a heuristic justification for the formula

$$P(B) = \int_T \left[\int_{B_t} p_{\text{meas}}(w|t)\lambda(dw) \right] \mu_{\text{prej}}(dt), \tag{87}$$

which we now simply adopt as the definition of P.

Of course $p_{\text{meas}} \geqq 0$, and we will assume that $p_{\text{meas}}(w|t)$ is integrable on $W \times T$ with respect to the product measure of λ and μ_{prej}. Therefore Equation (87) does define a probability measure P on (S, Ω). Bayesians regard the probability measure space (S, Ω, P) as describing our subjective probability distribution for the double event that we measure weight w for the Hope diamond and that its true weight is t.

Suppose the measurement is performed, and we are told only that its outcome, w, lies in a Borel subset $\tilde{W}$ of W. Then, according to Bayesians, we should revise our subjective probability distribution for t in T to a new measure $\mu_{\text{post}}^{\tilde{W}}$ which uses this information. For any Borel subset $\tilde{T}$ of T we should take $\mu_{\text{post}}^{\tilde{W}}(\tilde{T})$ to be the conditional probability that t lies in $\tilde{T}$, given that w lies in $\tilde{W}$. That is,

$$\mu_{\text{post}}^{\tilde{W}}(\tilde{T}) = P(W \times \tilde{T} | \tilde{W} \times T) = P(\tilde{W} \times \tilde{T}) / P(\tilde{W} \times T),$$

or

(88) $$\mu_{\text{post}}^{\tilde{W}}(\tilde{T}) = \int_{\tilde{W}} f_{\tilde{T}}(w)\lambda(dw) \Big/ \int_{\tilde{W}} f_T(w)\lambda(dw)$$

where, from (87),

$$f_{\tilde{T}}(w) = \int_{\tilde{T}} p_{\text{meas}}(w|t)\mu_{\text{prej}}(dt).$$

Now we assume that $p_{\text{meas}}(w|t)$ depends on w and t in such a way that $f_{\tilde{T}}(w)$ is a continuous function of w for every Borel subset $\tilde{T} \subseteqq T$. Then for any fixed w in W we can take the limit of (88) (using the Lebesgue bounded convergence theorem) over intervals $\tilde{W}$ which contain and shrink down to w. Thus the limit of $\mu_{\text{post}}^{\tilde{W}}(\tilde{T})$ is $\mu_{\text{post}}^{w}(\tilde{T}) = f_{\tilde{T}}(w)/f_T(w)$, or

(89) $$\mu_{\text{post}}^{w}(\tilde{T}) = \int_{\tilde{T}} \bar{p}_{\text{meas}}(w|t)\mu_{\text{prej}}(dt)$$

where

(90) $$\bar{p}_{\text{meas}}(w|t) = p_{\text{meas}}(w|t) \Big/ \int_T p_{\text{meas}}(w|t)\mu_{\text{prej}}(dt).$$

Equation (89) with (90) is evidently the generalization of (85).

As a function of t, $\bar{p}_{\text{meas}}(w|t)$ is integrable with respect to μ_{prej}, and $\bar{p}_{\text{meas}} \geqq 0$, so (89) does define a new probability measure μ_{post}^{w} on $(T, B(T))$. Bayesians claim that we ought to adopt it as our a posteriori subjective probability distribution for the true weight t of the Hope diamond, given that our measuring process produced a value w for the weight.

If μ_{prej} has a probability density $p_{\text{prej}}(t)$ then μ_{post}^{w} has, for each fixed w, a probability density in t given by

(91) $$p_{\text{post}}(t|w) = \bar{p}_{\text{meas}}(w|t)p_{\text{prej}}(t).$$

If $p_{\text{prej}}(t)$ has a much larger variance in t than $\bar{p}_{\text{meas}}(w|t)$ then $p_{\text{post}}(t|w)$ is very nearly $\bar{p}_{\text{meas}}(w|t)$, or at least this is true of the measures they generate on T. Then $p_{\text{post}}(t|w)$ is nearly independent of my a priori prejudices as expressed in $p_{\text{prej}}(t)$.

In the foregoing discussion we assumed that $p_{\text{meas}}(w|t)$ was known. In practice we might try to estimate it from the distribution of the actual weighings $w^{(1)}, \ldots, w^{(N)}$ of the Hope diamond by assuming that it was a member of some finite-parameter family of distributions and using objectivist methods to estimate the parameters. For example, we might assume that there was a normal distribution $n_v(w)$ with mean 0 and unknown variance v such that for any t, $p_{\text{meas}}(w|t) = n_v(w - t)$. Then an unbiased estimate for t (in fact the maximum likelihood estimate) is (81), the mean $w^{(0)}$ of the N weighings, and an unbiased estimate for v is given by $(N-1)v = \sum_{\nu=1}^{N} [w^{(\nu)} - w^{(0)}]^2$ [Cramer, 1946]. Moreover, $p_{\text{meas}}(w|t$ is normal in w with mean t and variance $V = v/N$, so an estimate of V is

$$V = \left(\sum_{\nu=1}^{N} [w^{(\nu)} - w^{(0)}]^2 \right) \Big/ N(N-1). \tag{92}$$

If W is not the real line but simply a topological space, and $B(W)$ is the σ-ring generated by the open subsets of W and λ is σ-finite, while $(T, B(T), \mu_{\text{prej}})$ is any probability measure space, the foregoing discussion can be repeated verbatim. We might regard W as the data space and T as the model space.

2. *Bayesian inference on* $\mathfrak{H}$. Let us now return to our problem of inference on the Hilbert space $\mathfrak{H}$ containing an unknown vector h_E which correctly models the physical object in which we are interested. We have D data vectors $\{d_1, \ldots, d_D\}$ *which may be linearly dependent* and which span the data space $\mathfrak{D} \subseteq \mathfrak{H}$. We have P prediction vectors $\{p_1, \ldots, p_P\}$ in $\mathfrak{H}$ which may be linearly dependent and which span the prediction space $\mathfrak{P}$. We want to estimate the P-tuple $(p_1 \cdot h_E, \ldots, p_P \cdot h_E)$, using measurements γ_i of the D quantities $d_i \cdot h_E$, $i = 1, \ldots, D$. These measurements may be in error by amounts whose statistics we know or can estimate (for example, by repeated measurement).

Let $G = R^D$ be the space of real D-tuples $\gamma = (\gamma_1, \ldots, \gamma_D)$. Our attempt to measure $d_1 \cdot h_E, \ldots, d_D \cdot h_E$ produces a D-tuple γ in G. We assume that for each h in $\mathfrak{H}$ we know a probability measure μ^h_{meas} on $(G, B(G))$. For any Borel subset $\tilde{G}$ of G, $\mu^h_{\text{meas}}(\tilde{G})$ is the probability that if h_E is h our measurements will produce a D-tuple which lies in $\tilde{G}$. We assume that μ^h_{meas} has a density function $p_{\text{meas}}(\gamma|h)$, so that

$$\mu^h_{\text{meas}}(\tilde{G}) = \int_{\tilde{G}} p_{\text{meas}}(\gamma|h)\lambda(d\gamma) \tag{93}$$

where λ is Lebesgue measure on G. And we assume that $\mu^h_{\text{meas}}(\tilde{G})$ as a function of h depends only on $d_1 \cdot h, \ldots, d_D \cdot h$, i.e. only on $\mathscr{D} \cdot h$. Thus

$$p_{\text{meas}}(\gamma|h) = p_{\text{meas}}(\gamma|\mathscr{D} \cdot h). \tag{94}$$

For any fixed γ, $p_{\text{meas}}(\gamma|h)$ is a cylinder function of h based on $\mathfrak{D}$.

What should we take for T, our space of models, and for μ_{prej}, our a priori probability distribution on T? There is an analogue on $\mathfrak{H}$ of the normal distribution with mean *0* and variance $M^2\mathscr{H}$, and choosing this for μ_{prej} would nicely reflect a belief that probably $\|h_E\| \leq M$. This μ_{prej}, however, is only a cylinder measure, not a measure. We can be sure that μ_{prej} is a measure by taking T to be a finite-dimensional subspace $\mathfrak{S}$ of $\mathfrak{H}$. Then $\mu_{\text{prej}\mathscr{S}}$ is our a priori subjective probability measure for

$\mathscr{S} \cdot h_E$ in $\mathfrak{S}$. From knowledge of $\mathscr{S} \cdot h_E$, we want to be able to predict the outcome of our measurements, i.e. to calculate $p_{\text{meas}}(\gamma|h_E)$. Therefore $\mathscr{S} \cdot h$ must determine $\mathscr{D} \cdot h$. That is, if $\mathscr{S} \cdot h = \mathscr{S} \cdot h'$ then $\mathscr{D} \cdot h = \mathscr{D} \cdot h'$; if $\mathscr{S} \cdot (h - h') = 0$ then $\mathscr{D} \cdot (h - h') = 0$. In other words, $\mathfrak{S}^{\perp} \subseteqq \mathfrak{D}^{\perp}$, whence $\mathfrak{D} \subseteqq \mathfrak{S}$.

Once we accept an a priori subjective probability measure $\mu_{\text{prej}\,\mathscr{S}}$ on $(\mathfrak{S}, B(\mathfrak{S}))$ we can calculate from (89) an a posteriori subjective probability measure $\mu_{\text{post}\,\mathscr{S}}$ for $\mathscr{S} \cdot h_E$ in $\mathfrak{S}$ which takes account of the data. From $\mu_{\text{post}\,\mathscr{S}}$ we want to obtain our a posteriori probability measure $\mu_{\text{post}\,\mathscr{P}}$ for $\mathscr{P} \cdot h_E$ in $\mathfrak{P}$. Therefore we choose $\mathfrak{S}$ so that $\mathfrak{P} \subseteqq \mathfrak{S}$, and we set $\mu_{\text{post}\,\mathscr{P}} = \mu_{\text{post}\,\mathscr{S}\mathscr{P}}$, the marginal distribution of $\mu_{\text{post}\,\mathscr{S}}$ on $\mathfrak{P}$.

Let $\mathfrak{S}$ be any finite-dimensional subspace of $\mathfrak{H}$ such that $\mathfrak{D} + \mathfrak{P} \subseteqq \mathfrak{S} \subseteqq \mathfrak{H}$. Suppose that we have an a priori subjective probability distribution for $\mathscr{S} \cdot h_E$ in $\mathfrak{S}$, i.e. a probability measure $\mu_{\text{prej}\,\mathscr{S}}$ on $(\mathfrak{S}, B(\mathfrak{S}))$. (We do not assume here that we have an a priori probability distribution for h_E in $\mathfrak{H}$.) The Bayesian recommendation is that after learning the D-tuple γ which is the outcome of our attempt to measure $d_1 \cdot h_E, \ldots, d_D \cdot h_E$ we adopt an a posteriori subjective probability measure $\mu^{\gamma}_{\text{post}\,\mathscr{S}}$ on $\mathfrak{S}$ which assigns to any Borel subset S of $\mathfrak{S}$ the probability

$$(95) \qquad \mu^{\gamma}_{\text{post}\,\mathscr{S}}(S) = \int_S p_{\text{meas}}(\gamma|s)\mu_{\text{prej}\,\mathscr{S}}(ds) \Big/ \int_{\mathfrak{S}} p_{\text{meas}}(\gamma|s)\mu_{\text{prej}\,\mathscr{S}}(ds).$$

From $\mu^{\gamma}_{\text{post}\,\mathscr{S}}$ we find, as its marginal distribution on $\mathfrak{P}$, our a posteriori subjective probability distribution $\mu^{\gamma}_{\text{post}\,\mathscr{P}}$ for $\mathscr{P} \cdot h_E$ in $\mathfrak{P}$. From this we can immediately obtain our a posteriori subjective probability distribution for $\tilde{\gamma} = (p_1 \cdot h_E, \ldots, p_P \cdot h_E)$ in R^P. For example, if $h_{\mathscr{S}}$ and $V_{\mathscr{S}}$ are the mean and variance tensor of $\mu^{\gamma}_{\text{post}\,\mathscr{S}}$ in $\mathfrak{S}$, then $h_{\mathscr{P}} = \mathscr{P} \cdot h_{\mathscr{S}}$ and $V_{\mathscr{P}} = \mathscr{P} \cdot V_{\mathscr{S}} \cdot \mathscr{P}$ are the mean and variance tensor of $\mu^{\gamma}_{\text{post}\,\mathscr{P}}$, and $(p_1 \cdot h_{\mathscr{P}}, \ldots, p_P \cdot h_{\mathscr{P}})$ and $p_i \cdot V_{\mathscr{P}} \cdot p_j$ are the mean and variance matrix of the a posteriori subjective probability distribution of $\tilde{\gamma}$ in R^P. Note that this latter will be singular if $\{p_1, \ldots, p_P\}$ are linearly dependent, since then all the mass in R^P lies on $\tilde{\Gamma}(R^P)$ where $\tilde{\Gamma}: \mathfrak{P} \to R^P$ is defined by requiring for each p in $\mathfrak{P}$ that $\tilde{\Gamma}(p) = (p_1 \cdot p, \ldots, p_P \cdot p)$.

One question remains: suppose that instead of $\mathfrak{S}$ we had chosen a different finite-dimensional space $\mathfrak{T}$ such that $\mathfrak{D} + \mathfrak{P} \subseteqq \mathfrak{T} \subseteqq \mathfrak{H}$ and another a priori subjective probability measure $\mu_{\text{prej}\,\mathscr{T}}$ on $\mathfrak{T}$. Would we obtain the same result for $\mu^{\gamma}_{\text{post}\,\mathscr{P}}$? Obviously not, unless $\mu_{\text{prej}\,\mathscr{S}}$ and $\mu_{\text{prej}\,\mathscr{T}}$ are consistent, by which we mean that they have the same marginal distribution on $\mathfrak{S} \cap \mathfrak{T}$. If they are consistent, we can replace $\mathfrak{S}$ by $\mathfrak{S} \cap \mathfrak{T}$, so it suffices to consider the case that $\mathfrak{D} + \mathfrak{P} \subseteqq \mathfrak{S} \subseteqq \mathfrak{T} \subseteqq \mathfrak{H}$, with $\mu_{\text{prej}\,\mathscr{S}} = \mu_{\text{prej}\,\mathscr{T}\mathscr{S}}$, and to show that $(\mathfrak{S}, B(\mathfrak{S}), \mu_{\text{prej}\,\mathscr{S}})$ and $(\mathfrak{T}, B(\mathfrak{T}), \mu_{\text{prej}\,\mathscr{T}})$ lead to the same

$\mu^{\gamma}_{\text{post}\,\mathscr{P}}$. We will show more, that they lead to the same $\mu^{\gamma}_{\text{post}\,\mathscr{S}}$. Let S be any Borel subset of $\mathfrak{S}$. Since $p_{\text{meas}}(\gamma|t)$ is a cylinder function of t based on $\mathfrak{S}$, Theorem 16 applied to $\mathfrak{T}$ says that the integrals in (95) are unchanged if in them we replace $\mathscr{S}$ by $\mathscr{T}$, s by t, and S by $T = (\mathscr{S}|\mathfrak{T})^{-1}(S)$. In other words, $\mu^{\gamma}_{\text{post}\,\mathscr{S}}(S) = \mu^{\gamma}_{\text{post}\,\mathscr{T}}(T)$. But this is precisely the assertion that $\mu^{\gamma}_{\text{post}\,\mathscr{S}} = \mu^{\gamma}_{\text{post}\,\mathscr{T}\mathscr{S}}$. We have proved

REMARK 28. *If $\mathfrak{S}$ and $\mathfrak{T}$ are finite-dimensional subspaces of Hilbert space $\mathfrak{H}$ and $\mathfrak{D} + \mathfrak{P} \subseteqq \mathfrak{S} \subseteqq \mathfrak{H}$ and $\mathfrak{D} + \mathfrak{P} \subseteqq \mathfrak{T} \subseteqq \mathfrak{H}$ then if $\mu_{\text{prej}\,\mathscr{S}}$ and $\mu_{\text{prej}\,\mathscr{T}}$ are consistent so are the $\mu^{\gamma}_{\text{post}\,\mathscr{S}}$ and $\mu^{\gamma}_{\text{post}\,\mathscr{T}}$ obtained from* (95).

If we want to be able to compute a consistent system of a posteriori subjective probability distributions $\mu^{\gamma}_{\text{post}\,\mathscr{P}}$ for $\mathscr{P} \cdot h_E$ in every finite-dimensional subspace $\mathfrak{P}$ of $\mathfrak{H}$, then we must have a consistent system of a priori subjective probability measure spaces $(\mathfrak{S}, B(\mathfrak{S}), \mu_{\text{prej}\,\mathscr{S}})$ for all the finite-dimensional subspaces $\mathfrak{S}$ of $\mathfrak{H}$ which contain $\mathfrak{D}$. We will have such a system if all the measures $\mu_{\text{prej}\,\mathscr{S}}$ are marginal distributions of a single cylinder measure μ_{prej} on $(\mathfrak{H}, \boldsymbol{C}(\mathfrak{H}))$. According to Remark 15 the converse is also true. Thus we are led in a natural way to choose a cylinder measure on $\mathfrak{H}$ to describe our a priori subjective probability distribution for h_E in $\mathfrak{H}$. This fact makes it worth while to formulate the result of the foregoing discussion as

THEOREM 29. *If μ_{prej} is a cylinder measure on Hilbert space $\mathfrak{H}$, and all the a priori subjective probability distributions $\mu_{\text{prej}\,\mathscr{S}}$ on finite-dimensional subspaces $\mathfrak{S}$ satisfying $\mathfrak{D} + \mathfrak{P} \subseteqq \mathfrak{S} \subseteqq \mathfrak{H}$ are marginal distributions of μ_{prej}, then all the a posteriori subjective probability distributions $\mu^{\gamma}_{\text{post}\,\mathscr{S}}$ obtained by the Bayesian Equation* (95) *are marginal distributions of a single cylinder measure $\mu^{\gamma}_{\text{post}}$ on $\mathfrak{H}$.* (*Here we assume that $p_{\text{meas}}(\gamma|d)$ is integrable on $R^D \times \mathfrak{D}$ with respect to the product measure of Lebesgue measure on R^P and $\mu_{\text{prej}\,\mathscr{D}}$ on $\mathfrak{D}$.*)

According to Theorem 29, if we want to use an a priori subjective probability distribution μ_{prej} for h_E in $\mathfrak{H}$, together with a known or estimated probability density $p_{\text{meas}}(\gamma|d)$ for the measurement D-tuple $\gamma = (\gamma_1, \ldots, \gamma_D)$, to obtain an a posteriori subjective probability distribution $\mu^{\gamma}_{\text{post}\,\mathscr{P}}$ on the prediction space $\mathfrak{P}$, we can work in any finite-dimensional space $\mathfrak{S}$ such that $\mathfrak{D} + \mathfrak{P} \subseteqq \mathfrak{S} \subseteqq \mathfrak{H}$. The probability measure $\mu^{\gamma}_{\text{post}\,\mathscr{S}\mathscr{P}}$ will be independent of $\mathfrak{S}$. Of course, the smaller $\mathfrak{S}$ is the easier the computation is, so usually we would take $\mathfrak{S} = \mathfrak{D} + \mathfrak{P}$.

In all the foregoing discussion we have supposed that we could agree on an a priori subjective probability distribution for h_E in $\mathfrak{H}$, such as, for example, a cylinder measure μ_{prej} on $\mathfrak{H}$ which somehow reflects the belief that probably $\|h_E\| \leqq M$. But we may not be willing, a priori, to assume more than that probably $\|\mathscr{F}\cdot h_E\| \leqq M$, where $\mathfrak{F}$ is a closed subspace of $\mathfrak{H}$ such that $\mathfrak{F}^\perp$ is finite dimensional. In this situation we have a cylinder measure $\mu_{\text{prej}\,\mathscr{F}}$ on $\mathfrak{F}$ which describes our a priori subjective probability distribution for $\mathscr{F}\cdot h_E$, but we do not have a cylinder measure μ_{prej} on all of $\mathfrak{H}$.

We can extend our Bayesian inference to this case by introducing on $\mathfrak{H}$ a one-parameter family of cylinder measures $\mu^{\varepsilon}_{\text{prej}}$ which are product measures of $\mu_{\text{prej}\mathscr{F}}$ on $\mathfrak{F}$ and measures $\mu^{\varepsilon}_{\text{prej}\mathscr{F}^\perp}$ on $\mathfrak{F}^\perp$ whose variances tend to infinity times $\mathscr{F}^\perp$ as ε tends to 0. Thus as ε approaches 0, $\mu^{\varepsilon}_{\text{prej}}$ applies a weaker and weaker constraint to $\|\mathscr{F}^\perp\cdot h_E\|$ but does not change the distribution of $\mathscr{F}\cdot h_E$. In the limit $\varepsilon = 0$ there is no constraint on $\|\mathscr{F}^\perp\cdot h_E\|$. Although the cylinder measures $\mu^{\varepsilon}_{\text{prej}}$ do not approach a cylinder measure on $\mathfrak{H}$ as $\varepsilon\to 0$, the cylinder measures $\mu^{\gamma,\varepsilon}_{\text{post}}$ do so under suitable assumptions on $p_{\text{meas}}(\gamma|d)$. We take this limiting cylinder measure on $\mathfrak{H}$ as $\mu^{\gamma}_{\text{post}}$ our a posteriori subjective probability distribution for h_E in $\mathfrak{H}$, determined by Bayesian inference from $\mu_{\text{prej}\,\mathscr{F}}$ and $p_{\text{meas}}(\gamma|d)$.

The details are as follows. Let $n:\mathfrak{F}^\perp\to R$ be a continuous, real valued function on $\mathfrak{F}^\perp$ satisfying $0\leqq n(f^\perp)\leqq n(0)$ for every $f^\perp$ in $\mathfrak{F}^\perp$, and satisfying $\int_{\mathfrak{F}^\perp} n(f^\perp)\lambda(df^\perp) = 1$ where λ is Lebesgue measure on $\mathfrak{F}^\perp$. Let $\mu^{\varepsilon}_{\text{prej}\mathscr{F}^\perp}$ be the probability measure on $(\mathfrak{F}^\perp, B(\mathfrak{F}^\perp))$ whose density function with respect to Lebesgue measure is $\varepsilon n(\varepsilon f^\perp)$. Let $\mu^{\varepsilon}_{\text{prej}}$ be the product cylinder measure on $\mathfrak{H}$ of $\mu_{\text{prej}\,\mathscr{F}}$ on $\mathfrak{F}$ and $\mu^{\varepsilon}_{\text{prej}\,\mathscr{F}^\perp}$ on $\mathfrak{F}^\perp$. Our knowledge of the outcome γ of our measurements transforms our subjective a priori probability distribution $\mu^{\varepsilon}_{\text{prej}}$ for h_E into a subjective a posteriori probability distribution $\mu^{\varepsilon}_{\text{post}}$. Let $\mathfrak{S}$ be any finite-dimensional subspace of $\mathfrak{H}$ such that

(96) $$\mathfrak{F}^\perp \oplus \mathscr{F}\cdot\mathfrak{D} \subseteqq \mathfrak{S} \subseteqq \mathfrak{H}.$$

Then $\mathfrak{D}\subseteqq\mathfrak{S}$. If S is any Borel subset of $\mathfrak{S}$, we have from (95)

(97) $$\mu^{\gamma,\varepsilon}_{\text{post}\,\mathscr{S}}(S) = \int_S p_{\text{meas}}(\gamma|s)\mu^{\varepsilon}_{\text{prej}\,\mathscr{S}}(ds) \Big/ \int_{\mathfrak{S}} p_{\text{meas}}(\gamma|s)\mu^{\varepsilon}_{\text{prej}\,\mathscr{S}}(ds).$$

Now $\mathfrak{S} = \mathfrak{F}^\perp \oplus \mathfrak{T}$ where $\mathfrak{T}$ is a finite-dimensional subspace of $\mathfrak{F}$, so $\mu^{\varepsilon}_{\text{prej}\,\mathscr{T}}$ is the marginal distribution of $\mu_{\text{prej}\,\mathscr{F}}$ on $\mathfrak{T}$, and is independent of ε. According to Lemma 22 $\mu^{\varepsilon}_{\text{prej}\mathscr{S}}$ is the product measure of $\mu^{\varepsilon}_{\text{prej}\,\mathscr{F}^\perp}$ and $\mu_{\text{prej}\,\mathscr{T}}$. If we write $\lambda_{\mathscr{F}^\perp}\mu_{\text{prej}\,\mathscr{T}}$ for the measure on $\mathfrak{S}$ which is the product of Legesgue measure $\lambda_{\mathscr{F}^\perp}$ on $\mathfrak{F}^\perp$ and $\mu_{\text{prej}\,\mathscr{T}}$ on $\mathfrak{T}$, then (97) can be written

(98) $$\mu^{\gamma,\varepsilon}_{\text{post}\,\mathscr{S}}(S) =$$

$$\int_S p_{\text{meas}}(\gamma|s)n(\varepsilon\mathscr{F}^\perp\cdot s)\lambda_{\mathscr{F}^\perp}\mu_{\text{prej}\mathscr{T}}(ds) \Big/ \int_{\mathfrak{S}} p_{\text{meas}}(\gamma|s)n(\varepsilon\mathscr{F}^\perp\cdot s)\lambda_{\mathscr{F}^\perp}\mu_{\text{prej}\mathscr{T}}(ds).$$

Now let $\mathfrak{C} = \mathfrak{D} \ominus (\mathfrak{D} \cap \mathfrak{F})$ and let $\lambda_{\mathscr{C}}$ be Lebesgue measure on $\mathfrak{C}$. Suppose that $p_{\text{meas}}(\gamma|d)$ is integrable on $\mathfrak{D}$ with respect to the product measure $\lambda_{\mathscr{C}}\mu_{\text{prej}\,\mathscr{D}\wedge\mathscr{F}}$. Since $p_{\text{meas}}(\gamma|s) = p_{\text{meas}}(\gamma|\mathscr{D}\cdot s)$, it follows that $p_{\text{meas}}(\gamma|s)$ is integrable on $\mathfrak{S}$ with respect to $\lambda_{\mathscr{F}^\perp}\mu_{\text{prej}\,\mathscr{T}}$ (a detailed computation appears in Lemma 32). Therefore the Lebesgue bounded convergence theorem applied to (98) asserts that $\mu^{\gamma}_{\text{post}\,\mathscr{S}}(S) = \lim_{\varepsilon\to 0} \mu^{\gamma,\varepsilon}_{\text{post}\,\mathscr{S}}(S)$ exists, and that

$$(99)\quad \mu^{\gamma}_{\text{post}\,\mathscr{S}}(S) = \int_S p_{\text{meas}}(\gamma|s)\lambda_{\mathscr{F}^\perp}\mu_{\text{prej}\,\mathscr{T}}(ds) \Big/ \int_{\mathfrak{S}} p_{\text{meas}}(\gamma|s)\lambda_{\mathscr{F}^\perp}\mu_{\text{prej}\,\mathscr{T}}(ds).$$

The integrability of $p_{\text{meas}}(\gamma|s)$ over $\mathfrak{S}$ with respect to $\lambda_{\mathscr{F}^\perp}\mu_{\text{prej}\,\mathscr{T}}$ assures us that $\mu^{\gamma}_{\text{post}\,\mathscr{S}}$, as defined by (99), is a probability measure on $\mathfrak{S}$. If $\mathfrak{S}$ and $\mathfrak{S}'$ are two finite-dimensional subspaces of $\mathfrak{H}$ such that

$$\mathfrak{F}^\perp \oplus \mathscr{F}\cdot\mathfrak{D} \subseteqq \mathfrak{S} \subseteqq \mathfrak{S}' \subseteqq \mathfrak{H},$$

then we can write $\mathfrak{S} = \mathfrak{F}^\perp \oplus \mathfrak{T}$, $\mathfrak{S}' = \mathfrak{F}^\perp \oplus \mathfrak{T}'$, with $\mathfrak{T} \subseteqq \mathfrak{T}' \subseteqq \mathfrak{F}$, and $\mu_{\text{prej}\,\mathscr{T}}$ is the marginal distribution of $\mu_{\text{prej}\,\mathscr{T}'}$ on $\mathfrak{T}$. Since $p_{\text{meas}}(\gamma|\mathscr{S}s')$ is a cylinder function in $\mathfrak{S}'$ based on $\mathfrak{S}$, it follows from (99) that $\mu^{\gamma}_{\text{post}\,\mathscr{S}}$ is the marginal distribution of $\mu^{\gamma}_{\text{post}\,\mathscr{S}'}$ on $\mathfrak{S}$. Therefore, by Remark 15 there is a unique cylinder measure $\mu^{\gamma}_{\text{post}}$ on $\mathfrak{H}$ such that all the measures $\mu^{\gamma}_{\text{post}\,\mathscr{S}}$ defined by (99) are marginal distributions of $\mu^{\gamma}_{\text{post}}$. We summarize these results as

THEOREM 30. *Suppose that $\mathfrak{F}$ is a closed subspace of Hilbert space $\mathfrak{H}$ and $\mathfrak{F}^\perp$ is finite dimensional. Suppose that $\mathfrak{D}$ is a finite-dimensional subspace of $\mathfrak{H}$ and $\mathfrak{H} = \mathfrak{D} + \mathfrak{F}$. Let $\mathfrak{C} = \mathfrak{D} \ominus (\mathfrak{D} \cap \mathfrak{F})$ and $G = R^D$. Let $\lambda_{\mathscr{C}}$ and λ_G be Lebesgue measure on $\mathfrak{C}$ and G respectively. Let $\mu_{\text{prej}\mathscr{F}}$ be a cylinder measure on $\mathfrak{F}$ and denote by $\mu_{\text{prej}\mathscr{T}}$ its marginal distribution on any finite-dimensional subspace $\mathfrak{T} \subseteqq \mathfrak{F}$. Let $p_{\text{meas}}(\gamma|d)$ be a continuous, real-valued nonnegative function on $G \times \mathfrak{D}$, integrable with respect to the product measure $\lambda_G(\lambda_{\mathscr{C}}\mu_{\text{prej}\mathscr{D}\wedge\mathscr{F}})$. Then for any γ in G there is a unique cylinder measure $\mu^{\gamma}_{\text{post}}$ on $\mathfrak{H}$ whose marginal distributions $\mu^{\gamma}_{\text{post}\mathscr{S}}$ are given by (99) for all finite-dimensional subspaces $\mathfrak{S}$ of $\mathfrak{H}$ satisfying $\mathfrak{F}^\perp \oplus \mathscr{F}\cdot\mathfrak{D} \subseteqq \mathfrak{S} \subseteqq \mathfrak{H}$.*

The problem of Bayesian inference on $\mathfrak{H}$ is now solved in principal, but there remain nontrivial computations. Sometimes useful in these computations is

COROLLARY 31. *Under the hypotheses of Theorem 30, let $\mathfrak{T}$ be any finite-dimensional subspace of $\mathfrak{F}$ such that $\mathscr{F}\cdot\mathfrak{D} \subseteqq \mathfrak{T} \subseteqq \mathfrak{F}$, and let $\mathfrak{S} = \mathfrak{F}^\perp \oplus \mathfrak{T}$. Then $\mathfrak{H} = \mathfrak{C} \dotplus \mathfrak{F}$, $\mathfrak{S} = \mathfrak{C} \dotplus \mathfrak{T}$, and for any Borel subset*

S of $\mathfrak{S}$ we can write (99) *as*

(100) $$\mu^{\gamma}_{\text{post}\mathscr{S}}(S) = \int_S p_{\text{meas}}(\gamma|s)\lambda_{\mathscr{C}}\mu_{\text{prej}\,\mathscr{T}}(ds) \Big/ \int_{\mathfrak{S}} p_{\text{meas}}(\gamma|s)\lambda_{\mathscr{C}}\mu_{\text{prej}\,\mathscr{T}}(ds)$$

where $\lambda_{\mathscr{C}}\mu_{\text{prej}\,\mathscr{T}}$ is the product measure on $\mathfrak{C}$ of $\lambda_{\mathscr{C}}$ on $\mathfrak{S}$ and $\mu_{\text{prej}\mathscr{T}}$ on $\mathfrak{T}$.

PROOF. Let $\xi_S(s)$ be the indicator function of the Borel set S, so that for any vector s in $\mathfrak{S}$, $\xi_S(s) = 1$ or 0 according as s is in or outside of S. Writing $\mu_{\mathscr{S}}$ for $\lambda_{\mathscr{F}^{\perp}}\mu_{\text{prej}\,\mathscr{T}}$, and $p(s)$ for $\xi_S(s)p_{\text{meas}}(\gamma|s)$, we have

$$\int_S p_{\text{meas}}(\gamma|s)\lambda_{\mathscr{F}^{\perp}}\mu_{\text{prej}\,\mathscr{T}}(ds) = \int_{\mathfrak{S}} p(s)\mu_{\mathscr{S}}(ds).$$

Now $s = \mathscr{F}^{\perp}\cdot s + \mathscr{T}\cdot s$ if s is in $\mathfrak{S}$, so according to Remark 21 (extended in an obvious way because $\lambda_{\mathscr{F}^{\perp}}$ is not a finite measure)

$$\int_{\mathfrak{S}} p(s)\mu_{\mathscr{S}}(ds) = \int_{\mathfrak{T}} \left[\int_{\mathfrak{F}^{\perp}} p(f^{\perp} + t)\lambda_{\mathscr{F}^{\perp}}(df^{\perp})\right]\mu_{\text{prej}\,\mathscr{T}}(dt).$$

The mapping $(\mathscr{C}|\mathfrak{F}^{\perp}):\mathfrak{F}^{\perp} \to \mathfrak{C}$ is one-to-one onto and has inverse $(\mathscr{C}^*_{\mathscr{F}}|\mathfrak{C}): \mathfrak{C} \to \mathfrak{F}^{\perp}$. Let $J(\mathscr{C}^*_{\mathscr{F}}|\mathfrak{C})$ be the Jacobian of this inverse. For any fixed t in $\mathfrak{T}$, the mapping $A:\mathfrak{C} \to \mathfrak{F}^{\perp}$ defined by $A(c) = \mathscr{C}^*_{\mathscr{F}}\cdot(c - t)$ is one-to-one onto, and its Jacobian is also $J(\mathscr{C}^*_{\mathscr{F}}|\mathfrak{C})$. If t is fixed, then for any $f^{\perp}$ in $\mathfrak{F}^{\perp}$ there is exactly one c in $\mathfrak{C}$ such that $f^{\perp} = \mathscr{C}^*_{\mathscr{F}}\cdot(c - t)$. Thus, for any fixed t in $\mathfrak{T}$,

$$\int_{\mathfrak{F}^{\perp}} p(f^{\perp} + t)\lambda_{\mathscr{F}^{\perp}}(df^{\perp}) = J(\mathscr{C}^*_{\mathscr{F}}|\mathfrak{C}) \int_{\mathfrak{C}} p[\mathscr{C}^*_{\mathscr{F}}\cdot(c - t) + t]\,\lambda_{\mathscr{C}}(dc).$$

Now $\mathscr{C}^*_{\mathscr{F}}\cdot(c - t) + t = \mathscr{C}^*_{\mathscr{F}}\cdot c + \mathscr{F}^*_{\mathscr{C}}\cdot t$, and so according to Remark 21,

$$\int_{\mathfrak{S}} p(s)\mu_{\mathscr{S}}(ds) = J(\mathscr{C}^*_{\mathscr{F}}|\mathfrak{C}) \int_{\mathfrak{S}} p[\mathscr{C}^*_{\mathscr{F}}\cdot\mathscr{C}\cdot s + \mathscr{F}^*_{\mathscr{C}}\cdot\mathscr{T}\cdot s]\lambda_{\mathscr{C}}\mu_{\text{prej}\,\mathscr{T}}(ds).$$

But for any s in $\mathfrak{S}$, $\mathscr{T}\cdot s = \mathscr{F}\cdot s$, so $\mathscr{C}^*_{\mathscr{F}}\cdot\mathscr{C}\cdot s + \mathscr{F}^*_{\mathscr{C}}\cdot\mathscr{T}\cdot s = s$. Therefore

$$\int_S p_{\text{meas}}(\gamma|s)\lambda_{\mathscr{F}^{\perp}}\mu_{\text{prej}\,\mathscr{T}}(ds) = J(\mathscr{C}^*_{\mathscr{F}}|\mathfrak{C}) \int_S p_{\text{meas}}(\gamma|s)\lambda_{\mathscr{C}}\mu_{\text{prej}\,\mathscr{T}}(ds).$$

This equation is true for any Borel subset S of $\mathfrak{S}$ and in particular for $S = \mathfrak{S}$. If this change of variables is carried out in both numerator and denominator of (99), the Jacobians cancel, and the result is (100).

A last result aimed at easing computation can be obtained from

LEMMA 32. *Suppose that $\mathfrak{K}$, $\mathfrak{L}$ and $\mathfrak{M}$ are closed subspaces of Hilbert space $\mathfrak{H}$ and that $\mathfrak{H} = (\mathfrak{K} \dotplus \mathfrak{L}) \dotplus \mathfrak{M}$. Suppose that $\mu_{\mathscr{K}}, \mu_{\mathscr{L}}$ and $\mu_{\mathscr{M}}$ are cylinder measures on $\mathfrak{K}$, $\mathfrak{L}$ and $\mathfrak{M}$. Then $\mathfrak{H} = \mathfrak{K} \dotplus (\mathfrak{L} \dotplus \mathfrak{M})$ and $\mu_{\mathscr{K}}(\mu_{\mathscr{L}}\mu_{\mathscr{M}}) = (\mu_{\mathscr{K}}\mu_{\mathscr{L}})\mu_{\mathscr{M}}$ or, in an earlier notation, $\mu_{\mathscr{K}\vee(\mathscr{L}\vee\mathscr{M})} = \mu_{(\mathscr{K}\vee\mathscr{L})\vee\mathscr{M}}$.*

PROOF. From $\mathfrak{H} = (\mathfrak{K} \dotplus \mathfrak{L}) \dotplus \mathfrak{M}$ it follows that for any h in $\mathfrak{H}$ there are unique vectors k, l, m in $\mathfrak{K}, \mathfrak{L}, \mathfrak{M}$ such that $h = k + l + m$, and that there is a constant Q such that for any h, $\|k + l\| \leqq Q\|h\|$ and $\|m\| \leqq Q\|h\|$. It also follows that there is a constant P such that $\|k\| \leqq P\|k + l\|$ and $\|l\| \leqq P\|k + l\|$. But then $\|k\| \leqq PQ\|h\|$ and $\|l + m\| \leqq (P + 1)Q\|h\|$, so by Theorem 3, $\mathfrak{L} \dotplus \mathfrak{M}$ is closed and $\mathfrak{H} = \mathfrak{K} \dotplus (\mathfrak{L} \dotplus \mathfrak{M})$. Now suppose that K, L and M are cylinder subsets of $\mathfrak{K}$, $\mathfrak{L}$, and $\mathfrak{M}$. We claim that the two cylinder measures $\mu_{(\mathscr{K} \vee \mathscr{L}) \vee \mathscr{M}}$ and $\mu_{\mathscr{K} \vee (\mathscr{L} \vee \mathscr{M})}$ assign the same number to $\mathscr{K}^{-1}(K) \cap \mathscr{L}^{-1}(L) \cap \mathscr{M}^{-1}(M)$. To see this, we note that

$$\mathscr{K}^{-1}(K) = (\mathscr{K} \vee \mathscr{L})^{-1}(\mathscr{K}|\mathfrak{K} \dotplus \mathfrak{L})^{-1}(K),$$

$$\mathscr{L}^{-1}(L) = (\mathscr{K} \vee \mathscr{L})^{-1}(\mathscr{L}|\mathfrak{K} \dotplus \mathfrak{L})^{-1}(L),$$

$$\mathscr{K}^{-1}(K) \cap \mathscr{L}^{-1}(L) = (\mathscr{K} \vee \mathscr{L})^{-1}[(\mathscr{K}|\mathfrak{K} \dotplus \mathfrak{L})^{-1}(K) \cap (\mathscr{L}|\mathfrak{K} \dotplus \mathfrak{L})^{-1}(L)].$$

(Here $\mathscr{K} \vee \mathscr{L}$ is the operator of orthogonal projection from $\mathfrak{H}$ onto $\mathfrak{K} + \mathfrak{L}$.) Therefore

$$\begin{aligned} &\mu_{(\mathscr{K} \vee \mathscr{L}) \vee \mathscr{M}}(\mathscr{K}^{-1}(K) \cap \mathscr{L}^{-1}(L) \cap \mathscr{M}^{-1}(M)) \\ &\quad = \mu_{\mathscr{K} \vee \mathscr{L}}[(\mathscr{K}|\mathfrak{K} \dotplus \mathfrak{L})^{-1}(K) \cap (\mathscr{L}|\mathfrak{K} \dotplus \mathfrak{L})^{-1}(L)]\mu_{\mathscr{M}}(M) \\ &\quad = \mu_{\mathscr{K}}(K)\mu_{\mathscr{L}}(L)\mu_{\mathscr{M}}(M). \end{aligned}$$

Similarly $\mu_{\mathscr{K} \vee (\mathscr{L} \vee \mathscr{M})}(\mathscr{K}^{-1}(K) \cap \mathscr{L}^{-1}(L) \cap \mathscr{M}^{-1}(M)) = \mu_{\mathscr{K}}(K)\mu_{\mathscr{L}}(L)\mu_{\mathscr{M}}(M)$. Now let H be an arbitrary cylinder subset of $\mathfrak{H}$. Then there is a finite-dimensional subspace $\mathfrak{T}$ of $\mathfrak{H}$ such that H is in $c_{\mathscr{H}}(\mathfrak{T})$. Let $\mathfrak{S} = \mathscr{K}_{\mathscr{L} \vee \mathscr{M}}(\mathfrak{T}) + \mathscr{L}_{\mathscr{K} \vee \mathscr{M}}(\mathfrak{T}) + \mathscr{M}_{\mathscr{K} \vee \mathscr{L}}(\mathfrak{T})$. Then $\mathfrak{T} \subseteqq \mathfrak{S}$ and $\mathfrak{S}$ is finite dimensional, so H is in $c_{\mathscr{H}}(\mathfrak{S})$. But $c_{\mathscr{H}}(\mathfrak{S})$ is the σ-ring generated by sets of the form $\mathscr{K}^{-1}(K) \cap \mathscr{L}^{-1}(L) \cap \mathscr{M}^{-1}(M)$ where K is in $c_{\mathscr{K}}(\mathscr{K}_{\mathscr{L} \vee \mathscr{M}}(\mathfrak{T}))$, L is in $c_{\mathscr{L}}(\mathscr{L}_{\mathscr{K} \vee \mathscr{M}}(\mathfrak{T}))$ and M is in $c_{\mathscr{M}}(\mathscr{M}_{\mathscr{K} \vee \mathscr{L}}(\mathfrak{T}))$. The measures $\mu_{(\mathscr{K} \vee \mathscr{L}) \vee \mathscr{M}}$ and $\mu_{\mathscr{K} \vee (\mathscr{L} \vee \mathscr{M})}$ are countably additive on $c_{\mathscr{H}}(\mathfrak{S})$ and agree on its generators, so they agree on $c_{\mathscr{H}}(\mathfrak{S})$. Thus they agree on arbitrary cylinder subsets H of $\mathfrak{H}$.

From Lemma 32 and Corollary 31 we obtain a result very useful in computations with certain kinds of a priori subjective probability distributions.

COROLLARY 33. *In addition to the hypotheses of Theorem* 30, *suppose also that* $\mu_{\mathrm{prej}\,\mathscr{F}}$ *is the product of its marginal distributions* $\mu_{\mathrm{prej}\,\mathscr{D} \wedge \mathscr{F}}$ *on* $\mathfrak{D} \cap \mathfrak{F}$ *and* $\mu_{\mathrm{prej}\,\mathscr{B}}$ *on* $\mathfrak{B} = \mathfrak{F} \ominus (\mathfrak{D} \cap \mathfrak{F})$. *Then for any* γ *in* R^D, $\mu^{\gamma}_{\mathrm{post}}$ *is the product cylinder measure on* $\mathfrak{H} = \mathfrak{D} \dotplus \mathfrak{B}$ *of* $\mu^{\gamma}_{\mathrm{post}\,\mathscr{D}}$ *on* $\mathfrak{D}$ *and* $\mu_{\mathrm{prej}\,\mathscr{B}}$ on $\mathfrak{B}$. *Moreover, for any Borel subset* D *of* $\mathfrak{D}$,

$$\mu^{\gamma}_{\mathrm{post}\,\mathscr{D}}(D) = \int_D p_{\mathrm{meas}}(\gamma|d)\lambda_{\mathscr{C}}\mu_{\mathscr{D} \wedge \mathscr{F}}(dd) \Big/ \int_{\mathfrak{D}} p_{\mathrm{meas}}(\gamma|d)\lambda_{\mathscr{C}}\mu_{\mathscr{D} \wedge \mathscr{F}}(dd) \tag{101}$$

PROOF. Let $\mathfrak{T}$ be any finite-dimensional subspace of $\mathfrak{F}$ such that $\mathscr{F} \cdot \mathfrak{D} \subseteqq \mathfrak{T} \subseteqq \mathfrak{F}$, and let $\mathfrak{U} = \mathfrak{T} \ominus (\mathfrak{D} \cap \mathfrak{F})$ Let $\mathfrak{S} = \mathfrak{F}^{\perp} \oplus \mathfrak{T}$. Then $\mathfrak{S} = \mathfrak{D} \dotplus \mathfrak{U}$. By hypothesis, $\mu_{\text{prej}\,\mathscr{T}} = \mu_{\text{prej}\,\mathscr{D} \wedge \mathscr{T}} \mu_{\text{prej}\,\mathscr{U}}$, and by Lemma 32 we have that $\lambda_{\mathscr{C}}(\mu_{\text{prej}\,\mathscr{D} \wedge \mathscr{F}} \mu_{\text{prej}\,\mathscr{U}}) = (\lambda_{\mathscr{C}} \mu_{\text{prej}\,\mathscr{D} \wedge \mathscr{F}}) \mu_{\text{prej}\,\mathscr{U}}$. If D is any Borel subset of $\mathfrak{D}$ and U is any Borel subset of $\mathfrak{U}$, let $\tilde{D} = (\mathscr{D}|\mathfrak{S})^{-1}(D)$ and $\tilde{U} = (\mathscr{U}|\mathfrak{S})^{-1}(U)$. Then from Equation (100) we have that

$$\mu^{\gamma}_{\text{post}\,\mathscr{S}}(\tilde{D} \cap \tilde{U}) = \frac{\int_{\tilde{D} \cap \tilde{U}} p_{\text{meas}}(\gamma|s)[\lambda_{\mathscr{C}} \mu_{\text{prej}\,\mathscr{D} \wedge \mathscr{F}} \mu_{\text{prej}\,\mathscr{U}}](ds)}{\int_{\mathfrak{S}} p_{\text{meas}}(\gamma|s)[\lambda_{\mathscr{C}} \mu_{\text{prej}\,\mathscr{D} \wedge \mathscr{F}} \mu_{\text{prej}\,\mathscr{U}}](ds)}.$$

But $p_{\text{meas}}(\gamma|s)$ is based on $\mathfrak{D}$, and the indicator function $\xi_{\tilde{D} \cap \tilde{U}}(s) = \xi_D(\mathscr{D} \cdot s)\xi_U(\mathscr{U} \cdot s)$. Therefore from Remark 21,

$$\mu^{\gamma}_{\text{post}\,\mathscr{S}}(\tilde{D} \cap \tilde{U}) = \frac{\left[\int_D p_{\text{meas}}(\gamma|d)\lambda_{\mathscr{C}} \mu_{\text{prej}\,\mathscr{D} \wedge \mathscr{F}}(dd)\right] \mu_{\text{prej}\,\mathscr{U}}(U)}{\int_{\mathfrak{D}} p_{\text{meas}}(\gamma|d)\lambda_{\mathscr{C}} \mu_{\text{prej}\,\mathscr{D} \wedge \mathscr{F}}(dd)}. \tag{102}$$

If we set $U = \mathfrak{U}$, then $\tilde{D} \cap \tilde{U} = \tilde{D} = (\mathscr{D}|\mathfrak{S})^{-1}(D)$, and $\mu^{\gamma}_{\text{post}\mathscr{S}}(\tilde{D}) = \mu^{\gamma}_{\text{post}\mathscr{D}}(D)$, which proves Equation (101). But then Equation (102) can be written $\mu^{\gamma}_{\text{post}\,\mathscr{S}}(\tilde{D} \cap \tilde{U}) = \mu^{\gamma}_{\text{post}\,\mathscr{D}}(D)\mu^{\gamma}_{\text{prej}\,\mathscr{U}}(U)$; according to Lemma 22 this proves Corollary 33.

The effect of Corollary 33 is to reduce the calculation of $\mu^{\gamma}_{\text{post}}$ from $\mu_{\text{prej}\,\mathscr{F}}$ to a single calculation of $\mu^{\gamma}_{\text{post}\,\mathscr{D}}$ on $\mathfrak{D}$. For example, if $h_{\mathscr{D}}$ and $V_{\mathscr{D}}$ are the mean and variance tensor of $\mu^{\gamma}_{\text{post}\,\mathscr{D}}$ on $\mathfrak{D}$, and $h_{\mathscr{B}}$ and $V_{\mathscr{B}}$ are the mean and variance tensor of $\mu_{\text{prej}\,\mathscr{B}}$ on $\mathfrak{B}$, then the mean $h_{\mathscr{H}}$ and variance tensor $V_{\mathscr{H}}$ of $\mu^{\gamma}_{\text{post}}$ on $\mathfrak{H}$ are given by

$$h_{\mathscr{H}} = \mathscr{D}^*_{\mathscr{B}} \cdot h_{\mathscr{D}} + \mathscr{B}^*_{\mathscr{D}} \cdot h_{\mathscr{B}},$$

$$V^{\text{inv}}_{\mathscr{H}} = V^{\text{inv}}_{\mathscr{D}} + V^{\text{inv}}_{\mathscr{B}}.$$

3. *Normal, or Gaussian, prejudices and data.* A cylinder measure μ on a Hilbert space $\mathfrak{H}$ is "normal" or "Gaussian" with mean m and variance $V: \mathfrak{H} \to \mathfrak{H}$ if m is a vector in $\mathfrak{H}$, V is a selfadjoint, positive semidefinite linear operator on $\mathfrak{H}$, and for any finite-dimensional subspace $\mathfrak{S} \subseteqq \mathfrak{H}$ the marginal distribution $\mu_{\mathscr{S}}$ is normal (Gaussian) on $\mathfrak{S}$ with mean $\mathscr{S} \cdot m$ and variance $\mathscr{S} \cdot V \cdot \mathscr{S}$. For each pair (m, V), this prescription defines a unique cylinder measure μ on $\mathfrak{H}$, and evidently μ has bounded moments of all orders, mean m, and variance V. If $\mathfrak{K}$ and $\mathfrak{L}$ are closed subspaces of $\mathfrak{H}$ such that $\mathfrak{H} = \mathfrak{K} \dotplus \mathfrak{L}$, and if $\mu_{\mathscr{K}}$ and $\mu_{\mathscr{L}}$ are normal cylinder measures on $\mathfrak{K}$ and $\mathfrak{L}$, then their product cylinder measure, $\mu_{\mathscr{K} \vee \mathscr{L}}$, is normal on $\mathfrak{H}$.

Conversely, if μ is a cylinder measure on $\mathfrak{H}$ such that $\mu = \mu_{\mathscr{K}}\mu_{\mathscr{K}^\perp}$ for all closed subspaces $\mathfrak{K} \subseteqq \mathfrak{H}$, or even for all one-dimensional subspaces $\mathfrak{K}$, then μ is a normal cylinder measure whose variance tensor is $M^2\mathscr{H}$ for some nonnegative real M. When $\mathfrak{H}$ is two dimensional, this converse is proved by studying the characteristic function of μ, which must be separable in every Cartesian axis system. The general result follows immediately from the two-dimensional case.

In our problem of inference, one very convenient way of expressing the belief that probably $\|h_E\|$ is not much larger than M is to adopt for μ_{prej} the normal cylinder measure on $\mathfrak{H}$ with mean 0 and variance tensor $V = M^2\mathscr{H}$. If $\mathfrak{K}$ is any closed subspace of $\mathfrak{H}$ then μ_{prej} is the product measure of its marginal distributions on $\mathfrak{K}$ and $\mathfrak{K}^\perp$; that is, the random variables $\mathscr{K} \cdot h_E$ and $\mathscr{K}^\perp \cdot h_E$ are independent.

When $\dim \mathfrak{H} = \infty$, it should be noted that in adopting the normal cylinder measure μ with mean 0 and variance $M^2\mathscr{H}$ as our a priori subjective probability distribution for h_E on $\mathfrak{H}$ we are assuming something much weaker than that probably $\|h_E\| \leqq M$. For any real K, the ball $B_K = \{h : h \text{ in } \mathfrak{H} \text{ and } \|h\| \leqq K\}$ is not a cylinder set, so $\mu(B_K)$ is not defined. There is, however, a sense in which μ assigns probability 0 to B_K. We can find a sequence of cylinder sets C_K^n with $C_K^1 \supset C_K^2 \supset \cdots \supset B_K$ such that $\lim_{n\to\infty} \mu(C_K^n) = 0$. Let $\{\mathfrak{S}_1, \mathfrak{S}_2, \ldots\}$ be a denumerable sequence of subspaces of $\mathfrak{H}$ such that $\dim \mathfrak{S}_n = n$ and $\mathfrak{S}_n \subseteqq \mathfrak{S}_{n+1}$. Let $B_K^n = \mathfrak{S}_n \cap B_K$ and let $C_K^n = \mathscr{S}_n^{-1}(B_K^n)$. Then $\mu(C_K^n) = \mu_{\mathscr{S}_n}(B_K^n) = I_n(K^2/2M^2)/I_n(\infty)$ where $I_n(x) = \int_0^x t^{n/2-1}e^{-t}\,dt$. Thus $\lim_{n\to\infty} \mu_{\mathscr{S}_n}(B_K^n) = 0$. In adopting for μ_{prej} the normal cylinder measure on $\mathfrak{H}$ with mean 0 and variance $M^2\mathscr{H}$ we are not asserting that probably $\|h_E\| \leqq M$ but only that probably $\|\mathscr{S} \cdot h_E\| \leqq M$ for every finite-dimensional subspace $\mathfrak{S} \subseteqq \mathfrak{H}$. However, if $\|\ \ \|_1$ is any measurable pseudonorm on $\mathfrak{H}$ [Gross, 1962], we are assigning positive a priori subjective probability to the event $\|h_E\|_1 \leqq M$.

If instead of $\|h_E\| \leqq M$ we believe only that it is probable that $\|\mathscr{F} \cdot h_E\| \leqq M$, we can adopt for $\mu_{\text{prej}\,\mathscr{F}}$ the normal cylinder measure on $\mathfrak{F}$ with mean 0 and variance tensor $M^2\mathscr{F}$. Then the hypotheses of Corollary 33 are satisfied. This choice of $\mu_{\text{prej}\,\mathscr{F}}$ is, of course, particularly convenient when the errors in our data are also normally distributed.

Suppose that the D-tuple of experimental errors $(\delta\gamma_1, \ldots, \delta\gamma_D)$ in our measurement of $\boldsymbol{\gamma} = (\gamma_1, \ldots, \gamma_D)$ is normally distributed in R^D with mean $(0, \ldots, 0)$ and variance matrix

$$V_{ij} = E[(\delta\gamma_i)(\delta\gamma_j)] \tag{103}$$

where E is expected value. The $D \times D$ matrix V_{ij} is symmetric and positive semidefinite. We will assume that it is strictly positive definite, i.e. that there is no linear combination of errors $\delta\gamma_1, \ldots, \delta\gamma_D$ which vanishes with probability 1.

Let $\Gamma : \mathfrak{D} \to R^D$ be defined by requiring for any d in $\mathfrak{D}$ that $\Gamma(d) = (d_1 \cdot d, \ldots, d_D \cdot d)$ where $\{d_1, \ldots, d_D\}$ are our data vectors. Then the observed D-tuple $\boldsymbol{\gamma}'$ is an unbiased estimate of $\boldsymbol{\gamma}_E = \Gamma(\mathscr{D} \cdot h_E)$, and $V_{ij} = E\{[\gamma_i - (\gamma_E)_i][\gamma_j - (\gamma_E)_j]\}$. As an example of how $\boldsymbol{\gamma}'$ and V_{ij} might be obtained in practice, we might make a large number N of independent measurements $\boldsymbol{\gamma}^{(v)} = (\gamma_1^{(v)}, \ldots, \gamma_D^{(v)})$, $v = 1, \ldots, N$, and then use for $\boldsymbol{\gamma}'$ the mean

$$\boldsymbol{\gamma}^{(0)} = \sum_{v=1}^{N} \boldsymbol{\gamma}^{(v)} \Big/ N$$

and for V_{ij} the estimate

(104) $$V_{ij} = \left(\sum_{v=1}^{N} [\gamma_i^{(v)} - \gamma_i^{(0)}][\gamma_j^{(v)} - \gamma_j^{(0)}] \right) \Big/ N(N-1).$$

We write $p_{\text{meas}}(\boldsymbol{\gamma})$ for the normal distribution on R^D with mean $(0, \ldots, 0)$ and variance matrix V_{ij}. Then the density for the probability distribution of our observed $\boldsymbol{\gamma}$ is $p_{\text{meas}}(\boldsymbol{\gamma} - \boldsymbol{\gamma}_E) = p_{\text{meas}}(\boldsymbol{\gamma} - \Gamma(\mathscr{D} \cdot h_E))$. In our Bayesian inference we have

(105) $$p_{\text{meas}}(\boldsymbol{\gamma}|d) = p_{\text{meas}}(\boldsymbol{\gamma} - \Gamma(d)).$$

We take for $\mu_{\text{prej}\,\mathscr{F}}$ the normal cylinder measure on $\mathfrak{F}$ with mean 0 and variance tensor $M^2\mathscr{F}$. Then from Corollary 33, $\mu^{\boldsymbol{\gamma}}_{\text{post}\,\mathscr{D}}$ has a probability density $p^{\boldsymbol{\gamma}}_{\text{post}\,\mathscr{D}}$ which satisfies

(106) $$\begin{aligned} -2 \ln p^{\boldsymbol{\gamma}}_{\text{post}\,\mathscr{D}}(d) &= \text{constant} + M^{-2} \| \mathscr{D} \wedge \mathscr{F} \cdot d \|^2 \\ &\quad + \sum_{i,j=1}^{D} (\gamma_i - d_i \cdot d)(\gamma_j - d_j \cdot d)(V^{-1})^{ij} \end{aligned}$$

where $(V^{-1})^{ij}$ is the $D \times D$ symmetric, positive-definite matrix inverse to V_{ij}. We define

(107) $$d(\boldsymbol{\gamma}) = \sum_{i,j=1}^{D} \gamma_i (V^{-1})^{ij} d_j$$

and

(108) $$Q = \sum_{i,j=1}^{D} d_i d_j (V^{-1})^{ij}.$$

Since $(V^{-1})^{ij}$ is positive-definite and $\{d_1, \ldots, d_D\}$ span $\mathfrak{D}$, $Q|\mathfrak{D}$ is positive-definite so $(Q|\mathfrak{D})^{-1} : \mathfrak{D} \to \mathfrak{D}$ exists. Thus Q lives on $\mathfrak{D}$ and Q^{inv} is well defined and lives on $\mathfrak{D}$. We define $V:\mathfrak{H}\to\mathfrak{H}$ as $V = Q^{\text{inv}}$, so that by the definition of V

$$V^{\text{inv}} = \sum_{i,j=1}^{D} d_i d_j (V^{-1})^{ij}. \tag{109}$$

Then (106) can be written

$$-2\ln p^{\gamma}_{\text{post}\,\mathscr{D}}(d) = \text{constant} + M^{-2}\, d\cdot\mathscr{D}\wedge\mathscr{F}\cdot d + d\cdot V^{\text{inv}}\cdot d - 2d\cdot d(\gamma).$$

It follows that $p^{\gamma}_{\text{post}\,\mathscr{D}}$ is Gaussian on $\mathfrak{D}$ with variance tensor $V_{\mathscr{D}}$ given by

$$V^{\text{inv}}_{\mathscr{D}} = M^{-2}(\mathscr{D}\wedge\mathscr{F}) + V^{\text{inv}} \tag{110}$$

and with mean $h_{\mathscr{D}}(\gamma)$ given by

$$h_{\mathscr{D}}(\gamma) = V_{\mathscr{D}}\cdot d(\gamma). \tag{111}$$

Then according to Corollary 33 and Equations (41) and (42), $\mu^{\gamma}_{\text{post}}$ is normal on $\mathfrak{H}$ with mean $h_{\mathscr{H}}(\gamma)$ and variance tensor $V_{\mathscr{H}}$ where $V^{\text{inv}}_{\mathscr{H}} = M^{-2}\mathscr{B} + V^{\text{inv}}_{\mathscr{D}} = M^{-2}(\mathscr{B} + \mathscr{F}\wedge\mathscr{D}) + V^{\text{inv}}$, or

$$V^{\text{inv}}_{\mathscr{H}} = M^{-2}\mathscr{F} + V^{\text{inv}} \tag{112}$$

and

$$h_{\mathscr{H}}(\gamma) = \mathscr{D}^{*}_{\mathscr{B}}\cdot V_{\mathscr{D}}\cdot d(\gamma). \tag{113}$$

It may happen that the experimental errors are normally distributed in R^D, as discussed above, but that we prefer to adopt for $\mu_{\text{prej}\,\mathscr{F}}$ the normal cylinder measure on $\mathfrak{F}$ with variance $M^2\mathscr{F}$ and with mean h_0, where h_0 is some preferred member of $\mathfrak{H}$ other than 0. This amounts to assuming, roughly, that $\|\mathscr{F}\cdot(h_E - h_0)\|$ is probably not much larger than M. As an example, if $\mathfrak{H}$ is the Hilbert space of square integrable functions $h(x)$ on $0 \leqq x \leqq 1$ and $h_E(x)$ is the density of the earth, then $h_E(x)$ is positive everywhere, and $\int_0^1 x^2 h_E(x)\,dx$ and $\int_0^1 x^4 h_E(x)\,dx$ are both rather accurately known. We might then take for $h_0(x)$ some smooth positive function (perhaps with a jump at the core-mantle boundary) which gave the correct values to $\int_0^1 x^2 h_0(x)dx$ and $\int_0^1 x^4 h_0(x)dx$. Then the hypothesis $\|h_E - h_0\| \leqq M$ would become appealing at smaller values of M than the hypothesis $\|h_E\| \leqq M$.

We can deal with this situation by applying translation operators to $\mathfrak{H}$ and R^D. For any h in $\mathfrak{H}$ let $\tilde{h} = h - h_0$. For any γ in R^D let $\tilde{\gamma} = \gamma - \Gamma(h_0)$. Then $\mathscr{F}\cdot\tilde{h}_E$ has as its a priori subjective probability distribution the normal cylinder measure on $\mathfrak{F}$ with mean 0 and variance $M^2\mathscr{F}$ and $\tilde{\gamma}$ has in R^D the normal distribution with variance matrix V_{ij} and mean $\Gamma(\tilde{h}_E)$. Therefore, according to the discussion we have already given, $\tilde{h}_E$ has as its a posteriori subjective probability distribution the normal cylinder measure on $\mathfrak{H}$ with variance $V_{\mathscr{H}}$ given by (112) and with mean $h_{\mathscr{H}}(\tilde{\gamma}) = \mathscr{D}^*_{\mathscr{B}} \cdot V_{\mathscr{D}} \cdot d(\tilde{\gamma})$. Then $h_E = \tilde{h}_E + h_0$ has as its a posteriori subjective probability distribution on $\mathfrak{H}$ the normal cylinder measure $\mu^{\gamma}_{\text{post}}$ whose variance is $V_{\mathscr{H}}$, given by (112), and whose mean is $m_{\mathscr{H}}(\gamma) = h_{\mathscr{H}}(\tilde{\gamma}) + h_0 = h_{\mathscr{H}}(\gamma - \Gamma(h_0)) + h_0$, so

$$m_{\mathscr{H}}(\gamma) = h_0 + \mathscr{D}^*_{\mathscr{B}} \cdot V_{\mathscr{D}} \cdot d(\gamma - \Gamma(h_0)) \tag{114}$$

where $d(\gamma)$ is given by (107).

At this point we recall that once we know $\mu^{\gamma}_{\text{post}}$ then our problem of inference is solved. If $\mathfrak{P}$ is the prediction space, spanned by the prediction vectors $\{p_1, \ldots, p_P\}$, then in the foregoing example involving normal distributions, $\mathscr{P} \cdot h_E$ is normally distributed in $\mathfrak{P}$ with mean $\mathscr{P} \cdot h(\gamma)$ or $\mathscr{P} \cdot m_{\mathscr{H}}(\gamma)$ and with variance tensor $\mathscr{P} \cdot V_{\mathscr{H}} \cdot \mathscr{P}$. Then $\tilde{\gamma} = (p_1 \cdot h_E, \ldots, p_P \cdot h_E)$ is normally distributed in R^P with mean $(p_1 \cdot h_{\mathscr{H}}(\gamma), \ldots, p_P \cdot h_{\mathscr{H}}(\gamma))$ or $(p_1 \cdot m_{\mathscr{H}}(\gamma), \ldots, p_P \cdot m_{\mathscr{H}}(\gamma))$, and with variance matrix $p_i \cdot V_{\mathscr{H}} \cdot p_j$.

4. *Comparison with optimal functionals.* When there is only one prediction vector, p, and when the data vectors $\{d_1, \ldots, d_D\}$ are linearly independent, we have two ways of estimating $p \cdot h_E$ from knowledge of the data and their error statistics. The first method involves approximating p by a member of $\mathfrak{D}$, and the second has just been described. Here we compare the results of the two methods when the experimental errors are normally distributed and when $\mu_{\text{prej}\,\mathscr{F}}$ is normal with mean 0 and variance $M^2\mathscr{F}$.

Let $d^1, \ldots, d^D$ be the basis for $\mathfrak{D}$ dual to $\{d_1, \ldots, d_D\}$. If V is the tensor defined by (57), then V lives on $\mathfrak{D}$ and $V^{\text{inv}} = \sum (V^{-1})^{ij} d_i d_j$, so the V defined by (57) is the same as that defined by (109). Therefore the $\tilde{V}_{\mathscr{B} \vee \mathscr{D}}$ defined by (68) is the same as the $V_{\mathscr{H}}$ defined by (112). It follows that the mean square error (66) is the same as the variance of the a posteriori subjective probability distribution for $p \cdot h_E$ on R which we deduce from the variance of $\mu^{\gamma}_{\text{post}\,\mathscr{P}}$ on $\mathfrak{P}$, as given by our Bayesian inference.

To complete the demonstration that the two methods agree, we will show that $p \cdot h_{\mathscr{H}}(\gamma)$, the mean of the a posteriori subjective probability distribution for $p \cdot h_E$ on R, as given by (113), is the same as the value $p \cdot \tilde{h}_{\mathscr{B} \vee \mathscr{D}}$ given by (65) for the best available estimate of $p \cdot h_E$. In fact, we will show that $\tilde{h}_{\mathscr{B} \vee \mathscr{D}} = h_{\mathscr{H}}(\gamma)$. To do so, it suffices to show that

$$h' - U \cdot W^{\text{inv}} \cdot h' = V_{\mathscr{D}} \cdot d(\gamma) \tag{115}$$

where $U = M^{-2}V$ and $W = \mathscr{D} \wedge \mathscr{F} + \mathscr{D} \wedge \mathscr{F} \cdot U \cdot \mathscr{D} \wedge \mathscr{F}$. Here h' is given by (56) and $d(\gamma)$ by (107). Therefore $\gamma_i = d_i \cdot h'$, and $d(\gamma) = h' \cdot \sum_{i,j=1}^{D} d_i d_j (V^{-1})^{ij} = h' \cdot V^{\text{inv}} = V^{\text{inv}} \cdot h'$. Hence to prove (115), since h' is in $\mathfrak{D}$, it suffices to prove that $\mathscr{D} - U \cdot W^{\text{inv}} = V_{\mathscr{D}} \cdot V^{\text{inv}}$ or, what is equivalent according to (110),

$$\mathscr{D} - U \cdot W^{\text{inv}} = [\mathscr{D} + U \cdot \mathscr{D} \wedge \mathscr{F}]^{\text{inv}}. \tag{116}$$

Now if $\mathfrak{C} = \mathfrak{D} \ominus (\mathfrak{D} \cap \mathfrak{F})$ then $\mathscr{D} = \mathscr{C} + \mathscr{D} \wedge \mathscr{F}$ so

$$\mathscr{D} + U \cdot \mathscr{D} \wedge \mathscr{F} = \mathscr{C} + \mathscr{C} \cdot U \cdot \mathscr{D} \wedge \mathscr{F} + W.$$

Therefore, according to (17),

$$\begin{aligned}(\mathscr{D} + U \cdot \mathscr{D} \wedge \mathscr{F})^{\text{inv}} &= \mathscr{C} - \mathscr{C} \cdot U \cdot \mathscr{D} \wedge \mathscr{F} \cdot W^{\text{inv}} + W^{\text{inv}} \\ &= \mathscr{D} - \mathscr{D} \wedge \mathscr{F} - \mathscr{C} \cdot U \cdot W^{\text{inv}} + W^{\text{inv}} \\ &= \mathscr{D} - U \cdot W^{\text{inv}} - \mathscr{D} \wedge \mathscr{F} \\ &\quad + \mathscr{D} \wedge \mathscr{F} \cdot U \cdot \mathscr{D} \wedge \mathscr{F} \cdot W^{\text{inv}} + W^{\text{inv}} \\ &= \mathscr{D} - U \cdot W^{\text{inv}} + W^{\text{inv}} \\ &\quad + (W - \mathscr{D} \wedge \mathscr{F}) \cdot W^{\text{inv}} - \mathscr{D} \wedge \mathscr{F}.\end{aligned}$$

Since W lives on $\mathfrak{D} \cap \mathfrak{F}$, this proves (116) and establishes the equivalence of the two methods for estimating $p \cdot h_E$, at least as far as the results of the two methods overlap, and under the assumption that the experimental errors are normally distributed in R^D and that $\mu_{\text{prej}\,\mathscr{F}}$ is normal on $\mathfrak{F}$ with mean 0 and variance $M^2\mathscr{F}$.

C. Linear quelling.

1. *General description.*

(a) *Non-Hilbert model spaces.* In trying to make predictions $\gamma_1, \ldots, \gamma_P$ about a physical object E from measurements $\gamma_1, \ldots, \gamma_D$ of some of its numerical properties, we have assumed so far that E was modelled by some member h_E of a Hilbert space $\mathfrak{H}$, and that the data functionals $g_1, \ldots, g_D$ and the prediction functionals $\tilde{g}_1, \ldots, \tilde{g}_P$ were continuous linear functionals on $\mathfrak{H}$.

In many applications the model space which first presents itself as the place to look for a model m_E of E is a real vector space $\mathfrak{M}$ on which there is no obviously appropriate inner product or on which some of the data and prediction functionals are discontinuous with respect to the obvious inner product. For example, suppose that $\mathfrak{M}$ is all continuous, real-valued functions $m(x)$ on the closed unit interval $0 \leqq x \leqq 1$. Suppose that for each i, $i = 1, \ldots, D$, there is a function $G_i(x)$ in $\mathfrak{M}$ such that the value of the ith data functional g_i at an arbitrary vector m in $\mathfrak{M}$ is

$$g_i(m) = \int_0^1 G_i(x) m(x)\, dx. \tag{117}$$

Suppose that for each j, $j = 1, \ldots, P$, there is a point x_j in $0 \leqq x_j \leqq 1$ such that the value of the jth prediction functional $\tilde{g}_j$ at an arbitrary vector m in $\mathfrak{M}$ is

$$\tilde{g}_j(m) = m(x_j). \tag{118}$$

Then our prediction problem consists in trying to estimate the values of $\gamma_j = m_E(x_j)$, $j = 1, \ldots, P$, from the values of

$$\gamma_i = \int_0^1 G_i(x) m_E(x)\, dx, \qquad i = 1, \ldots, D.$$

This sort of finite generalized moment problem is exactly what faces us in estimating the frictional lossiness of the earth's mantle at various depths, using the observed damping rates of a finite number of the earth's elastic-gravitational normal modes. If we were trying to estimate the $g_i(m_E)$ from the $\tilde{g}_j(m_E)$, then our problem of inference would be the problem of numerical integration.

In the example the obvious inner product on $\mathfrak{M}$ is

$$\langle m_1, m_2 \rangle = \int_0^1 m_1(x) m_2(x)\, dx.$$

With respect to this inner product, the functionals g_i are continuous, but the functionals $\tilde{g}_j$ are not. Indeed, we can have a sequence $\{m_1, m_2, \ldots\}$ of functions in $\mathfrak{M}$ with $\langle m_n, m_n \rangle \to 0$ and $\tilde{g}_1(m_n) \to \infty$ as $n \to \infty$. In many such examples we can find an unobvious inner product on $\mathfrak{M}$ (or enough of $\mathfrak{M}$ to be useful) which makes the data functionals and prediction functionals continuous. Our goal in §II.C.1 is to try to systematize the search for such unobvious inner products.

We start with a real vector space $\mathfrak{M}$ of models which we believe contains an unknown vector m_E which adequately describes the physical object E, at least as far as our measurements and predictions are concerned. We also have another vector space, $\mathfrak{M}^§$, consisting of those linear functionals $g : \mathfrak{M} \to R$ such that we are interested in $g(m_E)$. The data functionals and prediction functionals are, of course, in $\mathfrak{M}^§$, but they may not span $\mathfrak{M}^§$. As later examples will indicate, we may want to include other functionals in $\mathfrak{M}^§$. The members of $\mathfrak{M}^§$ are defined only as functionals on $\mathfrak{M}$, so two members of $\mathfrak{M}^§$ which behave alike are equal; that is, if g and g' are members of $\mathfrak{M}^§$ and $g(m) = g'(m)$ for every m in $\mathfrak{M}$, then $g = g'$.

If g_1 and g_2 are in $\mathfrak{M}^§$ and α_1 and α_2 are real numbers, then $\alpha_1 g_1 + \alpha_2 g_2$ denotes the linear functional $g : \mathfrak{M} \to R$ defined by requiring that for each m in $\mathfrak{M}$, $g(m) = \alpha_1 g_1(m) + \alpha_2 g_2(m)$. The assumption that $\mathfrak{M}^§$ is a vector space is the requirement that whenever g_1 and g_2 are in $\mathfrak{M}^§$ and α_1 and α_2 are real numbers then $\alpha_1 g_1 + \alpha_2 g_2$ is in $\mathfrak{M}^§$. If g is in $\mathfrak{M}^§$ and m is in $\mathfrak{M}$, we introduce for $g(m)$ the notation $[g, m]$. The real number $[g, m]$ depends linearly on m when g is fixed, because g is a linear functional. And $[g, m]$ depends linearly on g when m is fixed because of the definition of a linear combination of linear functionals. The resemblance of $[g, m]$ to an inner product can be exploited further. If M is any subset of $\mathfrak{M}$, we define $M^\perp$ as the subset of $\mathfrak{M}^§$ consisting of all those functionals g such that $[g, m] = 0$

for every vector m in M. If G is any subset of $\mathfrak{M}^\S$, we define $G^\perp$ as the subset of $\mathfrak{M}$ consisting of all those vectors m such that $[g, m] = 0$ for every functional g in G. Even if the sets G and M are not subspaces of $\mathfrak{M}^\S$ and $\mathfrak{M}$, $G^\perp$ is always a subspace of $\mathfrak{M}$ and $M^\perp$ is always a subspace of $\mathfrak{M}^\S$. If $\mathfrak{K}$ and $\mathfrak{L}$ are subspaces of $\mathfrak{M}$, $\mathfrak{K}^\S$ and $\mathfrak{L}^\S$ are subspaces of $\mathfrak{M}^\S$, and A and B are subsets of either $\mathfrak{M}$ or $\mathfrak{M}^\S$, then

$$(\mathfrak{K} + \mathfrak{L})^\perp = \mathfrak{K}^\perp \cap \mathfrak{L}^\perp, \tag{119}$$

$$(\mathfrak{K}^\S + \mathfrak{L}^\S)^\perp = \mathfrak{K}^{\S\perp} \cap \mathfrak{L}^{\S\perp}, \tag{120}$$

$$A \subseteqq B \quad \text{implies} \quad B^\perp \subseteqq A^\perp. \tag{121}$$

In a particular problem we may have the feeling that $\mathfrak{M}$ is unnecessarily large and that we can restrict the search for m_E to a certain subspace $\mathfrak{N} \subseteqq \mathfrak{M}$. For example, $\mathfrak{M}$ might be the space of real square integrable functions on the unit interval, and $m_E(x)$ might be the density of the earth at dimensionless radius x. Most geophysicists would be willing to assume that m_E was piecewise continuous and would restrict their search to the linear space $\mathfrak{N}$ consisting of such functions. Such a restriction of the search is comparable to accepting the hypothesis, in bounded linear inference on a Hilbert space, that we know a number M such that $\|h_E\|$ is probably not much larger than M. Both decisions are partly matters of judgment and intuition, and both should be tested by seeing how far variations in M or $\mathfrak{N}$ influence the predictions.

There are some quantitative requirements on the subspaces $\mathfrak{N} \subseteqq \mathfrak{M}$ to which it is reasonable to restrict our search for m_E. If, for two different functionals g_1 and g_2 in $\mathfrak{M}^\S$, we are certain that $[g_1, m_E] = [g_2, m_E]$, then there is no reason to consider any models m except those for which $[g_1, m] = [g_2, m]$, or $[g_1 - g_2, m] = 0$. That is, we can restrict our search for m_E to the subspace $\{g_1 - g_2\}^\perp \subseteqq \mathfrak{M}$. More generally, if $\mathfrak{K}^\S$ is the set of all functionals k in $\mathfrak{M}^\S$ for which we are certain that $[k, m_E] = 0$, we can restrict our search for m_E to $\mathfrak{K}^{\S\perp} \subseteqq \mathfrak{M}$, and we can replace any functional g in $\mathfrak{M}^\S$ by $g|\mathfrak{K}^{\S\perp}$. Thus we choose a new $\mathfrak{M}$ equal to $\mathfrak{K}^{\S\perp}$ and a new $\mathfrak{M}^\S$ consisting of the restrictions to $\mathfrak{K}^{\S\perp}$ of the functionals in the old $\mathfrak{M}^\S$. Some functionals which were different in the old $\mathfrak{M}^\S$ become indistinguishable in the new $\mathfrak{M}^\S$. The result of this shrinkage of the problem is that we no longer have in $\mathfrak{M}^\S$ any nonzero functionals k such that we are sure that $[k, m_E] = 0$. We will always suppose that this particular shrinkage has been carried out.

Suppose we try to restrict the search for m_E even further, to a subspace $\mathfrak{N} \subseteqq \mathfrak{M}$. If $\mathfrak{N}$ is so small that there are two different functionals g_1 and g_2 in $\mathfrak{M}^\S$ with $g_1|\mathfrak{N} = g_2|\mathfrak{N}$ then restricting the search for m_E to $\mathfrak{N}$ amounts to claiming that we are sure that $[g_1 - g_2, m_E] = 0$. But we suppose that

$\mathfrak{M}$ and $\mathfrak{M}^\S$ have already been shrunk so much that there are no nonzero members k of $\mathfrak{M}^\S$ about which we are sure that $[k, m_E] = 0$. Therefore we cannot restrict the search for m_E to such a small subspace $\mathfrak{N}$.

The foregoing discussion is best carried out with the help of a new term.

DEFINITION 34. Let $\mathfrak{N}$ be a subspace of $\mathfrak{M}$ and $\mathfrak{K}^\S$ a subspace of $\mathfrak{M}^\S$. We say that $\mathfrak{N}$ "separates" $\mathfrak{K}^\S$ iff $\mathfrak{K}^\S \cap \mathfrak{N}^\perp = \{0\}$, and $\mathfrak{K}^\S$ "separates" $\mathfrak{N}$ iff $\mathfrak{N} \cap \mathfrak{K}^{\S\perp} = \{0\}$.

Equivalent definitions are that $\mathfrak{N}$ separates $\mathfrak{K}^\S$ iff whenever k is in $\mathfrak{K}^\S$ and $[k, n] = 0$ for every n in $\mathfrak{N}$ then $k = 0$, and iff whenever k_1 and k_2 are in $\mathfrak{K}^\S$ and $[k_1, n] = [k_2, n]$ for all n in $\mathfrak{N}$ then $k_1 = k_2$. The space $\mathfrak{M}$ always separates $\mathfrak{M}^\S$ (because if it does not initially, we shrink it till it does) but $\mathfrak{M}^\S$ may not separate $\mathfrak{M}$. Any subspace $\mathfrak{N} \subseteq \mathfrak{M}$ to which we restrict our search for m_E should separate $\mathfrak{M}^\S$. Accepting this limitation on $\mathfrak{N}$ gives us a tool very useful in the applications. We have

THEOREM 35. *Suppose $\{k_1, \ldots, k_N\}$ is any finite set of linearly independent linear functionals on $\mathfrak{M}$. Suppose subspace $\mathfrak{N} \subseteq \mathfrak{M}$ separates $\operatorname{sp}\{k_1, \ldots, k_N\}$. Then there are vectors $\{n_1, \ldots, n_N\}$ in $\mathfrak{N}$ such that*

$$[k_i, n_j] = \delta_{ij}, \qquad i, j = 1, \ldots, N. \tag{122}$$

(Here $\delta_{ij} = 0$ if $i \neq j$ and $\delta_{ij} = 1$ if $i = j$.)

Moreover, if $\{n_1, \ldots, n_N\}$ is any set of vectors in $\mathfrak{N}$ satisfying Equation (122) then

(i) *$\{n_1, \ldots, n_N\}$ are linearly independent modulo $\{k_1, \ldots, k_N\}^\perp$, i.e. if $\alpha_1, \ldots, \alpha_N$ are real numbers such that $\sum_{i=1}^N \alpha_i n_i$ is in $\{k_1, \ldots, k_N\}^\perp$ then $\alpha_1 = \cdots = \alpha_N = 0$;*

(ii) *$\operatorname{sp}\{n_1, \ldots, n_N\}$ and $\operatorname{sp}\{k_1, \ldots, k_N\}$ separate each other;*

(iii) *$\mathfrak{M} = \operatorname{sp}\{n_1, \ldots, n_N\} + \{k_1, \ldots, k_N\}^\perp$;*

(iv) *$\operatorname{sp}\{n_1, \ldots, n_N\} \cap \{k_1, \ldots, k_N\}^\perp = \{0\}$.*

PROOF. First we prove that Equation (122) implies (i), (ii), (iii) and (iv). If $\alpha_1, \ldots, \alpha_N$ are real numbers such that $\sum_{j=1}^N \alpha_j n_j$ is in $\{k_1, \ldots, k_N\}^\perp$, then $0 = [k_i, \sum_{j=1}^N \alpha_j n_j] = \sum_{j=1}^N \alpha_j [k_i, n_j] = \sum_{j=1}^N \alpha_j \delta_{ij} = \alpha_i$, $i = 1, \ldots, N$. Hence (i). If $[k, n] = 0$ for every k in $\operatorname{sp}\{k_1, \ldots, k_N\}$ and if n is in $\operatorname{sp}\{n_1, \ldots, n_N\}$, the same argument shows that $n = 0$. And if $[k, n] = 0$ for every n in $\operatorname{sp}\{n_1, \ldots, n_N\}$, and if k is in $\operatorname{sp}\{k_1, \ldots, k_N\}$, the same argument shows that $k = 0$. Hence (ii). For (iii), suppose that m is in $\mathfrak{M}$. Let $m^\perp = m - \sum_{j=1}^N [k_j, m] n_j$. Then $[k_i, m^\perp] = [k_i, m] - \sum_{j=1}^N [k_j, m][k_i, n_j] = 0$, $i = 1, \ldots, N$, which proves (iii). Finally, (iv) follows from (ii) and Definition 34. It remains to show that we can find $\{n_1, \ldots, n_N\}$ satisfying Equation (122). We proceed by induction on N. For $N = 1$ we can find an n_1' in $\mathfrak{N}$ with $[k_1, n_1'] \neq 0$ since $\mathfrak{N}$ separates $\operatorname{sp}\{k_1, \ldots, k_N\}$. We take $n_1 =$

$n'_1/[k_1, n'_1]$. Now suppose the result has been proved for N and we want to extend it to $N + 1$. We have $N + 1$ linearly independent linear functionals $\{k_1, \ldots, k_{N+1}\}$ on $\mathfrak{M}$ and, by induction, N vectors $\{n'_1, \ldots, n'_N\}$ in $\mathfrak{N}$ such that $[k_i, n'_j] = \delta_{ij}, i, j = 1, \ldots, N$. If n is any vector in $\mathfrak{N}$ we write

$$n = n^{\perp} + \sum_{i=1}^{N} [k_i, n]n'_i.$$

Then $n^{\perp}$ is in $\{k_1, \ldots, k_N\}^{\perp} \cap \mathfrak{N}$, and

$$\begin{aligned}[k_{N+1}, n] &= [k_{N+1}, n^{\perp}] + \sum_{i=1}^{N} [k_i, n][k_{N+1}, n'_i] \\ &= [k_{N+1}, n^{\perp}] + \left[\sum_{i=1}^{N} [k_{N+1}, n'_i]k_i, n\right].\end{aligned}$$

If $[k_{N+1}, n^{\perp}] = 0$ for every $n^{\perp}$ in $\{k_1, \ldots, k_N\}^{\perp} \cap \mathfrak{N}$, then

$$[k_{N+1}, n] = \left[\sum_{i=1}^{N} [k_{N+1}, n'_i]k_i, n\right]$$

for every n in $\mathfrak{N}$. Since $\mathfrak{N}$ separates $\mathrm{sp}\{k_1, \ldots, k_{N+1}\}$ it follows that $k_{N+1} = \sum_{i=1}^{N} [k_{N+1}, n'_i]k_i$, contrary to the assumed linear independence of $\{k_1, \ldots, k_{N+1}\}$. Therefore we can find an $n^{\perp}$ in $\{k_1, \ldots, k_N\}^{\perp} \cap \mathfrak{N}$ such that $[k_{N+1}, n^{\perp}] \neq 0$. Then we define $n_{N+1} = n^{\perp}/[k_{N+1}, n^{\perp}]$. We have $[k_{N+1}, n_{N+1}] = 1$, and $[k_i, n_{N+1}] = 0$ if $i = 1, \ldots, N$. Now we define $n_j = n'_j - [k_{N+1}, n'_j]n_{N+1}, j = 1, \ldots, N$. Then $[k_{N+1}, n_j] = 0$, and for $i = 1, \ldots, N$, $[k_i, n_j] = [k_i, n'_j] = \delta_{ij}$ because $[k_i, n_{N+1}] = 0$. This completes the induction and the proof of Theorem 35.

(b) *Definition of a linear quelling and its adjoint.* Suppose that $\mathfrak{M}$ is a real vector space and $\mathfrak{M}^{\S}$ is a vector space of linear functionals on $\mathfrak{M}$ which is separated by $\mathfrak{M}$. A linear quelling of $(\mathfrak{M}, \mathfrak{M}^{\S})$ is an ordered pair $(\mathfrak{H}^0, Q)$ with these properties:

(i) $\mathfrak{H}^0$ is a real pre-Hilbert space, i.e. a real vector space with a real, bilinear, positive-definite inner product (the completion of $\mathfrak{H}^0$ in the norm given by its inner product is a Hilbert space which we will call $\mathfrak{H}$);

(ii) $Q : \mathfrak{H}^0 \to \mathfrak{M}$ is a linear injection, i.e. a linear, one-to-one mapping of $\mathfrak{H}^0$ into $\mathfrak{M}$;

(iii) for any g in $\mathfrak{M}^{\S}$ the real number $[g, Qh]$ depends continuously on h in $\mathfrak{H}^0$.

For any particular member g of $\mathfrak{M}^{\S}$ we can define a functional $g^* : \mathfrak{H}^0 \to R$ by requiring for each h in $\mathfrak{H}^0$ that $g^*(h) = [g, Qh]$. Then g^* is a linear functional because both Q and g are linear, and g^* is continuous because of part (iii) of the definition of a quelling. Therefore g^* can be extended by continuity in exactly one way so as to be a continuous linear functional on $\mathfrak{H}$, the completion of $\mathfrak{H}^0$. It follows that there is a unique vector Q^*g in $\mathfrak{H}$ such that for every h in $\mathfrak{H}$, $\langle Q^*g, h\rangle = g^*(h)$. (Here we continue to write the inner product of two vectors h_1 and h_2 in $\mathfrak{H}$ either as $\langle h_1, h_2\rangle$ or as $h_1 \cdot h_2$, whichever is convenient at the moment.) Therefore, for every h in $\mathfrak{H}^0$,

(123) $$\langle Q^*g, h\rangle = [g, Qh],$$

and the validity of this equation for all h in $\mathfrak{H}^0$ uniquely determines the vector Q^*g in $\mathfrak{H}$ once g is given in $\mathfrak{M}^\S$. Therefore $Q^*: \mathfrak{M}^\S \to \mathfrak{H}$ is a well-defined mapping. It is called the "adjoint" of Q.

The adjoint is a linear mapping. To see this, let α_1 and α_2 be real numbers and let g_1 and g_2 be functionals in $\mathfrak{M}^\S$. For any h in $\mathfrak{H}^0$ we have

$$\langle Q^*(\alpha_1 g_1 + \alpha_2 g_2), h\rangle = [\alpha_1 g_1 + \alpha_2 g_2, Qh] = \alpha_1[g_1, Qh] + \alpha_2[g_2, Qh]$$
$$= \alpha_1\langle Q^*g_1, h\rangle + \alpha_2\langle Q^*g_2, h\rangle.$$

Therefore, for every h in $\mathfrak{H}^0$ we have

(124) $$\langle Q^*(\alpha_1 g_1 + \alpha_2 g_2), h\rangle = \langle \alpha_1 Q^*g_1 + \alpha_2 Q^*g_2, h\rangle.$$

Since both sides of Equation (124) are continuous in h and $\mathfrak{H}^0$ is dense in $\mathfrak{H}$, it follows that the two vectors $Q^*(\alpha_1 g_1 + \alpha_2 g_2)$ and $\alpha_1 Q^*g_1 + \alpha_2 Q^*g_2$ are equal.

Part of the definition of a quelling requires that the inverse mapping $Q^{-1}: Q(\mathfrak{H}^0) \to \mathfrak{H}^0$ exist. Therefore another way to write the defining equation for Q^* is this: if m is any member of $Q(\mathfrak{H}^0)$ and g is any functional in $\mathfrak{M}^\S$ then

(125) $$[g, m] = \langle Q^*g, Q^{-1}m\rangle.$$

This equation is simply (123) with h set equal to $Q^{-1}m$.

(c) *Application of adequate linear quellings.*

DEFINITION 36. A linear quelling $(\mathfrak{H}^0, Q)$ of $(\mathfrak{M}, \mathfrak{M}^\S)$ will be called "adequate" iff $Q(\mathfrak{H}^0)$ separates $\mathfrak{M}^\S$.

Thus $(\mathfrak{H}^0, Q)$ is adequate iff $Q(\mathfrak{H}^0)^\perp = \{0\}$. We have

REMARK 37. *If $(\mathfrak{H}^0, Q)$ is a linear quelling of $(\mathfrak{M}, \mathfrak{M}^\S)$ then*

(126) $$(Q^*)^{-1}(\{0\}) = Q(\mathfrak{H}^0)^\perp.$$

Therefore $(\mathfrak{H}^0, Q)$ is adequate iff $Q^*: \mathfrak{M}^\S \to \mathfrak{H}$ is an injection.

PROOF. Suppose that g is in $\mathfrak{M}^\S$ and h is in $\mathfrak{H}^0$. Then $[g, Qh] = \langle Q^*g, h\rangle$, so $Q^*g = 0$ iff $[g, Qh] = 0$ for all h in $\mathfrak{H}^0$, i.e. iff g is a member of $Q(\mathfrak{H}^0)^\perp$.

If $(\mathfrak{H}^0, Q)$ is an adequate linear quelling of $(\mathfrak{M}, \mathfrak{M}^\S)$, then the assumption that m_E lies in $Q(\mathfrak{H}^0)$ does not force $[g, m_E] = 0$ for any nonzero g in $\mathfrak{M}^\S$.

If $Q(\mathfrak{H}^0)$ is large enough that there are no obvious physical objections to assuming that m_E is in $Q(\mathfrak{H}^0)$, then we can make that assumption. Its effect is immediately to reduce our problem of linear inference to bounded linear inference on a Hilbert space.

We take for our Hilbert space the completion of $\mathfrak{H}^0$, which we call $\mathfrak{H}$. The mapping $Q^{-1}: Q(\mathfrak{H}^0) \to \mathfrak{H}^0$ establishes a one-to-one correspondence between a subspace $Q(\mathfrak{H}^0)$ of models in $\mathfrak{M}$ which contains m_E and a subspace $\mathfrak{H}^0 \subseteqq \mathfrak{H}$ which contains $h_E = Q^{-1}m_E$. The mapping $Q^*: \mathfrak{M}^\S \to \mathfrak{H}$ establishes a one-to-one correspondence between the functionals g for which we might want to know $[g, m_E]$ and a subspace of vectors in $\mathfrak{H}$. In particular the data functionals g_i are mapped into data vectors $d_i = Q^* g_i$, $i = 1, \ldots, D$, and the prediction functionals $\tilde{g}_j$ are mapped into prediction vectors $p_j = Q^*\tilde{g}_j, j = 1, \ldots, P$. According to (125) we have $\gamma_i = [g_i, m_E] = d_i \cdot h_E$, $i = 1, \ldots, D$, and $\tilde{\gamma}_j = [\tilde{g}_j, m_E] = p_j \cdot h_E, j = 1, \ldots, P$. Thus we have a problem of bounded linear inference on a Hilbert space, and all the apparatus of §II.B is available to us if we think we know an M such that $\|h_E\|$ is probably not much larger than M, i.e. $\|Q^{-1}m_E\|$ is probably not much larger than M.

In the applications of quelling, $\mathfrak{M}$ is usually a function space and some members of $\mathfrak{M}^\S$ are discontinuous in an obvious topology on $\mathfrak{M}$. Then, loosely speaking, Q^* converts vectors of infinite length into vectors of finite length, so Q^* is a sort of smoothing operator, and so is Q. Hence Q^{-1} roughens, and the hypothesis $\|Q^{-1}m_E\| \leqq M$ is a hypothesis about the smoothness of m_E as well as its general size.

In a sense the new bounded linear inference on $\mathfrak{H}$ is more appropriate to the physical problem that was the old linear inference on $\mathfrak{M}$. There was no assumption that the data $[g_i, m_E]$ or the predictions $[\tilde{g}_j, m_E]$ depend continuously on m_E, so very similar models could produce very different data or predictions. In the new formulation, $d_i \cdot h_E$ and $p_j \cdot h_E$ do depend continuously on h_E, so $\mathfrak{H}$ is a more natural place than $\mathfrak{M}$ to look for models of E. Therefore we ought to be willing to accept any member of $\mathfrak{H}$ as a possible model for E. The space $\mathfrak{H}^0$ to which $Q^{-1}m_E$ belongs is dense in $\mathfrak{H}$, so there is no way of determining by measurements with nonzero errors, whether a given h belongs to $\mathfrak{H}^0$ or only to $\mathfrak{H}$. The question is like asking whether the mass of the earth in grams is a rational number.

(d) *Construction of adequate quellings.* A quelling $(\mathfrak{H}^0, Q)$ of $(\mathfrak{M}, \mathfrak{M}^\S)$ will be adequate iff $(Q^*)^{-1}(\{0\}) = \{0\}$, so the inadequacy of a quelling is measured either by the largeness of $(Q^*)^{-1}(\{0\})$ or the smallness of $Q(\mathfrak{H}^0)$. In addition to Equation (126) we have, from (121),

$$Q(\mathfrak{H}^0) \subseteqq [(Q^*)^{-1}(\{0\})]^\perp. \tag{127}$$

If $(\mathfrak{H}_1^0, Q_1)$ and $(\mathfrak{H}_2^0, Q_2)$ are both quellings of $(\mathfrak{M}, \mathfrak{M}^{\S})$, we will write $(\mathfrak{H}_1^0, Q_1) \subseteqq (\mathfrak{H}_2^0, Q_2)$ and say that $(\mathfrak{H}_2^0, Q_2)$ is an extension of $(\mathfrak{H}_1^0, Q_1)$ if we have $\mathfrak{H}_1^0 \subseteqq \mathfrak{H}_2^0$ and $Q_1 = Q_2|\mathfrak{H}_1^0$. An immediate consequence of Equations (121) and (126) is

REMARK 38. *If $(\mathfrak{H}_1^0, Q_1)$ and $(\mathfrak{H}_2^0, Q_2)$ are both linear quellings of $(\mathfrak{M}, \mathfrak{M}^{\S})$ and $(\mathfrak{H}_1^0, Q_1) \subseteqq (\mathfrak{H}_2^0, Q_2)$, then $(Q_2^*)^{-1}(\{0\}) \subseteqq (Q_1^*)^{-1}(\{0\})$.*

If $(\mathfrak{H}^0, Q)$ is an inadequate quelling of $(\mathfrak{M}, \mathfrak{M}^{\S})$, we would like to find ways to extend it, and if possible to extend it to an adequate quelling. One procedure which sometimes works is to construct a direct sum of two inadequate quellings. Suppose that $(\mathfrak{H}_1^0, Q_1)$ and $(\mathfrak{H}_2^0, Q_2)$ are both linear quellings of $(\mathfrak{M}, \mathfrak{M}^{\S})$. The set of all ordered pairs (h_1, h_2) with h_1 in $\mathfrak{H}_1^0$ and h_2 in $\mathfrak{H}_2^0$ can be made into a real vector space by defining, for any real α and α',

$$\alpha(h_1, h_2) + \alpha'(h_1', h_2') = (\alpha h_1 + \alpha' h_1', \alpha h_2 + \alpha' h_2').$$

This vector space can be made into a pre-Hilbert space by introducing the dot product

$$(h_1, h_2) \cdot (h_1', h_2') = h_1 \cdot h_1' + h_2 \cdot h_2'.$$

The pre-Hilbert space so obtained we will denote by $\mathfrak{H}_1^0 \oplus \mathfrak{H}_2^0$. Loosely speaking, it contains both $\mathfrak{H}_1^0$ and $\mathfrak{H}_2^0$, and its completion is $\mathfrak{H}_1 \oplus \mathfrak{H}_2$, where $\mathfrak{H}_i$ is the completion of $\mathfrak{H}_i^0$, $i = 1, 2$.

We define the mapping $Q_1 \oplus Q_2 : \mathfrak{H}_1^0 \oplus \mathfrak{H}_2^0 \to \mathfrak{M}$ by requiring for any h_1 in $\mathfrak{H}_1^0$ and any h_2 in $\mathfrak{H}_2^0$ that

$$Q_1 \oplus Q_2(h_1, h_2) = Q_1 h_1 + Q_2 h_2. \tag{128}$$

Then $Q_1 \oplus Q_2$ is linear. If g is a member of $\mathfrak{M}^{\S}$ we have

$$[g, Q_1 \oplus Q_2(h_1, h_2)] = [g, Q_1 h_1 + Q_2 h_2] = [g, Q_1 h_1] + [g, Q_2 h_2]$$

so $[g, Q_1 \oplus Q_2(h_1, h_2)]$ depends continuously on (h_1, h_2). Therefore $(\mathfrak{H}_1^0 \oplus \mathfrak{H}_2^0, Q_1 \oplus Q_2)$ is a linear quelling of $(\mathfrak{M}, \mathfrak{M}^{\S})$ iff $Q_1 \oplus Q_2$ is an injection. From (128), this is the requirement that if $Q_1 h_1 + Q_2 h_2 = 0$ then $h_1 = 0$ and $h_2 = 0$, which is equivalent to the demand that $Q_1(\mathfrak{H}_1^0) \cap Q_2(\mathfrak{H}_2^0) = \{0\}$. Thus we have

THEOREM 39. *Let $(\mathfrak{H}_1^0, Q_1)$ and $(\mathfrak{H}_2^0, Q_2)$ be linear quellings of $(\mathfrak{M}, \mathfrak{M}^{\S})$. Then $(\mathfrak{H}_1^0 \oplus \mathfrak{H}_2^0, Q_1 \oplus Q_2)$ is a linear quelling of $(\mathfrak{M}, \mathfrak{M}^{\S})$ iff $Q_1(\mathfrak{H}_1^0) \cap Q_2(\mathfrak{H}_2^0) = \{0\}$. When this condition is satisfied, $(\mathfrak{H}_1^0 \oplus \mathfrak{H}_2^0, Q_1 \oplus Q_2)$ is an extension of both $(\mathfrak{H}_1^0, Q_1)$ and $(\mathfrak{H}_2^0, Q_2)$, and*

$$Q_1 \oplus Q_2(\mathfrak{H}_1^0 \oplus \mathfrak{H}_2^0) = Q_1(\mathfrak{H}_1^0) + Q_2(\mathfrak{H}_2^0). \tag{129}$$

Moreover, the adjoint of $Q_1 \oplus Q_2$ *is given by requiring for every* g *in* $\mathfrak{M}^{\S}$ *that*

$$(130) \qquad (Q_1 \oplus Q_2)^* g = (Q_1^* g, Q_2^* g)$$

whence

$$(131) \qquad [(Q_1 \oplus Q_2)^*]^{-1}(\{0\}) = (Q_1^*)^{-1}(\{0\}) \cap (Q_2^*)^{-1}(\{0\}).$$

PROOF. The only part of Theorem 39 not yet proved concerns the adjoints. Since Equation (131) follows immediately from (130), we need prove only (130). Let $(Q_1 \oplus Q_2)^{\dagger}: \mathfrak{M}^{\S} \to \mathfrak{H}_1 \oplus \mathfrak{H}_2$ be defined by requiring for each g in $\mathfrak{M}^{\S}$ that $(Q_1 \oplus Q_2)^{\dagger} g = (Q_1^* g, Q_2^* g)$. Then for any (h_1, h_2) in $\mathfrak{H}_1^0 \oplus \mathfrak{H}_2^0$ and any g in $\mathfrak{M}^{\S}$ we have

$$\langle (Q_1 \oplus Q_2)^{\dagger} g, (h_1, h_2) \rangle = (Q_1^* g, Q_2^* g) \cdot (h_1, h_2) = \langle Q_1^* g, h_1 \rangle + \langle Q_2^* g, h_2 \rangle$$
$$= [g, Q_1 h_1] + [g, Q_2 h_2] = [g, Q_1 \oplus Q_2(h_1, h_2)].$$

Since Equation (123) determines Q^* uniquely, it follows that $(Q_1 \oplus Q_2)^{\dagger} = (Q_1 \oplus Q_2)^*$. This completes the proof of Theorem 39.

Even if $Q_1(\mathfrak{H}_1^0) \cap Q_2(\mathfrak{H}_2^0) \neq \{0\}$ it is sometimes still possible to find a quelling $(\mathfrak{H}^0, Q)$ with $Q(\mathfrak{H}^0) = Q_1(\mathfrak{H}_1^0) + Q_2(\mathfrak{H}_2^0)$, and hence $(Q^*)^{-1}(\{0\}) = (Q_1^*)^{-1}(\{0\}) \cap (Q_2^*)^{-1}(\{0\})$. We have

REMARK 40. *Suppose that* $(\mathfrak{H}_1^0, Q_1)$ *and* $(\mathfrak{H}_2^0, Q_2)$ *are linear quellings of* $(\mathfrak{M}, \mathfrak{M}^{\S})$. *Let* $\mathfrak{T} = Q_2^{-1}[Q_1(\mathfrak{H}_1^0) \cap Q_2(\mathfrak{H}_2^0)]$. *Suppose that there is a subspace* $\mathfrak{H}_3^0 \subseteq \mathfrak{H}_2^0$ *such that* $\mathfrak{H}_3^0 \cap \mathfrak{T} = \{0\}$ *and* $\mathfrak{H}_2^0 = \mathfrak{H}_3^0 + \mathfrak{T}$. *Let* $Q_3 = Q_2 | \mathfrak{H}_3^0$. *Then* $(\mathfrak{H}_3^0, Q_3)$ *is a linear quelling of* $(\mathfrak{M}, \mathfrak{M}^{\S})$ *and* $Q_1(\mathfrak{H}_1^0) \cap Q_3(\mathfrak{H}_3^0) = \{0\}$, *so* $(\mathfrak{H}_1^0 \oplus \mathfrak{H}_3^0, Q_1 \oplus Q_3)$ *is also a linear quelling of* $(\mathfrak{M}, \mathfrak{M}^{\S})$. *Moreover,* $Q_1 \oplus Q_3(\mathfrak{H}_1^0 \oplus \mathfrak{H}_3^0) = Q_1(\mathfrak{H}_1^0) + Q_2(\mathfrak{H}_2^0)$.

PROOF. It is clear that $(\mathfrak{H}_3^0, Q_3)$ is a quelling of $(\mathfrak{M}, \mathfrak{M}^{\S})$. If there are vectors h_1 in $\mathfrak{H}_1^0$ and h_3 in $\mathfrak{H}_3^0$ such that $Q_1 h_1 = Q_3 h_3$, then $Q_1 h_1 = Q_2 h_3$ so $Q_2 h_3$ is in $Q_1(\mathfrak{H}_1^0) \cap Q_2(\mathfrak{H}_2^0)$, and h_3 is in $\mathfrak{T}$ as well as $\mathfrak{H}_3^0$. Therefore, by hypothesis, $h_3 = 0$. Thus $Q_3 h_3 = 0$, so $Q_1(\mathfrak{H}_1^0) \cap Q_3(\mathfrak{H}_3^0) = \{0\}$. Next, for any vectors h_1 in $\mathfrak{H}_1^0$ and h_2 in $\mathfrak{H}_2^0$, we can find an h_3 in $\mathfrak{H}_3^0$ and a t in $\mathfrak{T}$ such that $h_2 = h_3 + t$. But there is an h_1' in $\mathfrak{H}_1^0$ such that $Q_1 h_1' = Q_2 t$, so $Q_1 h_1 + Q_2 h_2 = Q_1 h_1 + Q_2 h_3 + Q_2 t = Q_1(h_1 + h_1') + Q_3 h_3$. Thus $Q_1(\mathfrak{H}_1^0) + Q_2(\mathfrak{H}_2^0) \subseteq Q_1(\mathfrak{H}_1^0) + Q_3(\mathfrak{H}_3^0)$. Containment in the other direction being obvious, the two sets are equal.

The question of when Theorem 39 produces an adequate linear quelling is answered by

REMARK 41. *Suppose that* $(\mathfrak{H}_1^0, Q_1)$, $(\mathfrak{H}_2^0, Q_2)$ *and* $(\mathfrak{H}_1^0 \oplus \mathfrak{H}_2^0, Q_1 \oplus Q_2)$ *are linear quellings of* $(\mathfrak{M}, \mathfrak{M}^{\S})$. *Then* $(\mathfrak{H}_1^0 \oplus \mathfrak{H}_2^0, Q_1 \oplus Q_2)$ *is adequate iff* $Q_2(\mathfrak{H}_2^0)$ *separates* $(Q_1^*)^{-1}(\{0\})$.

PROOF. From Definition 34, $Q_2(\mathfrak{H}_2^0)$ separates $(Q_1^*)^{-1}(\{0\})$ iff $(Q_1^*)^{-1}(\{0\}) \cap Q_2(\mathfrak{H}_2^0)^\perp = \{0\}$. From Remark 37 this is equivalent to the condition $(Q_1^*)^{-1}(\{0\}) \cap (Q_2^*)^{-1}(\{0\}) = \{0\}$. From equation (131) this condition is $[(Q_1 \oplus Q_2)^*]^{-1}(\{0\}) = \{0\}$, and according to Remark 37 the condition is $Q_1 \oplus Q_2(\mathfrak{H}_1^0 \oplus \mathfrak{H}_2^0)^\perp = \{0\}$, which is precisely the Definition 36 that $(\mathfrak{H}_1^0 \oplus \mathfrak{H}_2^0, Q_1 \oplus Q_2)$ be adequate.

If $(\mathfrak{H}^0, Q)$ is a linear quelling of $(\mathfrak{M}, \mathfrak{M}^\S)$ such that $(Q^*)^{-1}(\{0\})$ is finite dimensional, then Remark 41 makes it possible to extent $(\mathfrak{H}^0, Q)$ to an adequate linear quelling. We have

THEOREM 42. *Suppose that* $(\mathfrak{H}^0, Q)$ *is a quelling of* $(\mathfrak{M}, \mathfrak{M}^\S)$ *and that* $\{m_1, \ldots, m_N\}$ *is a finite set of members of* $\mathfrak{M}$ *which are linearly independent modulo* $Q(\mathfrak{H}^0)$ *and such that* $\operatorname{sp}\{m_1, \ldots, m_N\}$ *separates* $(Q^*)^{-1}(\{0\})$. *Define a mapping* $Q^N : R^N \to \mathfrak{M}$ *by requiring for any N-tuple of real numbers* $(\alpha_1, \ldots, \alpha_N)$ *that* $Q^N(\alpha_1, \ldots, \alpha_N) = \alpha_1 m_1 + \cdots + \alpha_N m_N$. *Then* (R^N, Q^N) *is a linear quelling of* $(\mathfrak{M}, \mathfrak{M}^\S)$ *and* $(\mathfrak{H}^0 \oplus R^N, Q \oplus Q^N)$ *is an adequate linear quelling of* $(\mathfrak{M}, \mathfrak{M}^\S)$. *The adjoint* $Q^{N*} : \mathfrak{M}^\S \to R^N$ *is the mapping which assigns to any functional g in* $\mathfrak{M}^\S$ *the N-tuple*

$$Q^{N*}g = ([g, m_1], \ldots, [g, m_N]). \tag{132}$$

Thus the mapping $Q \oplus Q^N : \mathfrak{H}^0 \oplus R^N \to \mathfrak{M}$ *assigns to any* $(N+1)$*-tuple* $(h, \alpha_1, \ldots, \alpha_N)$, *with h in* $\mathfrak{H}^0$ *and* $\alpha_1, \ldots, \alpha_N$ *real, the vector*

$$Q \oplus Q^N(h, \alpha_1, \ldots, \alpha_N) = Qh + \sum_{i=1}^{N} \alpha_i m_i \tag{133}$$

and the adjoint $(Q \oplus Q^N)^* : \mathfrak{M}^\S \to \mathfrak{H} \oplus R^N$ *assigns to any functional g in* $\mathfrak{M}^\S$ *the* $(N+1)$*-tuple*

$$(Q \oplus Q^N)^* g = (Q^* g, [g, m_1], \ldots, [g, m_N]), \tag{134}$$

which is in $\mathfrak{H} \oplus R^N$.

PROOF. First we must show that $Q(\mathfrak{H}^0) \cap Q^N(R^N) = \{0\}$. If there is an h in $\mathfrak{H}^0$ and an N-tuple $(\alpha_1, \ldots, \alpha_N)$ in R^N such that $Qh = Q^N(\alpha_1, \ldots, \alpha_N)$, then $Qh = \sum_{i=1}^N \alpha_i m_i$. By hypothesis, $\{m_1, \ldots, m_N\}$ are linearly independent modulo $Q(\mathfrak{H}^0)$, so $\alpha_1 = \cdots = \alpha_N = 0$. Thus $Qh = 0$. Hence, by Theorem 39, $(\mathfrak{H}^0 \oplus R^N, Q \oplus Q^N)$ is indeed a linear quelling of $(\mathfrak{M}, \mathfrak{M}^\S)$. Its adequacy follows immediately from Remark 41. To show that Q^{N*} is given by Equation (132), observe that for any $(\alpha_1, \ldots, \alpha_N)$ in R^N and any g in $\mathfrak{M}^\S$,

$$[g, Q^N(\alpha_1, \ldots, \alpha_N)] = \sum_{i=1}^{N} \alpha_i [g, m_i] = ([g, m_1], \ldots, [g, m_N]) \cdot (\alpha_1, \ldots, \alpha_N).$$

Comparison with Equation (123) proves Equation (132). Equation (133) is simply the definition of $Q \oplus Q^N$, and Equation (134) is an immediate consequence of Equations (130) and (132). This completes the proof.

COROLLARY 43. *If* $(\mathfrak{H}^0, Q)$ *is a linear quelling of* $(\mathfrak{M}, \mathfrak{M}^{\S})$ *and* $\dim(Q^*)^{-1}(\{0\}) = N < \infty$, *we can always find a set of members* $\{m_1, \ldots, m_N\}$ *of* $\mathfrak{M}$ *satisfying the hypotheses of Theorem* 42, *so we can always construct an adequate linear quelling* $(\mathfrak{H}^0 \oplus R^N, Q \oplus Q^N)$ *which is an extension of* $(\mathfrak{H}^0, Q)$.

PROOF. Let $\{k_1, \ldots, k_N\}$ be a basis for $(Q^*)^{-1}(\{0\})$, and let $\{m_1, \ldots, m_N\}$ be a set of vectors in $\mathfrak{M}$ such that $[k_i, m_j] = \delta_{ij}$, $i, j = 1, \ldots, N$. Because $\mathfrak{M}$ separates $\mathfrak{M}^{\S}$, Theorem 35 assures the existence of such a set $\{m_1, \ldots, m_N\}$. According to that theorem, $\{m_1, \ldots, m_N\}$ are linearly independent modulo $[(Q^*)^{-1}(\{0\})]^{\perp}$ so, from Equation (127), they are linearly independent modulo $Q(\mathfrak{H}^0)$. Also, according to Theorem 35, $\mathrm{sp}\{m_1, \ldots, m_N\}$ separates $(Q^*)^{-1}(\{0\})$. This proves Corollary 43.

A use of Corollary 43 which will appear often in the applications is the following: Suppose that $(\mathfrak{H}^0, Q)$ is an adequate quelling of $(\mathfrak{M}, \mathfrak{M}^{\S})$. Suppose there are a finite number of linear functionals $\{k_1, \ldots, k_N\}$ on $\mathfrak{M}$ such that $Q(\mathfrak{H}^0) \subseteq \{k_1, \ldots, k_N\}^{\perp}$. We may think it unlikely that $[k_i, m_E] = 0$, $i = 1, \ldots, N$. In that case, we do not want to restrict the search for m_E to $Q(\mathfrak{H}^0)$, even though $(\mathfrak{H}^0, Q)$ is adequate. We can choose a new $\mathfrak{M}^{\S}$ equal to $\mathfrak{M}^{\S} + \mathrm{sp}\{k_1, \ldots, k_N\}$. With this new $\mathfrak{M}^{\S}$, $(\mathfrak{H}^0, Q)$ is still a quelling of $(\mathfrak{M}, \mathfrak{M}^{\S})$, but is no longer adequate. It can be extended to an adequate quelling $(\mathfrak{H}_1^0, Q_1)$ by Corollary 43 and restricting the search for m_E to $Q_1(\mathfrak{H}_1^0)$ does not entail assuming that $[k_i, m_E] = 0$, $i = 1, \ldots, N$.

2. *Quelling derivatives of the Dirac delta function by multiplication.*

(a) *Statement of the problem.* As an illustration of the foregoing general discussion, we consider the following problem of inference. We choose a positive integer P and a point x_0 in the closed unit interval $[0, 1]$. Our model space $\mathfrak{M}$ consists of all functions $m(x)$ which are continuous everywhere on $[0, 1]$ and which have $(P - 1)$st derivatives at x_0. To be precise, we assume that there are numbers $\tilde{\gamma}_1, \ldots, \tilde{\gamma}_P$ such that

$$\lim_{x \to x_0} \frac{m(x) - \sum_{j=1}^{P} \tilde{\gamma}_j (x - x_0)^{j-1}}{(x - x_0)^{P-1}} = 0. \tag{135}$$

Clearly, the function m uniquely determines its $\tilde{\gamma}_1, \ldots, \tilde{\gamma}_P$, and we regard $\tilde{\gamma}_j (j - 1)!$ heuristically as $m^{(j-1)}(x_0)$, the $(j - 1)$st derivative of m at x_0.

We assume that the D data functionals all have the form (117) with data kernels $G_i(x)$ which are continuous on $[0, 1]$. We take only one prediction functional, $\tilde{g}$. For any m in $\mathfrak{M}$, $[g, m]$ is $\tilde{\gamma}_P$ as defined by (135); roughly speaking, $[g, m] = m^{(P-1)}(x_0)/(P-1)!$. In terms of the theory of generalized functions (distributions) $\tilde{g}$ has the form (117) with kernel $\tilde{G}(x) = (-1)^{P-1}\delta^{(P-1)}(x - x_0)/(P-1)!$, where $\delta(x)$ is the Dirac delta function. At the outset we take $\mathfrak{M}^{\S} = \text{sp}\{g_1, \ldots, g_D, \tilde{g}\}$.

(b) *A quelling and its extensions.* Our hope is to convert the very singular "function" $\delta^{(P-1)}(x - x_0)$ into a square-integrable function by applying a linear mapping $Q^*: \mathfrak{M}^{\S} \to \mathfrak{H}^0$ which is the adjoint of an injection Q of some pre-Hilbert space $\mathfrak{H}^0$ onto a subspace $Q(\mathfrak{H}^0) \subseteqq \mathfrak{M}$ sufficiently large that is is plausible to seek m_E in $Q(\mathfrak{H}^0)$.

The simplest way to make $\delta^{(P-1)}(x - x_0)$ square integrable is to multiply it by 0, but the range of the adjoint of the 0 operator contains only the single model $m = 0$, and we do not want to assume $m_E = 0$. The next simplest procedure is to multiply $\delta^{(P-1)}(x - x_0)$ by $(x - x_0)^P$. This still transforms $\delta^{(P-1)}(x - x_0)$ into 0, but there is some hope that the range of its adjoint is not uncomfortably small. We take for $\mathfrak{H}^0$ the space of all continuous functions on [0, 1], with inner product

$$\langle h_1, h_2 \rangle = \int_0^1 h_1(x) h_2(x)\, dx.$$

The completion of $\mathfrak{H}^0$ is $\mathfrak{H} = L_2[0, 1]$, the space of all square-integrable functions on [0, 1] modulo the null functions.

We take for $Q: \mathfrak{H}^0 \to \mathfrak{M}$ the mapping which assigns to any function h in $\mathfrak{H}^0$ a function Qh whose value at x in [0, 1] is given by

$$(Qh)(x) = (x - x_0)^P h(x). \tag{136}$$

Then Q is a linear injection of $\mathfrak{H}^0$ into $\mathfrak{M}$. To prove that $(\mathfrak{H}^0, Q)$ is a linear quelling of $(\mathfrak{M}, \mathfrak{M}^{\S})$ we must also show that for any g in $\mathfrak{M}^{\S}$, $[g, Qh]$ depends continuously on h. If g is in $\mathfrak{M}^{\S}$ there are real numbers $\alpha_0, \alpha_1, \ldots, \alpha_D$ such that $g = \alpha_0 \tilde{g} + \alpha_1 g_1 + \cdots + \alpha_D g_D = \alpha_0 \tilde{g} + g'$ where $[g', m] = \int_0^1 G'(x) m(x)\, dx$ with $G'(x) = \alpha_1 G_1(x) + \cdots + \alpha_D G_D(x)$. If $m(x) = (x - x_0)^P h(x)$ with h in $\mathfrak{H}^0$ then m satisfies (135) with $\tilde{\gamma}_1 = \cdots = \tilde{\gamma}_P = 0$, so $[g, m] = [g', m]$. Thus $[g, Qh] = \int_0^1 G'(x)(x - x_0)^P h(x)\, dx$, and, by Schwarz's inequality in $L_2[0, 1]$,

$$|[g, Qh]|^2 \leqq \|h\|^2 \int_0^1 G'(x)^2 (x - x_0)^{2P}\, dx.$$

Therefore $[g, Qh]$ defines a bounded linear functional on $\mathfrak{H}^0$ and is continuous in h. It follows that $(\mathfrak{H}^0, Q)$ is a quelling of $(\mathfrak{M}, \mathfrak{M}^{\S})$.

The adjoint, $Q^*:\mathfrak{M}^\S \to \mathfrak{H}^0$, is found as follows: for any g in $\mathfrak{M}^\S$ we write $g = \alpha_0 \tilde{g} + g'$ as before. Then

$$[g, Qh] = [g', Qh] = \int_0^1 [G'(x)(x - x_0)^P]h(x)\,dx = \langle Q^*g, h\rangle.$$

Thus Q^*g is that function in $\mathfrak{H}^0$ whose value at x in $[0, 1]$ is the real number $(Q^*g)(x) = (x - x_0)^P G'(x)$. If we write all linear functionals g on $\mathfrak{M}$ in the notation of distribution theory, so that $[g, m] = \int_0^1 G(x)m(x)\,dx$, then $(Q^*g)(x) = (x - x_0)^P G(x)$, because $(x - x_0)^P \delta^{(P-1)}(x - x_0) = 0$.

We note immediately that $(\mathfrak{H}^0, Q)$ is an inadequate quelling of $(\mathfrak{M}, \mathfrak{M}^\S)$, since $Q^*\tilde{g} = 0$. We can use Corollary 43 to extend $(\mathfrak{H}^0, Q)$ to an adequate quelling $(\mathfrak{H}^{0\prime}, Q')$, but if $P \geqq 2$ this extended quelling, even though adequate, suffers a serious defect: assuming that m_E is in $Q'(\mathfrak{H}^{0\prime})$ prejudges the values of certain interesting physical properties of E. For $j = 1, \ldots, P$, define $\tilde{g}_j:\mathfrak{M} \to R$ by requiring for any m in $\mathfrak{M}$ that $[\tilde{g}_j, m] = \tilde{\gamma}_j$, as given by (135). Then $\tilde{g} = \tilde{g}_P$. The assumption that m_E is in $Q'(\mathfrak{H}^{0\prime})$ forces us to accept that $[\tilde{g}_j, m_E] = 0$, $j = 1, \ldots, P - 1$. That is, $Q'(\mathfrak{H}^{0\prime})$ contains only functions in $\mathfrak{M}$ whose first $P - 2$ derivatives vanish at x_0. We will suppose that the original physical problem did not make this a reasonable assumption about m_E.

As remarked at the end of §II.C.1.d, the easiest way to enlarge $Q(\mathfrak{H}^0)$ to an acceptable size is first to enlarge $\mathfrak{M}^\S$ to include the offending functionals; we take $\mathfrak{M}^\S = \text{sp}\{g_1, \ldots, g_D, \tilde{g}_1, \ldots, \tilde{g}_P\}$. Since $\tilde{g}_j|Q(\mathfrak{H}^0) = 0$, $j = 1, \ldots, P$, it follows that $[g, Qh]$ is continuous in h for every g in the enlarged $\mathfrak{M}^\S$. Now, however, $(Q^*)^{-1}(\{0\}) = \text{sp}\{\tilde{g}_1, \ldots, \tilde{g}_P\}$, so $\dim(Q^*)^{-1}(\{0\}) = P$.

The functions $m_1, \ldots, m_P$ whose existence is asserted in Corollary 43 can be chosen in any convenient way. We will take $m_j(x) = (x - x_0)^{j-1}$, $j = 1, \ldots, P$. If $\sum_{j=1}^P \alpha_j m_j$ is in $Q(\mathfrak{H}^0)$ for some real numbers $\alpha_1, \ldots, \alpha_P$, then there is an h in $\mathfrak{H}^0$ such that $\sum_{j=1}^P \alpha_j(x - x_0)^{j-1} = (x - x_0)^P h(x)$, so $\alpha_1 = \cdots = \alpha_P = 0$. Thus $\{m_1, \ldots, m_P\}$ are linearly independent modulo $Q(\mathfrak{H}^0)$, and we can take $\tilde{\mathfrak{H}}^0 = \mathfrak{H}^0 \oplus R^P$, $\tilde{Q} = Q \oplus Q^P$, where $Q^P(\alpha_1, \ldots, \alpha_P) = \sum_{j=1}^P \alpha_j m_j$. Then for any $(P + 1)$-tuple $\tilde{h} = (h, \alpha_1, \ldots, \alpha_P)$ in $\tilde{\mathfrak{H}}^0$, $\tilde{Q}\tilde{h}$ is that function on $[0, 1]$ which assigns to any x the value

$$(\tilde{Q}\tilde{h})(x) = \sum_{j=1}^P \alpha_j(x - x_0)^{j-1} + (x - x_0)^P h(x). \tag{137}$$

How big is $\tilde{Q}(\tilde{\mathfrak{H}}^0)$? For any function m in $\mathfrak{M}$, define a function $h:[0, 1] \to R$ as

$$h(x) = \left(m(x) - \sum_{j=1}^P \tilde{\gamma}_j(x - x_0)^{j-1}\right)\Big/(x - x_0)^P. \tag{138}$$

In general, if m is in $\mathfrak{M}$, $h(x)$ may not be defined at x_0, and as $x \to x_0$ $h(x)$ may approach infinity or oscillate wildly. The function m is in $\tilde{Q}(\tilde{\mathfrak{H}}^0)$ iff h as defined by (138) has a limit as $x \to x_0$. Therefore, assuming that m_E is in $\tilde{Q}(\tilde{\mathfrak{H}}^0)$ amounts to assuming that m_E has a Pth derivative at x_0, at least in the sense of (138). If, as suggested in §II.C.1.b, we shrink our model space for the physical object E from $\mathfrak{M}$ to $\tilde{Q}(\tilde{\mathfrak{H}}^0)$, then convert $\tilde{Q}(\tilde{\mathfrak{H}}^0)$ to $\tilde{\mathfrak{H}}^0$ by means of the mapping $\tilde{Q}^{-1}:\tilde{Q}(\tilde{\mathfrak{H}}^0) \to \tilde{\mathfrak{H}}^0$, and finally enlarge the model space $\tilde{\mathfrak{H}}^0$ to its completion $\tilde{\mathfrak{H}}$, what we are doing is to start with all the functions m in $\mathfrak{M}$, then to restrict attention to those which have a continuous h as defined in (138), and finally to expand our vision to those for which the h defined by (138) is square integrable on $[0, 1]$.

The adjoint $\tilde{Q}^*:\mathfrak{M}^\S \to \tilde{\mathfrak{H}}$ is given by

$$\tilde{Q}^*g = (Q^*g, [g, m_1], \ldots, [g, m_P]), \tag{139}$$

where Q^*g has already been defined. If we regard g as a distribution with a singular kernel G, so that $[g, m] = \int_0^1 G(x)m(x)\,dx$, then $(Q^*g)(x) = (x - x_0)^P G(x)$.

To apply the results of §II.B, we ask whether there is a real M such that we think that $\|Q^{-1}m_E\|$ is probably not much larger than M. From (135) and (138), for any m in $\tilde{Q}(\tilde{\mathfrak{H}}^0)$, $\tilde{Q}^{-1}m = (h, \tilde{\gamma}_1, \ldots, \tilde{\gamma}_P)$ where $\tilde{\gamma}_j = [\tilde{g}_j, m] = m^{(j-1)}(x_0)/(j-1)!$ and h is given by (138). Then

$$\|Q^{-1}m\|^2 = \sum_{j=1}^P \tilde{\gamma}_j^2 + \int_0^1 \left| \left(m(x) - \sum_{j=1}^P \tilde{\gamma}_j (x - x_0)^{j-1}\right) \Big/ (x - x_0)^P \right|^2 dx.$$

To assume that $\|\tilde{Q}^{-1}m\| \leqq M$ is to limit not only the average size of m but the values of its first $P - 1$ derivatives at x_0 and also the extent of its failure to have a Pth derivative at x_0. All this is a demand on the smoothness of m at x_0.

(c) *Weaker assumptions on the smoothness of* m_E. Suppose we are willing to assume that m_E satisfies

$$\int_0^1 \left[\left(m(x) - \sum_{j=1}^P \tilde{\gamma}_j (x - x_0)^{j-1}\right) \Big/ (x - x_0)^P \right]^2 dx \leqq M^2 \tag{141}$$

where $\tilde{\gamma}_1, \ldots, \tilde{\gamma}_P$ are given by (135), but suppose we are unwilling to make a priori assumptions about probable upper bounds for $m_E(x_0), \ldots, m_E^{(P-1)}(x_0)$. Then we define a closed subspace $\mathfrak{F} \subseteqq \tilde{\mathfrak{H}}$ as

$$\mathfrak{F} = \{Q^*\tilde{g}_1, \ldots, Q^*\tilde{g}_P\}^\perp.$$

Then $\mathfrak{F}$ consists of all the $(P + 1)$-tuples $(h, \alpha_1, \ldots, \alpha_P)$ in $\mathfrak{H} \oplus R^P$ which have $\alpha_1 = \cdots = \alpha_P = 0$. The integral on the left in (141) is $\|\mathscr{F} \cdot \tilde{Q}^{-1}m\|^2$, so the problem is reduced to a form already considered in §II.B.

We might want to consider a completely different type of possible roughness in m_E: we might want to admit the possibility that, at some known point a in the unit interval, m_E has a discontinuity of known type. To be sure that the integrals (117) exist, we will assume that m_E is Lebesgue integrable on $[0, 1]$. Therefore we can admit singularities of the form $|x - x_0|^{-\beta}$ with $0 < \beta < 1$. If $\frac{1}{2} \leqq \beta < 1$ then the function $h(x)$ defined by (138) cannot be square integrable if the discontinuity is present. If m_E is in $\tilde{Q}(\tilde{\mathfrak{H}}^0)$ and $\frac{1}{2} \leqq \beta < 1$ then m_E is continuous at $x = a$. To overcome this difficulty, let $q:[0, 1] \to R$ be any function satisfying (135) and continuous in $[0, 1]$ except that it has at $x = a$ a discontinuity of the assumed form. We take for $\mathfrak{M}$ the space of all functions $m = \beta q + m'$ where β is real and m' is continuous on $[0, 1]$ and satisfies (135). For any such function m we define $[\tilde{g}_{P+1}, m] = \beta$, so $\tilde{g}_{P+1}:\mathfrak{M} \to R$ is a linear functional. Then m_E is in $\mathfrak{M}$. We take $\bar{\mathfrak{H}}^0 = \mathfrak{H}^0 \oplus R^{P+1}$ and define $\bar{Q}:\bar{\mathfrak{H}}^0 \to \mathfrak{M}$ by requiring for any $(P + 2)$-tuple $\bar{h} = (h, \alpha_1, \ldots, \alpha_P, \beta)$ in $\bar{\mathfrak{H}}^0$ that $\bar{Q}\bar{h}$ is the function whose value at x is

$$(142) \qquad (\bar{Q}\bar{h})(x) = \beta q(x) + \sum_{j=1}^{P} \alpha_j (x - x_0)^j + (x - x_0)^P h(x).$$

We take $\mathfrak{M}^{\S} = \mathrm{sp}\{g_1, \ldots, g_D, \tilde{g}_1, \ldots, \tilde{g}_P, \tilde{g}_{P+1}\}$. Then $(\bar{\mathfrak{H}}^0, \bar{Q})$ is an adequate quelling of $(\mathfrak{M}, \mathfrak{M}^{\S})$, and the argument proceeds as before.

A subtler type of control on the smoothness of m_E appears in the following problem. Suppose we believe that m_E has $P - 1$ continuous derivatives in $[0, 1]$ except possibly at $x = a$, where $m_E(x), m_E^{(1)}(x), \ldots, m_E^{(P-1)}(x)$ may all have jump discontinuities. We let $\mathfrak{M}$ be the space of all such functions. Then every member m of $\mathfrak{M}$ satisfies (135) with

$$\tilde{\gamma}_j = m^{(j-1)}(x_0)/(j - 1)!, \qquad j = 1, \ldots, P.$$

We take $\mathfrak{M}^{\S} = \mathrm{sp}\{g_1, \ldots, g_D, \tilde{g}_1, \ldots, \tilde{g}_P\}$.

As a first attempt at quelling we take $\mathfrak{H}^0 = \mathfrak{M}$, with the inner product $\langle h_1, h_2 \rangle = \int_0^1 h_1(x) h_2(x)\, dx$. We take $\tilde{\mathfrak{H}}^0 = \mathfrak{H}^0 \oplus R^P$. We define $\tilde{Q}:\tilde{\mathfrak{H}}^0 \to \mathfrak{M}$ by requiring for any $\tilde{h} = (h, \alpha_1, \ldots, \alpha_P)$ in $\tilde{\mathfrak{H}}^0$ that $\tilde{Q}\tilde{h}$ be the function given by (137). Then if $x_0 \neq a$, $(\tilde{\mathfrak{H}}^0, \tilde{Q})$ is an adequate quelling of $(\mathfrak{M}, \mathfrak{M}^{\S})$, and $\tilde{Q}(\tilde{\mathfrak{H}}^0)$ consists of those functions m in $\mathfrak{M}$ for which $m^{(P)}(x_0)$ exists in the sense of (135). It may be quite reasonable to assume that m_E is in $\tilde{Q}(\tilde{\mathfrak{H}}^0)$, and yet $(\tilde{\mathfrak{H}}^0, \tilde{Q})$ is an unsatisfactory quelling if x_0 is close to a. To see why, consider the case $P = 1$. Then, from (140),

$$(143) \qquad \|\tilde{Q}^{-1} m\|^2 = m(x_0)^2 + \int_0^1 \left[\frac{m(x) - m(x_0)}{x - x_0}\right]^2 dx.$$

If m has a jump discontinuity at $x = a$, then $\|\tilde{Q}^{-1}m\| \to \infty$ as $x_0 \to a$, so if x_0 is close to a the inequality $\|\tilde{Q}^{-1}m_E\| \leqq M$ is a reasonable a priori assumption only for extremely large values of M. But for any prediction vector p with $\mathcal{D}^\perp \cdot p \neq 0$, the variance of $\langle p, \tilde{Q}^{-1}m_E\rangle$ approaches ∞ with M in all of the methods described in §II.B.

To overcome this difficulty for arbitrary P, we take $\mathfrak{H}^0$ to be the space of all functions $h(x)$ which are $(P-1)$ times continuously differentiable on $[0, 1]$. For $j = 1, \ldots, P$ we define a linear functional $\tilde{g}_{P+j}:\mathfrak{M} \to R$ by requiring for each m in $\mathfrak{M}$ that $[\tilde{g}_{P+j}, m] = m^{(j-1)}(a+) - m^{(j-1)}(a-)$. For each positive integer j we define a function $\delta^{(-j)}:R \to R$ by requiring that $\delta^{(-j)}(x) = 0$ if $x \leqq 0$ and $\delta^{(-j)}(x) = x^{j-1}/(j-1)!$ if $x > 0$. Take $\bar{\mathfrak{H}}^0 = \mathfrak{H}^0 \oplus R^{2P}$ and define $\bar{Q}:\bar{\mathfrak{H}}^0 \to \mathfrak{M}$ by requiring for each $\bar{h} = (h, \alpha_1, \ldots, \alpha_P, \beta_1, \ldots, \beta_P)$ in $\bar{\mathfrak{H}}^0$ that $\bar{Q}\bar{h}$ is the function which assigns to any real x in $[0, 1]$ the value

$$(144)\quad (\bar{Q}\bar{h})(x) = \sum_{j=1}^{P} [\alpha_j(x - x_0)^{j-1} + \beta_j\delta^{(-j)}(x - a)] + (x - x_0)^P h(x).$$

Take $\mathfrak{M}^\S = \mathrm{sp}\{g_1, \ldots, g_D, \tilde{g}_1, \ldots, \tilde{g}_{2P}\}$. Then $(\bar{\mathfrak{H}}^0, \bar{Q})$ is an adequate linear quelling of $(\mathfrak{M}, \mathfrak{M}^\S)$. If m is in $\bar{Q}(\bar{\mathfrak{H}}^0)$, m has the form (144) for some h in $\mathfrak{H}^0$, and

$$(145)\qquad \|\bar{Q}^{-1}m\|^2 = \sum_{j=1}^{P} (\alpha_j^2 + \beta_j^2) + \int_0^1 h(x)^2\, dx.$$

Under this new quelling, the inequality $\|\bar{Q}^{-1}m\| \leqq M$ for a fixed M does not become less plausible as x_0 approaches a.

(d) *Approximating prediction functionals.* In §§II.C.2.a, b, c above, we have adopted the point of view that the data and our initial prejudice limit h_E in $\mathfrak{H}$, and that we estimate $p \cdot h_E$ by finding the range of possible variation of $p \cdot h$ for all h which satisfy those limits. When there is only one prediction functional $p = Q^*\tilde{g}$, we can adopt the alternative point of view that we estimate $p \cdot h_E$ as $d \cdot h_E$ where d is that vector in $\mathfrak{D} = Q^*\,\mathrm{sp}\{g_1, \ldots, g_D\}$ which best approximates p subject to certain constraints. In this section, §II.C.2.d, we want to examine quelling from this alternative point of view.

We return to our first example: $\mathfrak{M}$ is all functions $m(x)$ continuous on $[0, 1]$ and satisfying (135). We are trying to estimate $m_E^{(P-1)}(x_0)$ from $\gamma_i = [g_i, m_E]$, $i = 1, \ldots, D$, with g_i given by (117) and G_i continuous on $[0, 1]$. For simplicity we assume that the γ_i are measured without error; errors are treated as in §II.B.2.b or c.

We take $\mathfrak{M}^\S = \mathrm{sp}\{g_1, \ldots, g_D, \tilde{g}_1, \ldots, \tilde{g}_P\}$, and take $\mathfrak{H}^0$ and $\tilde{\mathfrak{H}}^0$ as in §II.C.2.b. We take $\tilde{Q}:\tilde{\mathfrak{H}}^0 \to \mathfrak{M}$ as given by (137). Then $[\tilde{g}_P, m] = \langle \tilde{Q}^*\tilde{g}_P,$

$\tilde{Q}^{-1}m\rangle$. We estimate $\langle \tilde{Q}^* g_P, \tilde{Q}^{-1} m_E \rangle$ as $\langle d, \tilde{Q}^{-1} m_E \rangle$ where d is that vector in $\mathfrak{D} = \tilde{Q}^* \operatorname{sp}\{g_1, \ldots, g_D\}$ which best approximates $p = \tilde{Q}^* \tilde{g}_P$. Thus we estimate $[\tilde{g}_P, m_E]$ as $[g, m_E]$ where g is that functional in $\operatorname{sp}\{g_1, \ldots, g_D\}$ which makes $\tilde{Q}^* g$ as close as possible to $\tilde{Q}^* \tilde{g}_P$.

Any g in $\operatorname{sp}\{g_1, \ldots, g_D\}$ can be written $g = \sum_{i=1}^D \alpha^i g_i$ where $\alpha^1, \ldots, \alpha^D$ are real. Then

$$[g, m] = \int_0^1 G(x) m(x)\, dx \tag{146}$$

where

$$G(x) = \sum_{i=1}^{D} \alpha^i G_i(x). \tag{147}$$

Then in $\mathfrak{H}^0 \oplus R^P$, $\tilde{Q}^* g$ is the following $(P + 1)$-tuple:

$$\tilde{Q}^* g = \left((x - x_0)^P G(x), \int_0^1 G(x)\, dx, \int_0^1 (x - x_0) G(x)\, dx, \ldots, \int_0^1 (x - x_0)^{P-1} G(x)\, dx \right);$$

and $\tilde{Q}^* \tilde{g}_P = (0, 0, \ldots, 0, 1)$. Therefore

$$\begin{aligned} \|\tilde{Q}^* g - \tilde{Q}^* \tilde{g}_P\|^2 = \sum_{j=1}^{P-1} \left[\int_0^1 G(x)(x - x_0)^{j-1}\, dx \right]^2 \\ + \left[\int_0^1 G(x)(x - x_0)^{P-1}\, dx - 1 \right]^2 \\ + \int_0^1 (x - x_0)^{2P} G(x)^2\, dx. \end{aligned} \tag{148}$$

In case $P = 1$ this equation reduces to

$$\|\tilde{Q}^* g - \tilde{Q}^* \tilde{g}_1\|^2 = \left[\int_0^1 G(x)\, dx - 1 \right]^2 + \int_0^1 (x - x_0)^2 G(x)^2\, dx. \tag{149}$$

First let us examine the case $P = 1$ in detail. The data permit us to calculate $\int_0^1 G(x)\, m_E(x)\, dx$ only for functions $G(x)$ of the form (147). We would like to know $\int_0^1 \delta(x - x_0) m_E(x)\, dx$. Therefore we take that $G(x)$ in $\operatorname{sp}\{G_1, \ldots, G_D\}$ which is the best possible approximation to $\delta(x - x_0)$, in the sense that it minimizes (149) as a function of $\alpha^1, \ldots, \alpha^D$. What does such minimization mean in terms of the function G? We would like $G(x)$ to be nearly 0 except when x is close to x_0 and to have a tall peak near $x = x_0$. The center of such a peak could be defined as x_G, that value of y which minimizes $\int_0^1 (x - y)^2 G(x)^2\,dx$. Then $\int_0^1 (x - x_G)G(x)^2\,dx = 0$, so

$$\int_0^1 (x - x_0)^2 G(x)^2\,dx = \int_0^1 (x - x_G)^2 G(x)^2\,dx + (x_G - x_0)^2 \int_0^1 G(x)^2\,dx.$$

Now let $A_{\varepsilon,\zeta}(x) = \zeta/2\varepsilon$ if $|x - x_G| < \varepsilon$ and $=0$ otherwise. Then

$$\int_0^1 A_{\varepsilon,\zeta}(x)\,dx = \zeta, \qquad \int_0^1 A_{\varepsilon,\zeta}(x)^2\,dx = \zeta^2/2\varepsilon,$$

and

$$\int_0^1 (x - x_0)^2 A_{\varepsilon,\zeta}(x)^2\,dx = (2\varepsilon)\zeta^2/12.$$

The function $A_{\varepsilon,\xi}$ has a peak width equal to 2ε. To understand heuristically what it means to minimize (149), we define $\zeta_G = \int_0^1 G(x)\,dx$ and $\boldsymbol{W}_G = 12\zeta_G^{-2}\int_0^1 (x - x_G)^2 G(x)^2\,dx$. If $G = A_{\varepsilon,\zeta}$, then $\boldsymbol{W}_G = 2\varepsilon$, the peak width of G, so we regard $\boldsymbol{W}_G$ as the peak width, or resolving length, of any kernel G. If G is strongly peaked near x_0 and nearly 0 elsewhere, this definition coincides with our geometrical notion. When $G = A_{\varepsilon,\zeta}$, then $\int_0^1 G^2 dx = \zeta_G^2/\boldsymbol{W}_G$, and when G is strongly peaked near x_0, the dimensionless quantity $\Psi_G = \boldsymbol{W}_G\zeta_G^{-2}\int_0^1 G^2\,dx$ will not differ from 1 by a large factor. Therefore we can write

$$\int_0^1 (x - x_0)^2 G(x)^2\,dx = \zeta_G^2[\boldsymbol{W}_G/12 + \Psi_G(x_0 - x_G)^2/\boldsymbol{W}_G]$$

and

$$(150) \qquad \|\tilde{Q}^*g - \tilde{Q}^*\tilde{g}_1\|^2 = (\zeta_G - 1)^2 + \zeta_G^2[\boldsymbol{W}_G/12 + \Psi_G(x_0 - x_G)^2/\boldsymbol{W}_G].$$

Thus minimizing $\|\tilde{Q}^*g - \tilde{Q}^*\tilde{g}_1\|$ forces $\zeta_G = \int_0^1 G(x)\,dx$ to be close to 1, forces the peak width $\boldsymbol{W}_G$ to be small, and forces $|x_G - x_0|$ to be small. When $\|\tilde{Q}^*g - \tilde{Q}^*\tilde{g}_1\|$ is small, $[g, m_E]$ is not precisely $m_E(x_0)$, but is at any rate nearly a localized average of the values of $m_E(x)$ for x near x_0. The resolving length of this average is $\boldsymbol{W}_G$, and its point of localization (the "center" in Backus and Gilbert [1970]) is x_G.

For aesthetic or other reasons we may want to minimize (149) subject to other constraints besides (147). For example, we might want to demand that $\int_0^1 G(x)\,dx = 1$, so that $[g, m_E]$ is exactly a localized average. If we believe that m_E has a jump discontinuity at $x = a$, then to avoid a Gibbs phenomenon there we want to restrict attention to kernels G which treat $\delta^{(-1)}(x - a)$ exactly, i.e. which have

$$\int_0^1 \delta^{(-1)}(x - a)G(x)\,dx = \int_0^1 \delta^{(-1)}(x - a)\delta(x - x_0)\,dx = \delta^{(-1)}(x_0 - a).$$

Then we have a finite number of functions $m_1, \ldots, m_F$ such that we demand $[g, m_k] = [\tilde{g}_1, m_k]$, $k = 1, \ldots, F$. That is, we demand that $\tilde{Q}^*g - \tilde{Q}^*\tilde{g}_1$ be in $\mathfrak{F} = \{Q^{-1}m_1, \ldots, Q^{-1}m_F\}^{\perp}$. This situation has already been discussed in §II.B.2.b and c.

When $P > 1$ a similar discussion can be carried out for (148), showing the sense in which the smallness of $\|\tilde{Q}^*g - \tilde{Q}^*\tilde{g}_P\|$ forces g to look like $(-1)^{P-1}\delta^{(P-1)}(x - x_0)/(P - 1)!$.

3. *Quelling derivatives of the Dirac delta function by integration.*

(a) *Statement of the problem.* Suppose we think $m_E(x)$ is a continuous function on $[0, 1]$, and suppose we know $\gamma_i = [g_i, m_E]$ where g_i are functionals of the form (117) with $G_i(x)$ continuous on $[0, 1]$, $i = 1, \ldots, D$. Suppose we want to use these data to estimate the values of m_E at several points, $x_1, \ldots, x_P$. Then we take $\tilde{g}_j : \mathfrak{M} \to R$ as $[\tilde{g}_j, m] = m(x_j)$, $j = 1, \ldots, P$, and we take $\mathfrak{M}^{\S} = \mathrm{sp}\{g_1, \ldots, g_D, \tilde{g}_1, \ldots, \tilde{g}_P\}$. For any g in $\mathfrak{M}^{\S}$ we write $[g, m] = \int_0^1 G(x)m(x)\,dx$, where G is interpreted as a distribution if necessary. Thus $\tilde{G}_j(x) = \delta(x - x_j)$, $j = 1, \ldots, P$. A quelling $(\mathfrak{H}^0, Q)$ will make $Q^*\tilde{g}_j$ a square integrable function for $j = 1, \ldots, P$. If we quell by multiplication, we take $\mathfrak{H}^0$ to be all continuous functions on $[0, 1]$, with $\langle h_1, h_2 \rangle = \int_0^1 h_1(x)h_2(x)\,dx$, and we define $Q : \mathfrak{H}^0 \to \mathfrak{M}$ by requiring $(Qh)(x) = q(x)h(x)$ where $q(x) = \prod_{j=1}^P (x - x_j)$. This quelling being inadequate, we take $\tilde{\mathfrak{H}}^0 = \mathfrak{H}^0 \oplus R^P$, $n_j = q(x)/(x - x_j)$, $m_j(x) = n_j(x)/n_j(x_j)$, and for any $\tilde{h} = (h, \alpha_1, \ldots, \alpha_P)$ in $\tilde{\mathfrak{H}}^0$ we define $(\tilde{Q}\tilde{h})(x) = q(x)h(x) + \sum_{j=1}^P \alpha_j m_j(x)$. Then $(\tilde{\mathfrak{H}}^0, \tilde{Q})$ is an adequate quelling of $(\mathfrak{M}, \mathfrak{M}^{\S})$, but if P is large the procedure involves heavy numerical computation and, what is worse, we must change the quelling every time we want to add a new point x_{P+1} at which to evaluate m_E.

To overcome these disadvantages of quelling by multiplication, we consider a second method for quelling derivatives of Dirac delta functions. If $[\tilde{g}_j, m] = m^{(j-1)}(x_0)/(j - 1)!$, then $[\tilde{g}_j, m] = \int_0^1 \tilde{G}_j(x)m(x)\,dm$ with $\tilde{G}_j(x) = (-1)^{j-1}\delta^{(j-1)}(x - x_0)/(j - 1)!$. This generalized function can be made square integrable by integrating it j times from x to 1. If we want to estimate $[\tilde{g}_j, m_E]$ for $j = 1, \ldots, P$, then we can take $Q^* : \mathfrak{M}^{\S} \to \mathfrak{H}$ as the operation of integrating the distributions in $\mathfrak{M}^{\S} P$ times from x to 1. The adjoint of this operation will be to integrate P times from 0 to x. To see this we note that if f and g are both continuous functions on $[0, 1]$ then

$$(151)\qquad \int_0^1 dx\, f(x)\int_x^1 d\xi_1\int_{\xi_1}^1 d\xi_2\cdots\int_{\xi_{P-1}}^1 d\xi_P g(\xi_P) = \int_0^1 dx\, g(x)\int_0^x d\xi_1\int_0^{\xi} d\xi_2\cdots\int_0^{\xi_{P-1}} d\xi_P f(\xi_P),$$

an equation which can be deduced by P integrations by parts. The definition of a distribution is such that if f is a distribution for which one side of (151) makes sense, then the other side makes sense and the two sides are equal.

In what follows, we will find the following well-known identities very useful:

$$(152)\ (P-1)!\int_0^x d\xi_1\int_0^{\xi_1} d\xi_2\cdots\int_0^{\xi_{P-1}} d\xi_P f(\xi_P) = \int_0^x (x-\xi)^{P-1} f(\xi)\, d\xi,$$

$$(153)\ (P-1)!\int_x^1 d\xi_1\int_{\xi}^1 d\xi_2\cdots\int_{\xi_{P-1}}^1 d\xi_P g(\xi_P) = \int_x^1 (\xi-x)^{P-1} g(\xi)\, d\xi.$$

(b) *A quelling and its extensions.* We take $\mathfrak{M}$ to be all functions m on $[0, 1]$ which have continuous $(P-1)$st derivatives, $m^{(P-1)}(x)$, for all x in $[0, 1]$. We define $\tilde{g}_j : \mathfrak{M} \to R$ by requiring $[\tilde{g}_j, m] = m^{(j-1)}(x_0)/(j-1)!$. Thus $[\tilde{g}_j, m] = \int_0^1 \tilde{G}_j(x)m(x)\, dx$ where

$$\tilde{G}_j(x) = (-1)^{j-1}\delta^{(j-1)}(x-x_0)/(j-1)!.$$

We take $\mathfrak{M}^{\S} = \operatorname{sp}\{g_1, \ldots, g_D, \tilde{g}_1, \ldots, \tilde{g}_P\}$. We take $\mathfrak{H}^0$ to be the space of all continuous functions on $[0, 1]$, with inner product $\langle h_1, h_2\rangle = \int_0^1 h_1(x)h_2(x)\, dx$. Then for any h in $\mathfrak{H}^0$ we define Qh as the function in $\mathfrak{M}$ which assigns to an arbitrary x in $[0, 1]$ the value

$$(154)\qquad (Qh)(x) = (P-1)!\int_0^x d\xi_1\int_0^{\xi_1} d\xi_2\cdots\int_0^{\xi_{P-1}} d\xi_P h(\xi_P).$$

According to (152), we can write this also as

$$(155)\qquad (Qh)(x) = \int_0^x (x-\xi)^{P-1} h(\xi)\, d\xi.$$

Clearly $Q:\mathfrak{H}^0 \to \mathfrak{M}$ is a linear injection. We want to prove that for any g in $\mathfrak{M}^{\S}$ the number $[g, Qh] = \int_0^1 G(x)[(Qh)(x)]\, dx$ depends continuously on h, regarded as a vector in $\mathfrak{H}^0$. That is, we want to prove that if g is fixed in $\mathfrak{M}^{\S}$ then there is a constant K such that $|[g, Qh]| \leqq K\|h\|$ for all h in $\mathfrak{H}^0$. On account of (151), (152) and (153), we have

$$(156)\qquad [g, Qh] = \int_0^1 dx\, h(x)\int_x^1 (\xi-x)^{P-1} G(\xi)\, d\xi.$$

If $g = g_i$, $i = 1, \ldots, D$, then $\int_x^1 (\xi - x)^{P-1} G_i(\xi)\, d\xi$ is continuous for all x in $[0, 1]$. If $g = \tilde{g}_j$, $j = 1, \ldots, P$, then

$$(157) \qquad \int_x^1 (\xi - x)^{P-1} \tilde{G}_j(\xi)\, d\xi = \frac{(P-1)!}{(j-1)!} \delta^{-(P+1-j)}(x_0 - x).$$

For $j = 1, \ldots, P$, this is also a continuous function of x in $[0, 1]$. Therefore, for any g in $\mathfrak{M}^{\S}$ there is a function $\Gamma(x)$ continuous on $[0, 1]$ such that $[g, Qh] = \int_0^1 h(x)\Gamma(x)\, dx$ for all h in $\mathfrak{H}^0$. Therefore, by Schwarz's inequality, $|[g, Qh]| \leqq \|h\|\, \|\Gamma\|$. It follows that $(\mathfrak{H}^0, Q)$ is a linear quelling of $(\mathfrak{M}, \mathfrak{M}^{\S})$.

From (151), (152), and (153), the adjoint mapping $Q^* : \mathfrak{M}^{\S} \to \mathfrak{H}$ assigns to any g in $\mathfrak{M}^{\S}$ a function Q^*g whose value at x is

$$(158) \qquad (Q^*g)(x) = \int_x^1 (\xi - x)^{P-1} G(\xi)\, d\xi.$$

Thus $(Q^*)^{-1}(\{0\}) = \{0\}$, and $(\mathfrak{H}^0, Q)$ is an adequate quelling of $(\mathfrak{M}, \mathfrak{M}^{\S})$. Nevertheless it is an unsatisfactory quelling, because if m is in $Q(\mathfrak{H}^0)$ then $m(0) = m^{(1)}(0) = \cdots = m^{(P-1)}(0) = 0$. We extend the quelling in the manner suggested in §II.C.1.d. We take

$$(159) \qquad m_j(x) = x^{j-1}, \qquad j = 1, \ldots, P.$$

Then $m_1, \ldots, m_P$ are linearly independent modulo $Q(\mathfrak{H}^0)$, so we take $\tilde{h}^0 = \mathfrak{H}^0 \oplus R^P$ and, for any $\tilde{h} = (h, \alpha_1, \ldots, \alpha_P)$ in $\tilde{\mathfrak{H}}^0$, we define $\tilde{Q}\tilde{h}$ as that function which assigns to x in $[0, 1]$ the value

$$(160) \qquad (\tilde{Q}\tilde{h})(x) = \sum_{j=1}^{P} \alpha_j x^{j-1} + \int_0^x (x - \xi)^{P-1} h(\xi)\, d\xi.$$

Then $\tilde{Q}^* : \mathfrak{M}^{\S} \to \tilde{\mathfrak{H}}$ is given by

$$(161) \qquad \tilde{Q}^* g = \left(\int_x^1 (\xi - x)^{P-1} G(\xi)\, d\xi, [g, m_1], \ldots, [g, m_P] \right).$$

If m is in $\tilde{Q}(\tilde{\mathfrak{H}}^0)$, then $m(x)$ is given by (160) with $\alpha_j = m^{(j-1)}(0)/(j-1)!$ and $h(x) = m^{(P)}(x)/(P-1)!$. Thus m is P times continuously differentiable on $[0, 1]$ and

$$(162) \qquad \|\tilde{Q}^{-1} m\|^2 = \sum_{j=1}^{P} \left[\frac{m^{(j-1)}(0)}{(j-1)!} \right]^2 + \int_0^1 \left[\frac{m^{(P)}(x)}{(P-1)!} \right]^2 dx.$$

If we look for h_E not only in $\tilde{\mathfrak{H}}^0$ but in its completion, $\tilde{\mathfrak{H}}$, this amounts to assuming that $m_E^{(P)}(x)$ is square integrable on $[0, 1]$, or, more precisely, that $m_E^{(P-1)}(x)$ is the integral of a function square integrable on $[0, 1]$.

This is, of course, a stronger demand than that $m_E^{(P-1)}(x)$ simply be continuous, so $\tilde{Q}(\mathfrak{H})$ is a proper subset of $\mathfrak{M}$.

(c) *Weaker assumptions on the smoothness of* m_E. If we are willing to assume a priori that probably

$$\int_0^1 \left[\frac{m_E^{(P)}(x)}{(P-1)!}\right]^2 dx \leqq M^2 \tag{163}$$

but we prefer to make no a priori assumptions about $m^{(j-1)}(0), j = 1, \ldots, P$, then we can take $\mathfrak{F} = \{\tilde{Q}^{-1}m_1, \ldots, \tilde{Q}^{-1}m_P\}^{\perp}$, a closed subspace of $\mathfrak{H}$. Equation (163) is precisely the condition that $\|\mathscr{F} \cdot Q^{-1}m_E\| \leqq M$, so we can proceed as in §II.B.3.

If we want to permit m_E to have infinities or jump discontinuities at a finite number of known locations in [0, 1] we proceed as in §II.C.2.c.

(d) *Approximating prediction functionals.* Here we take the point of view that we want to estimate $[\tilde{g}_P, m_E]$ as $[g, m_E]$ where g is that functional in $\mathrm{sp}\{g_1, \ldots, g_D\}$ which makes $\tilde{Q}^*g$ as close as possible to $\tilde{Q}^*\tilde{g}_P$. From (161) we have

$$\tilde{Q}^*\tilde{g}_P = (\delta^{(-1)}(x_0 - x), 0, \ldots, 0, 1).$$

Therefore

$$\begin{aligned}\|\tilde{Q}^*g - \tilde{Q}^*\tilde{g}_P\|^2 = {}& \int_0^1 dx\left[\int_x^1 d\xi(\xi - x)^{P-1}G(\xi) - \delta^{(-1)}(x_0 - x)\right]^2 \\ &+ \sum_{j=1}^{P}\left[\int_0^1 G(x)x^{j-1}\,dx - \delta_{jP}\right]^2.\end{aligned} \tag{164}$$

In case $P = 1$ this equation reduces to

$$\begin{aligned}\|\tilde{Q}^*g - \tilde{Q}^*\tilde{g}_1\|^2 = {}& \int_0^1 dx\left[\int_x^1 d\xi G(\xi) - \delta^{(-1)}(x_0 - x)\right]^2 \\ &+ \left[\int_0^1 G(x)\,dx - 1\right]^2.\end{aligned} \tag{165}$$

Equation (165) is another way, different from (149), of measuring the deviation of $G(x)$ from $\delta(x_0 - x)$. If we want to impose no constraints on g other than (147), we simply minimize (165) subject to (147), i.e. as a function of $\alpha^1, \ldots, \alpha^D$. If we have aesthetic or practical reasons for imposing constraints on G, such as $\int_0^1 G(x)\,dx = 1$, then we minimize (165) subject to those constraints as well as (147). The discussion proceeds as in §§II.C.2.d and II.B.2.

4. *Quelling derivatives of the Dirac delta function by convolution.*

(a) *Statement of the problem.* We have discussed two quellings of $\delta^{(P-1)}(x - x_0)$, based on the square-integrability of the functions $(x - x_0)^P \delta^{(P-1)}(x - x_0)$ and $\int_0^x d\xi_1 \int_0^{\xi_1} d\xi_2 \cdots \int_0^{\xi_{P-1}} d\xi_P \delta^{(P-1)}(\xi_P - x_0)$. A third way to make $\delta^{(P-1)}(x - x_0)$ square-integrable is to convolute it with a function $s(x)$ whose $(P - 1)$st derivative is square-integrable.

As a first and simplest illustration, suppose $\mathfrak{M}$ consists of all functions $m(x)$ defined and absolutely integrable on the whole real line R, bounded there, and having $P - 1$ continuous derivatives everywhere. Suppose $[g_i, m] = \int_{-\infty}^{\infty} G_i(x)m(x)\,dx$, $i = 1, \ldots, D$, where G_i is bounded and continuous. Suppose $[g_j, m] = m^{(j-1)}(x_0)/(j - 1)!$, $j = 1, \ldots, P$. Let $\mathfrak{M}^\S = \mathrm{sp}\{g_1, \ldots, g_D, \tilde{g}_1, \ldots, \tilde{g}_P\}$. Let $\mathfrak{H}^0$ consists of all functions $h(x)$ defined, bounded, continuous, and absolutely integrable on all of R. Take for the inner product on $\mathfrak{H}^0$ the value $\langle h_1, h_2 \rangle = \int_{-\infty}^{\infty} h_1(x)h_2(x)\,dx$.

For any h in $\mathfrak{H}^0$ we define the Fourier transform Fh as that complex-valued function which assigns to any real number t the value $(Fh)(t) = \int_{-\infty}^{\infty} h(x)\exp(-2\pi itx)\,dx$. If h_1 and h_2 are in $\mathfrak{H}^0$ we define $h_1 * h_2$, their convolution, as that function which assigns to any real x the value $(h_1 * h_2)(x) = \int_{-\infty}^{\infty} h_1(x - \xi)h_2(\xi)\,d\xi$. As is well known, $F(h_1 * h_2) = F(h_1)F(h_2)$.

Now let $s(x)$ be a function in $\mathfrak{H}^0$ whose first $P - 1$ derivatives are also in $\mathfrak{H}^0$ and whose Fourier transform never vanishes. Such functions do exist, one example being $\exp(-x^2)$. For any h in $\mathfrak{H}^0$ define $Qh = s * h$. Then $Q : \mathfrak{H}^0 \to \mathfrak{M}$ is a linear mapping. To see that Q is an injection, observe that $F(Qh) = (Fs)(Fh)$. Since Fs never vanishes, $Fh = F(Qh)/Fs$, and thus if $Qh = 0$ then $F(Qh) = 0$ so $Fh = 0$ so $h = 0$. To prove that $(\mathfrak{H}^0, Q)$ is a quelling of $(\mathfrak{M}, \mathfrak{M}^\S)$, it remains only to show that for any g in $\mathfrak{M}^\S$, $[g, Qh]$ depends continuously on h as h varies in $\mathfrak{H}^0$. For any g in $\mathfrak{M}^\S$ and any m in $\mathfrak{M}$ we write $[g, m] = \int_{-\infty}^{\infty} G(x)m(x)\,dx$. The kernel $G(x)$ will be a distribution rather than a function, unless g is in $\mathrm{sp}\{g_1, \ldots, g_D\}$. If $m = Qh = s * h$ then, proceeding formally for a moment, we have

$$
\begin{aligned}
[g, Qh] &= \int_{-\infty}^{\infty} dx\, G(x) \int_{-\infty}^{\infty} d\xi\, s(x - \xi)h(\xi) \\
&= \int_{-\infty}^{\infty} d\xi\, h(\xi) \int_{-\infty}^{\infty} dx\, s(x - \xi)G(x) \qquad (166) \\
&= \int_{-\infty}^{\infty} d\xi\, h(\xi)(s_- * G)(\xi)
\end{aligned}
$$

where $s_-(x) = s(-x)$. If g is in $\mathrm{sp}\{g_1, \ldots, g_D\}$ then G is absolutely integrable on R, and Fubini's theorem justifies the formal calculation (166). If g is in $\mathrm{sp}\{\tilde{g}_1, \ldots, \tilde{g}_P\}$ then $[g, Qh]$ is a linear combination of derivatives of $s*h$, and Fubini's theorem assures us that $(s*h)^{(j-1)} = s^{(j-1)}*h$, $j = 1, \ldots, P$. This observation justifies (166) in case g is in $\mathrm{sp}\{\tilde{g}_1, \ldots, \tilde{g}_P\}$. For any g in $\mathfrak{M}^{\S}$, s_-*G is a function in $\mathfrak{H}^0$, so $|[g, Qh]| \leq \|h\| \, \|s_-*G\|$. This proves that $[g, Qh]$ depends continuously on h, so $(\mathfrak{H}^0, Q)$ is indeed a quelling of $(\mathfrak{M}, \mathfrak{M}^{\S})$. From Equation (166) we can immediately write down the adjoint $Q^*: \mathfrak{M}^{\S} \to \mathfrak{H}$. For any g in $\mathfrak{M}^{\S}$, $Q^*g = s_-*G$, where G is the kernel of g. Since $Fs_- = (Fs)^*$, $F(Q^*g) = (Fs)^*FG$, so if $Q^*g = 0$ then $F(Q^*g) = 0$, so $FG = 0$, so $G = 0$ so $g = 0$. Therefore Q^* is an injection, and $(\mathfrak{H}^0, Q)$ is an adequate quelling of $(\mathfrak{M}, \mathfrak{M}^{\S})$.

If m is in $Q(\mathfrak{H}^0)$ then $m = s*h$ for some h in $\mathfrak{H}^0$, and $h = Q^{-1}m$. But $Fh = Fm/Fs$, and, by the Parseval theorem, $\|h\| = \|Fh\|$. Therefore

$$\|Q^{-1}m\|^2 = \int_{-\infty}^{\infty} \left| \frac{(Fm)(t)}{(Fs)(t)} \right|^2 dt.$$

In short, the a priori prejudice that $\|Q^{-1}m\|$ is probably not much larger than M amounts to the belief that there is a restricted amount of energy at high wave numbers in the Fourier spectrum of m.

(b) *A quelling and its extensions.* The foregoing discussion was particularly simple because the quelling needed no extension. If we are interested in functions $m(x)$ defined only on $[0, 1]$, convolution quelling is not so simple. We take for $\mathfrak{M}$ the set of all functions with $P - 1$ continuous derivatives on $[0, 1]$. We assume that there are functions $G_1(x), \ldots, G_D(x)$ piecewise continuous on $[0, 1]$ such that for any m in $\mathfrak{M}$, $[g_i, m] = \int_0^1 G_i(x)m(x)\,dx$, $i = 1, \ldots, D$. We take

$$[\tilde{g}_j, m] = m^{(j-1)}(x_0)/(j-1)!, \qquad j = 1, \ldots, P,$$

and take $\mathfrak{M}^{\S} = \mathrm{sp}\{g_1, \ldots, g_D, \tilde{g}_1, \ldots, \tilde{g}_P\}$.

We take $\mathfrak{H}^0$ to be the space of all real functions defined and continuous on the whole real line R and periodic with period 1. We define $\langle h_1, h_2 \rangle = \int_0^1 h_1(x)h_2(x)\,dx$. We write the nth Fourier coefficient of any such function h as $(Fh)(n)$, so for any integer n

$$(Fh)(n) = \int_0^1 h(x) \exp(-2\pi inx)\,dx.$$

The Fourier coefficients $(Fm)(n)$ of any function m in $\mathfrak{M}$ are defined similarly. For any g in $\mathfrak{M}^{\S}$ we write $[g, m] = \int_0^1 G(x)m(x)\,dx$. The kernel G may be a distribution rather than an ordinary function. We define the Fourier coefficients of g as $(Fg)(n) = [g, \exp(-2\pi inx)]$, or

$$(Fg)(n) = \int_0^1 G(x) \exp(-2\pi inx)\,dx.$$

If h_1 and h_2 are in $\mathfrak{H}^0$ we define their convolution, $h_1 * h_2$, as the function of x given by

$$(h_1 * h_2)(x) = \int_0^1 h_1(x - \xi) h_2(\xi)\, d\xi.$$

Then $h_1 * h_2$ is in $\mathfrak{H}^0$, and $F(h_1 * h_2) = (Fh_1)(Fh_2)$.

Let $s(x)$ be any real function such that $s^{(j)}(x)$ is in $\mathfrak{H}^0$ for $j = 0, 1, \ldots P - 1$ and such that $(Fs)(n)$ is different from 0 for every integer n. Define $Q : \mathfrak{H}^0 \to \mathfrak{M}$ by requiring for any h in $\mathfrak{H}^0$ that $Qh = (s*h)|[0, 1]$. Since $s*h$ is periodic with period 1 it is completely determined by Qh. And since $[F(s*h)](n) = [(Fs)(n)][(Fh)(n)]$, and $(Fs)(n) \neq 0$, it follows that Fh is completely determined by $F(s*h)$. Thus h is determined by Qh. That is, $Q : \mathfrak{H}^0 \to \mathfrak{M}$ is an injection. If g is in $\mathfrak{M}^\S$ then $[g, Qh]$ depends continuously on h for essentially the same reason that it did in the preceding §II.C.4.a. Therefore, $(\mathfrak{H}^0, Q)$ is a quelling of $(\mathfrak{M}, \mathfrak{M}^\S)$. As in §II.C.4.a, the adjoint mapping $Q^* : \mathfrak{M}^\S \to \mathfrak{H}$ is given by $Q^*g = s*G$. That is, for any x in R

$$(Q^*g)(x) = \int_0^1 s(\xi - x) G(\xi)\, d\xi.$$

By looking at Fourier coefficients, we see that g is uniquely determined by Q^*g, so $Q^* : \mathfrak{M}^\S \to \mathfrak{H}$ is an injection. Thus $(\mathfrak{H}^0, Q)$ is an adequate quelling of $(\mathfrak{M}, \mathfrak{M}^\S)$.

Nevertheless, the quelling must be extended. If m is in $Q(\mathfrak{H}^0)$ then $m^{(j-1)}(1) = m^{(j-1)}(0)$, $j = 1, \ldots, P$. We define functionals $\tilde{g}_{P+j} : \mathfrak{M} \to R$ by requiring for any m in $\mathfrak{M}$ that $[\tilde{g}_{P+j}, m] = [^{(j-1)}(1) - m^{(j-1)}(0)]/(j-1)!$, $j = 1, \ldots, P$. Every m in $Q(\mathfrak{H}^0)$ satisfies $[\tilde{g}_{P+j}, m] = 0, j = 1, \ldots, P$. Unless we have reason to believe that m_E satisfies these P constraints we must extend the quelling. We enlarge $\mathfrak{M}^\S$ to be $\mathrm{sp}\{g_1, \ldots, g_D, \tilde{g}_1, \ldots, \tilde{g}_{2P}\}$, so that $(\mathfrak{H}^0, Q)$ is no longer an adequate quelling of $(\mathfrak{M}, \mathfrak{M}^\S)$. We choose functions $m_1, \ldots, m_P$ in $\mathfrak{M}$ such that $[\tilde{g}_{P+j}, m_i] = \delta_{ij}, i, j = 1, \ldots, P$. The most convenient choice is to demand that $m_i(x)$ be a polynomial in x of degree i satisfying $m_i(0) = 0$. Then $m_i(x)$ is uniquely determined as

$$m_i(x) = (i-1)! \sum_{k=1}^{i} c_{i+1-k} x^k / k!$$

where the infinite sequence $\{c_1, c_2, \ldots\}$ is determined by the recursion relations $c_1 = 1$, $\sum_{k=1}^{i} c_{i+1-k}/k! = 0$ for $i = 2, 3, \ldots$.

We take $\tilde{\mathfrak{H}}^0 = \mathfrak{H}^0 \oplus R^P$ and define $\tilde{Q} : \tilde{\mathfrak{H}}^0 \to \mathfrak{M}$ by requiring that if $\tilde{h} = (h, \alpha_1, \ldots, \alpha_P)$ is any member of $\tilde{\mathfrak{H}}^0$ then

$$\tilde{Q}\tilde{h} = Qh + \alpha_1 m_1 + \cdots + \alpha_P m_P = s*h + \alpha_1 m_1 + \cdots + \alpha_P m_P. \tag{167}$$

If m is in $\tilde{Q}(\tilde{\mathfrak{H}}^0)$ then m has the form (167) with $(h, \alpha_1, \ldots, \alpha_P)$ uniquely determined by m, and in fact

$$\alpha_j = [\tilde{g}_{p+j}, m] = [m^{(j-1)}(1) - m^{(j-1)}(0)]/(j-1)!, \qquad j = 1, \ldots, P.$$

Thus

$$(168) \qquad \|\tilde{Q}^{-1}m\|^2 = \int_0^1 h(x)^2\,dx + \sum_{j=1}^{P} \left[\frac{m^{(j-1)}(1) - m^{(j-1)}(0)}{(j-1)!}\right]^2.$$

Another way of looking at $\|\tilde{Q}^{-1}m\|$ is as follows. We let $\tilde{m}(x) = s*h$ and $p_P(x) = \alpha_1 m_1(x) + \cdots + \alpha_P m_P(x)$. Then $\tilde{m}$ can be extended to all of R so as to be $P - 1$ times continuously differentiable for all real x and periodic with period 1, while $p_P(x)$ is a polynomial of degree P, defined by requiring that $p_P(0) = 0$ and $p_P^{(j-1)}(1) - p_P^{(j-1)}(0) = \tilde{m}^{(j-1)}(1) - \tilde{m}^{(j-1)}(0)$, $j = 1, \ldots, P$. We have both $\|h\|^2 = \sum_{n=-\infty}^{\infty} (Fh)(n)^2$ and $(Fh)(n) = (F\tilde{m})(n)/(Fs)(n)$, so

$$(169) \quad \|\tilde{Q}^{-1}m\|^2 = \sum_{n=-\infty}^{\infty} \left|\frac{F\tilde{m}(n)}{Fs(n)}\right|^2 + \sum_{j=1}^{P} \left[\frac{m^{(j-1)}(1) - m^{(j-1)}(0)}{(j-1)!}\right]^2.$$

Thus the hypothesis $\|Q^{-1}m\| \leqq M$ limits the size of the polynomial $p_P(x)$ of degree P which we must subtract from $m(x)$ in order to make $\tilde{m}(x) = m(x) - p_P(x)$ $(P - 1)$ times continuously differentiable when extended periodically with period 1; and the hypothesis also limits the amount of energy at high wave numbers in the Fourier spectrum of $\tilde{m}$.

(c) *Weaker assumptions on the smoothness of* m_E. The possibility that m_E has infinities or jump discontinuities at known locations in [0, 1] can be dealt with just as it was in §§II.C.2.c and II.C.3.c.

(d) *Approximating prediction functionals.* For any g in $\mathfrak{M}^{\S}$, $\tilde{Q}^*g = (s*G, [g, m_1], \ldots, [g, m_P])$. In particular,

$$\tilde{Q}^*\tilde{g}_P = (s^{(P-1)}(x_0 - x)/(P-1)!, 0, \ldots, 0, x_0).$$

Therefore if g is in $\mathrm{sp}\{g_1, \ldots, g_D\}$, with kernel G in $\mathrm{sp}\{G_1, \ldots, G_D\}$, then

$$\begin{aligned} \|\tilde{Q}^*g - \tilde{Q}^*\tilde{g}_P\|^2 &= \int_0^1 \left[\frac{s^{(P-1)}(x_0 - x)}{(P-1)!} - s*G\right]^2 dx \\ (170) \qquad &+ \sum_{j=1}^{P-1} \left[\int_0^1 G(x)m_j(x)\,dx\right]^2 \\ &+ \left[\int_0^1 G(x)m_P(x)\,dx - x_0\right]^2. \end{aligned}$$

If we adopt the point of view of §II.B.2, we will try to make $\tilde{Q}^*g$ as close to $\tilde{Q}^*\tilde{g}_P$ as possible, subject to the constraint that g be in $\mathrm{sp}\{g_1, \ldots, g_D\}$, and to any other constraints which seem appropriate. Then we simply minimize (170) as a function of G, subject to those constraints.

5. *Quelling Dirac delta functions in more than one dimension.* Evidently problems of the foregoing sort can be studied when the models m are functions defined on an open subset U of R^T where T is any positive integer. Multiplication quelling extends to such higher-dimensional problems without modification, as does convolution quelling in case $U = R^T$. I know of no way to generalize convolution quelling for arbitrary bounded open sets U, but there probably is one.

There are interesting generalizations of integration quelling. Suppose for simplicity that we are trying to estimate $m_E(x_0)$ from $\int_U G_i(x)m_E(x)\, d^Tx$, where x_0 is a given point in U. Suppose that ∂U, the boundary of U, has a continuous outward unit normal $\hat{n}$. Take for $\mathfrak{M}$ the space of all real functions m bounded and continuous on cU, the closure of U. Let $\mathfrak{H}_1^0 = \mathfrak{M}$ with inner product $\langle h_1, h_2 \rangle = \int_U h_1(x)h_2(x)\, d^Tx$, and let $\mathfrak{H}_2^0$ be the space of all real functions $k(x)$ defined and continuous for x on ∂U, with $\langle k_1, k_2 \rangle = \int_{\partial U} k_1(x)k_2(x)\, d^{T-1}x$, where $d^{T-1}x$ is the element of "area" on ∂U. Define $Q_1 : \mathfrak{H}_1^0 \to \mathfrak{M}$ by requiring for any h in $\mathfrak{H}_1^0$ that $Q_1 h$ be that function $m(x)$ which satisfies $\Delta m = h$ in U and $m(x) = 0$ if x is on ∂U. Here Δ is the T-dimensional Laplacian operator. Let $Q_2 : \mathfrak{H}_2^0 \to \mathfrak{M}$ be defined by requiring for any k in $\mathfrak{H}_2^0$ that $Q_2 k$ be that function $m(x)$ which satisfies $\Delta m = 0$ in U and $m(x) = k(x)$ if x is on ∂U. Then every function in $Q_1(\mathfrak{H}_1^0)$ vanishes on ∂U, while every function in $Q_2(\mathfrak{H}_2^0)$ is harmonic in U, so neither $(\mathfrak{H}_1^0, Q_1)$ nor $(\mathfrak{H}_2^0, Q_2)$ is an adequate quelling of $(\mathfrak{M}, \mathfrak{M}^{\S})$ if $\mathfrak{M}^{\S}$ includes any functionals g which evaluate m on ∂U or evaluate derivatives of m in U. However, $Q_1 \oplus Q_2(\mathfrak{H}_1^0 \oplus \mathfrak{H}_2^0)$ contains all twice continuously differentiable functions whose Laplacians satisfy a Hölder condition, so $(\mathfrak{H}_1^0 \oplus \mathfrak{H}_2^0, Q_1 \oplus Q_2)$ may be an acceptable quelling of $(\mathfrak{M}, \mathfrak{M}^{\S})$.

On the surface of a sphere, both multiplication quelling and convolution quelling proceed essentially as on the real line. Integration quelling may be less useful.

6. *Other applications of linear quelling.* In all of the preceding examples, the data functionals were of the form $[g, m] = \int G(x)m(x)\, dx$ with G a continuous function. Now we want to permit the data functionals themselves to be distributions. We will take the simplest case. $\mathfrak{M}$ is all continuous functions on $[0, 1]$, and for any m in $\mathfrak{M}$, $[g_i, m] = m(x_i)$, $i = 1, \ldots, D$, where $x_1, x_2, \ldots, x_D$ are points in $[0, 1]$. If $\tilde{g} : \mathfrak{M} \to R$ is given by $[\tilde{g}, m] = m(x_0)$ for all m in $\mathfrak{M}$, then estimating $[\tilde{g}, m]$ from $[g_i, m]$, $i = 1, \ldots, D$, is the problem of numerical interpolation or extrapolation, depending on whether x_0 is inside or outside the convex hull of $x_1, \ldots, x_P$. If $\tilde{g} : \mathfrak{M} \to R$

is given by $[\tilde{g}, m] = \int_0^1 G(x)m(x)\,dx$ for all m in $\mathfrak{M}$, then estimating $[\tilde{g}, m]$ from $[g_i, m]$, $i = 1, \ldots, D$, is the problem of numerical integration. If all members of $\mathfrak{M}$ are required to be differentiable and $[\tilde{g}, m] = m'(x_0)$ for all m in $\mathfrak{M}$, then estimating $[\tilde{g}, m]$ from $[g_i, m]$, $i = 1, \ldots, D$, is the problem of numerical differentiation.

If we use multiplication quelling, then our method of inference will lead simply to fitting a polynomial (or some other function selected from a D-dimensional family) to the data points, and evaluating it or its derivative at x_0 or integrating G times it from 0 to 1. The estimates of the integral, derivative, and interpolated value are not new, but we do get a new sort of error estimate for them from §II.B.3.a.

Either integration quelling or convolution quelling seems to lead to new methods of numerical integration, differentiation, interpolation and extrapolation. Whether these new methods have any advantages over the usual methods I do not know, but it seems possible that they may be useful in smoothing functions for which limitations on the spectrum are more natural than limitations on (or, as is often assumed, the absence of) higher order terms in the Taylor series. Thus they could be used as an alternative to a polynomial or trigonometric fit in interpolating topographic data.

III. LOCALIZED NONLINEAR INFERENCE

A. **Nonlinear quelling.** If our model m_E is a member of a linear space $\mathfrak{M}$ but some of the data functionals or some of the prediction functionals are nonlinear, it may be profitable to consider nonlinear methods of converting $\mathfrak{M}$ to a Hilbert space, $\mathfrak{H}$, and even for linear mappings $Q:\mathfrak{H}^0 \to \mathfrak{M}$ the definition of a quelling must be generalized.

If $g:\mathfrak{M} \to R$ is any functional, linear or not, we write $g(m)$, the value of g at a vector m in $\mathfrak{M}$, as $[g, m]$. Since g need not be linear, we may no longer have $[g, \alpha_1 m_1 + \alpha_2 m_2] = \alpha_1[g, m_1] + \alpha_2[g, m_2]$, but we always have $[\alpha_1 g_1 + \alpha_2 g_2, m] = \alpha_1[g_1, m] + \alpha_2[g_2, m]$ because this equation defines $\alpha_1 g_1 + \alpha_2 g_2$. The set $\mathfrak{M}^{\S}$, consisting of all linear combinations of the functionals on $\mathfrak{M}$ which interest us, is still a vector space, even though its members may be nonlinear functionals. The data functionals $g_1, \ldots, g_D$ and the prediction functionals $\tilde{g}_1, \ldots, \tilde{g}_P$ are in $\mathfrak{M}^{\S}$.

Suppose that $Q:\mathfrak{H}^0 \to \mathfrak{M}$ is a not necessarily linear injection of pre-Hilbert space $\mathfrak{H}^0$ into $\mathfrak{M}$. For any fixed g in $\mathfrak{M}^{\S}$ we can define a functional $(Q^*g):\mathfrak{H}^0 \to R$ by requiring for each h in $\mathfrak{H}^0$ that

$$[Q^*g, h] = [g, Qh]. \tag{171}$$

The functional $Q^*g:\mathfrak{H}^0 \to R$ may be nonlinear if $Q:\mathfrak{H}^0 \to \mathfrak{M}$ is a nonlinear mapping or $g:\mathfrak{M} \to R$ is a nonlinear functional. However, Q^*g depends linearly on g. We have

$$\begin{aligned}[Q^*(\alpha_1 g_1 + \alpha_2 g_2), h] &= [\alpha_1 g_1 + \alpha_2 g_2, Qh] = \alpha_1[g_1, Qh] + \alpha_2[g_2, Qh] \\ &= \alpha_1[Q^* g_1, h] + \alpha_2[Q^* g_2, h] \\ &= [\alpha_1 Q^* g_1 + \alpha_2 Q^* g_2, h]\end{aligned} \tag{172}$$

for every h in $\mathfrak{H}^0$. The linear mapping $Q^* : \mathfrak{M}^{\S} \to$ (linear space of all functionals on $\mathfrak{H}^0$) is the adjoint of Q.

We will call a possibly nonlinear functional $g : \mathfrak{H}^0 \to R$ on a pre-Hilbert space $\mathfrak{H}^0$ "Cauchy continuous" if for every Cauchy sequence $\{h_1, h_2, \ldots\}$ of vectors in $\mathfrak{H}^0$, $\{[g, h_1], [g, h_2], \ldots\}$ is a Cauchy sequence of real numbers. The Cauchy continuous functionals on $\mathfrak{H}^0$ are precisely those which can be extended in exactly one way to continuous functionals on $\mathfrak{H}$, the completion of $\mathfrak{H}^0$. We denote by $C(\mathfrak{H})$ the linear space of all continuous functionals on $\mathfrak{H}$. Then the Cauchy continuous functionals on $\mathfrak{H}^0$ can be regarded as the members of $C(\mathfrak{H})$.

Suppose that $Q : \mathfrak{H}^0 \to \mathfrak{M}$ is an injection of pre-Hilbert space $\mathfrak{H}^0$ into vector space $\mathfrak{M}$ and that for each g in $\mathfrak{M}^{\S}$, $Q^* g : \mathfrak{H}^0 \to R$ is Cauchy continuous. Then we will call $(\mathfrak{H}^0, Q)$ a quelling of $(\mathfrak{M}, \mathfrak{M}^{\S})$. The functional $Q^* g$ can be thought of as a member of $C(\mathfrak{H})$, so $Q^* : \mathfrak{M}^{\S} \to C(\mathfrak{H})$ and this mapping is linear because, by continuity in h, (172) holds not only for all h in $\mathfrak{H}^0$ but for all h in $\mathfrak{H}$.

If $(\mathfrak{H}^0, Q)$ is a quelling of $(\mathfrak{M}, \mathfrak{M}^{\S})$, and m is any member of $Q(\mathfrak{H}^0)$, we can set $h = Q^{-1} m$ in (171) and obtain

$$[g, m] = [Q^* g, Q^{-1} m]. \tag{173}$$

If $Q^* : \mathfrak{M}^{\S} \to C(\mathfrak{H})$ is an injection, then we can replace our original problem of inference on $(\mathfrak{M}, \mathfrak{M}^{\S})$ by a problem of inference on $(\mathfrak{H}, Q^* \mathfrak{M}^{\S})$. We replace m_E by $h_E = Q^{-1} m_E$ and replace every g in $\mathfrak{M}^{\S}$ by $Q^* g$ in $Q^* \mathfrak{M}^{\S} \subseteq C(\mathfrak{H})$. The data then take the form $\gamma_i = [Q^* g_i, h_E]$, $i = 1, \ldots, D$, and the predictions take the form $\tilde{\gamma}_j = [Q^* \tilde{g}_j, h_E]$, $j = 1, \ldots, P$. Both the data and the predictions depend continuously on the model. Of course, to be able to accept this new problem of inference as an adequate substitute for the old one we must think it reasonable to assume that m_E is in $Q(\mathfrak{H}^0)$.

If $g^* : \mathfrak{H} \to R$ is a functional on Hilbert space $\mathfrak{H}$ and h_0 is a particular point in $\mathfrak{H}$, it may happen that there is a vector h^* in $\mathfrak{H}$ such that

$$\lim_{\|h\| \to 0} \frac{[g^*, h_0 + h] - [g^*, h_0] - \langle h^*, h \rangle}{\|h\|} = 0. \tag{174}$$

If this happens, then h^* is uniquely determined by g^* and is called the Fréchet derivative of g^* at h_0, written $h^* = \partial g^* / \partial h_0$. And if this happens, g^* is called Fréchet differentiable at h_0. Another way to state (174) is to demand that there be a function $r:\mathfrak{H} \to R$ such that

(175)
(i) $\lim_{\|h\| \to 0} r(h) = 0$ and
(ii) $[g^*, h_1] - [g^*, h_0] = (\partial g^*/\partial h_0)\cdot(h_1 - h_0) + \|h_1 - h_0\| r(h_1 - h_0)$ for all h_1 sufficiently close to h_0.

Thus (174) simply states that if we know $[g^*, h_0]$ and $\|h\|$ is not too large we can calculate $[g^*, h_0 + h]$ with good accuracy by means of a first-order perturbation theory.

If $(\mathfrak{H}^0, Q)$ is a quelling of $(\mathfrak{M}, \mathfrak{M}^\S)$ and if, for every g in $\mathfrak{M}^\S$, $Q^*g:\mathfrak{H} \to R$ is Fréchet differentiable at every point in some open subset U of $\mathfrak{H}$ which we think contains h_E (U is independent of g) then we will call $(\mathfrak{H}^0, Q)$ a smooth quelling of $(U, \mathfrak{M}, \mathfrak{M}^\S)$. If $(\mathfrak{H}^0, Q)$ is a smooth, adequate quelling of $(U, \mathfrak{M}, \mathfrak{M}^\S)$ then the functionals Q^*g in $C(\mathfrak{H})$ which replace the functionals g in $\mathfrak{M}^\S$ are all Fréchet differentiable everywhere in U.

B. **Differentiable nonlinear inference on Hilbert spaces.** We suppose that by the use of a smooth, adequate quelling we have reduced our problem of inference to one of the following sort: we have an open subset U of a Hilbert space $\mathfrak{H}$, and U contains h_E, the unknown vector which describes the physical object E in which we are interested. We have D Fréchet differentiable functionals $g_i:U \to R$, $i = 1, \ldots, D$, and P Fréchet differentiable functionals $\tilde{g}_j:U \to R, j = 1, \ldots, P$. We have attempted to measure the values of $[g_i, h_E]$, $i = 1, \ldots, D$, and have obtained a D-tuple of results $\gamma' = (\gamma'_1, \ldots, \gamma'_D)$, with $\gamma'_i = [g_i, h_E] + \delta\gamma_i$. The experimental errors $\delta\gamma_i$ are unknown, but we assume that their joint probability distribution in R^D is known for every possible choice of h_E. That is, for any h in U we know $p_{\text{meas}}(\gamma|h)$, the probability density in R^D that, if h_E is h, our attempt to measure $([g_1, h_E], \ldots, [g_D, h_E])$ will produce the D-tuple γ. From the measured γ' and the known function $p_{\text{meas}}(\gamma|h)$ we want to estimate $[\tilde{g}_j, h_E]$, $j = 1, \ldots, P$.

In nonlinear problems it is usually a nontrivial task to find even one model which fits all the observations, and indeed the experimental errors, though small, may be such that there is no h in U, or in $\mathfrak{H}$ for that matter, with $\gamma'_i = [g_i, h]$, $i = 1, \ldots, D$. Therefore we seek an h_0 in U which satisfies the data only approximately, within experimental error. We must make this idea precise. We suppose that the expected values of the experimental errors, $E(\delta\gamma_i)$, all vanish, for $i = 1, \ldots, D$. We also suppose that the $D \times D$ variance matrix $V_{ij} = E[(\delta\gamma_i)(\delta\gamma_j)]$ is known or estimated, and we write its inverse as $(V^{-1})^{ij}$. Then the probability that our measured γ satisfies

$$\sum_{i,j=1}^{D} (\gamma_i - [g_i, h_E])(\gamma_j - [g_j, h_E])(V^{-1})^{ij} \leqq \Psi^2$$

is a function $p(\Psi)$ which increases monotonically from 0 to 1 as Ψ increases from 0 to ∞. Given that we observe γ', then for any vector h in U we define the "credibility" of h as $c(h|\gamma') = 1 - p(\Psi)$ where $\Psi = \Psi(h|\gamma')$ is defined by

$$\Psi(h|\gamma')^2 = \sum_{i,j=1}^{D} (\gamma_i' - [g_i, h])(\gamma_j' - [g_j, h])(V^{-1})^{ij}. \tag{176}$$

If the credibility of h is 1, then $[g_i, h] = \gamma_i', i = 1, \ldots, D$. If the credibility of h is close to 0, then the D-tuple $([g_1, h], \ldots, [g_D, h])$ is far from γ', as measured in units equal to the experimental errors. It is easier to work with $\Psi(h|\gamma')$, which is nearly 0 when h fits the data well, and is very large when h fits the data poorly.

One procedure for finding an h_0 which fits the data within experimental error would be to minimize $\|h\|$ subject to the requirement $\Psi(h|\gamma') \leqq q$ for various values of q, and to choose $h_0 = h(q_0)$ where q_0 is itself acceptably small and makes $\|h(q_0)\|$ acceptably small. "Acceptability" here is a vague notion. Acceptable sizes for $\|h(q)\|$ involve our a priori prejudice, if any, that $\|h_E\| \leqq M$. An acceptable size for q would perhaps make $p(q) \leqq 0.9$, or perhaps even $p(q) \leqq 0.5$. When V_{ij} is small enough, the h which minimizes $\|h\|$ subject to $\Psi(h|\gamma') \leqq q$ will usually lie on the hypersurface $\Psi(h|\gamma') = q$, so h can be found by introducing a Lagrange multiplier. Then the problem is equivalent to minimizing $\|h\|^2 + M^2\Psi(h|\gamma')^2$ as a function of h for various fixed M. As far as I know, this particular scheme has never been tried, but it seems likely to be convenient since it lends itself well to Newton's method and other numerical techniques for minimizing functions of many variables. A method roughly like the above was found to work well in the problem of learning the earth's internal density from the frequencies of oscillation of a finite number of its elastic-gravitational normal modes [Backus and Gilbert, 1967].

Once we have a solution h_0 with high credibility and acceptably low norm, we can try to explore the locus of all vectors h in U whose credibility is at least $c(h_0|\gamma')$. The part of this locus which lies close to h_0 can be explored by perturbation theory. If $\|h - h_0\|$ is sufficiently small, then for any g in $\mathfrak{M}^{\S}$ the remainder term can be neglected in the equation

$$[g, h] - [g, h_0] = (\partial g/\partial h_0) \cdot (h - h_0) + \|h - h_0\| r(h - h_0).$$

If we neglect the remainder term then we have a problem of bounded linear inference. All the differences $[g, h] - [g, h_0]$ for g in $\mathfrak{M}^{\S}$ depend linearly on $h - h_0$, and we can take $\gamma - ([g_1, h_0], \ldots, [g_D, h_0])$ as our D-tuple of observations. Then the problem is that considered in §II. If there are no experimental errors, then h_0 will lie on the manifold of simultaneous solutions h of the D equations $[g_i, h] = \gamma_i'$, $i = 1, \ldots, D$.

Solving the linearized problem will give the linear manifold tangent to the solution manifold at h_0. The effect of experimental errors is to "thicken" both manifolds to open subsets of $\mathfrak{H}$.

There remains the problem of exploring the part of the locus of vectors with high credibility which is so far from h_0 that first-order perturbation theory is inapplicable. So far as I know, no progress has been made on this problem in general, but certain particular problems have been discussed by exploiting their peculiarities. Even in these particular problems, usually all that has been proved is that a certain infinite collection of data, if known without experimental error, uniquely determines the model. We do not usually know how closely a given finite set of data confines the model, except in all problems of linear inference and in Gerver's [1970] recent work on the inverse problem in one-dimensional wave propagation.

In §IV we will discuss some examples of inverse problems in which some results are known about whether the model is uniquely determined by a certain infinite class of data. These results are of interest even in problems of linear inference, since finding the extent to which a finite collection of data determines the model usually involves considerable numerical computation. If it is known that this finite collection is a subset of an infinite collection of data which does not uniquely determine the model, we might hesitate to spend computer time on the finite problem.

IV. Nonlocalized Nonlinear Inference

A. **Formulation of the problem.** Here we will restrict attention to idealized inverse problems in which an infinite amount of perfectly accurate data is available. Since the measured data are contaminated by experimental errors, the existence question is of interest to us: Is there any model h at all which will produce the observed data? The uniqueness problem is equally important and sometimes easier to solve. If we can prove uniqueness for a given infinite set of data then we are encouraged to compute the resolving power of finite subsets of those data. And if we find that a particular finite collection of data does not limit h_E in certain interesting directions, uniqueness theorems for infinite data sets may suggest what new kinds of data we should collect to constrain h_E in those directions.

B. **Examples from geophysics.**

1. *Seismic velocities.* Perhaps the most famous geophysical inverse problem is that of finding the seismic velocity as a function of depth, knowing the time required for the seismic signal from an impulsive surface source to travel to any other surface point.

We will consider this problem for a flat, isotropic, perfectly elastic earth in which v, the seismic velocity being studied (the compressional velocity or the shear velocity) depends only on the depth z. The method is due originally to Herglotz and Wiechert, who observed that if v increases monotonically with depth then the problem amounts simply to solving Abel's integral equation. The case of the spherical earth, with v a function of radius r alone, is treated by [Bullen, 1965], and is like the flat earth except that $v(r)/r$ rather than $v(r)$ must increase with depth.

Let $\hat{x}$, $\hat{y}$ and $\hat{z}$ be unit vectors in the coordinate directions, and assume that the earth occupies the half-space $z > 0$. We consider only the short-wave approximation (geometrical optics). If $\boldsymbol{k} = l\hat{x} + m\hat{y} + n\hat{z}$ is the local wave vector (the spatial gradient of the phase) of a seismic wave group, then the circular frequency of that wave group (the negative partial derivative of the phase with respect to time) is $w = kv(z)$ where $v(z)$ is the phase velocity at the depth z of the group. Here $k^2 = l^2 + m^2 + n^2$. Therefore wave groups behave like particles with momentum $\boldsymbol{k}$ and Hamiltonian $H(l, m, n, x, y, z, t) = kv(z)$. Since $\partial_t H = 0$, the value of H (i.e., the circular frequency w) is constant at a wave group. The Hamiltonian equations of motion are

$$\begin{aligned} dx/dt &= \partial H/\partial l = (l/k)v(z), & dl/dt &= -\partial H/\partial x = 0, \\ dy/dt &= \partial H/\partial m = (m/k)v(z), & dm/dt &= -\partial H/\partial y = 0, \\ dz/dt &= \partial H/\partial n = (n/k)v(z), & dn/dt &= -\partial H/\partial z = -kv'(z). \end{aligned}$$

During its motion, any wave group keeps constant its values of l, m and dy/dx, so it moves in a vertical plane. We choose our axes so that this is the xz plane. Then $m = 0$, $y = 0$, and l and H are both constants of the motion, as is $p = l/H$. But $p = l/kv(z)$, so

$$p = (\sin i)/v(z) \tag{177}$$

where i is the angle between the vertical and the ray (the trajectory of the wave group). The constancy of p in (177) is, of course, Snell's law.

Define $s(z) = 1/v(z)$. Then by hypothesis $s(z)$ decreases as z increases, and from (177), $s(0) > p$. Also, at every point on the ray $l/k = p/s(z)$ and $n/k = (s(z)^2 - p^2)^{1/2}/s(z)$. When $n = 0$, $dz/dt = 0$. For any fixed p, let z_p denote the unique solution of

$$s(z_p) = p. \tag{178}$$

A wave packet leaving the surface impulsive source at angle i to the vertical will have $p = (\sin i)/v(0)$. As the packet travels, its p will remain constant, and after time $T/2$ the packet will reach the depth z_p, at a horizontal distance $X/2$ from the source. There it will begin to turn upward, and will

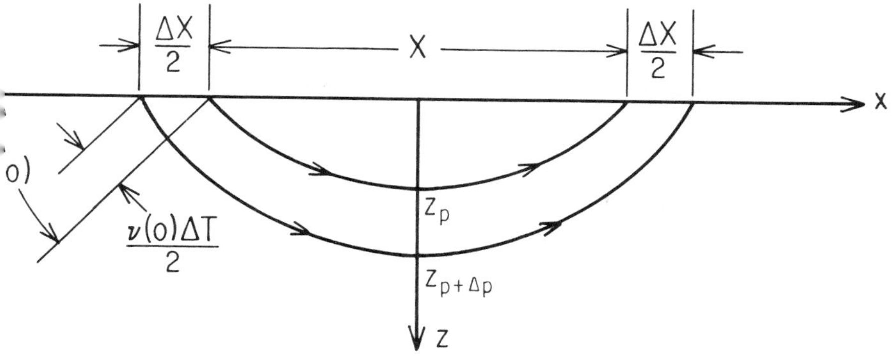

FIGURE 4

arrive again at the surface after time T, at a horizontal distance X from the source. From Hamilton's equations,

$$T(p) = 2\int_0^{z_p} s(z)^2(s(z)^2 - p^2)^{-1/2}\,dz \tag{179}$$

and

$$X(p) = 2\int_0^{z_p} p(s(z)^2 - p^2)^{-1/2}\,dz. \tag{180}$$

We assume that T has been measured as a function of X. From this we can find X as a function of p, and finally, by solving (180), we find z_p as a function of p. Then we know v as a function of z, because $v(z_p) = 1/p$.

To find X as a function of p we observe that $dT/dX = p$, a fact which can be proved either by integrating (179) and (180) by parts and differentiating both with respect to p, or by inspecting Figure 4. From this diagram, at $z = 0$ we have $\sin i = v(0)\Delta T/\Delta X$, or $p = \Delta T/\Delta X$.

We differentiate the observed function $T(X)$, and write X as a function of dT/dX. This function is $X(p)$. Next we replace p by q in (180) and change the variable of integration, obtaining

$$X(q) = -2\int_q^{s(0)} q\frac{dz}{ds}(s^2 - q^2)^{-1/2}\,ds.$$

We multiply both sides of this equation by $(q^2 - p^2)^{-1/2}$, integrate with respect to q from p to $s(0)$, and change the order of the double integration on the right. The result is

$$\int_p^{s(0)} \frac{X(q)\,dq}{(q^2 - p^2)^{1/2}} = -\int_p^{s(0)} ds\frac{dz}{ds}\int_p^s \frac{2q\,dq}{(s^2 - q^2)^{1/2}(q^2 - p^2)^{1/2}}.$$

The second integral on the right is π, so

$$\pi z_p = \int_p^{s(0)} X(q)(q^2 - p^2)^{-1/2}\, dq.$$

There are two difficulties with the foregoing inversion scheme. It requires us to differentiate the data curve $T(X)$ numerically, and it requires that $v(z)$ be monotone. The former requirement sometimes strains the data, and the earth appears not to satisfy the latter requirement. At any rate no one has yet found a shear wave velocity profile which explains the observed dispersion relation for Rayleigh waves and does not have a low velocity zone 50 to 150 km thick at a depth between 100 and 200 km.

The foregoing inversion technique has one advantage. It establishes the existence of a profile $v(z)$ to explain any travel time curve $T(X)$ for which dT/dX is a continuous, monotone increasing function of X. Errors in the data will prevent the existence of a solution only if they make dT/dX nonmonotone.

2. *Surface wave geography on a sphere.* If two seismographs, P and Q are located on a single great circle passing through the epicenter of a large earthquake, then the unknown variation of phase with azimuth at the source is of no consequence because both seismographs are at the same azimuth relative to the source. Therefore, Fourier analysis of the seismograms at P and Q will give a very accurate value for the average slowness (reciprocal of the phase velocity) of Rayleigh waves of any frequency along the great circular arc joining P and Q.

Rayleigh waves with a one-minute period have a wavelength of about 250 km and penetrate the upper mantle to a depth of about 40 km; 5-minutes waves have a wavelength of 2000 km and penetrate to a depth of about 300 km. Rayleigh waves with such long periods and great penetration depths are generated only by very large or very deep earthquakes, both of which are rare and make the geometrical coincidence envisioned in the first paragraph most unlikely. Toksöz and Ben-Menahem [1963] overcame this difficulty in a very clever way. They noticed that P and Q could be the same seismograph, because the Rayleigh waves generated by large earthquakes often travel 7 times around the earth before being lost in the noise. Then every large earthquake would generate a datum for every seismograph, namely the average slowness of Rayleigh waves of any frequency, averaged around the great circle passing through the seismograph and the epicenter of the earthquake. For Rayleigh waves with periods between one and five minutes Toksöz and Ben-Menahem succeeded in measuring such great circular averages of slowness accurate to about 0.1 per cent. They found differences of as much as one per cent between different great circles, after correcting for the polar caustics and the ellipticity of the earth. It is possible that the observed differences in average slowness reflect the geography of the outer 300 km of the earth. Therefore it becomes interesting to know whether the great circular averages of a function defined on the surface of a sphere determine that function.

They do not [Backus, 1964]. If $\hat{w}$ denotes a point on S^2, the surface of the unit sphere, and if $f(\hat{w}) = -f(-\hat{w})$ for all $\hat{w}$, then the functions $g(\hat{w})$ and $g(\hat{w}) + f(\hat{w})$ will have the same averages over all great circles. If, however, $g(\hat{w}) = g(-\hat{w})$ for all $\hat{w}$ on S^2, then g is determined by its great circular averages. This follows from a remarkable property of spherical harmonics. Let $C(S^2)$ be the space of all functions g defined and continuous on S^2. Let $A: C(S^2) \to C(S^2)$ be defined as follows: for any g in $C(S^2)$, Ag is that function which assigns to any point $\hat{w}$ on S^2 the average value of g around the great circle whose pole is $\hat{w}$. Then A is a linear operator, and it turns out that the surface spherical harmonics are eigenfunctions of A. The harmonic Y_l^m has eigenvalue $P_l(0)$ where P_l is the Legendre polynomial of degree l. Thus if $g(\hat{w}) = g(-\hat{w})$ and we know Ag, we can find g simply by expanding Ag in surface spherical harmonics.

We do not want to know the even part of the upper mantle geography; we want the geography itself. All is not lost, however. Suppose that g is in $C(S^2)$ and that we know all the great *semi*circular averages of g. Then we know the great circular averages, so we know $g(\hat{w}) + g(-\hat{w})$ for every $\hat{w}$. Now consider two nearly coincident great semicircles which are part of the same great circle G. Suppose one great semicircle starts at $\hat{w}$ and ends at $-\hat{w}$, and the other starts a small distance δs along G away from $\hat{w}$ and ends δs short of $-\hat{w}$. Subtracting the first great semicircular integral of g from the second gives $\delta s[g(\hat{w}) - g(-\hat{w})]$, so $g(\hat{w}) - g(-\hat{w})$ can be obtained by differentiating the great semicircular integrals of g. Therefore g is uniquely determined by its great semicircular averages.

If seismographs always occurred in antipodal pairs, then every large earthquake would produce two great semicircular averages of phase slowness for every antipodal pair of seismographs, because the great circle passing through one member of the pair always passes through the other member. Therefore antipodal pairs of seismographs accumulate averages of slowness on great semicircles as rapidly as individual seismographs accumulate averages of slowness on great circles. Since the former data determine the slowness geography uniquely while the latter do not, there is a case for antipodal pairing of long period seismographs.

3. *Motions of the surface of the earth's fluid core.* One of the applications to which we can put our inversion scheme is to obtain estimates of the geomagnetic field and its secular variation at the surface of the core from observations at the surface of the mantle. The problem is ill-posed in the

sense of Hadamard; arbitrarily small errors in the exterior measurements can produce arbitrarily large errors in the core field if they occur at small enough horizontal wavelengths. If we are willing to accept a limit on the amount of energy in the geomagnetic field at high wave numbers, then convolution quelling will estimate the core field. So far, such calculations have not been done, but it will be surprising if the results are qualitatively different from those obtained by downward extrapolation of a truncated spherical harmonic expansion of the surface field [Booker, 1969].

If the radial component of the geomagnetic field at the surface of the core is known, and if we accept current fashions in core resistivity and vertical length scale for the magnetic field inside the core, then for the periods involved in the secular variation, the core beneath its magnetic and viscous boundary layers advects the magnetic field as would a perfect conductor. From this hypothesis and the secular variation data we can obtain interesting information about the fluid velocity at the top of the "free stream" just under the boundary layers in the core.

Let B be the radial component of the magnetic field on the surface of the core. A curve on that surface satisfying $B = 0$ will be called a null-flux line. If the core fluid is perfectly conducting, null-flux lines move with it and act as tracers of its motion. Booker [1969] finds four closed null-flux lines, one being a wiggly magnetic equator, one a loop about 20° across, east of Japan, and two about 40° across under Argentina and south of Madagascar. The small loop in the northern hemisphere is at, and perhaps beyond, the resolution of the data. All three small loops are being distorted as well as translated, but all three do have westward components to their velocities, of the order of 0.2° per year.

If the approximation of perfect conductivity in the core is a good one in discussing the secular variation, then the surface integral of $\partial_t B$ over the interior of any closed null-flux loop ought to vanish. Within experimental error, Booker [1969] finds this to be true for his three small null-flux loops but not for the wiggly magnetic equator. It remains to be seen whether this last anomaly will be resolved by a more precise technique for calculating the core field from the observations or whether it represents a failure of the hypothesis that in the secular variation the core behaves like a perfect conductor.

Incidentally, if v is the velocity of the core at the top of the free stream and ∇_s is the surface gradient, the equation which governs B is $\partial_t B + \nabla_s \cdot (Bv) = 0$ if the core is a perfect conductor. It is clear from this equation that westbound waves in v will produce heterodyning in B rather than westward waves. The westward drift of the nondipole field cannot be explained as a result of westbound Alfvén-inertial waves [Backus, 1968]. This is not to underestimate the importance of those waves to the dynamics of the core.

4. *Determining the external geomagnetic field from intensity measurements.* The strength of the earth's steady gravitational field has always been easier to measure than its direction relative to the surface of the earth because atmospheric refraction interferes with accurate measurement of the shape of the surface. Recently, introduction of the proton-precession magnetometer has produced a similar situation in geomagnetism. This magnetometer gives a very accurate measurement of the local magnetic field strength, but no indication of its direction. Precession magnetometers are ideally suited for situations where the measuring instrument's orientation relative to the earth cannot easily be measured, as in satellites and when towed far enough behind a steel oceanographic vessel to be outside the ship's magnetic field.

We are led to ask whether such data determine the earth's external magnetic field. The relevant existence and uniqueness questions can be formulated as follows: Let V be an open, bounded, simply connected subset of R^N. Let ∂V, the boundary of V, have a continuous outward unit normal. Let $H_N(V)$ be the set of all continuous real-valued scalar functions u defined on $R^N - V$ (the region outside V), harmonic there ($\nabla^2 u = 0$), and vanishing uniformly at infinity ($\lim_{r\to\infty} u(r) = 0$). Let $\partial H_N(V)$ be the set of all continuous, nonnegative real-valued functions g defined on ∂V and such that there exists a function u in $H_N(V)$ with $g = |\nabla u|$ everywhere on ∂V. The existence question is to decide which nonnegative, real-valued functions on ∂V are in $\partial H_N(V)$ (i.e., what restrictions, if any, are imposed on the boundary data by the requirement that they be boundary data). The uniqueness question is to decide, for a given g in $\partial H_N(V)$, how many different functions u in $H_N(V)$ satisfy $|\nabla u| = g$ everywhere on ∂V. Obviously, if u works so does $-u$, and the question is whether there is a less trivial nonuniqueness as well.

For $N = 2$, these questions are elementary problems in analytic function theory [Backus, 1968]. It can be shown that $\partial H_2(V)$ contains all functions positive and continuous on ∂V and that if g is in $\partial H_2(V)$ then there is an infinite-dimensional family of functions u in $H_2(V)$ all of which satisfy $|\nabla u| = g$ everywhere on ∂V. These results are of interest in interpreting the ship magnetometer records of magnetic stripes produced on the ocean floor parallel to mid-ocean ridges by sea-floor spreading and the occasional reversals of the geomagnetic dipole.

For $N \geqq 3$ nothing seems to be known about the existence question, but [Backus, 1968] has given some partial answers to the uniqueness question. Specifically, for $N \geqq 3$,

(i) If V is convex and u is the external gravity field of a nonnegative mass distribution in V, then u in $H_N(V)$ is uniquely determined by $|\nabla u|$ on ∂V. (This result also holds for $N = 2$.)

(ii) If V is a sphere and u is known to be the sum of finitely many homogeneous exterior spherical harmonics, then u is determined, except for sign, throughout $R^N - V$ by the values of $|\nabla u|$ on ∂V. (This result fails for $N = 2$.)

(iii) If u is in $H_N(V)$ and $|\nabla u|$ is known everywhere in a spherical shell containing V inside the inner sphere, then u is determined throughout $R^N - V$ except for sign. (This result fails for $N = 2$.)

The three foregoing results suggest the conjecture that when $N \geqq 3$, if u is in $H_N(V)$, then u is determined throughout. $R^N - V$, except for sign, by the values of $|\nabla u|$ on ∂V. This conjecture can be formulated as follows: if u and v are both in $H_N(V)$ and $|\nabla u| = |\nabla v|$ on ∂V then either $u + v$ or $u - v$ vanishes throughout $R^N - V$. The equation $|\nabla u| = |\nabla v|$ can be written $\nabla u \cdot \nabla u - \nabla v \cdot \nabla v = 0$, or $\nabla(u - v) \cdot \nabla(u + v) = 0$. If we define $\phi = u + v$ and $\psi = u - v$, then the conjecture can be reformulated thus: if ϕ and ψ are in $H_N(V)$ and $\nabla\phi \cdot \nabla\psi = 0$ on ∂V then either ϕ or ψ vanishes identically in $R^N - V$.

Unfortunately, I will now show that this conjecture is false if $N = 3$ and V is a sphere of radius a. Let r, θ, λ be radius, colatitude and longitude in R^3. Let $\mu = \cos\theta$. Let $P_l^m(\mu)$ be an associated Legendre polynomial, and let $Y_l^m(\theta, \lambda) = C_l^m P_l^{|m|}(\cos\theta)e^{im\lambda}$, the normalizing constant C_l^m being chosen so that $\int_0^\pi d\theta \int_0^{2\pi} d\lambda \sin\theta |Y_l^m|^2 = 1$ and $Y_l^{-m} = (-1)^m (Y_l^m)^*$. The addition theorem for spherical harmonics says that if $\hat{w}$ and $\hat{w}'$ are any unit vectors then

$$4\pi \sum_{m=-l}^{l} Y_l^m(\hat{w}) Y_l^m(\hat{w}')^* = (2l + 1)P_l(\hat{w} \cdot \hat{w}').$$

If we set $\hat{w}' = \hat{w}$ we obtain $4\pi \sum_{m=-l}^{l} |Y_l^m(\hat{\omega})|^2 = 2l + 1$. It follows that for any θ and λ,

$$|Y_l^m(\theta, \lambda)|^2 \leqq (2l + 1)/4\pi. \tag{181}$$

Any member of $H_3(V)$ can be written in the form

$$\psi(r, \theta, \lambda) = \sum_{l=0}^{\infty} \sum_{m=-l}^{l} \psi_l^m \left(\frac{a}{r}\right)^{l+1} Y_l^m(\theta, \lambda) \tag{182}$$

where ψ is real if for all l and m

$$\psi_l^{-m} = (-1)^m (\psi_l^m)^*. \tag{183}$$

We take

$$\phi = \mu(a/r)^2 \tag{184}$$

and try to find coefficients ψ_l^m satisfying (183) such that the ψ defined by (182) satisfies $\nabla\phi \cdot \nabla\psi = 0$ on $r = a$. Since $\partial_\lambda \phi = 0$, this equation can be written $\partial_r\phi\partial_r\psi + a^{-2}\partial_\theta\phi\partial_\theta\psi = 0$. In case ϕ is given by (184), this condition becomes, on $r = a$,

(185) $$(1 - \mu^2)\,\partial_\mu\psi - 2\mu a\,\partial_r\psi = 0.$$

Substituting (182) in (185) gives

(186) $$\sum_{l=0}^{\infty}\sum_{m=-l}^{l} \psi_l^m[(1 - \mu^2)\partial_\mu Y_l^m + 2(l + 1)\mu Y_l^m] = 0.$$

If we define

$$g_l^m = [(l + m)(l - m)/(2l + 1)(2l - 1)]^{1/2}$$

and if we agree that $Y_l^m = 0$ unless $|m| \leqq l$, then for all integral l and m we have

$$\mu Y_l^m = g_l^m Y_{l-1}^m + g_{l+1}^m Y_{l+1}^m,$$

$$(1 - \mu^2)\partial_\mu Y_l^m = (l + 1)g_l^m Y_{l-1}^m - l g_{l+1}^m Y_{l+1}^m.$$

If we substitute these expressions in (186) we obtain

$$\sum_{l=0}^{\infty}\sum_{m=-l}^{l} \psi_l^m[3(l + 1)g_l^m Y_{l-1}^m + (l + 2)g_{l+1}^m Y_{l+1}^m] = 0.$$

If we define $\psi_{|m|-1}^m = 0$, this last equation can be written

(187) $$\sum_{m=-\infty}^{\infty}\sum_{l=|m|}^{\infty} Y_l^m[3(l + 2)g_{l+1}^m\psi_{l+1}^m + (l + 1)g_l^m\psi_{l-1}^m] = 0.$$

The spherical harmonics are orthonormal, so (187) is equivalent to

(188) $$3(l + 2)g_{l+1}^m\psi_{l+1}^m + (l + 1)g_l^m\psi_{l-1}^m = 0$$

for all $l \geqq 0$ and all m between $-l$ and l. For any fixed m, $\psi_{|m|-1}^m = 0$, so $\psi_{|m|-1+2n}^m = 0$ for every positive integer n. However, we can choose $\psi_{|m|}^m$ arbitrarily and then (188) determines $\psi_{|m|+2n}^m$ for all positive integers n.

It remains to verify that we can choose the $\psi_{|m|}^m$ in such a way that the foregoing formal calculations are justified. From (181) and (188), if $l \geqq 1$ then $|\psi_{l+1}^m| \leqq |\psi_{l-1}^m|/3$. Suppose we choose positive constants C and β with $\beta < 1$. Then we choose any infinite sequence of complex constants $(\psi_0^0, \psi_1^1, \psi_2^2, \ldots)$ satisfying $|\psi_m^m| \leqq C\beta^m$, $m = 0, 1, 2, \ldots$, and we define $\psi_m^{-m} = (-1)^m(\psi_m^m)^*$, $m = 0, 1, 2, \ldots$. (Thus ψ_0^0 must be real.) Then the series (182) and all its derivatives converge absolutely and uniformly outside any sphere centered on the origin and having radius larger than $a/3$. If we want the magnetic fields $\mathbf{V}(\phi - \psi)$ and $\mathbf{V}(\phi + \psi)$ to be physically realizable, they must contain no monopole term ψ_0^0. Therefore, from (188), $\psi_l^0 = 0$ for all $l \geqq 0$.

It should be noted that $\psi = u - v$ can be arbitrarily small, so the non-uniqueness is not resolvable by considering only fields v so close to a given field u that one can linearize the problem. The "generalized Neumann problem," to find v in $H_N(V)$ if $f \cdot \nabla v$ is given on ∂V for some specified vector field f on ∂V, appears not to have been solved in general.

It should be pointed out that our result does not preclude the existence of some magnetic fields ∇u which are uniquely determined by $|\nabla u|$ on ∂V. We have proved only that the set of fields not so determined is an infinite-dimensional family.

Bibliography

L. E. Alsop, G. Sutton, and M. Ewing, *Free oscillations of the Earth observed on strain and pendulum seismographs*, J. Geophys. Res. **66** (1961), 631–642.

G. Backus and F. Gilbert, *Numerical applications of a formalism for geophysical inverse problems*, Geophys. J. R. Astr. Soc. **13** (1967), 247–276.

———, *The resolving power of gross Earth data*, Geophys. J. R. Astr. Soc. **16** (1968), 169–205.

———, *Uniqueness in the inversion of inaccurate gross Earth data*, Philos. Trans. Roy. Soc. London Ser. A **266** (1970), 123–192.

G. Backus, *Inference from inadequate and inaccurate data*. I, II, III, Proc. Nat. Acad. Sci. U.S.A. **65** (1970), 1–7, 281–287; III **67** (1970), 282–289.

———, *Application of a non-linear boundary-value problem for Laplace's equation to gravity and geomagnetic intensity surveys*, Quart. J. Mech. Appl. Math. **21** (1968), 195–221. MR **37** #3028.

———, *Non-uniqueness of the external geomagnetic field determined by surface intensity measurements*, J. Geophys. Res. **75** (1970), 6339–6341.

———, *Kinematics of geomagnetic secular variation in a perfectly conducting core*, Philos. Trans. Roy. Soc. London Ser. A **263** (1968), 239–266.

———, *Geographical interpretation of measurements of average phase velocities of surface waves over great circular and semi-circular paths*, Bull. Seism. Soc. Am. **54** (1964), 571–610.

H. Benioff, F. Press, and S. Smith, *Excitation of the free oscillations of the Earth by earthquakes*, J. Geophys. Res. **66** (1961), 605–620.

S. Bochner, *Harmonic analysis and the theory of probability*, Univ. of California Press, Berkeley, Calif., 1955. MR **17**, 273.

J. Booker, *Geomagnetic data and core motions*, Proc. Roy. Soc. Ser. A **309** (1969), 27–40.

K. Bullen, *An introduction to the theory of seismology*, Cambridge Univ. Press, Cambridge, 3rd ed., 1965.

N. Dunford and J. Schwartz, *Linear operators*. I: *General theory*, Pure and Appl. Math., vol. 7, Interscience, New York, 1958. MR **22** #8302.

M. Gerver, *Inverse problem for the one-dimensional wave equation*, Geophys. J. R. Astr. Soc. **20** (to appear).

R. K. Getoor, *On characteristic functions of Banach space valued random variables*, Pacific J. Math. **7** (1957), 855–896. MR **19**, 584.

W. Gibbs and E. B. Wilson, *Vector analysis*, Yale Univ. Press, New Haven, Conn., 1901.

L. Gross, *Measurable functions on Hilbert space*, Trans. Amer. Math. Soc. **105** (1962), 372–390. MR **26** #5121.

———, *Harmonic analysis on Hilbert space*, Mem. Amer. Math. Soc. No. 46 (1963). MR **28** #4304.

———, *Classical analysis on a Hilbert space*, Analysis in Function Space, M.I.T. Press, Cambridge, Mass., 1964, pp. 51–68. MR **29** #4423.

———, *Abstract Wiener spaces*, Proc. Fifth Berkeley Sympos. Math. Statist. and Probability (Berkeley, Calif., 1965/66), vol. II, part I, Univ. of California Press, Berkeley, Calif., 1967, pp. 31–42, MR **35** #3027.

P. R. Halmos, *Measure theory*, Van Nostrand, Princeton, N.J., 1950. MR **11**, 504.

———, *Introduction to Hilbert space and theory of spectral multiplicity*, Chelsea, New York, 1951. MR **13**, 563.

———, *Finite-dimensional vector spaces*, 2nd ed., University Series in Undergraduate Math., Van Nostrand, Princeton, N.J., 1958. MR **19**, 725.

A. N. Kolmogorov, *Foundations of the theory of probability*, Springer, Berlin, 1933; English transl., Chelsea, New York, 1956. MR **18**, 155.

K. Loewner, *Grundzüge liner Inhaltslehre im Hilbertschen Raume*, Ann. of Math. (2) **40** (1939), 816–833. MR **1**, 48.

E. R. Lorch, *On a calculus of operators in reflective vector spaces*, Trans. Amer. Math. Soc. **45** (1939), 217–234.

———, *Spectral theory*, University Texts in Math. Sci., Oxford Univ. Press, New York, 1962. MR **25** #427.

E. H. Moore, *On the reciprocal of the general algebraic matrix*, Bull. Amer. Math. Soc. **26** (1920), 394–395. (This is a report on a talk given at the Fourteenth Western Meeting of the Amer. Math. Soc.)

G. Morris, R. Raitt, and G. Shor, *Velocity anisotropy and delay-time maps of the mantle near Hawaii*, J. Geophys. Res. **74** (1969), 4300–4316.

N. F. Ness, J. Harrison, and L. B. Slichter, *Observations of the free oscillations of the Earth*, J. Geophys. Res. **66** (1961), 621–630.

E. Parzen, *Modern probability theory and its applications*, Wiley, New York, 1960. MR **22** #3021.

R. Penrose, *A generalized inverse for matrices*, Proc. Cambridge Philos. Soc. **51** (1955), 406–413. MR **16**, 1082.

F. Reisz and B. Sz.-Nagy, *Functional analysis*, Akad. Kiadó, Budapest, 1953; English transl., Ungar, New York, 1955. MR **15**, 132; MR **17**, 175.

L. J. Savage, *The foundations of statistical inference*, Methuen, London, 1962.

I. E. Segal, *Tensor algebras over Hilbert spaces*. I, Trans. Amer. Math. Soc. **81** (1956), 106–134. MR **17**, 880.

M. N. Toksöz and A. Ben-Menahem, *Velocities of Mantle, Love and Rayleigh waves over multiple paths*, Bull. Seism. Soc. Am. **53** (1963), 741–764.

University of California, San Diego

Inverse Problems for the Earth's Normal Modes

Freeman Gilbert

Whenever there is a large earthquake the Earth vibrates for days afterwards. The vibrations consist of the superposition of the elastic-gravitational normal modes of the Earth that are excited by the earthquake. Dissipation causes the modes to decay, typical values of $\mathcal{Q}$ falling in the range 100–400. The lowest observed frequencies belong to a quintet of modes at nearly 1.1 cycles/hour (c/h). The $\mathcal{Q}$ for this quintet is close to 350, so after ten days the amplitude is diminished by a factor of ten. By far the highest $\mathcal{Q}$ belongs to the fundamental radial mode. With a period of 20.46 min. this mode has a motion where every point moves in or out along a radius, all points at the same radius having the same amplitude. There is a small amount of shear in this mode, but it is nearly purely dilational and has a $\mathcal{Q}$ of about 6000 so that its amplitude is diminished by a factor of ten in 2 months [**1**].

Until recently instruments were not sufficiently sensitive and quiet to permit the observation of the lower frequency normal modes for earthquakes with Richter magnitude below 7-3/4. In the past twenty years there have been 9 earthquakes with Richter magnitude equal to or greater than 7-3/4, the latest being the Peruvian shock of 31 May 1970.

In the past year newly developed accelerometers have allowed geophysicists to lower the magnitude threshold for low frequency modes from 7-3/4 to 6-1/2. Since October 1969, when the first of the new accelerometers became operational [**2**], there have been more than 20 earthquakes of magnitude 6-1/2 or greater. In principle, then, we could have accumulated more normal mode data in the past 10 months with the new instruments than was accumulated in the past two decades with more traditional instruments. In practice, however, operational difficulties with the systems for data acquisition and recording have prevented our achieving such a goal. Nevertheless, the deployment and operation of the new instruments promises to give a great improvement in both the quality and the amount of seismological data at our disposal.

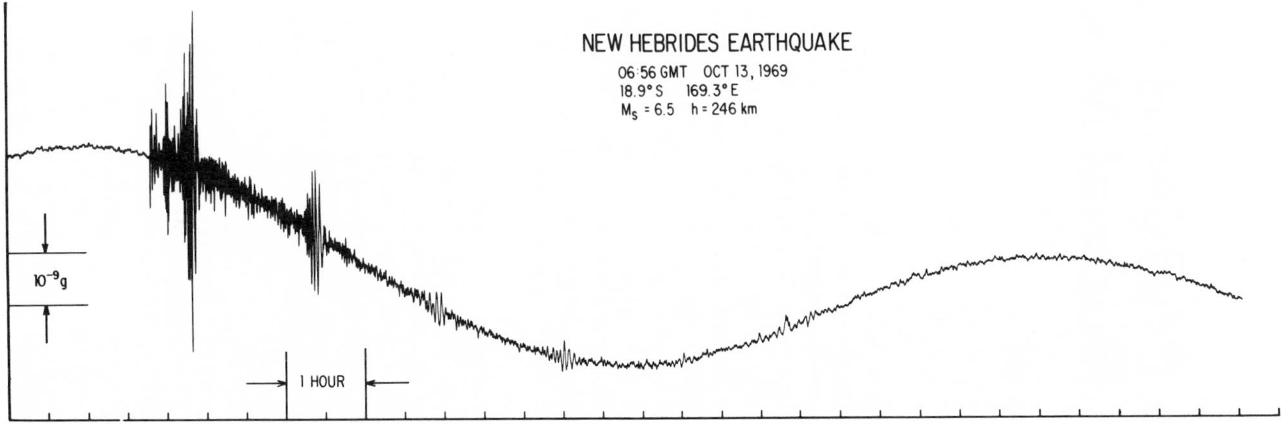

FIGURE 1. Accelerogram for New Hebrides earthquake. The frequency band 1-30 c/h is amplified 40 dB with respect to the bodily tidal frequencies. The vertical scale, 10^{-9} g, is to be used for the band 1-30 c/h.

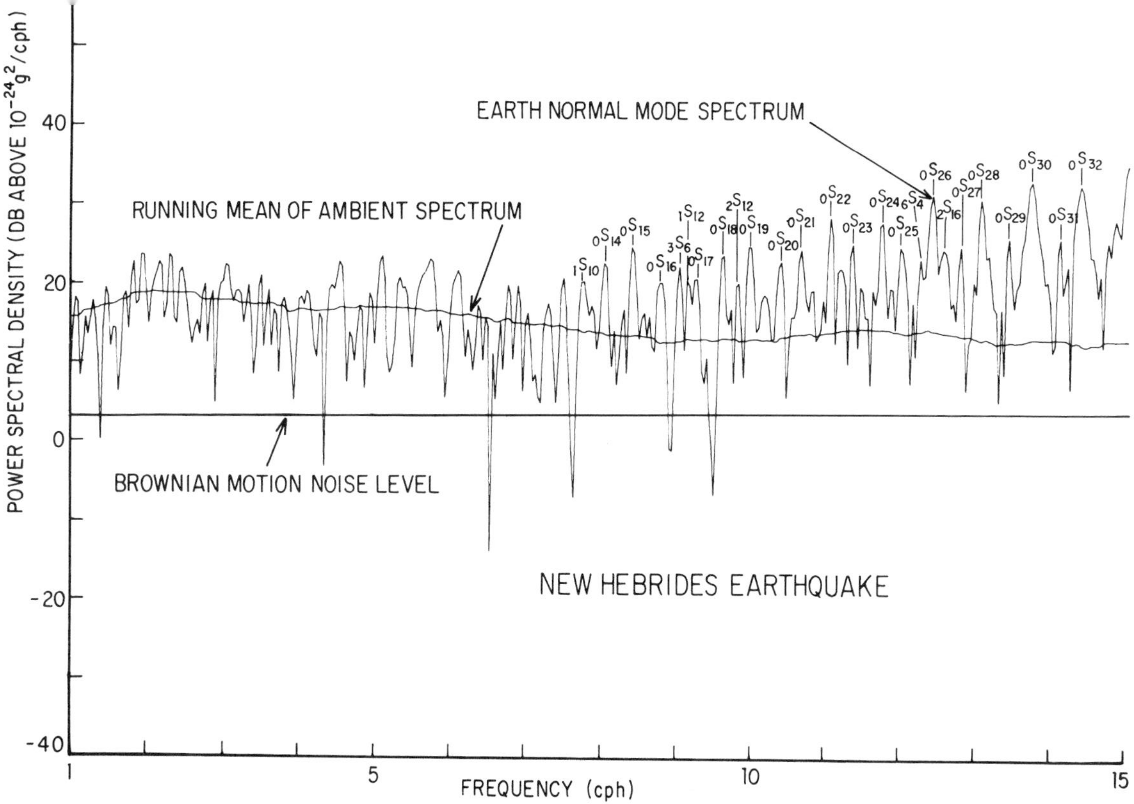

FIGURE 2. Spectra below 14 c/h. The instrumental Brownian noise is shown as the heavy solid line. The 0 dB point refers to a spectral density of 7.4×10^{-25} $g^2/c/h$.

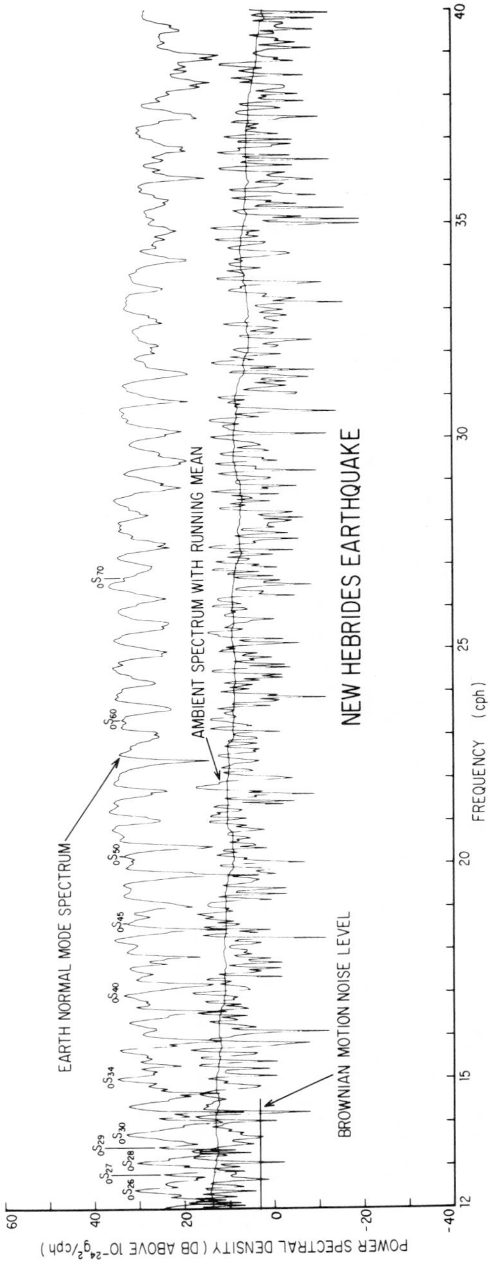

FIGURE 3. Spectra above 12 c/h.

As an example of what can be achieved [**3**], we refer to Figure 1. This is a recording of the New Hebrides earthquake of 13 October 1969 made in San Diego by a Block–Moore vertical accelerometer. In the frequency band 1-30 c/h the signal is amplified 100 times with respect to other frequency bands. We see the earthquake signal superimposed on the bodily tide, whose amplitude is about 10^{-7} g. A standard numerical fourier analysis of the earthquake signal is shown in Figure 2. For comparison an averaged fourier spectrum of background noise, recorded prior to the earthquake, is shown as the ambient spectrum. Clearly there are several peaks, such as the one labeled ${}_0S_{30}$, that stand significantly above the ambient spectrum. The continuation of the spectrum to higher frequencies is shown in Figure 3. We interpret the positions of spectral peaks in Figures 2 and 3 as the eigenfrequencies of the normal modes of the Earth.

If we knew the mechanical properties of the Earth, we could calculate its eigenfrequencies and the observations would merely serve to corroborate our theory. We do believe that our knowledge, although incomplete, is good enough to make such a calculation meaningful. Seismologists have become convinced that the interpretation of travel time data has led to models of the Earth that are grossly correct; that differ from the real Earth only in small, but important, details. Consequently, the calculated eigenfrequencies may not agree exactly with the observed ones, but the differences could reasonably be expected to be small. One of the inverse problems for the Earth's normal modes is to find perturbations to a standard Earth model so that the differences between calculated and observed eigenfrequencies are minimized [**4**]. We shall use the nomenclature and symbolism of the classical theory of small oscillations in our exposition.

Consider a conservative system of N particles in small oscillation. Let the linearized equation for the conservation of linear momentum for the αth particle be

$$m_\alpha \frac{d^2}{dt^2}\boldsymbol{u}_\alpha + \sum_{\beta=1}^{N} \boldsymbol{V}_{\alpha\beta}\cdot\boldsymbol{u}_\beta = \boldsymbol{f}_\alpha \tag{1}$$

where m is the mass, $\boldsymbol{u}$ the displacement, $\boldsymbol{V}$ the symmetric positive definite potential energy matrix, and $\boldsymbol{f}$ the applied force. The solution to (1) can be represented as a superposition of the normal modes, $\boldsymbol{s}_{\alpha,n}$, $n = 1,\ldots,3N$, that satisfy the equation

$$-m_\alpha\omega_n^2\boldsymbol{s}_{\alpha,n} + \sum_{\beta} \boldsymbol{V}_{\alpha\beta}\cdot\boldsymbol{s}_{\beta,n} = 0 \tag{2}$$

where the eigenfrequencies are ω_n.

The Laplace transform of (1) is

$$m_\alpha p^2 \bar{\boldsymbol{u}}_\alpha + \sum_\beta \boldsymbol{V}_{\alpha\beta} \cdot \bar{\boldsymbol{u}}_\beta = \bar{\boldsymbol{f}}_\alpha \tag{3}$$

where

$$\bar{\boldsymbol{u}}_\alpha(p) = \int_0^\infty \boldsymbol{u}_\alpha(t)\, e^{-pt}\, dt, \qquad \operatorname{Re} p > 0$$

and

$$\boldsymbol{u}_\alpha(0) = \left(\frac{d}{dt}\boldsymbol{u}_\alpha(t)\right)_{t=0} = 0.$$

The normal modes in (2) are orthonormal (* denotes complex conjugate)

$$\sum_\alpha m_\alpha \boldsymbol{s}^*_{\alpha,n} \cdot \boldsymbol{s}_{\alpha,l} = \delta_{nl} \tag{4}$$

and form a complete basis for the finite-dimensional vector space of the $3N$ degrees of freedom. Consequently $\bar{\boldsymbol{u}}_\alpha$ has the representation

$$\bar{\boldsymbol{u}}_\alpha = \sum_n a_n \boldsymbol{s}_{\alpha,n} \tag{5}$$

where

$$a_n = \sum_\alpha m_\alpha \boldsymbol{s}^*_{\alpha,n} \cdot \bar{\boldsymbol{u}}_\alpha. \tag{6}$$

If we take the scalar product of (3) with $\boldsymbol{s}^*_{\alpha,n}$ and sum over α we get

$$p^2 a_n + \sum_{\alpha\beta} \boldsymbol{s}^*_{\alpha,n} \cdot \boldsymbol{V}_{\alpha\beta} \cdot \left(\sum_l a_l \boldsymbol{s}_{\beta,l}\right) = \sum_\alpha \boldsymbol{s}^*_{\alpha,n} \cdot \bar{\boldsymbol{f}}_\alpha. \tag{7}$$

If we take the scalar product of (2) with $\boldsymbol{s}^*_{\alpha,l}$, sum over α, and use (4) we get

$$\sum_{\alpha\beta} \boldsymbol{s}^*_{\alpha,l} \cdot \boldsymbol{V}_{\alpha\beta} \cdot \boldsymbol{s}_{\beta,n} = \omega_n^2 \delta_{nl}, \tag{8}$$

which can be used to reduce (7) to

$$a_n(p^2 + \omega_n^2) = \sum_\alpha \boldsymbol{s}^*_{\alpha,n} \cdot \bar{\boldsymbol{f}}_\alpha. \tag{9}$$

Thus the representation (5) becomes

$$\bar{\boldsymbol{u}}_\alpha = \sum_n \left[\sum_\beta \boldsymbol{s}^*_{\beta,n} \cdot \bar{\boldsymbol{f}}_\beta\right] \boldsymbol{s}_{\alpha,n}/(p^2 + \omega_n^2). \tag{10}$$

Most earthquakes are modeled as step functions so that $\bar{f}_\beta = f_\beta p^{-1}$. Then the Laplace inversion of (10) is, for $t > 0$,

$$u_\alpha(t) = \sum_n s_{\alpha,n}\left[\sum_\beta s^*_{\beta,n}\cdot f_\beta\right]\frac{1-\cos\omega_n t}{\omega_n^2}. \tag{11}$$

When there is a small amount of dissipation ($Q \gg 1$) we have

$$u_\alpha(t) = \sum_n s_{\alpha,n}\left[\sum_\beta s^*_{\beta,n}\cdot f_\beta\right]\frac{1-\cos\omega_n t\, e^{-\omega_n t/2Q_n}}{\omega_n^2} \tag{12}$$

so that after a long time all that remains is the static displacement

$$\lim_{t\to\infty} u_\alpha(t) = \sum_n s_{\alpha,n}\left[\sum_\beta s^*_{\beta,n}\cdot f_\beta\right]\frac{1}{\omega_n^2}. \tag{13}$$

The study of static displacements (strain offsets) following large earthquakes has become an important research subject.

For the Earth, which we regard as a classical continuum, a sum such as

$$\sum_\beta s^*_{\beta,n}\cdot f_\beta \quad \text{becomes} \quad \int dV\, s^*_n(r)\cdot f(r)$$

and $f(r)$ is the body force per unit volume. Almost all earth models that we use are radially stratified spheres. Spherical symmetry, plus the properties of perfect elasticity and isotropy, permit a normal mode eigenfunction to be written in spherical coordinates as $s(r) = s(r, \theta, \phi) = {}_n\sigma_l^m$ or ${}_n\tau_l^m$ where

$$\begin{aligned} {}_n\sigma_l^m &= r\,{}_nU_l(r)Y_l^m(\theta,\phi) + r\,{}_nV_l(r)\nabla Y_l^m(\theta,\phi), \\ {}_n\tau_l^m &= -{}_nW_l(r)r \times \nabla Y_l^m(\theta,\phi). \end{aligned} \tag{14}$$

The modes denoted by σ are spheroidal or poloidal modes. The curl of σ is nonradial. The modes denoted by τ are torsional or toroidal modes. The curl of τ is radial. The scalars U, V, and W are solutions to ordinary differential equations with prescribed boundary conditions [**5**]. The eigenvalues are the squared eigenfrequencies.

For a particular l and m the smallest eigenfrequency is labeled $n = 0$, the next $n = 1$, etc. We call l the total angular order, m the azimuthal order ($-l \leqq m \leqq l$) and n the radial order. The azimuthal order m does not appear in the differential equations or boundary conditions so that, for each l and n, there are $2l + 1$ normal mode eigenfunctions, all of which belong to the same eigenfrequency. This is the familiar phenomenon of degeneracy. Small perturbation due to the rotation of the Earth and to the ellipticity of the Earth's figure completely remove the degeneracy and lead to a split multiplet [**6**], the splitting formula having the form

$$(15) \qquad {}_n\omega_l^m = {}_n\omega_l^0(1 + {}_n\alpha_l + m\,{}_n\beta_l + m^2\,{}_n\gamma_l) \quad -l \leqq m \leqq l.$$

The splitting effect is on the order of 1 in 1000 and can interfere with measurements of dissipation. Symmetrical splitting, due to the Corioli force and represented by β in (15), has been observed for $n = 0, l = 2, 3$, bu there have been no reports of the asymmetries due to γ. For the toroida modes ${}_n\beta_l = 1/l(l + 1)$ which provides a mechanical analog of the Zeema effect. For the spheroidal modes all the splitting parameters depend on th Earth's mechanical structure so that their observation, with the exceptio of α, would provide new data for the inverse problem.

Because of the near degeneracy in m, the azimuthal order is customaril not used in identifying a normal mode. Thus a spheroidal spectral line i denoted ${}_nS_l$ and a toroidal line ${}_nT_l$. The line labeled ${}_0S_{30}$ in Figure 2 reall consists of all 61 members of the spheroidal multiplet that has the smalles unperturbed eigenfrequency ($n = 0$) and total angular order $l = 30$. Th amplitude, but, of course, not the frequency, of each member of th multiplet varies over the Earth's surface according to its spherical harmoni pattern (14); consequently the whole multiplet's shape will also vary wit position. That is, the apparent peak of the line ${}_0S_{30}$ will not be the same i San Diego as in, say, New York. If we knew the splitting formula an splitting parameters we could conceivably "super resolve" the multiplet but so far that has not been attempted.

Naturally there is ambiguity in the identification of spectral peaks, bu for many observed spectral lines there is a very close correlation, both i frequency and amplitude, with theoretical calculations, and the best fittin theoretical normal mode is assigned to the observed one. In fact that is how the line in Figure 2 near 13.7 c/h is identified as ${}_0S_{30}$.

We assume that it is possible to identify spectral lines, to measure thei eigenfrequencies and Q's, and to assign experimental errors to the observa tions. Such data are raw material for the inverse problem.

Let us return to Rayleigh's principle (8) in the form

$$(16) \qquad \omega_l^2 \sum_\alpha m_\alpha \mathbf{s}^*_{\alpha,l} \cdot \mathbf{s}_{\alpha,l} = \sum_{\alpha\beta} \mathbf{s}^*_{\alpha,l} \cdot \mathbf{V}_{\alpha\beta} \cdot \mathbf{s}_{\beta,l}.$$

We suppose, for the purposes of demonstration, that the mass of eac particle is changed by a small amount δm_α but that $\mathbf{V}_{\alpha\beta}$ is not changed. W seek the perturbation in ω^2 due to the perturbation in mass. A firs variation of (16) gives

$$
\begin{aligned}
&\delta\omega_l^2 \sum_\alpha m_\alpha \boldsymbol{s}_{\alpha,l}^* \cdot \boldsymbol{s}_{\alpha,l} + \omega_l^2 \sum_\alpha \delta m_\alpha \boldsymbol{s}_{\alpha,l}^* \cdot \boldsymbol{s}_{\alpha,l} \\
&\quad + \omega_l^2 \sum_\alpha \left[m_\alpha \boldsymbol{s}_{\alpha,l}^* - \sum_\beta \boldsymbol{V}_{\alpha\beta} \cdot \boldsymbol{s}_{\beta,l}^* \right] \cdot \delta \boldsymbol{s}_{\alpha,l} \\
&\quad + \omega_l^2 \sum_\alpha \left[m_\alpha \boldsymbol{s}_{\alpha,l} - \sum_\beta \boldsymbol{V}_{\alpha\beta} \cdot \boldsymbol{s}_{\beta,l} \right] \cdot \delta \boldsymbol{s}_{\alpha,l}^* = 0.
\end{aligned} \tag{17}
$$

The use of (2) and its complex conjugate allows us to simplify (17)

$$\delta\omega_l^2 \sum_\alpha m_\alpha \boldsymbol{s}_{\alpha,l}^* \cdot \boldsymbol{s}_{\alpha,l} = -\omega_l^2 \sum_\alpha \delta m_\alpha \boldsymbol{s}_{\alpha,l}^* \cdot \boldsymbol{s}_{\alpha,l}. \tag{18}$$

If we knew the small perturbations δm_α we could calculate the small perturbation $\delta\omega_l^2$. Using the normalization (4) and the notation

$$M_{\alpha,l} = -\omega_l^2 \boldsymbol{s}_{\alpha,l}^* \cdot \boldsymbol{s}_{\alpha,l},$$

we simplify (18)

$$\delta\omega_l^2 = \sum_\alpha \delta m_\alpha M_{\alpha,l}, \tag{19}$$

and we suppose now that $\delta\omega_l^2$ is known and that we want to find δm_α. That is

$$\delta\omega_l^2 = (\omega_{\text{obs}}^2 - \omega_{\text{comp}}^2)_l$$

where ω_{obs} is the observed value of ω_l and ω_{comp} is the computed value. Now there are N values of δm_α and $3N$ values of $\delta\omega_l^2$ so the inverse problem appears to be overdetermined. In reality N is very large, infinite for continuum mechanics, and it is usual that the number of observations is much less than N. Thus we assume that we know $\delta\omega_l^2$, $l = 1, \ldots, L \ll N$, and we want to find δm_α. Formulated in this way the inverse problem appears to be hopelessly nonunique, and it is customary to give up and pursue some other line of work. However, if we are satisfied to have any δm_α satisfying (19) we can exploit the nonuniqueness to our advantage. For example, suppose we demand $\sum_\alpha (\delta m_\alpha)^2$ be a minimum [**4**]. We regard the central problem to be the minimization of $\sum_\alpha (\delta m_\alpha)^2$ with constraints (19). This is the classical isoperimetric problem in the calculus of variations and we introduce Lagrange multipliers v_l to find the solution

$$\delta m_\alpha = \sum_l v_l M_{\alpha,l}. \tag{20}$$

Substituting (20) into (19) gives

$$\sum_l \left[\sum_\alpha M_{\alpha,l} M_{\alpha,n} \right] v_l = \delta\omega_n^2 . \tag{21}$$

The matrix $\sum_\alpha M_{\alpha,l} M_{\alpha,n}$ is an inner product matrix, so it is symmetric and positive definite. Consequently, there is a unique solution, v_l, to (21) such that (20) gives a perturbation δm_α that satisfies (19). Now we have a new mass distribution, $m_\alpha + \delta m_\alpha$ with which we can return to (2) and calculate new eigenfrequencies and normal modes. The process (19)–(21) can be repeated. Like a modification of Newton's method, the process we have outlined should converge provided the initial values of $\delta\omega_n^2$ are not too large. (It would prove embarrassing if $\delta m_\alpha < -m_\alpha$ in one of the iterations.) This is exactly the process that has been used in finding solutions to the inverse normal mode problem for the free oscillations of the Earth [**7**]. The only difference in computational detail is that $\sum_\alpha \ldots$ is replaced by $\int dV \ldots$.

One obvious modification of this method is to include observational errors in solving (21). For example, if σ_l is the standard error in $(\omega_{\text{obs}}^2)_l$ we replace (21) by

$$\delta\omega_n^2 - \sigma_n \leqq \sum_l \left[\sum_\alpha M_{\alpha,l} M_{\alpha,n} \right] v_l \leqq \delta\omega_n^2 + \sigma_n \tag{22}$$

and we find that solution to (22) that minimizes $\sum_\alpha (\delta m_\alpha)^2$.

TABLE 1. Data used to obtain Figure 5 from Figure 4 in two iterations.

mass:	$\bar{\rho} = 5.517$ gm/cm^3
moment:	$z = .33089$
modes:	${}_0S_0, {}_1S_0, {}_2S_0, {}_3S_0, {}_0S_2, {}_1S_2, {}_2S_2, {}_1S_4, {}_0S_{25}, {}_0S_{49}, {}_0S_{73}, {}_0S_{97}, {}_0T_{27}, {}_0T_{53}, {}_0T_{105}$

As an example of the application of the method, we use the set of data listed in Table 1. Our initial model, used to start the iterative process, is shown in Figure 4. Its computed eigenfrequencies are compared to the observed ones in Table 2, and we can see that it is not a bad fit to the data, although a better fit is desirable. After two iterations we have the model shown in Figure 5. It is only slightly different from the initial model, but is in better agreement with the data (Table 3).

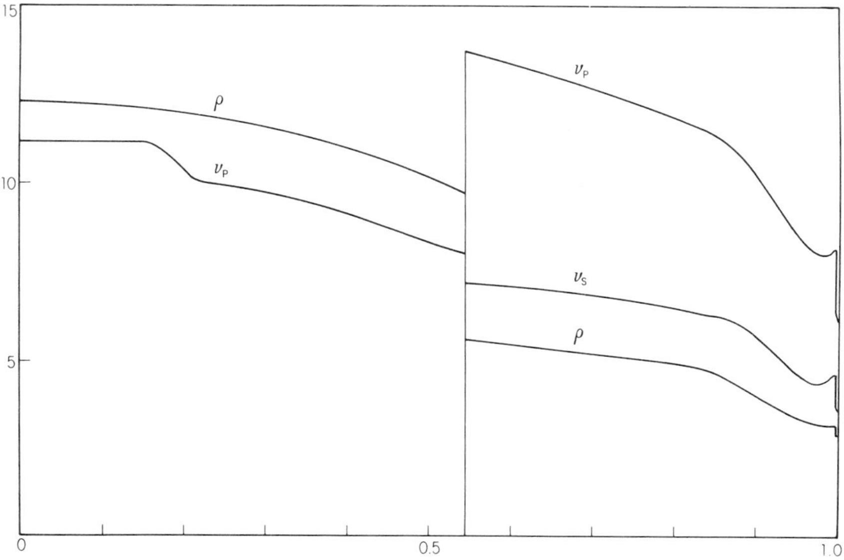

FIGURE 4. The Gutenberg Earth model. The density, ρ, is given in gm/cm^3. The S-wave velocity, v_S, and the P-wave velocity, v_P, are given in km/sec. The radius of the Earth is normalized to unity.

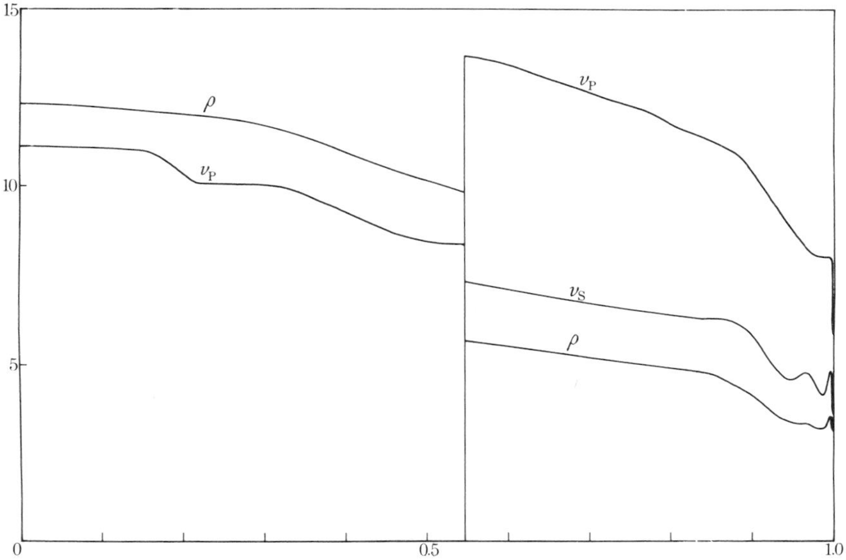

FIGURE 5. Model 8734 (computer job number) derived in two iterations from the Gutenberg model by using the data in Tables 1 and 2.

TABLE 2. Comparison of observed eigenfrequencies with ones computed for Figure 4. Values are given in radians/sec.

mode	ω (*observed*)	ω (*computed*)
	$\times 10^2$	$\times 10^2$
${}_0S_0$	.51162	.50734
${}_1S_0$	1.0403	1.0358
${}_2S_0$	1.5745	1.5876
${}_3S_0$	2.0735	2.0710
${}_0S_2$	.19411	.19610
${}_1S_2$	.42705	.43328
${}_2S_2$	.68771	.69039
${}_1S_4$	.73494	.74727
${}_0S_{25}$	2.1050	2.1466
${}_0S_{49}$	3.4614	3.5246
${}_0S_{73}$	4.8715	4.9331
${}_0S_{97}$	6.2641	6.3861
${}_0T_{27}$	2.2208	2.2516
${}_0T_{53}$	4.0005	4.0455
${}_0T_{105}$	7.5445	7.6443

r.m.s. relative error = .013

Now that we have a model that fits a chosen set of data, we must face the problem of uniqueness. Since the number of data we have is much smaller than the number of eigenfrequencies of the Earth, we cannot hope that our model is a unique one. Our difficulty is further compounded when we admit observational errors in the data.

To begin, we shall suppose that there is a *linear* relation between the data, γ_n, ($\delta\omega^2$ in our example) and the model, $m(r)$, (δm_α, $\delta\rho$, δv_p, δv_s in our example)

$$\gamma_n = \int_0^1 dr \mathscr{G}_n(r) m(r) \tag{23}$$

TABLE 3. Comparison of observed eigenfrequencies with ones computed for Figure 5. Values are given in radians/sec.

mode	ω (*observed*)	ω (*computed*)
	$\times 10^2$	$\times 10^2$
${}_0S_0$	.51162	.51165
${}_1S_0$	1.0403	1.0400
${}_2S_0$	1.5745	1.5744
${}_3S_0$	2.0735	2.0727
${}_0S_2$	.19411	.19424
${}_1S_2$	.42705	.42713
${}_2S_2$	.68771	.68812
${}_1S_4$	.73494	.73508
${}_0S_{25}$	2.1050	2.1048
${}_0S_{49}$	3.4614	3.4607
${}_0S_{73}$	4.8715	4.8700
${}_0S_{97}$	6.2641	6.2617
${}_0T_{27}$	2.2208	2.2208
${}_0T_{53}$	4.0005	3.9997
${}_0T_{105}$	7.5445	7.5432

r.m.s. relative error = .00031

and that $\mathcal{G}$ does not depend on m. For the moment all observed data are assumed to be exact. Thus we can interpret (23) by saying that each datum can be regarded as a linear average of the model. Consequently, any linear average that we can calculate must necessarily be a linear combination of the data that we have.

$$\sum_n a_n \gamma_n = \int_0^1 dr \left(\sum_n a_n \mathcal{G}_n(r) \right) m(r). \tag{24}$$

Let $A = \sum_n a_n \mathcal{G}_n(r)$. Then the averaging kernel, A, is a linear combination of the data kernels $\mathcal{G}_n(r)$. For the sake of argument suppose $\mathcal{G}_n(r) = \sin n\pi r$,

$n = 1, 2, \ldots, N$. Without necessarily being aware of this fortuitous circumstance, suppose we decide to choose $a_n = 2 \sin n\pi r_0$. Then

$$(25) \qquad \sum_{n=1}^{N} a_n \mathcal{G}_n(r) = 2 \sum_{n=1}^{N} \sin n\pi r_0 \sin n\pi r,$$

$$(26) \qquad \sum_{n=1}^{N} a_n \mathcal{G}_n(r) = \frac{\sin[(2N+1)(r-r_0)\pi/2]}{2\sin[(r-r_0)\pi/2]} - \frac{\sin[(2N+1)(r+r_0)\pi/2]}{2\sin[(r+r_0)\pi/2]}.$$

Thus when r_0 is not close to 0 or 1 and when r is close to r_0, A is a function with a tall central spike at $r = r_0$ and is small elsewhere; that is, A looks like a traditional approximation to the δ-function. In this special case (24) would become

$$(27) \qquad \sum_n a_n \gamma_n = \int_0^1 dr A m(r) \cong m(r_0).$$

It is conceivable, then, that a particular linear combination of the observed data can yield a good estimate of the model at a particular place, r_0 [**8**]. Our principle task is to determine whether A can be made to resemble a δ-function.

We could try to minimize the quadratic form

$$(28) \qquad \mathcal{J} = \int_0^1 dr(A(r) - \delta(r - r_0))^2.$$

Let us write $a = \sum_n a_n \mathcal{G}_n(r)$ as $A = \boldsymbol{a} \cdot \boldsymbol{\mathcal{G}}(r)$. Also, let $\mathcal{S} = \int_0^1 dr \boldsymbol{\mathcal{G}}(r)\boldsymbol{\mathcal{G}}(r)$ and denote the elements of $\boldsymbol{S}$ by S_{ij}. Then (28) becomes

$$(29) \qquad \mathcal{J} = \boldsymbol{a} \cdot \boldsymbol{S} \cdot \boldsymbol{a} - 2\boldsymbol{a} \cdot \boldsymbol{\mathcal{G}}(r_0) + \mathcal{J}_\delta$$

where $\mathcal{J}_\delta = \int_0^1 dr(\delta(r - r_0))^2$. Although $\mathcal{J}_\delta$ is infinite it does not depend on $\boldsymbol{a}$ so the minimum of $\mathcal{J}$ is the solution to

$$(30) \qquad \boldsymbol{S} \cdot \boldsymbol{a} = \boldsymbol{\mathcal{G}}(r_0)$$

and

$$(31) \qquad A = \boldsymbol{\mathcal{G}}(r_0) \cdot \boldsymbol{S}^{-1} \cdot \boldsymbol{\mathcal{G}}(r).$$

For $\mathcal{G}_n(r) = \sin n\pi r$ the solution to (30) yields (25) for A. More generally, if we regard $\mathcal{G}_n(r)$, $n = 1, \ldots, N$, as a set of basis functions for the finite dimensional space of observed data, then $\sum_{l=1}^{N} S_{nl}^{-1} \mathcal{G}_l(r)$; $n = 1, \ldots, N$, is the dual basis and the averaging kernel is the scalar product of the two.

For those who dislike the presence of $\mathcal{J}_\delta$ in (29), another approach is to minimize

$$(32) \qquad \mathcal{K} = \int_0^1 dr(r - r_0)^2(A(r) - \delta(r - r_0))^2.$$

Obviously the minimum of (32) is $\boldsymbol{a} = 0$, so some constraint is needed. A simple and reasonable one is to demand unit area for A

$$\int_0^1 A(r)\,dr = 1.$$

If $\boldsymbol{G} = \int_0^1 \mathscr{G}(r)\,dr$, then we want to minimize (32) with the constraint $\boldsymbol{a}\cdot\boldsymbol{G} = 1$. We note that $\mathscr{K}$ has the same dimensions as r. Since $\mathscr{K} = 0$ for $A(r) = \delta(r - r_0)$, and $\mathscr{K} = 1/12$ for $A(r) = w^{-1}$ for $r_0 - w/2 \leqq r \leqq r_0 + w/2$, and $A = 0$ for other values of r, we take $12\,\mathscr{K}$ as a measure of the width of A when A is centered on r_0. Let $s = 12\,\mathscr{K}$. We call s the spread of A. When A is centered on r_0, s is the width of A. We could add another constraint $\int A(r)r\,dr = r_0$ to center A, but we don't want to become too fancy too soon. Let

$$\mathscr{S} = 12\int_0^1 dr(r - r_0)^2\mathscr{G}(r)\mathscr{G}(r). \tag{33}$$

Then we want to minimize $\boldsymbol{a}\cdot\mathscr{S}\cdot\boldsymbol{a}$ with the side condition $\boldsymbol{a}\cdot\boldsymbol{G} = 1$. Clearly

$$\boldsymbol{a} = \mathscr{S}^{-1}\cdot\boldsymbol{G}/\boldsymbol{G}\cdot\mathscr{S}^{-1}\cdot\boldsymbol{G}. \tag{34}$$

To compare minimizing $\mathscr{J}$ in (28) with $\mathscr{K}$ in (32) we refer to Figure 6.

We subjectively prefer the $\mathscr{K}$ criterion to the $\mathscr{J}$ criterion because the $\mathscr{K}$ criterion gives lower sidebands for A. The slight increase in the peak width of A is the price we have to pay for lower sidebands.

To summarize briefly, we minimize $s = \boldsymbol{a}\cdot\mathscr{S}\cdot\boldsymbol{a}$ with $\boldsymbol{a}\cdot\boldsymbol{G} = 1$. If $\boldsymbol{a}\cdot\mathscr{G}(r)$ resembles $\delta(r - r_0)$, then $\boldsymbol{a}\cdot\gamma \cong m(r_0)$ and $s = \boldsymbol{a}\cdot\mathscr{S}\cdot\boldsymbol{a}$ is the spread of A, or the resolving length of A.

For the problems where the data are nonlinear functionals of the model, we restrict attention to models that are "close." Suppose that m_1 and m_2 are "close." We mean

$$\delta\gamma = \int_0^1 dr(m_1(r) - m_2(r))\mathscr{G}(r) + \mathscr{O}[(m_1 - m_2)^2] \tag{35}$$

and $\mathscr{O}[(m_1 - m_2)^2]$ may be neglected. Then

$$\boldsymbol{a}\cdot\delta\gamma = \int_0^1 dr(m_1(r) - m_2(r))A(r)\,dr. \tag{36}$$

If both models agree with the data, $\delta\gamma = 0$ and

$$\int_0^1 drm_1(r)A(r) = \int_0^1 drm_2(r)A(r), \tag{37}$$

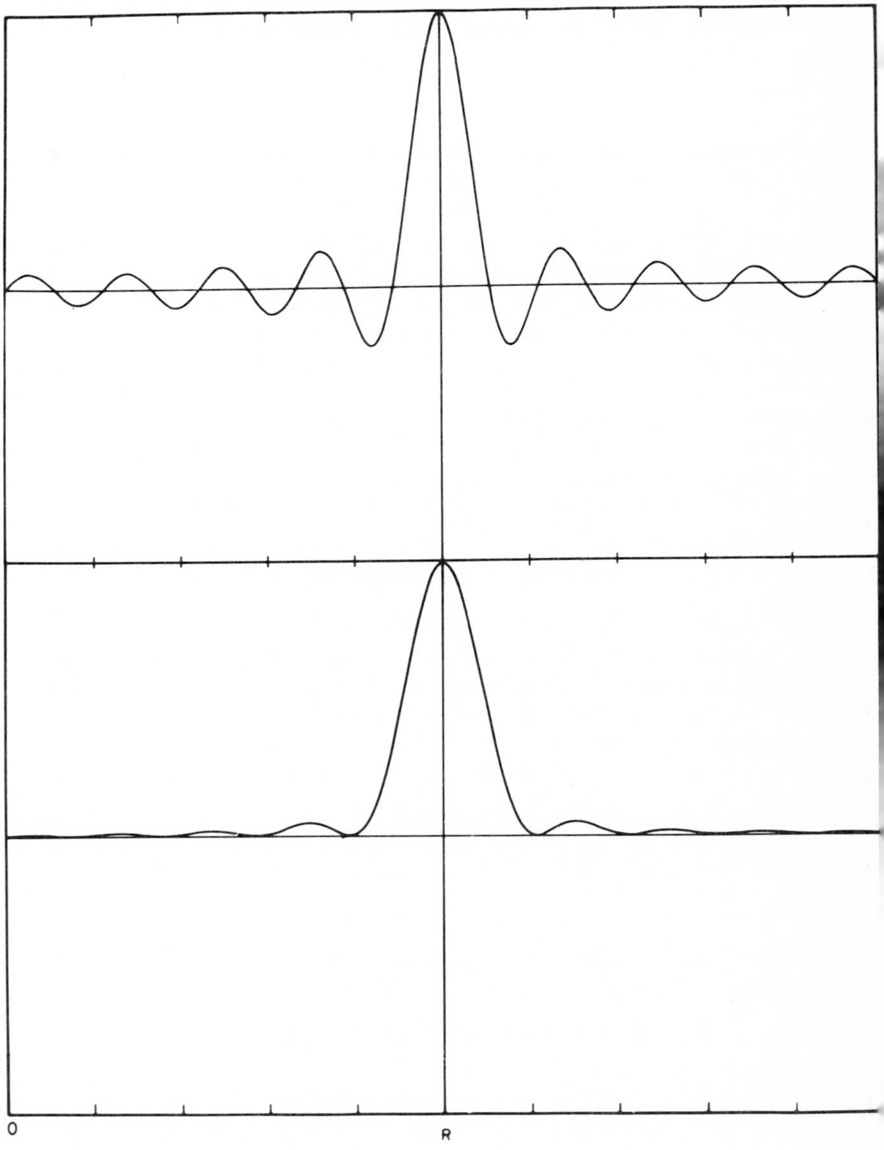

FIGURE 6. Averaging kernels, A, computed at $r_0 = 1/2$ from $\sin n\pi r$, $n = 1, \ldots, 17$. Above is the J kernel (28) and below is the κ kernel (32).

both models have the *same* linear averages [**8**]. If one model is the real Earth and another model is close to it, then certain linear averages of the structure of the real Earth can be found by (37). Such linear averages could be of interest if s is not too big.

TABLE 4. Data used in computations for Figures 7–9. Mass and moment are given in Table 1.

modes:	${}_0S_0, {}_1S_0, {}_2S_0, {}_3S_0, {}_1S_1, {}_2S_1, {}_0S_2, {}_2S_2, {}_1S_3, {}_0S_4, {}_1S_4, {}_2S_4, {}_4S_4, {}_0S_7, {}_1S_8,$ ${}_0S_{25}, {}_0S_{49}, {}_0S_{73}, {}_0S_{97}, {}_0T_7, {}_0T_{14}, {}_0T_{27}, {}_0T_{53}, {}_0T_{105}$

As an illustration of this technique we use the set of data listed in Table 4, pretending that both v_P and v_S are known exactly, and that ρ is "close" to the density of the real Earth. For various r_0 the averaging kernels, A, are shown in Figure 7. The density model, ρ, is shown as the piecewise continuous curve in Figure 8. Each dot represents the linear average of ρ (at a value of r_0) made with the appropriate A in Figure 7.

It is time to admit that if we use the $\mathscr{K}$ criterion (32) to find A, then the averaged m will not satisfy (23). In fact only the $\mathscr{J}$ criterion leads to an averaged m that satisfies (23). An obvious response is to be satisfied with an averaged m if it satisfies (23) to within one standard error in the data. However, if there are errors, $\delta\gamma$, in the data there will be associated with them uncertainties or errors in the averaged model

$$(38) \qquad \delta\gamma = \int_0^1 dr\mathscr{G}(r)\delta m(r), \qquad \boldsymbol{a}\cdot\delta\gamma = \int_0^1 dr A(r)\delta m(r) \cong \delta m(r_0).$$

Experimentally we should be able to find the variance of the data

$$(39) \qquad \boldsymbol{a}\cdot\overline{\delta\gamma\delta\gamma}\cdot\boldsymbol{a} = \overline{\delta m(r_0)\delta m(r_0)} = \varepsilon^2$$

where $\boldsymbol{V} = \overline{\delta\gamma\delta\gamma}$ is the experimentally estimated variance matrix. Given $\boldsymbol{V}$ we find $\boldsymbol{a}$ and A and determine $\boldsymbol{a}\cdot\gamma$ to find the linear average of m; we then calculate ε, the uncertainty or error in the average.

Clearly, we should like to minimize ε. That is, we want not only to minimize $s = \boldsymbol{a}\cdot\mathscr{S}\cdot\boldsymbol{a}$, but also to minimize $\varepsilon^2 = \boldsymbol{a}\cdot\boldsymbol{V}\cdot\boldsymbol{a}$, still satisfying the constraint $\boldsymbol{a}\cdot\boldsymbol{G} = 1$. Except under the most unusual and unrealistic circumstances it is obvious that ε^2 and s cannot be minimized simultaneously. It is equally obvious that some linear combination of ε^2 and s can be minimized [**9**].

A geometrical sketch of the basic idea may help one develop an intuitive understanding. For a proof of the assertions made here, one is referred to the literature [**9**].

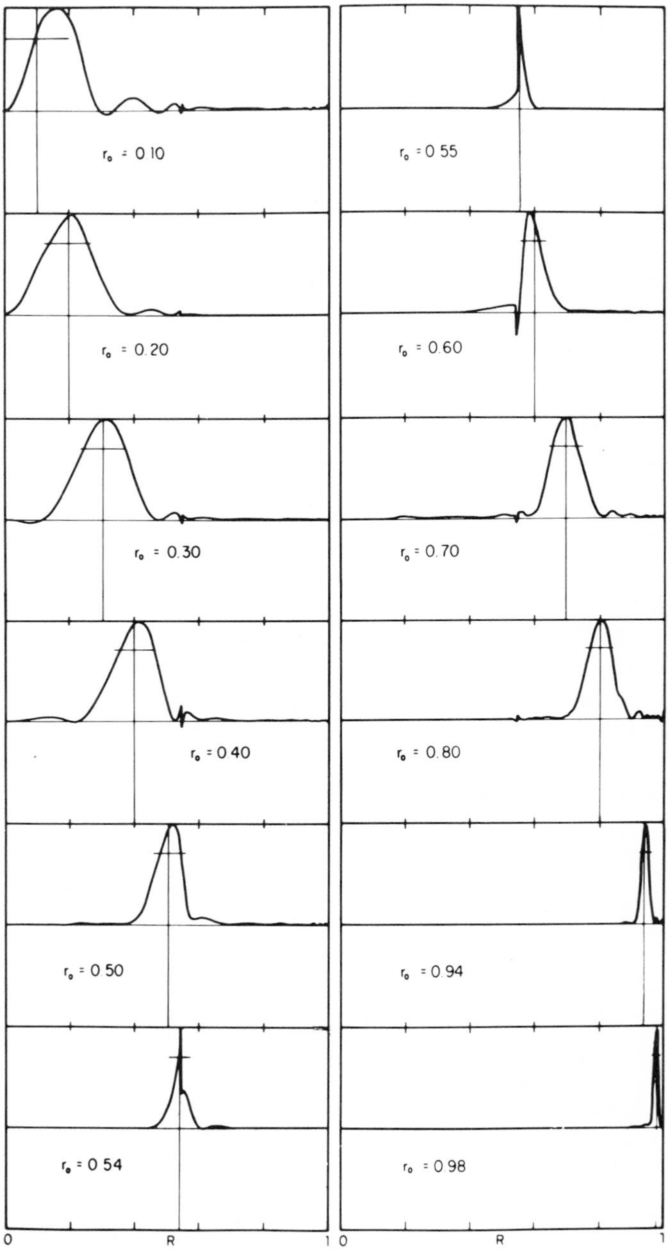

FIGURE 7. The averaging kernels, A, for ρ at selected r_0 (vertical lines). The κ criterion (32) is used and the data are given in Table 4. The spread, s, is represented by the horizontal bars.

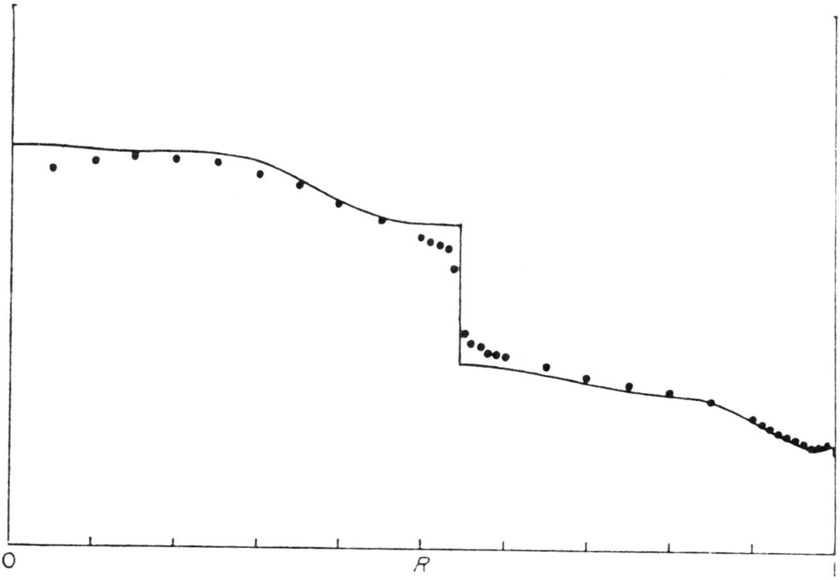

FIGURE 8. The solid curve is the density, ρ, and the dots are its local averages computed with the kernels in Figure 7.

The constraint $\boldsymbol{a} \cdot \boldsymbol{G} = 1$ can be regarded as a hyperplane in the N dimensional space of observed data. Both $\boldsymbol{a} \cdot \mathscr{S} \cdot \boldsymbol{a}$ and $\boldsymbol{a} \cdot \boldsymbol{V} \cdot \boldsymbol{a}$ are hyperspheres. When s is minimized

(40) $$\boldsymbol{a} = \mathscr{S}^{-1} \cdot \boldsymbol{G}/\boldsymbol{G} \cdot \mathscr{S}^{-1} \cdot \boldsymbol{G} = \boldsymbol{a}_s,$$

and $\boldsymbol{a}_s$ is in $\boldsymbol{a} \cdot \boldsymbol{G} = 1$ and represents a point (hyperpoint). Similarly when ε^2 is minimized

(41) $$\boldsymbol{a} = \boldsymbol{V}^{-1} \cdot \boldsymbol{G}/\boldsymbol{G} \cdot \boldsymbol{V}^{-1} \cdot \boldsymbol{G} = \boldsymbol{a}_\varepsilon,$$

and $\boldsymbol{a}_\varepsilon$ is another point on the hyperplane. It has been proved that there is a line in $\boldsymbol{a} \cdot \boldsymbol{G} = 1$ joining $\boldsymbol{a}_s$ and $\boldsymbol{a}_\varepsilon$ such that to each point on the line there corresponds an $\boldsymbol{a}$, and, therefore, an s and ε^2; and that the hypercircle $\mathscr{S}_G$, representing the intersection of $\boldsymbol{a} \cdot G = 1$ with $\boldsymbol{a} \cdot \mathscr{S} \cdot \boldsymbol{a}$, and the hypercircle ε_G^2, representing the intersection of $\boldsymbol{a} \cdot \boldsymbol{G} = 1$ with $\boldsymbol{a} \cdot \boldsymbol{V} \cdot \boldsymbol{a}$, are tangent at the point $\boldsymbol{a}$. In other words, for each value of s on the line joining $\boldsymbol{a}_s$ and $\boldsymbol{a}_\varepsilon$, ε^2 is minimized. Furthermore $\varepsilon^2(s)$ is a monotonically decreasing function of s.

Let

(42) $$\boldsymbol{W} = \mathscr{S}(1 - \alpha) + \boldsymbol{V}\alpha, \qquad 0 \leqq \alpha \leqq 1.$$

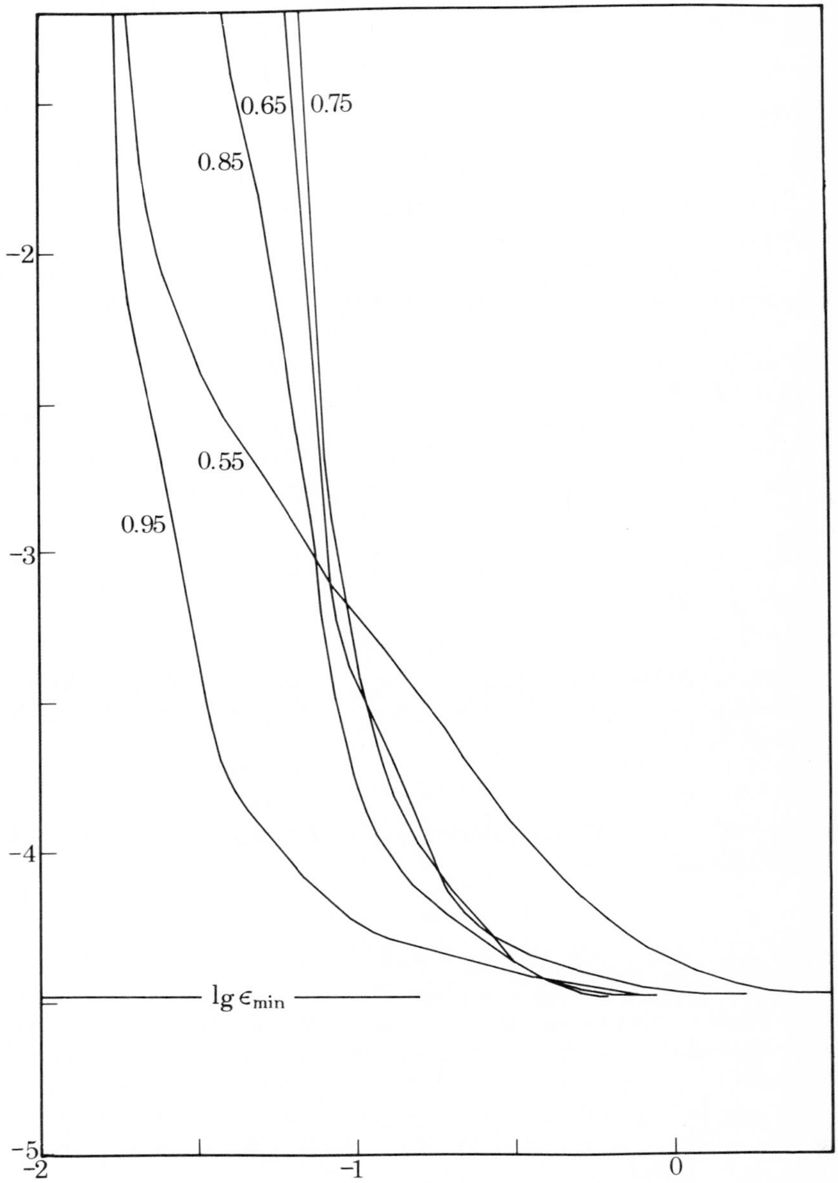

FIGURE 9. Tradeoff curves of absolute error, ε, as a function of spread, s, at selected values of r_0.

Then when $\alpha = 0$ we minimize s, and when $\alpha = 1$ we minimize ε^2. For some α, $0 \leqq \alpha \leqq 1$, let

$$\boldsymbol{a}_\alpha = \boldsymbol{W}^{-1} \cdot \boldsymbol{G} / \boldsymbol{G} \cdot \boldsymbol{W}^{-1} \cdot \boldsymbol{G}. \tag{43}$$

Then

$$s_\alpha = \boldsymbol{a}_\alpha \cdot \mathscr{S} \cdot \boldsymbol{a}_\alpha \quad \text{and} \quad \varepsilon_\alpha^2 = \boldsymbol{a}_\alpha \cdot \boldsymbol{V} \cdot \boldsymbol{a}_\alpha$$

and α parametrizes the curve $\varepsilon(s)$. This curve is called the tradeoff curve of absolute error vs. spread.

For nonlinear problems we appeal to (35)–(37). When $\delta\gamma$ is less than one standard error, both models have effectively the same tradeoff curves.

In most cases one is more interested in the relative error than the absolute error. The analysis is more involved and not even a sketch will be given here [**9**].

For an illustration of the application of tradeoff curves to geophysical data, the only example we give is an artificial one. A simple, even though physically unrealizable, model for dissipation (see (12)) is

$$\mathscr{Q}_n^{-1} = \int_0^1 dr \mathscr{G}_n(r) \mathscr{Q}^{-1}(r). \tag{44}$$

In the Earth $\mathscr{G}_n(r)$ is 0 for $r < r_c = 0.545$ where r_c is the radius of the fluid core. For $r > r_c$ we take $\mathscr{Q}^{-1}(r) = 0.004r$ so that $\mathscr{Q}(1) = 250$. For the 24 modes listed in Table 4 we calculate $\mathscr{Q}_n^{-1}$ in (44) and assume standard errors of 5%. Further we assume that the errors are uncorrelated so $\boldsymbol{V}$ is diagonal. For selected values of $r_0, r_c < r_0 < 1$, tradeoff curves are presented in Figure 9. In every case, a very slight increase in s above $s_{\min}$ leads to an enormous decrease of ε below $\varepsilon_{\max}$. If we had used a linear rather than logarithmic scale, the tradeoff curves would have closely resembled the shape ∟. The place to be is down at the corner.

References

1. L. B. Slichter, *Free oscillations of the earth*, Int'l. Dict. Geoph., edited by S. Runcorn, Pergamon Press, New York and London, 1967, pp. 331–343.
2. B. Block and R. D. Moore, J.G.R. **75** (1970), 1493–1505.
3. B. Block, J. Dratler and R. D. Moore, Nature **226** (1970), 343–344.
4. G. Backus and F. Gilbert, Geoph. J.R.A.S. **13** (1967), 247–276.
5. Z. Alterman, H. Jarosch and C. L. Pekeris, Proc. Roy. Soc. Ser. A **252** (1959), 80–95.
6. F. A. Dahlen, Geoph. J.R.A.S. **16** (1968), 329–367; ibid. **18** (1969), 397–436.
7. F. Gilbert and G. Backus, Bull. Seism. Soc. Amer. **58** (1968), 103–131.
8. G. Backus and F. Gilbert, Geoph. J.R.A.S. **16** (1968), 169–205.
9. ——, Philos. Trans. Roy. Soc. London Ser. A **266** (1970), 123–192.

INSTITUTE OF GEOPHYSICS AND PLANETARY PHYSICS AND
SCRIPPS INSTITUTION OF OCEANOGRAPHY

Dynamo Theory

P. H. Roberts

1. **Foundations.**

1.1. *Introduction.* How can a large cosmic body, such as the Earth, maintain (as the study of palaeomagnetism assures us it has) its magnetic field at much the same strength over all geological time? Ideas of permanent magnetism, or of intrinsic magnetism due to rotation, look extremely unattractive; and the consensus of opinion today is that the main geomagnetic field is created by electric currents flowing in the Earth's liquid, and presumed electrically conducting, core (and possibly also in the lower mantle).

Clearly the question now becomes: "How are these currents maintained?" For it is well known that, in the absence of electromotive forces, they would decay in a time of the order of 10^5 years at most (cf. equation (1.3.4) below[1]). Several possibilities have been suggested, but the most plausible (and aesthetically attractive) seems to be the dynamo theory, viz. that the fluid motions in the core create currents by electromagnetic induction as in a self-excited dynamo. Such a possibility, of self-regeneration, is commonly realized in engineering. The man-made dynamo is, however, multiply-connected and highly asymmetric, whereas the Earth is simply-connected and approximately spherical. It is far from obvious that the motions, $\boldsymbol{u}$, within a homogeneous body (such as the Earth) can, by their asymmetry, atone for the lack of asymmetry of the container. The first approach to this obstacle, one in which $\boldsymbol{u}$ is specified without reference to the dynamics, is known as *the kinematic dynamo problem*; it is linear and is the subject of much of the discussion below. It is now known, from many points of view, that kinematic dynamos exist. Indeed it seems that practically any motion, of "sufficient vigour and complexity," will act as a kinematic dynamo.

---

[1] (*L.M.N.*) denotes the *N*th numbered equation in §*L.M.* but will be referred to simply as "(*N*)" within the text of that subsection.

The initial question is now pushed back one stage further to: "Assuming a physically reasonable driving force, $\boldsymbol{F}$, will a flow of the right magnitude and character be created? And what will decide the strength and nature of the field created, bearing in mind that Lorentz forces will affect the flow?" In one sense, the answer to the first of these questions is "Yes, trivially." One can choose a known dynamo flow, $\boldsymbol{u}$, which creates a field, $\boldsymbol{B}$, of the right spatial structure. Having decided on a magnitude for $\boldsymbol{B}$, the Lorentz force can be computed and, together with the $\boldsymbol{u}$ selected, substituted into the Navier–Stokes equation, which then balances only for a particular $\boldsymbol{F}$. This inversion of the natural procedure holds scant prospects, however, of a proper understanding of the curious aperiodic reversals suffered by the geomagnetic field. More seriously, it gives no help whatever in the development of a full theory in which $\boldsymbol{F}$, rather than being assigned, is itself determined from other physical considerations, e.g. bouyancy in a convection theory. *The hydromagnetic dynamo problem* in which $\boldsymbol{B}$ and $\boldsymbol{u}$ are deduced from an assigned $\boldsymbol{F}$ is nonlinear and difficult, but some answers are beginning to emerge. The topic is discussed in §5.

1.2. *Electrodynamic equations and boundary conditions.* The basic equations governing the kinematic dynamo are Ohm's law and the pre-Maxwell equations. The latter are

$$\text{(1.2.1, 2)} \qquad \operatorname{curl} \boldsymbol{B} = \mu \boldsymbol{j}, \qquad \operatorname{div} \boldsymbol{B} = 0,$$

$$\text{(1.2.3, 4)} \qquad \operatorname{curl} \boldsymbol{E} = -\partial \boldsymbol{B}/\partial t, \qquad \operatorname{div} \boldsymbol{E} = \mathscr{I}/\varepsilon.$$

(Here $\boldsymbol{E}$ is electric field; $\boldsymbol{j}$, electric current density; $\mathscr{I}$, charge density; μ, permeability; ε, dielectric constant; μ and ε assume their free space values everywhere, and m.k.s. units are used.) Equation (4) does no more than determine $\mathscr{I}$ from $\boldsymbol{E}$ and will be discarded from now onwards. Ohm's law, in the form suitable for a conductor in nonrelativistic motion, is

$$\text{(1.2.5)} \qquad \boldsymbol{j} = \sigma(\boldsymbol{E} + \boldsymbol{u} \times \boldsymbol{B}),$$

where σ is the electrical conductivity. The flows, $\boldsymbol{u}$, specified will satisfy

$$\text{(1.2.6)} \qquad \operatorname{div} \boldsymbol{u} = 0.$$

At any surface, S, of discontinuity we must have

$$\text{(1.2.7)} \qquad \langle \boldsymbol{B} \rangle = \langle \boldsymbol{n} \times \boldsymbol{E} \rangle = 0, \quad \text{on } S,$$

where $\langle Q \rangle$ denotes the leap in any quantity, Q, across S, and $\boldsymbol{n}$ is the unit normal to S. It will be supposed that S is fixed and separates a conducting fluid volume, V, of finite spatial extent from a surrounding insulator, $\hat{V}$. In $\hat{V}$, we have, by (1–4),

$$\text{(1.2.}\hat{1},\hat{2}\text{)} \qquad \operatorname{curl} \hat{\boldsymbol{B}} = 0, \qquad \operatorname{div} \hat{\boldsymbol{B}} = 0,$$

$$\text{(1.2.}\hat{3},\hat{4}\text{)} \qquad \operatorname{curl} \hat{\boldsymbol{E}} = -\partial \hat{\boldsymbol{B}}/\partial t, \qquad \operatorname{div} \hat{\boldsymbol{E}} = 0.$$

The crucial demand, that $\hat{\boldsymbol{B}}$ and $\hat{\boldsymbol{E}}$ are *self-excited* (i.e. have no sources "at infinity"), requires

$$\text{(1.2.8)} \qquad \hat{\boldsymbol{B}} = O(r^{-3}), \qquad r \to \infty,$$

where r denotes distance from some origin in V.

From (1–3) and (5), we obtain

$$\text{(1.2.9)} \qquad \partial \boldsymbol{B}/\partial t = \operatorname{curl}(\boldsymbol{u} \times \boldsymbol{B}) + \eta \nabla^2 \boldsymbol{B},$$

where

$$\text{(1.2.10)} \qquad \eta = 1/\mu\sigma,$$

a quantity often called "the magnetic diffusivity" (dimensions L^2/T). Equation (9) is usually called "the induction equation." Taken together with (8), $(\hat{1})$, $(\hat{2})$, and the first of (7), it defines the kinematic dynamo problem; viz.: Do these equations, for the right choice of $\boldsymbol{u}$, possess a solution which does not decay to zero as $t \to \infty$?

The kinematic dynamo problem is clearly linear in $\boldsymbol{B}$ and, if we suppose $\boldsymbol{B}$ is proportional to $\exp(st)$, becomes a search for an eigenvalue, s, with a nonnegative real part. Apparently, a steady dynamo ($s = 0$) is less typical than a dynamo which oscillates at constant amplitude [$\mathscr{R}(s) = 0$, $\mathscr{I}(s) \neq 0$]. In the framework of the kinematic dynamo problem, it is quite possible to obtain physically senseless solutions [$\mathscr{R}(s) > 0$] which become infinite in amplitude as $t \to \infty$. The Lorentz force prevents the hydromagnetic dynamo from suffering this fate, and this nonlinearity also rules out solutions whose time dependence is purely sinusoidal.

It is worth noting, for future reference that, by (1), (5) and (10), Ohm's law may be written

$$\text{(1.2.11)} \qquad \boldsymbol{E} = -\boldsymbol{u} \times \boldsymbol{B} + \eta \operatorname{curl} \boldsymbol{B}.$$

1.3. *The electrodynamics of moving conductors.* Let $\mathscr{L}$ be a length characteristic of $\boldsymbol{B}$ and $\mathscr{U}$ a typical fluid velocity. The dimensionless number describing the relative magnitude of the terms on the right of (1.2.9) or (1.2.11) is

$$\text{(1.3.1)} \qquad R = \mathscr{U}\mathscr{L}/\eta = \mu\sigma\mathscr{U}\mathscr{L},$$

and is known as "the magnetic Reynolds number"; a related quantity is

$$R^* = \mathscr{L}^2/\eta\mathscr{T} = \mu\sigma\mathscr{L}^2/\mathscr{T}, \tag{1.3.2}$$

where $\mathscr{T}$ is the time scale over which $\mathscr{U}$ changes.

If $R \ll 1$, the first terms on the right of (1.2.9) and (1.2.11) may, in a first approximation, be omitted in comparison with the second, leading to

$$\partial \boldsymbol{B}/\partial t = \eta\nabla^2\boldsymbol{B}, \qquad \boldsymbol{E} = \eta \operatorname{curl} \boldsymbol{B}, \qquad (R \to 0), \tag{1.3.3}$$

as for a stationary conductor. These define two important quantities:

$$\tau_\eta = \mathscr{L}^2/\eta = \mu\sigma\mathscr{L}^2, \qquad \lambda_\eta = (\eta\mathscr{T})^{1/2} = (\mathscr{T}/\mu\sigma)^{1/2}. \tag{1.3.4}$$

The first of these, the "electromagnetic decay time," determines the time scale on which the flux through an element of fluid of size $\mathscr{L}$ can change: in particular, it gives an estimate for the time of its decay in the absence of applied electromotive forces. For the Earth, taking $\sigma = 3 \cdot 10^5$ mho/m ($\eta = 3\ \mathrm{m}^2/\mathrm{s}$) and $\mathscr{L} = 3.5 \cdot 10^6$m (the radius of the core) we obtain $\tau_\eta \approx 3 \cdot 10^5$ years (cf. §1.1). The second of (4), the "electromagnetic skin-depth," determines the distance to which an applied field can penetrate a stationary conductor in time $\mathscr{T}$.

If $R \gg 1$, or equivalently $\mathscr{L}/\mathscr{U} \ll \tau_\eta$, the terms in η in (1.2.9) and (1.2.11) may, in a first approximation, be neglected except in electromagnetic boundary layers on S whose thickness is $O(\lambda_\eta^*)$, where $\lambda_\eta^* = \eta/\mathscr{U}$ is a skin-depth based on $\mathscr{U}$. We then have the case of the perfect conductor, for which

$$\partial \boldsymbol{B}/\partial t = \operatorname{curl}(\boldsymbol{u} \times \boldsymbol{B}), \qquad \boldsymbol{E} = -\boldsymbol{u} \times \boldsymbol{B}, \qquad (R \to \infty). \tag{1.3.5}$$

These equations, obtained by comparing the last terms in (1.2.9) and (1.2.11) with the second, are also valid when they are small compared with the first terms, i.e. when $R^* \gg 1$, or equivalently $\mathscr{T} \ll \tau_\eta$. When (5) holds, the magnetic flux tubes may be pictured as being attached to the material it threads (cf. many references, e.g. P. H. Roberts [**76**, §2.3]). (During an interval as short as $\mathscr{T}$ or $\mathscr{L}/\mathscr{U}$, the flux does not have the time, τ_η, required for significant change.) The theory of (5) is simpler than that of its parent equation and, moreover, the frozen field picture is a useful aid to thought. It should be emphasized however that the essence of the dynamo problem is to understand how the body concerned has been able to keep its flux over a time ($\mathscr{T}$) *long* compared with that (τ_η) over which it would naturally decay. The idea of perfect conductivity can therefore be used only with caution. It is nevertheless useful even for finite R^* and R, as we may illustrate by the simple example of two-dimensional duct flow.

The arrangement is depicted in Figure 1. A field $\boldsymbol{B}_0$ is applied in the z-direction across a rectangular duct of width $\mathscr{L}$ whose side walls are, say, fixed insulators. In the kinematic problem, a flow $\boldsymbol{u}(z)$ in the x-direction, down the duct, is *assigned*. The frozen field picture leads, quite correctly, to the idea that the induced magnetic field is in the x-direction, but might also incorrectly give the impression that no steady-state solution is possible, the lines of force being "paid out" continuously from the walls as the flow takes them downstream. The steady-state form of (1.2.9) shows, however, that B_x will equilibrate when

$$B_x \simeq (\mathscr{L}^2\zeta/\eta)B_0, \tag{1.3.6}$$

where ζ is a typical shear $[u'(z)]$ in the flow. Another way of interpreting (6) is to say that, in a finitely conducting fluid, a line of force tends to slip, relative to the conductor it threads, at the typical speed $\eta/\tilde{\rho}$, where $\tilde{\rho}$ is the local radius curvature of the line. In equilibrium, this must balance a characteristic speed, $\zeta\mathscr{L}$, of motion down the duct. (It may be remarked that in the laboratory, τ_η is so small, that B_x/B_0 and $\tilde{\rho}$ are minute.) In the magnetohydrodynamic problem, a constant pressure gradient, p_0', is applied in the x-direction and both B_x *and* u are to be determined. Although these are still related by (6) this fact is not immediately useful. The steady-state is primarily determined, in the absence of viscosity, by a dynamical balance between Lorentz force and pressure gradient, leading to

$$B_x \simeq \mu p_0'\mathscr{L}/B_0, \tag{1.3.7}$$

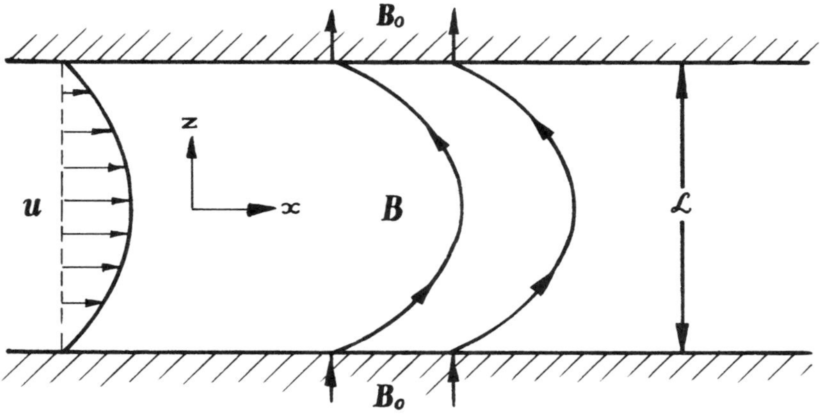

FIGURE 1

and *now* (6) shows that

(1.3.8) $$u \simeq p_0'/\sigma B_0^2,$$

(cf. e.g. P. H. Roberts [**76**, §7.3]).

The frozen field picture provides a useful qualitative sieve to divide the plausible kinematic dynamo motions from the implausible. For example, consider the cellular motion (G. O. Roberts [**75**])

(1.3.9) $$\boldsymbol{u} = (\cos y - \cos z, \sin z, \sin y),$$

whose streamlines are depicted by the yz-plane in Figure 2, the $\pm$ signs indicating maxima and minima of u_x. A field in the SW → NE direction is twisted in the clockwise (counterclockwise) sense, and also pushed below (lifted above) the plane of the paper, in the minus (plus) cells.

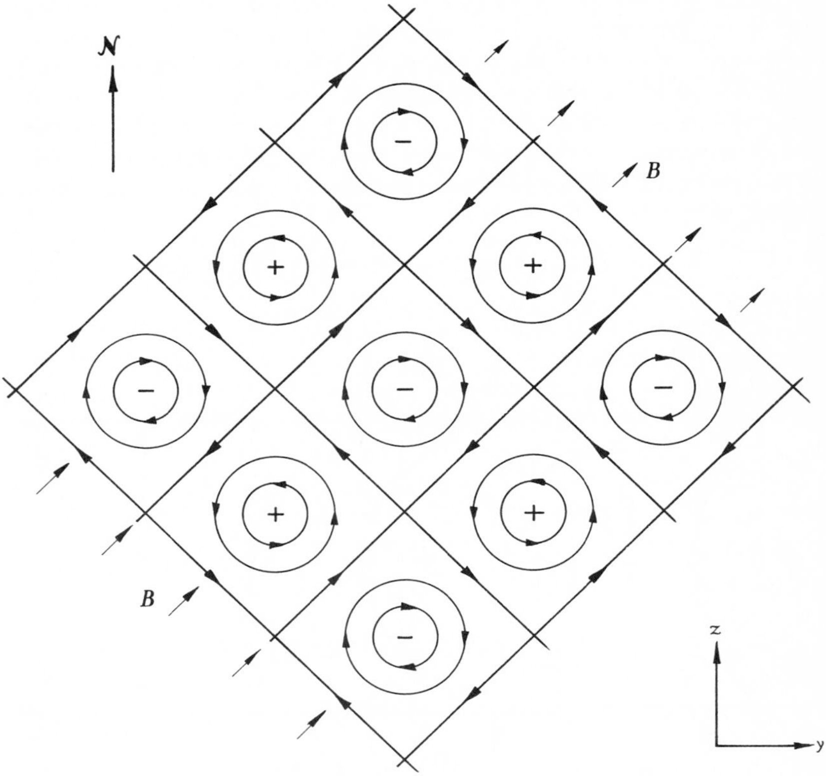

FIGURE 2

This tends to create a field in the NW → SE direction below the plane of the figure and in the SE → NW direction above the plane of the figure. This indicates fairly strongly that (9) is a candidate for producing a dynamo field predominantly of the form

$$\boldsymbol{B} = (0, \cos kx, \sin kx), \tag{1.3.10}$$

(together with other parts fluctuating on the scale of $\boldsymbol{u}$, cf. §3.3 below). Such indeed is the case. Had the sign in the x-component of (9) been different, the maximum and minimum of u_x would have occurred at stagnation points of the yz-motion, i.e. in positions much less favourable for imparting the helpful twists to the flux tubes, and it is not surprising that that motion is not such an efficient dynamo, although in fact it too turns out to be regenerative (cf. §3.3 below).

Finally, the frozen field idea also correctly indicates that self-sustaining dynamo action can occur only if the rate of stretching of flux tubes is sufficiently rapid to compensate for their diffusive drift relative to the fluid, i.e. it is necessary that a suitably defined magnetic Reynolds number is "sufficiently large." One such bound can be derived from the electromagnetic energy equation written in a form

$$\begin{aligned}\frac{\partial E}{\partial t} &\equiv \frac{\partial}{\partial t}\left[\frac{1}{2\mu}\int_{V+\hat{V}} \boldsymbol{B}^2\,dV\right]\\ &= -\frac{\eta}{\mu}\int_V (\operatorname{curl}\boldsymbol{B})^2\,dV + \frac{1}{\mu}\int_V \boldsymbol{u}\cdot(\boldsymbol{B}\times\operatorname{curl}\boldsymbol{B})\,dV,\end{aligned} \tag{1.3.11}$$

which can be readily deduced from $(1.2.\hat{1}, \hat{2})$ and (1.2.7–9). Thus, if $u_{\max}$ is the largest fluid velocity in V, we have

$$\begin{aligned}\mu\frac{\partial E}{\partial t} &\leqq -\eta\int_V (\operatorname{curl}\boldsymbol{B})^2\,dV + u_{\max}\int_V |\boldsymbol{B}|\,|\operatorname{curl}\boldsymbol{B}|\,dV\\ &\leqq -\eta\int_V (\operatorname{curl}\boldsymbol{B})^2\,dV + u_{\max}\left[\int_V \boldsymbol{B}^2\,dV \int_V (\operatorname{curl}\boldsymbol{B})^2\,dV\right]^{1/2}\\ &\leqq -\eta\int_V (\operatorname{curl}\boldsymbol{B})^2\,dV + u_{\max}\left[2\mu E\int_V (\operatorname{curl}\boldsymbol{B})^2\,dV\right]^{1/2}.\end{aligned} \tag{1.3.12}$$

Now, in the absence of fluid motions, any unmaintained field must decay to zero and, in fact, a lower bound, $k^2\eta/\mathscr{L}$ (say), for its decay rate can be established. (For example, for a sphere of radius $\mathscr{L}$, it follows easily from the formalism of §1.4 below that $k = \pi$.) And it is readily shown that any continuously differentiable field $\boldsymbol{B}$ which obeys $(1.2.\hat{1}, \hat{2})$ and (1.2.7–8), though not necessarily the induction equation (1.2.9), satisfies the inequality

$$\int_V (\text{curl } \boldsymbol{B})^2\, dV \geqq \frac{k^2}{\mathscr{L}^2} \int_{V+\hat{V}} \boldsymbol{B}^2\, dV. \tag{1.3.13}$$

On applying (13) to (12), it is seen that

$$\frac{\partial E}{\partial t} \leqq -\frac{1}{\mu}\left(\eta - \frac{u_{\max}\mathscr{L}}{k}\right) \int_V (\text{curl } \boldsymbol{B})^2\, dV. \tag{1.3.14}$$

Thus, for dynamo action, it is necessary that the Reynolds number,

$$R_1 \equiv u_{\max}\mathscr{L}/\eta, \tag{1.3.15}$$

based on the maximum fluid velocity $u_{\max}$ should exceed k. This result is due to Childress [**23**], who also observed that, by (12), $\partial E/\partial t \leqq u_{\max^2} E/2\eta$, a relation which, like (19) below, places an upper bound on the growth rate of field energy.

The inequality just derived would not, for example, rule out a fluid motion consisting of a solid body rotation, even though we know from the rotational invariance of Maxwell's equations that the field in such a case would decay at the same rate as it would in a stationary conductor. Another necessary condition which complements this inequality, and which does not suffer from this disadvantage, was derived earlier by Backus [**2**] (see also P. H. Roberts [**76**, §3.2]), by writing the energy equation (11) in the alternative form

$$\frac{\partial E}{\partial t} = -\frac{\eta}{\mu} \int_V (\text{curl } \boldsymbol{B})^2\, dV + \frac{1}{\mu} \int_V B_i B_j e_{ij}\, dV, \tag{1.3.16}$$

where

$$e_{ij} = \tfrac{1}{2}(\partial u_i/\partial x_j + \partial u_j/\partial x_i) \tag{1.3.17}$$

is the rate of strain tensor. By (1.2.6), the trace of $e_{ij}(\boldsymbol{x})$ is zero, and its largest eigenvalue, $\lambda(\boldsymbol{x})$, is therefore nonnegative. If Λ denotes the upper bound of λ in V, (16) now gives

(1.3.18)

$$\mu \frac{\partial E}{\partial t} \leqq -\eta \int_V (\text{curl } \boldsymbol{B})^2\, dV + \Lambda \int_V \boldsymbol{B}^2\, dV \leqq -\eta \int_V (\text{curl } \boldsymbol{B})^2\, dV + 2\mu\Lambda E.$$

On applying (13) to (18), it is seen that

$$\partial E/\partial t \leqq -2(\eta k^2/\mathscr{L}^2 - \Lambda)E. \tag{1.3.19}$$

Thus, for dynamo action, it is necessary that the Reynolds number,

$$R_2 \equiv \Lambda \mathscr{L}^2/\eta, \tag{1.3.20}$$

based on the maximum rate of shear, Λ, should exceed k^2.

Neither of these conditions on the Reynolds numbers are sufficient. In fact, by a historically early result of the subject, neither axially symmetric nor two-dimensional magnetic fields can be self-maintained, nor indeed can their decay time be substantially lengthened no matter how large R_1 and R_2 may be. (See discussion beneath (1.4.19) below; see also Backus [**1**]. By "two-dimensional" we mean fields whose components are independent of one cartesian coordinate.) Similarly, a toroidal flow (§1.4) cannot maintain a field no matter how vigourous or highly sheared it may be.

1.4. *Spherical dynamos.* The theory of dynamos in spherical conductors is of obvious relevance to solar, stellar and planetary magnetism; it has been systematized by Bullard and Gellman [**18**] in terms of a representation due to Lamb: A divergenceless vector, such as $\boldsymbol{B}$ or $\boldsymbol{u}$, may be written as

$$\boldsymbol{B} = \text{curl } T\boldsymbol{r} + \text{curl}^2 S\boldsymbol{r}, \tag{1.4.1}$$

where $\boldsymbol{r}$ is the radius vector from the centre, O, of the sphere, V (radius a, say). The vectors,

$$\boldsymbol{T} = \text{curl } T\boldsymbol{r} = -\boldsymbol{r} \times \nabla T, \qquad \boldsymbol{S} = \text{curl}^2 S\boldsymbol{r} = \nabla\left[\frac{\partial}{\partial r}(rS)\right] - \boldsymbol{r}\nabla^2 S, \tag{1.4.2}$$

are called the toroidal and poloidal parts of $\boldsymbol{B}$. In terms of spherical (r, θ, ϕ) coordinates, they are

$$\begin{aligned} \boldsymbol{T} &= \left[0, \frac{1}{\sin\theta}\frac{\partial T}{\partial\phi}, -\frac{\partial T}{\partial\theta}\right], \\ \boldsymbol{S} &= \left[\frac{L^2 S}{r}, \frac{1}{r}\frac{\partial}{\partial\theta}\left\{\frac{\partial}{\partial r}(rS)\right\}, \frac{1}{r\sin\theta}\frac{\partial}{\partial\phi}\left\{\frac{\partial}{\partial r}(rS)\right\}\right], \end{aligned} \tag{1.4.3}$$

where

$$L^2 = r^2\frac{\partial^2}{\partial r^2} + 2r\frac{\partial}{\partial r} - r^2\nabla^2 = -\left[\frac{1}{\sin\theta}\frac{\partial}{\partial\theta}\left(\sin\theta\frac{\partial}{\partial\theta}\right) + \frac{1}{\sin^2\theta}\frac{\partial^2}{\partial\phi^2}\right]. \tag{1.4.4}$$

Arbitrary functions of r can be added to T and S without affecting $\boldsymbol{T}$ and $\boldsymbol{S}$. This ambiguity is usually eliminated by subtracting from T and S their averages over θ and ϕ. It is clear that the radial component of a toroidal vector vanishes everywhere. If $\boldsymbol{B}$ vanishes, so do the defining scalars T and S.

The suitability of the division (1) arises from three main factors: (i) by (1), (2) and Ampère's law (1.2.1), the electric current is given by

(1.4.5) $$\mu \boldsymbol{j} = \text{curl}\, \boldsymbol{B} = \text{curl}(-\boldsymbol{r}\nabla^2 S) + \text{curl}^2\, T\boldsymbol{r},$$

i.e. the poloidal field is generated by a toroidal current *and* the toroidal field is produced by a poloidal current. By (1.2.$\hat{1}$) and (5),

(1.4.6) $$\hat{T} = \nabla^2 \hat{S} = 0,$$

in the insulator, $\hat{V}$, surrounding V. By the first of (6), the toroidal field is trapped within V. From (2) and (6), $\hat{\boldsymbol{B}}$ is the gradient of the scalar $\partial(r\hat{S})/\partial r$, which according to the second of (6) is harmonic [$r^2\nabla^2$ commutes with $\partial(r\cdot\ \ )/\partial r$]; and (ii) potential theory allows us to express that scalar in a form equivalent to

(1.4.7) $$\hat{S} = \sum_{n=1}^{\infty} \hat{S}_n(t)\left(\frac{a}{r}\right)^{n+1} Y_n(\theta, \phi),$$

which immediately incorporates the requirement (1.2.8) of no sources at ∞. Here Y_n denotes a surface harmonic of degree n. Finally (iii), the expansion of T and S in V in similar harmonic forms,

(1.4.8) $$T = \sum_{n=1}^{\infty} T_n(r, t) Y_n(\theta, \phi), \qquad S = \sum_{n=1}^{\infty} S_n(r, t) Y_n(\theta, \phi),$$

not only leads to direct simplifications, stemming from

(1.4.9) $$L^2 Y_n = -n(n+1) Y_n,$$

but also from (6) and (7) allows the conditions (1.2.7) to be written neatly as

(1.4.10) $$T_n = \frac{\partial S_n}{\partial r} + \frac{(n+1)}{r} S_n = 0, \quad \text{on } r = a.$$

Further properties of the expansion will be found in P. H. Roberts [**76**, §3.3b].

The difficulties of the scheme become apparent when $\boldsymbol{u}$ is expanded in similar form

(1.4.11) $$\boldsymbol{u} = \text{curl}\, \tilde{T}\boldsymbol{r} + \text{curl}^2\, \tilde{S}\boldsymbol{r},$$

and (1), (8), and (11) are substituted into the induction equation (1.2.9). Even if a simple $\boldsymbol{u}$ is chosen, its interactions with $\boldsymbol{B}$ in general excites an infinite set of the harmonics (T_n, S_n) of $\boldsymbol{B}$. It is necessary to truncate (8) after N terms and hope that numerical integrations for different N will give strong signs of convergence for $N \to \infty$. Bullard and Gellman [**18**], concentrating their search for steady $\boldsymbol{B}$ on motions of the type (in units scaled so that $\eta = 1/R$ and $a = 1$)

$$(1.4.12) \quad \tilde{S} = Rr^2(1-r)^2 P_2^2(\cos\theta)\cos 2\phi, \qquad \tilde{T} = \varepsilon Rr(1-r)P_1(\cos\theta),$$

found no compelling evidence of convergence either for $\varepsilon = 5$ or $\varepsilon = 10$. Gibson and P. H. Roberts [**32**] carried the $\varepsilon = 5$ case to larger N and found R_N increased systematically, a conclusion confirmed in general terms by Lilley [**52**], although the detailed agreement between his values of R_N and those of Gibson and P. H. Roberts was not very satisfactory. Prompted by insight provided by Braginskiĭ (see §4 below), Lilley also examined the motion

(1.4.13)

$$\tilde{S} = Rr^2 P_2^2(\cos\theta)[(1-r^2)^2\cos 2\phi + 1.6(1-4r^2)^2 H(\tfrac{1}{2}-r)\sin 2\phi],$$

$$\tilde{T} = 10Rr(1-r^2)P_1(\cos\theta),$$

(where H is the Heaviside unit function) and found fairly convincing indications of the existence of an eigenvalue near $R = 20$ (although, see general remarks about the oscillatory behaviour of the $\alpha\omega$ dynamos in §3.4 below).

Bullard and Gellman were able, from analytic arguments, to prove that a purely toroidal motion ($\tilde{S} \equiv 0$) cannot maintain a steady dynamo field (cf. P. H. Roberts [**76**, §3.3c]). Nevertheless it is very important to realize in what follows that, if any independent mechanism can be found to maintain the poloidal field, a toroidal field can easily be created from it by toroidal motions. Examples are given in Figure 3: these show how a simple differential rotation [$\tilde{T} \propto r^2 P_1(\cos\theta)$] would distort (a) an axial dipole, and (b) an axial quadrupole. The dotted lines indicate the undisturbed field lines, on their respective meridian planes, and the full lines illustrate their distortion, expected on the frozen field picture, from the shear. If the original poloidal multipoles are now subtracted, the toroidal remainder has the structure indicated in Figures 4. In 4(a), $T \propto P_2(\cos\theta)$; in 4(b), it is a combination of terms proportional to $P_1(\cos\theta)$ and $P_3(\cos\theta)$.

Another way of formalizing these arguments for axisymmetric fields is to replace (1) by the representation,

$$(1.4.14) \qquad \boldsymbol{B} = \left[-\frac{\partial A}{\partial z}, B, \frac{1}{\varpi}\frac{\partial}{\partial\varpi}(\varpi A)\right],$$

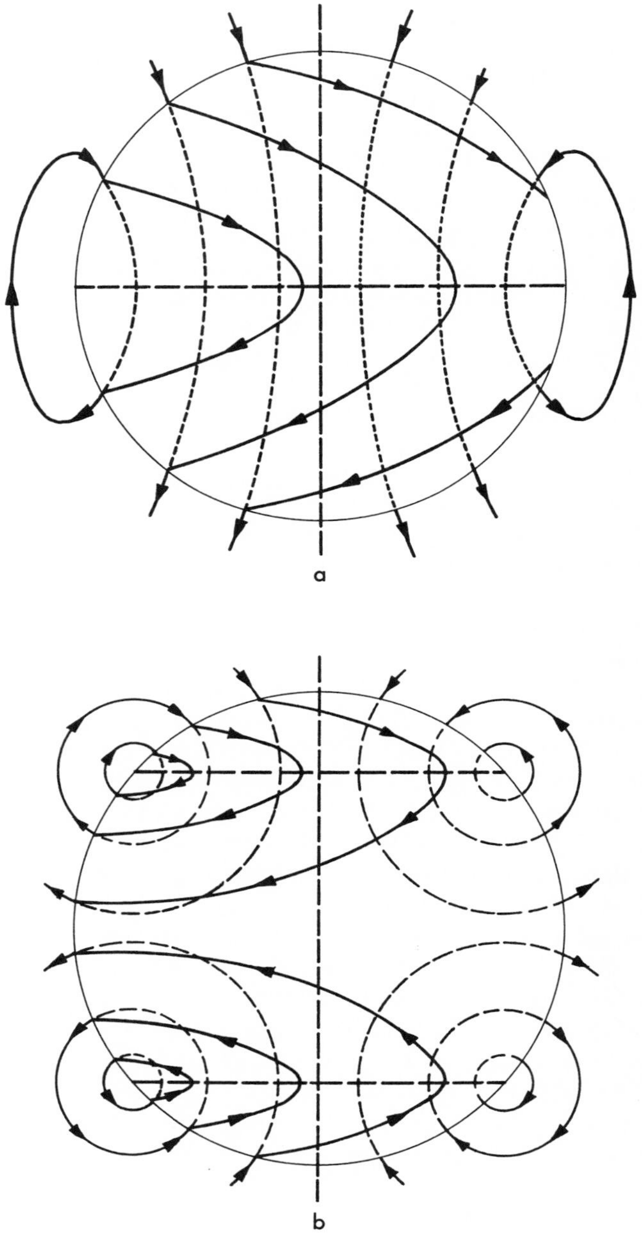

FIGURE 3

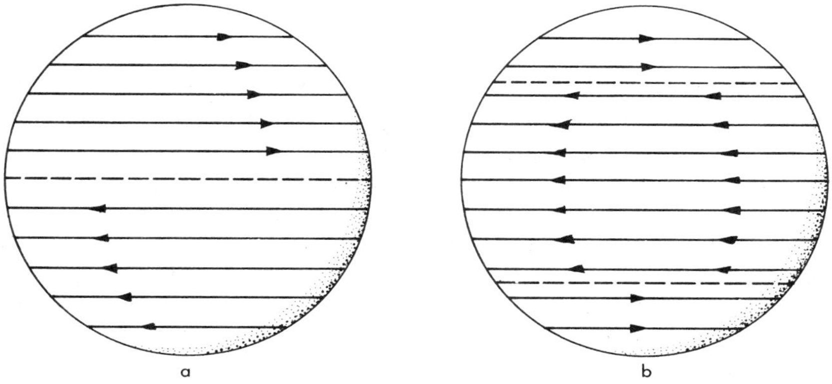

FIGURE 4

in cylindrical coordinates (ϖ, ϕ, z), where A, B and the components of $\boldsymbol{u}$ are independent of ϕ. On substituting into the induction equation (1.2.9), it is found that

$$\partial A/\partial t + (\boldsymbol{u}/\varpi)\cdot\nabla(\varpi A) = \eta\Delta A, \tag{1.4.15}$$

$$\partial B/\partial t + \varpi\boldsymbol{u}\cdot\nabla(B/\varpi) = \eta\Delta B + [\nabla(u_\phi/\varpi)\times\nabla(\varpi A)]_\phi, \tag{1.4.16}$$

where $\Delta = \nabla^2 - \varpi^{-2}$. These equations must be solved subject to

$$\langle A\rangle = \langle\partial A/\partial r\rangle = \langle B\rangle = 0, \quad \text{on } r = a, \tag{1.4.17}$$

where

$$\Delta\hat{A} = \hat{B} = 0; \qquad \hat{A} = O(r^{-2}), \quad r\to\infty. \tag{1.4.18}$$

The last term of (16) represents the creation of toroidal field (B) from the poloidal field, $\boldsymbol{B}_M$ [viz. that associated with A in (14)]. In agreement with (1.3.6) it indicates that, in a steady state,

$$B \sim (\mathscr{L}u_\phi/\eta)|\boldsymbol{B}_M|. \tag{1.4.19}$$

For example, for the Earth, if we take $|\boldsymbol{B}_M| \simeq 5$ gauss (on the core surface) and $u_\phi \doteqdot 10^{-4}$m/s (as inferred from the westward drift), we obtain $B \doteqdot 500$ gauss. Such a field would not, of course, be directly observable at the Earth's surface by magnetic measurements.

One might also argue, again from the frozen field picture, that it should be possible to create poloidal field by "pushing upwards" the axisymmetric toroidal field lines by a poloidal motion. One sees at once that to succeed such a motion would need to be nonaxisymmetric; for, by (15), the axisymmetric field, $\boldsymbol{B}_M$, has no source term [analogous to the last term of (16)] to regenerate it from B. Without such a source, the lines of poloidal field would "collapse inwards" to disappear, leaving the toroidal field, then also without support, to follow suit. [This result, that axisymmetric fields cannot be steadily self-maintained, has already been mentioned in §1.3. It was to avoid this theorem that Bullard and Gellman chose the complicated flow (12).] There is certainly no reason why locally an asymmetric poloidal flow should not counter the diffusive inward drift of field by an upward motion, but it is also necessary that it should balance their enhanced downward motion required elsewhere to satisfy continuity (1.2.6). The situation is, then, not clearcut. Summarizing, it is perfectly possible to create toroidal field from poloidal but there is, at this stage of the argument, no obvious process to do the reverse.

Before concluding this section, some further remarks on numerical integrations of the dynamo equations should be made. The beauty of the spherical harmonic representation of the dynamo equation lies principally in the simple form of the conditions (10) expressing the absence (1.2.8) of sources at infinity. In some other respects, particularly that of looking at the hydromagnetic dynamo at large Hartmann and Taylor numbers (§5), it is rather inconvenient. Thirlby (unpublished) has devised an alternative method which uses a rectangular (ϖ, z) grid which covers not only the interior of V but also the set of external points at which field values are required for the operations of vector calculus within V. These values (which represent the analytic extension of $\boldsymbol{B}$ rather than the field $\hat{\boldsymbol{B}}$ at the grid points) are obtained by a matrix multiplication on the interior values which automatically incorporates (1.2.8). The method has been tested on the stationary conductor, and the free decay modes have been recovered satisfactorily: the results obtained from Lilley's motion (13) also seem promising.

2. **Simple models.**

2.1. *The helical dynamo of Lortz.* The following dynamo functions in a conductor of infinite spatial extent: Let (ϖ, ϕ, z) be cylindrical coordinates, and introduce

$$\xi = m\phi + kz, \qquad q = 1/(m^2 + k^2\varpi^2), \tag{2.1.1}$$

where k and m, an integer, are constant. Define $\boldsymbol{w}$ by

$$\boldsymbol{w} = (0, -k\varpi q, mq), \tag{2.1.2}$$

which, it may be noted, satisfies

$$\text{(2.1.3)} \qquad \operatorname{div} \boldsymbol{w} = 0, \qquad \operatorname{curl} \boldsymbol{w} = -2kmq\boldsymbol{w}.$$

The field lines of $\boldsymbol{w}$ are helices, and the dynamo below is therefore called "helical." We seek steady dynamo fields of the form

$$\text{(2.1.4)} \qquad \boldsymbol{B} = h\boldsymbol{w} + \operatorname{grad} H \times \boldsymbol{w},$$

where h and H are functions of ξ and ϖ alone. Since the gradient of any such function is perpendicular to $\boldsymbol{w}$, (4) is divergenceless. If $m = 0$, the field (4) is axisymmetric; if $k = 0$, it is two dimensional: both possibilities are ruled out as potential dynamos by an early theorem of the subject (see §1.3). The electric current is given by a form similar to (4), viz.

$$\text{(2.1.5)} \qquad \mu \boldsymbol{j} = \operatorname{curl} \boldsymbol{B} = J\boldsymbol{w} - \operatorname{grad} h \times \boldsymbol{w},$$

where

$$\text{(2.1.6)} \qquad J = -q(2kmh - LH),$$

$$\text{(2.1.7)} \qquad L = \frac{1}{\varpi}\frac{\partial}{\partial \varpi}\left[q\varpi \frac{\partial}{\partial \varpi}\right] + \frac{1}{\varpi^2}\frac{\partial^2}{\partial \xi^2}.$$

Let us examine the possibility that (4) is supported by the flow

$$\text{(2.1.8)} \qquad \boldsymbol{u} = v\boldsymbol{w} + \operatorname{grad} V \times \boldsymbol{w}.$$

Substituting (4), (5) and (8) into the steady induction equation (1.2.9), we obtain

$$\text{(2.1.9)} \qquad \partial(V, H)/\partial(\varpi, \xi) = \eta\varpi[q^{-1}LH - 2kmqh],$$

$$\text{(2.1.10)} \qquad \partial(V, qh)/\partial(\varpi, \xi) + \partial(qv, H)/\partial(\varpi, \xi) = \eta\varpi Lh.$$

If one assumed

$$\text{(2.1.11)} \qquad v = v_1(\varpi)\sin\xi + v_2(\varpi)\sin 2\xi, \qquad V = V_1(\varpi)\sin\xi,$$

one would, a priori, expect to find that (9) and (10) required h and H to possess all terms in their Fourier cosine expansions in ξ. By careful choice of v_1, v_2 and V_1 however, it is possible to obtain an *exact* solution of the form

$$\text{(2.1.12)} \qquad h = h_0(\varpi) + h_1(\varpi)\cos\xi, \qquad H = f_0'(\varpi).$$

In fact, on substituting (11) and (12) into (9) and (10), v_1, v_2, V_1, h_0 and h_1 can be expressed *directly* in terms of f_0. Moreover, f_0 can be chosen in such a way that the solutions are self-excited and physically sensible both at $\varpi = 0$ and $\varpi = \infty$.

This ingenious example, due to Lortz [**53**], is the simplest case of a self-exciting homogeneous dynamo known at the present time.

2.2. *The vortex dynamos of Tverskoy and Gailitis.* Suppose V is unbounded, and let $(\varpi, z) = (c, 0)$ be the circular axis of a torus whose cross-sectional radius is a ($\ll c$). Introduce displaced polar coordinates (T, χ, ϕ) as illustrated in xz-projection in Figure 5, the axis intersecting this plane at C and C'. The surface, S, of the torus is now $T = a$. We first consider flows which are nonzero only in the ring, concentrating in fact on the incompressible motion whose components in displaced polars are

$$(2.2.1) \qquad \boldsymbol{u} = [0, u, 0], \qquad \text{where} \quad u = cv(T)/(c - T\cos\chi) \simeq v(T).$$

We will depart from the notation of §1.2 by using ˆ for the exterior of the ring, even though here this region is not an insulator but a conductor of the same resistivity as the interior of the ring (a region distinguished by the absence of a ˆ). We will temporarily suspend condition (1.2.8) which excludes sources at infinity, supposing that the ring lies in a field, $\boldsymbol{B}_E$, which, though nonuniform, does not vary appreciably over lengths of the order of a. Condition (1.2.8) is nevertheless still applicable to $\boldsymbol{B}_I$, the field induced from $\boldsymbol{B}_E$ by the motion. We will first consider the case of steady fields and motions. The argument is simplest when it is supposed

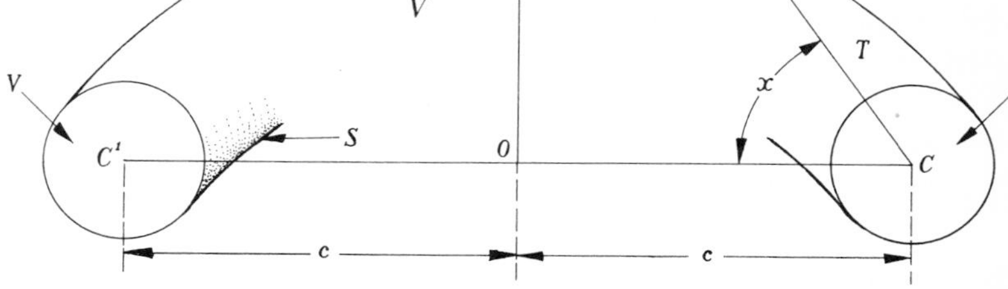

FIGURE 5

hat the magnetic Reynolds number,

2.2.2) $$R_m \equiv av(a)/\eta,$$

s large, but similar results hold provided that R, defined by (10) below, s large.

Consider a section, $\mathscr{S}$, of the ring between the planes ϕ and $\phi + \delta\phi$. The field $\boldsymbol{B}_E$ at $\mathscr{S}$ may be divided into two parts $\boldsymbol{B}_{E1}$ and $\boldsymbol{B}_{E2}$ which are espectively axisymmetric and asymmetric with respect to χ. Since a is small, $\boldsymbol{B}_{E2}$ does not vary rapidly in space over $\mathscr{S}$ but, in a frame moving with some surface element δS of $\mathscr{S}$, it will appear to vary rapidly in time with a frequency of $\omega = v(a)/a$, in fact), and will therefore penetrate beneath S only to a distance of [cf. equation (1.3.4)] order $\lambda_\eta = (\eta/\omega)^{1/2} \simeq aR_m^{-1/2}$ ($\ll a$), i.e. the interior of the ring will be almost completely shielded, electromagnetically, from $\boldsymbol{B}_{E2}$. In other words, if $\boldsymbol{B}_{I2}$ is the field induced by the motion from $\boldsymbol{B}_{E2}$, we must have (to leading order)

$$\boldsymbol{n} \cdot (\boldsymbol{B}_{E2} + \boldsymbol{B}_{I2})_T = 0, \quad \text{on } r = a.$$

This conclusion may also be obtained more formally by a boundary-layer argument: cf. e.g. P. H. Roberts [**77**, pp. 92–95].) Since $\boldsymbol{B}_{E2}$ is known, 3) provides a boundary condition on $\boldsymbol{B}_{I2}$ which, together with (1.2.8) and the equations

2.2.4) $$\nabla^2 \hat{\boldsymbol{B}} = 0, \qquad \operatorname{div} \hat{\boldsymbol{B}} = 0,$$

determine $\boldsymbol{B}_{I2}$ completely. It is clear from the form of (3) that the answer obtained will not depend on R_m in the limit $R_m \to \infty$. Although such an increase in v intensifies the skin currents, it also decreases λ_η, the net "surface" current being independent of R_m in the limit. In contrast, the field $\boldsymbol{B}_{I1}$ induced from $\boldsymbol{B}_{E1}$ increases linearly with R_m as $R_m \to \infty$. t follows that, to leading order, we may concentrate on $\boldsymbol{B}_{E1}$, and omit he suffix 1, as we will now do.

The axisymmetric field $\boldsymbol{B}_E$ is not subject to the skin depth argument given above, for, in the frame moving with δS, it will appear to be constant n time. It will therefore penetrate $\mathscr{S}$ completely. In general, $B_{E\phi}$ will depend on ϕ and, for definiteness, we will consider only one Fourier component, viz.

2.2.5) $$B_{E\phi} = \frac{(c - T\cos\chi)}{c} B_0\, e^{im\phi} \simeq B_0\, e^{im\phi},$$

where B_0 is a constant and m is an integer, and we have recognized that $B_{E\phi}$ will scarcely vary across $\mathscr{S}$. Since $B_{E\phi}$ varies with ϕ, flux conservation 1.2.2) requires B_{ET} to vary with T. In fact, according to (5) and (1.2.2),

2.2.6) $$B_{ET} \simeq -(imT/2c)B_0\, e^{im\phi}.$$

It is clear, from the frozen field picture, that the motion (1) will create from B_{ET} a large $B_{I\chi}$ component (and an associated current $j_{I\phi}$). Since, however, this field is parallel to $\boldsymbol{u}$ it has no further inductive effect, and we may compute $\hat{\boldsymbol{B}}_I$ from the Biot-Savart law

$$\hat{\boldsymbol{B}}_I(\hat{\boldsymbol{r}}) = \frac{1}{4\pi\eta}\int_V \frac{[\boldsymbol{u}(\boldsymbol{r}) \times \boldsymbol{B}_E(\boldsymbol{r})] \times (\hat{\boldsymbol{r}} - \boldsymbol{r})}{|\hat{\boldsymbol{r}} - \boldsymbol{r}|^3}\, dV. \tag{2.2.7}$$

Such a computation has been performed by Gailitis [**28**] who obtained, using (5) and (6),

$$\hat{B}_{I\phi}(c, \phi, z_0) = \frac{B_0\, e^{im\phi}}{\eta c^2 K_m}\int_0^a v(T)T^2\, dT, \tag{2.2.8}$$

where the arguments on the left refer to cylindrical coordinates (ϖ, ϕ, z), and $K_m = K_m(z_0/c)$ is given by

(2.2.9)

$$\frac{1}{K_m} = \tfrac{1}{2}mc^2 z_0\left[\int_0^{2\pi} \frac{\sin\phi \sin m\phi}{1\hat{\boldsymbol{r}} - \boldsymbol{r}|^3}\, d\phi\right]_{\varpi=\hat{\varpi}=c;\ z-\hat{z}=z_0}$$

$$= \frac{\pi^{1/2} m}{(m-1)!}\left(\frac{k}{2}\right)^{2m}(1 - k^2)^{1/2}\Gamma(m + \tfrac{1}{2})\ {}_2F_1(m + \tfrac{1}{2}, m + \tfrac{1}{2}; 2m + 1; k^2)$$

where $k = [1 + (z_0/2c)^2]^{-1/2}$, and ${}_2F_1$ denotes the hypergeometric function.

Now consider two identical coaxial vortex rings with identical (though opposite) internal motions, their exteriors being (as before) of equal resistivity and at rest; see Figure 6. The field (8) created by each can be the exciting field for the other, and a steady dynamo will clearly result if the magnetic Reynolds number,

$$R \equiv \frac{1}{\eta}\int_0^a \left(\frac{T}{a}\right)^2 v(T)\, dT, \tag{2.2.10}$$

is sufficiently large; if, in fact [using (8)],

$$R = (c/a)^2 K_m. \tag{2.2.11}$$

Gailitis concludes that the $m = 1$ mode is the easiest to excite. Although the motion (1) is axisymmetric, the field is not; indeed the analysis breaks down in the case $m = 0$.

The Gailitis dynamo is not the first example of an axisymmetric dynamo, i.e. a dynamo in which the *motions* are axisymmetric. That honour appears to be due to the Tverskoy dynamo, which involves induction by a single sporadic vortex ring in a conducting sphere

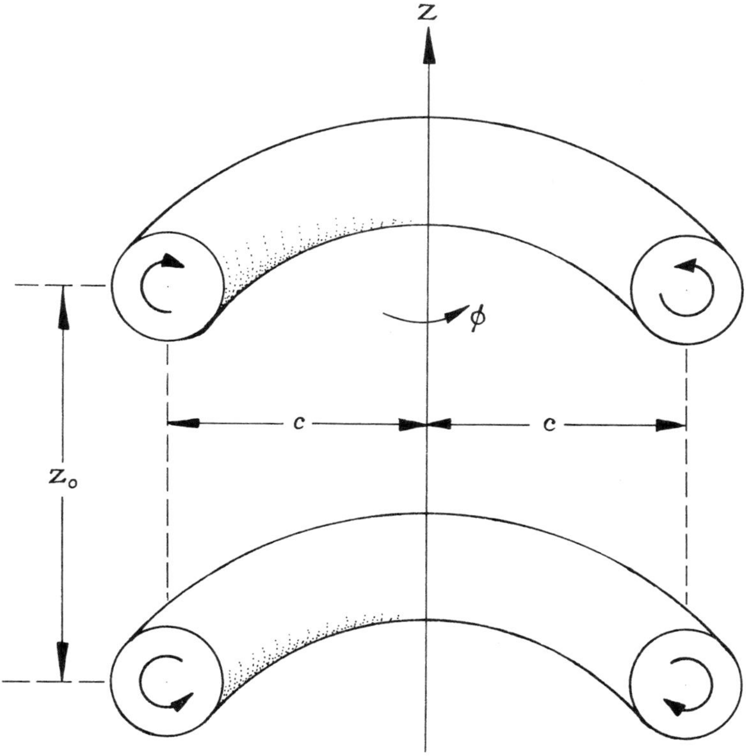

FIGURE 6. The Gailitis dynamo

surrounded by insulator (Tverskoy [**95**]). The basic idea, that of separating the effects of motion and diffusion in the induction process in a moving conductor, was originally suggested by Bade [**4**] and Parker [**60**], and was used with great effect by Backus [**2**] in one of the first existence proofs of the subject. It makes use of two facts. First, (A) if a motion is "sufficiently rapid" it is possible over a "sufficiently short time interval" to neglect resistivity almost entirely in following the evolution of the field: in other words, the frozen field equations can be applied. Second, (B) if all motion is stopped, the resulting ohmic diffusion quenches the fields of small-scale more rapidly than those of large-scale; for example, in a sphere, the decay time for the components of harmonic number, n, decreases as n^{-2}, for $n \to \infty$. (Cf. equation (1.3.4) and e.g. P. H. Roberts [**76**, pp. 97–98].) The "jerky dynamo" consists of an infinite alternation of identical periods of A and B. Even if the field consists initially of low harmonics (e.g., $n = 1$), it will acquire, during an episode of A, too great a degree of spatial complexity to be amenable to simple theoretical treatment. After a "sufficiently long period" of B, however, it will be virtually reduced again to the (more easily dealt with) harmonics $n = 1$. Naturally, the amplitude of these harmonics will also have suffered during this period of stasis, but this may be counteracted by a further "sufficiently vigourous" episode of A, and so on. This dynamo is, of course, not stationary, but it may be considered to be self-excited if, even when no sources exist at infinity, the field is periodic in time over the basic AB periodicity of the motion. Tverskoy applied this method to a single vortex situated symmetrically in a conducting sphere as illustrated in Figure 7 and found conditions under which the $m \neq 0$ modes could be self-maintained.

Before leaving the topic of axisymmetric dynamos, we should remark that we would not expect such models to be very realistic since, unless rather artificial body forces are assumed, the Lorentz forces of the fields would destroy the axisymmetry of the motions in any hydromagnetic models. We note in passing that a preliminary report has appeared [**79**] of an incomplete numerical study by G. O. Roberts of an axisymmetric kinematic dynamo in a sphere. This, unlike the Gailitis and Tverskoy dynamos, requires a ϕ-component of velocity, but the flow is not localized.

2.3. *The eddy dynamos of Herzenberg and Kropachev.* There is a parallel between the theory underlying this section and that described in the last. Since, however, the subject is perhaps less topical, and since it has, in any case, been discussed several times in recent literature (e.g. P. H. Roberts [**76**, pp. 95–104]; Gibson [**29**], [**30**], [**31**]), we will not attempt to give a full account here. The starting point is, again, an unbounded conductor, stationary apart from a region (here a sphere of radius a) which rotates with an angular velocity, $\boldsymbol{\omega}$, where

$$R_m \equiv a^2\omega/\eta \tag{2.3.1}$$

is large. Again it is convenient to consider the effect of this motion on an externally applied field, $\boldsymbol{B}_E$, whose spatial scale greatly exceeds a, and to divide this field into axisymmetric and asymmetric parts, $\boldsymbol{B}_{E1}$ and $\boldsymbol{B}_{E2}$ respectively, with respect to the direction of $\boldsymbol{\omega}$. Again the corresponding induced fields $\boldsymbol{B}_{I1}$ and $\boldsymbol{B}_{I2}$ are asymptotically proportional to R_m and 1, as $R_m \to \infty$, and in the first instance we may therefore again confine attention to the axisymmetric fields, ignoring the distinction between $\boldsymbol{B}_I$ and $\boldsymbol{B}_{I1}$.

We may expand $\boldsymbol{B}_E$ in a Taylor series about the centre, O, of the rotor, denoting by a superfix (0) a value taken at O itself. If the length scale

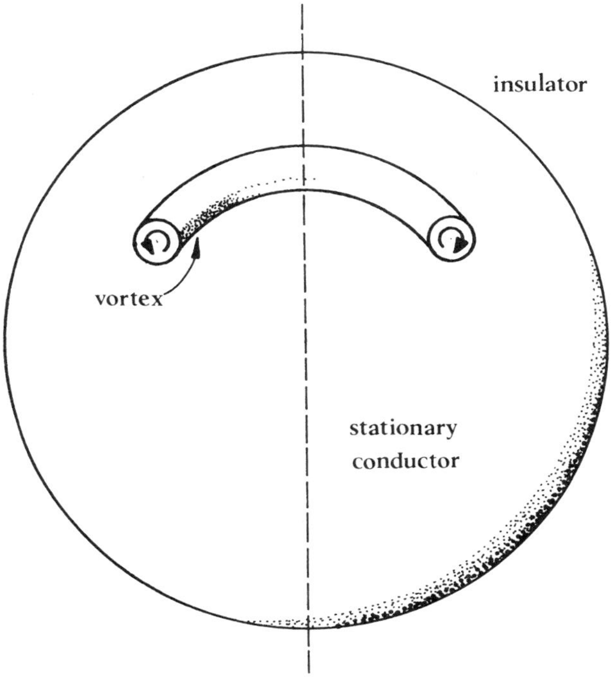

FIGURE 7. The Tverskoy dynamo

of $\boldsymbol{B}_E$ is $\mathscr{L}$ ($\gg a$), the $(n+1)$th term of this series will be of order $(r/\mathscr{L})^n \cdot |\boldsymbol{B}_E^{(0)}|$. After extracting the symmetric part from it, the induced field it creates is found to be of order $R_m(a/r)^{n+1}(a/\mathscr{L})^n|\boldsymbol{B}_E^{(0)}|$ for large r, except for the first ($n = 0$) term which is of order $R_m(a/r)^3|\boldsymbol{B}_E^{(0)}|$. Thus, for example, at a distance r of order $\mathscr{L}$, the $n = 0$ and $n = 1$ fields are both of order $R_m(a/\mathscr{L})^3|\boldsymbol{B}_E^{(0)}|$ while the remainder are (in a first approximation) negligibly smaller. Then, if $R_m(a/\mathscr{L})^3 = O(1)$, the fields at distance $\mathscr{L}$ are $O(\boldsymbol{B}_E^{(0)})$ and are given by

$$(2.3.2) \qquad \boldsymbol{B}_I = -\tfrac{1}{5}R_m(\hat{\boldsymbol{\omega}} \times \hat{\boldsymbol{r}})\left[(\hat{\boldsymbol{\omega}} \cdot \boldsymbol{B}_E^{(0)})(\hat{\boldsymbol{\omega}} \cdot \hat{\boldsymbol{r}})\left(\frac{a}{r}\right)^3 - \tfrac{1}{3}a(\hat{\boldsymbol{\omega}} \cdot \nabla \boldsymbol{B}_E^{(0)} \cdot \hat{\boldsymbol{\omega}})\left(\frac{a}{r}\right)^2\right],$$

a result obtained from the $n = 0$ and $n = 1$ harmonics alone. (Here $\hat{}$ signifies the unit vector.)

Now consider two identical rotors separated by a distance R ($\gg a$), where $R_m(a/R)^3 = O(1)$, whose exteriors are (as before) of equal resistivity and at rest; Figure 8 shows this "Herzenberg dynamo." The field (2)

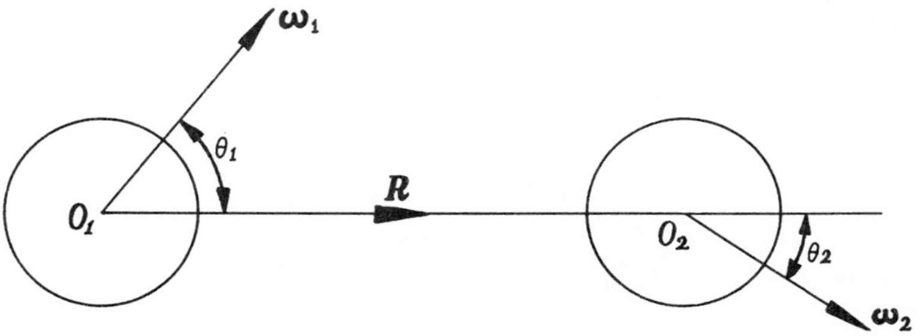

FIGURE 8

created by one can be the exciting field for the other. The condition for this, when stripped of a complication discussed fully in the references cited above, is (Herzenberg [**36**]), for $|\omega_1| = |\omega_2|$,

(2.3.3)

$$[\tfrac{1}{5}R_m(a/R)^3 \sin\theta_1 \sin\theta_2 \sin\Phi]^2(\cos\theta_1 \cos\theta_2 - \sin\theta_1 \sin\theta_2 \cos\Phi) = 3,$$

where θ_i is the angle between $\boldsymbol{\omega}_i$ and the line $\boldsymbol{R} = \overrightarrow{O_1O_2}$ joining their centres, and Φ is the angle between the planes defined by $\boldsymbol{\omega}_1$ and $\boldsymbol{R}$ and by $\boldsymbol{\omega}_2$ and $\boldsymbol{R}$. It is clear that, if (3) is obeyed, it will be violated if the direction of one (but not both) $\boldsymbol{\omega}_i$ is reversed; i.e. roughly half of the motions of this type can (for the right value of R_m) act as dynamos; in particular, $\theta_1 = \theta_2 = \frac{1}{4}\pi$, $\Phi = \frac{1}{2}\pi$ works if $R_m(a/R)^3 = 10\sqrt{6} \simeq 26.4$. The Herzenberg dynamo provided one of the first existence theorems of the subject: it has been successfully built in the laboratory (Lowes and Wilkinson [**54**], [**55**]).

Another application of (2), which is relevant to §3 below, is to an assembly of a large number of identical rotors, each embedded in a stationary infinite medium of the same conductivity and separated by a mean distance large compared with their radius, a. Let N ($\ll a^{-3}$) be the number of rotors per unit volume, and define an angular momentum per unit mass by

(2.3.4) $$\boldsymbol{m}(\boldsymbol{x}) = (8\pi/15)Na^5\bar{\boldsymbol{\omega}}(\boldsymbol{x}),$$

where $\bar{\boldsymbol{\omega}}(\boldsymbol{x})$ is the mean angular velocity of the rotors in the neighbourhood of the point $\boldsymbol{x}$. We suppose the length scale of $\boldsymbol{m}$, and therefore of the field it creates, is $\mathscr{L}$ ($\gg a$). The inducing field at any one rotor is the sum of that generated by every other rotor and (replacing this sum by an integral) the consistency condition for dynamo action is, according to (2),

(2.3.5) $\quad \boldsymbol{B}(\boldsymbol{x}) =$

$$-\frac{1}{8\pi\eta}\int\left\{\frac{3(\hat{\boldsymbol{m}}\cdot\boldsymbol{B})'[\hat{\boldsymbol{m}}'\cdot(\boldsymbol{x}-\boldsymbol{x}')]}{|\boldsymbol{x}-\boldsymbol{x}'|} + (\hat{\boldsymbol{m}}\cdot\nabla\boldsymbol{B}\cdot\hat{\boldsymbol{m}})'\right\}\frac{[\boldsymbol{m}'\times(\boldsymbol{x}-\boldsymbol{x}')]}{|\boldsymbol{x}-\boldsymbol{x}'|^3}\,dV'.$$

On applying the divergence theorem to the first term on the right, and discarding a surface integral (which, like that arising from the Φ of (7) below, vanishes when taken over the sphere at infinity), this reduces to

$$\boldsymbol{B}(\boldsymbol{x}) = \frac{\mu}{4\pi}\int\frac{\bar{\boldsymbol{j}}'\times(\boldsymbol{x}-\boldsymbol{x}')}{|\boldsymbol{x}-\boldsymbol{x}'|^3}\,dV', \tag{2.3.6}$$

where

$$\bar{j}_i = \sigma\left[-\frac{\partial\Phi}{\partial x_i} + \frac{1}{2}\frac{\partial}{\partial x_k}(\hat{m}_i\hat{m}_j m_k)B_j\right]. \tag{2.3.7}$$

Thus, according to the Biot-Savart law (6) and the Ohm's law (7), the average effect of the rotor assembly is the creation of an electromotive force $\mathscr{E}_i = a_{ij}B_j$ throughout the medium. This is a preview of mean field electrodynamics, the subject of §3. Further studies of rotor arrays, embedded in spheres, have been made by Kropachev [**48**], [**49**], in an effort to regenerate poloidal field from toroidal, and so complete the cycle of §1.4. Kropachev made use only of inducing fields asymmetric with respect to (in fact transverse to) the angular velocities of the rotors. In a further intricate analysis (Kropachev [**50**]), he was able to show that, if a single rotor is placed in, and near the boundary of, the conducting sphere (a situation in which the symmetric fields are not regenerative; cf. Gibson [**30**]), dynamo action could be expected when asymmetric fields are allowed for; see also Herzenberg and Lowes [**37**, cf. discussion of case C, p. 568].

3. **Mean field electrodynamics.**

3.1. *Introduction.* Already, in describing the development of Herzenberg's model to a calculus of many rotors, we have pointed out that small-scale motions can regenerate large-scale fields. This was recognized first by Parker [**60**], [**61**], who used the idea to complete the cycle proposed in §1.4; there it was demonstrated that a toroidal shear can create toroidal field from poloidal field, and Parker showed how small-scale motions could do the opposite. Consider a horizontal layer containing an approximately horizontal uniform field, $\boldsymbol{B}$, and in which the fluid moves in vigorous small-scale short-lived vertical eddies, excited perhaps by bouyancy forces in the manner sketched in Figure 9, where a meridian

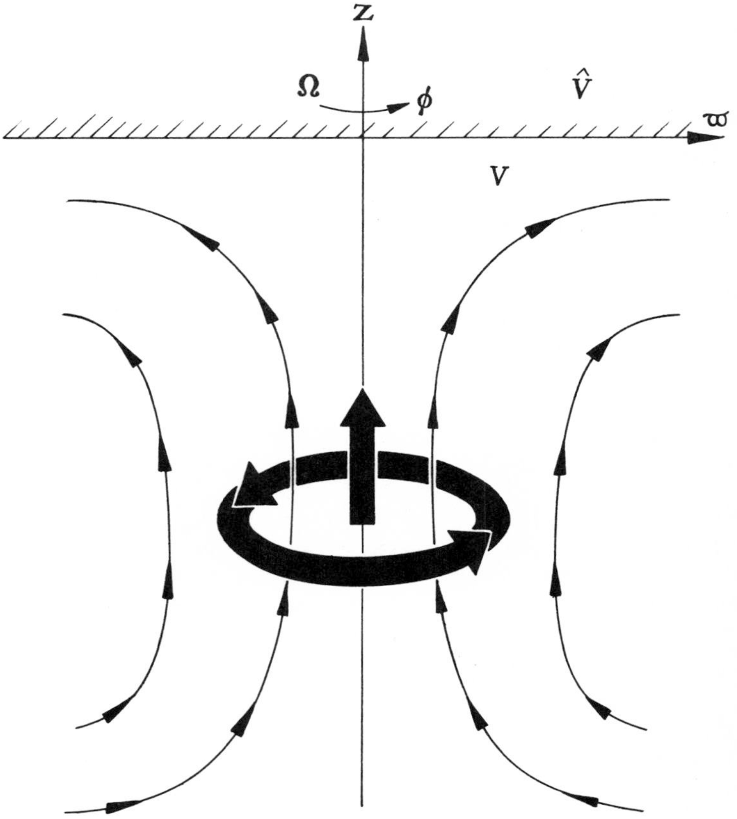

FIGURE 9

section is shown of such an upwelling which, for definiteness, is taken to be axisymmetric. Parker recognized that, as in the case of the Earth's atmosphere, Ferrel's rule would apply: the Coriolis force, $-2\boldsymbol{\Omega} \times \boldsymbol{u}$, associated with the radial (ϖ) motion is, viewed from above in the northern hemisphere of the core, counterclockwise for fluid flowing into the base of the rising column, and will deflect it, producing a corresponding counterclockwise flow. (To some extent this will be offset in the upper parts of the column by the Coriolis forces associated with the divergence of the flow, which are in the opposite sense.) To examine the effect of the upwelling on $\boldsymbol{B}$ we first note that, since the motion is small-scale, we can assume the field is uniform across it. Second, since the motion is vigourous and short-lived, we may assume perfect conductivity and picture

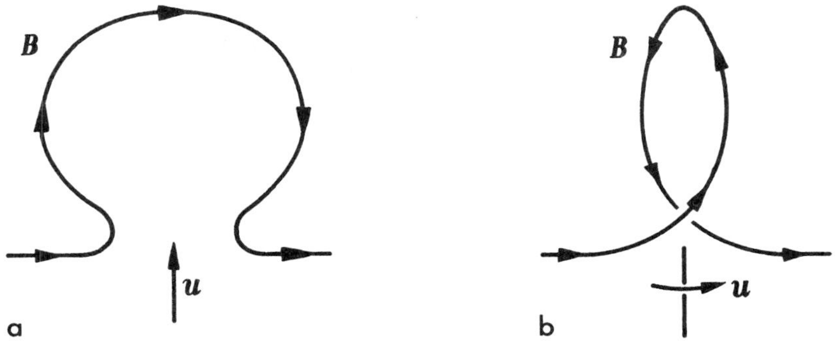

FIGURE 10

each field line as being frozen to the fluid. The vertical motion will bend the line into an Ω shape (Figure 10a), and the vorticity about it will twist the Ω out of its plane (Figure 10b). According to Ampère's law, the current associated with this out-of-plane loop will have a component antiparallel to the field. If we imagine a large number of these loops, and if we suppose after each motion a period of rest (§2.2) over which the fields created can diffuse over the mean distance separating loops, it seems plausible that these small-scale currents will produce an average large-scale current, $\boldsymbol{j}$, antiparallel to $\boldsymbol{B}$. As we have seen (§1.4), a toroidal current of this kind will generate a poloidal field, as Parker required. An even clearer example of the effect is the following.

Imagine a box (Figure 11a) containing homogeneous isotropic turbulence with the additional property that, like Parker's cyclonic motions above, the correlation between velocity and vorticity in any direction is positive i.e. "the statistical properties of the turbulence do not have a centre of symmetry," or "the turbulence is not mirror symmetric," or, following the terminology of Moffatt [**57**], "the turbulence possesses (positive) helicity" (German: Schraubensinn, cf. Krause [**42**]). If a uniform field $\boldsymbol{B}$ is applied (Figure 11b) and the ends of the walls are conducting, the box can act as a battery supplying electrical power at the expense of turbulent energy. If the box is insulating, a charge distribution will build up at the ends which will create a large-scale current equal and opposite to the average current created by the turbulence. It is, then, best to think of the effect of the small-scale motion as creating an e.m.f. of $\alpha\boldsymbol{B}$ where α (here negative) has the dimensions of velocity. Following Steenbeck and Krause [**87**], we refer to this as "the α-effect." It may be noticed in particular that, if the current from the ends is wound (in the correct sense) in a helix round the box as in a solenoid (Figure 11c),

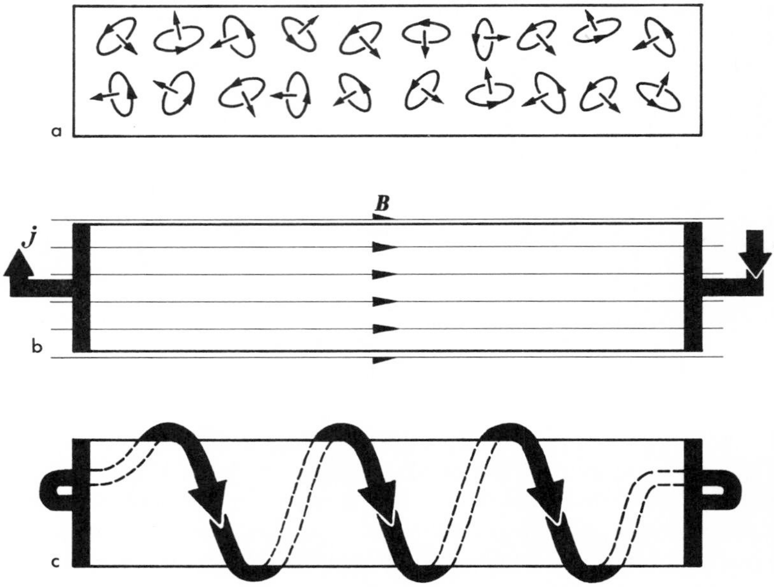

FIGURE 11

it will generate within the box a uniform field in the same direction as that initially applied. If this field is sufficiently strong, the original applied field may be dispensed with, i.e. we have dynamo action. In other words, the α-effect can alone regenerate field, without the shearing effect of the large-scale motion. This may also be demonstrated by another, self-explanatory, example of two rings containing turbulent conductor (Figure 12), and also by the more mathematical α^2-models of §3.4. Experimental demonstration of the α-effect has been given by Steenbeck et al. [**84**].

Although these examples make the creation of field from cyclonic events seem plausible, a closer look at the generation of the events themselves makes it seem unlikely that rotation *alone* will suffice to create helicity and its associated α-effect in the turbulence (see also Biermann, [**8**]). For example, what about the downwellings? Since these converge at the top, they will twist about the z-axis in the *same* sense as that shown in Figure 9, i.e. the helicity in the main part of the column will be opposite to that of an upwelling. The average helicity and α-effect apparently vanish *unless* there is a difference in the statistics of upwellings and downwellings, or in their internal flow-structure, or in *some* other attribute

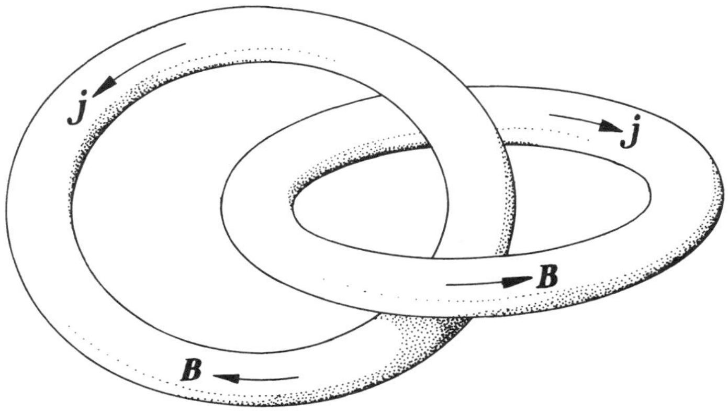

FIGURE 12

which singles out the sense of the vertical. In the case of a highly compressible material such as the sun's atmosphere there is no difficulty: the gradient, $\mathbf{g}$, of density suffices. The expansion of gas as it rises creates, when viewed from above in the northern hemisphere of the sun, a clockwise rotation, and the compression as it sinks produces counterclockwise rotation. In the case of the incompressible fluid in the Earth's core, it is not obvious which factor is the most potent. An allowance for the sphericity of the core will mean that rising (falling) bodies of the fluid will tend to spread (contract) horizontally, leading to helicity effects similar to those arising in the compressible gas. Alternatively Steenbeck, Krause and Rädler [**89**] have postulated a vertical gradient, $\mathbf{g}$, of turbulent intensity, a possibility which Steenbeck and Krause have examined both physically [**86**, p. 167] and through particular models [**88**]. There seems, incidentally, to have been little experimental work on turbulence in rotating fluids (but see [**35**], [**94**]).

3.2. *A more formal approach to turbulent dynamos.* We first describe turbulence in simple terms. Consider flow (velocity $\sim \mathscr{U}$) in the gap (width $\sim \mathscr{L}$) between two coaxial cylinders differentially rotating (Figure 13a). If the Reynolds number $R_e = \mathscr{U}\mathscr{L}/\nu$ is sufficiently small, this flow is laminar and in the azimuthal direction about the axis of symmetry. If, however, R_e is increased beyond a certain critical value, the flow becomes unstable, and secondary cellular motions, often called Taylor cells, result (Figure 13b). These are steady motions, having a length scale of order $\mathscr{L}$ and a preferred orientation (axisymmetric, in fact). On a further increase in R_e these become unstable and, superimposed

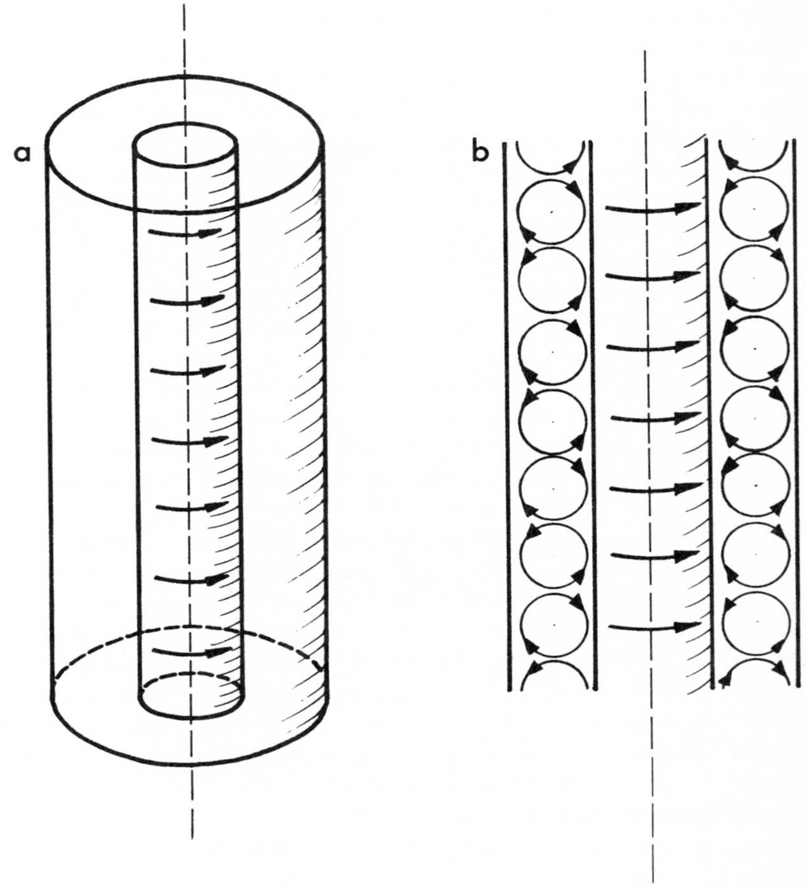

FIGURE 13

upon them, there appear irregularities of smaller length scale having a less organised orientation. On increasing R_e yet further, the degree of irregularity increases, the possible eddy size covering an even broader spectrum of scales and displaying an even greater randomness in orientation. In fact here, as in other examples of turbulence, one tendency can be detected for small eddies to develop from the "instability" of the flow in the larger eddies (a process generally called "energy cascade" or "cascade down the spectrum"), and another tendency for the orientation of the eddies to become increasingly random and to "forget" the (necessarily) preferred direction of the large-scale motions which gave them birth. The cascade down the spectrum continues until the size, λ, of the eddies is such that *their* Reynolds numbers are $O(1)$; then viscosity is, by definition, important and for smaller scales converts to heat the energy cascaded from the large eddies. The eddy sizes, between this "dissipation range" and the "energy containing range" already referred to, define "the inertial range." It should be remarked that the secondary motions and turbulence described in this example increase the torque, which must be applied to the cylinders to maintain R_e, far beyond the value which would have obtained had the flow remained laminar. The effect can be allowed for in a practical, though crude, way by the introduction of a turbulent viscosity, ν_T, which for violent turbulence greatly exceeds ν.

The irregularity and complexity of the flow just described suggests a statistical theory. For this purpose an "ensemble" of many realizations of the same experiment is imagined which, individually through the vagaries of turbulence, differ in detail; an average (denoted by an overbar) over this ensemble is introduced, and F' is used to denote the fluctuating part of any field, F, i.e. the difference between F and its average $\bar{F}$:

$$F = \bar{F} + F'. \tag{3.2.1}$$

It is clear that

$$\bar{F'} = 0, \qquad \bar{\bar{F}} = \bar{F}, \tag{3.2.2}$$

and

$$\overline{F + G} = \bar{F} + \bar{G}, \qquad \overline{\bar{F}G} = \bar{F}\bar{G}, \qquad \overline{\bar{F}G'} = 0, \qquad \text{etc.} \tag{3.2.3}$$

It is important to realize, however, that

$$\overline{FG} = \bar{F}\bar{G} + \overline{F'G'} \tag{3.2.4}$$

is not, in general, equal to $\bar{F}\bar{G}$. Cases of (4) of particular interest are provided by induced e.m.f. and by Lorentz force:

$$\overline{\boldsymbol{u} \times \boldsymbol{B}} = \bar{\boldsymbol{u}} \times \bar{\boldsymbol{B}} + \overline{\boldsymbol{u}' \times \boldsymbol{B}'}, \qquad \overline{\boldsymbol{j} \times \boldsymbol{B}} = \bar{\boldsymbol{j}} \times \bar{\boldsymbol{B}} + \overline{\boldsymbol{j}' \times \boldsymbol{B}'}, \tag{3.2.5}$$

and the main interest in the discussions below is in the estimation of

$$\overline{\boldsymbol{u}' \times \boldsymbol{B}'}$$

in terms of the mean field $\bar{\boldsymbol{B}}$.

When the ensemble average of any quantity depends on $\boldsymbol{x}$ but not on t, the turbulence is said to be "(statistically) steady," and the ensemble average may be replaced by a time average over any one member of the ensemble. Similarly, if the ensemble average depends on t but not $\boldsymbol{x}$, the turbulence is said to be "homogeneous," and a space average may be used instead of the ensemble average. If statistical properties are independent of the orientation of the coordinate frame (at a point), the turbulence is said to be "isotropic" (at that point); if they are independent of whether the frame is right-handed or left-handed, it is said to be "mirror-symmetric" or to "possess a centre of symmetry."

Tensors such as

$$Q_{ij}(\boldsymbol{x}, t; \boldsymbol{x}', t') = \overline{u_i'(\boldsymbol{x}, t)u_j'(\boldsymbol{x}', t')} \tag{3.2.6}$$

are called "correlation tensors." It is evident that in homogeneous turbulence they will depend on $\boldsymbol{x}$ and $\boldsymbol{x}'$ only in the combination $\boldsymbol{\xi} \equiv \boldsymbol{x} - \boldsymbol{x}'$, and that in steady turbulence they will depend on t and t' only in the combination $T = t - t'$, i.e. we may replace (6) by $Q_{ij}(\boldsymbol{\xi}, T)$; see [7]. It is also evident, for the case of (6), that

$$Q_{ij}(\boldsymbol{x}, t; \boldsymbol{x}', t') \equiv Q_{ji}(\boldsymbol{x}', t'; \boldsymbol{x}, t), \tag{3.2.7}$$

$$Q_{ij}(\boldsymbol{x}, t; \boldsymbol{x}, t) = v_i^2\delta_{ij}, \quad \text{(not summed)}, \tag{3.2.8}$$

where $v_i = (u_i^2)^{1/2}$ is the r.m.s. of the ith component of $\boldsymbol{u}$ at $(\boldsymbol{x}, t)$. It is clear that, if $|\boldsymbol{\xi}| \to \infty$ or $|T| \to \infty$ or both, $Q_{ij} \to 0$. In the case $t = t'$, a "typical" length $\lambda = |\boldsymbol{x} - \boldsymbol{x}'|$ over which Q_{ii} decays is known as "the correlation length." Similarly, in the case $\boldsymbol{x} = \boldsymbol{x}'$, a "typical" time $\tau = |t - t'|$ over which Q_{ii} decays is known as "the correlation time." In what follows we suppose that $\lambda \ll \mathscr{L}$ and $\tau \ll \mathscr{T}$. The exact form of Q_{ij}, even in the simplest case of steady homogeneous isotropic turbulence, is not precisely known, but a model such as

$$Q_{ij}(\boldsymbol{x}, t; \boldsymbol{x}', t') = \frac{1}{3}v^2\delta_{ij} \exp\left[-\left(\frac{T}{2\tau} + \frac{\xi^2}{2\lambda^2}\right)\right], \tag{3.2.9}$$

though crude, and misleading in detail, is analytically simple and has been used to estimate some of the scalars arising below; here $v^2 = 3v_i^2$ is the r.m.s. of $\boldsymbol{u}'^2$.

We consider the problems raised by the direct evaluation of the mean electromotive force,

$$\mathscr{E} = \overline{\boldsymbol{u}' \times \boldsymbol{B}'}, \tag{3.2.10}$$

created by the turbulence. The mean and fluctuating parts of the induction equation (1.2.9) are clearly

$$\partial\overline{\boldsymbol{B}}/\partial t = \operatorname{curl}(\overline{\boldsymbol{u}} \times \overline{\boldsymbol{B}} + \mathscr{E}) + \eta\nabla^2\overline{\boldsymbol{B}}, \tag{3.2.11}$$

$$\partial\boldsymbol{B}'/\partial t = \operatorname{curl}(\overline{\boldsymbol{u}} \times \boldsymbol{B}' + \boldsymbol{u}' \times \overline{\boldsymbol{B}} + \boldsymbol{G}') + \eta\nabla^2\boldsymbol{B}', \tag{3.2.12}$$

where

$$G' = u' \times B' - \overline{u' \times B'}. \tag{3.2.13}$$

Ignoring a possible difficulty, viz. that self-excited ($\bar{B} = 0$) solutions of (12) exist, this equation, together with the appropriate boundary conditions on B', determines B' as a linear functional of $\bar{B}$, and, returning to (10), $\mathscr{E}$ can in principle be evaluated, also as a linear functional of $\bar{B}$. To show how the process can work, we may first attempt to expand B' in, hopefully, a convergent power series in u', i.e. we may write

$$B' = \sum_{n=1}^{a} B'_n, \tag{3.2.14}$$

where

$$\frac{\partial B'_n}{\partial t} - \operatorname{curl}(\bar{u} \times B'_n) - \eta\nabla^2 B'_n = \operatorname{curl}\begin{cases} u' \times \bar{B}, & \text{if } n = 1, \\ G'_n, & \text{if } n > 1. \end{cases} \tag{3.2.15}$$

For the special cases $v\tau \ll \lambda$ or $v\lambda \ll \eta$ (or both), the leading term ($n = 1$) should suffice, i.e. the relevant solution of

$$\frac{\partial B'}{\partial t} + \bar{u} \cdot \nabla B' - B' \cdot \nabla \bar{u} - \eta\nabla^2 B' = \bar{B} \cdot \nabla u' - u' \cdot \nabla \bar{B}. \tag{3.2.16}$$

To solve (16), we first note that, since B' is created from $\bar{B}$ by u', the correlation between u' and B' will be nonzero only over distances of order λ and times of order τ, and that the boundaries of V should thus have a negligible effect in evaluating $\mathscr{E}$ [cf. (19) below]. It should therefore suffice to solve (16) in terms of the Green's function, $G_{ij}(x, t; x', t')$, vanishing for $|x - x'| \to \infty$, for the conductor of infinite extent:

$$B_i(x, t) = \int_{-\infty}^{t} dt' \iiint_{-\infty}^{\infty} dx' G_{ik}(x, t; x', t') \varepsilon_{k\alpha\beta} \frac{\partial}{\partial x'_\alpha} [\varepsilon_{\beta lm} u'_l(x', t') \bar{B}_m(x', t')], \tag{3.2.17}$$

or

$$B_i(x, t) = -\varepsilon_{k\alpha\beta}\varepsilon_{lm\beta} \int_{-\infty}^{t} dt' \iiint_{-\infty}^{\infty} dx' \frac{\partial G_{ik}(x, t; x', t')}{\partial x'_\alpha} u'_l(x', t) \bar{B}_m(x', t'), \tag{3.2.18}$$

whence, substituting into (10) and using (3) and (6),

(3.2.19) $\mathscr{E}_i(\boldsymbol{x}, t) =$

$$\varepsilon_{ijn}\varepsilon_{k\alpha\beta}\varepsilon_{lm\beta}\int_{-\infty}^{t} dt' \int\int\int_{-\infty}^{\infty} d\boldsymbol{x}' \frac{\partial G_{jk}(\boldsymbol{x}, t; \boldsymbol{x}', t')}{\partial x'_\alpha} Q_{nl}(\boldsymbol{x}, t; \boldsymbol{x}', t')\bar{B}_m(\boldsymbol{x}', t').$$

Suppose we expand $\bar{\boldsymbol{B}}(\boldsymbol{x}', t')$ in Taylor series about $(\boldsymbol{x}, t)$. Consider a term involving p space derivative and q time derivative. By supposition this will be of order $\bar{\boldsymbol{B}}/\mathscr{L}^p\mathscr{T}^q$ and, since Q_{ij} vanishes with $|\xi|$ and T over the length and time scales of λ and τ, the resulting integrated term will be of order $(\tau v^2/\lambda)\bar{\boldsymbol{B}}(\lambda/\mathscr{L})^p(\tau/\mathscr{T})^q$. In other words, the dominant terms should arise from the lowest derivatives, and (questions of helicity apart) the largest of all should arise from assuming $\bar{\boldsymbol{B}}$ is constant and ignoring the last term in (16). If we proceed as far as the first space derivative, we must replace $\bar{\boldsymbol{B}}(\boldsymbol{x}', t')$ in (19) by $\bar{\boldsymbol{B}} - \xi \cdot \nabla\bar{\boldsymbol{B}}$ evaluated at $(\boldsymbol{x}, t)$. Thus we will obtain, for the first two terms in the expansion of $\mathscr{E}$,

$$\mathscr{E}_i \equiv \overline{[\boldsymbol{u}' \times \boldsymbol{B}']}_i = a_{ij}\bar{B}_j + b_{ijk}(\partial\bar{B}_j/\partial x_k). \tag{3.2.20}$$

The tensors a_{ij} and b_{ijk}, which have dimensions of velocity and diffusivity respectively, depend on $\bar{\boldsymbol{u}}$ and on the statistical properties of $\boldsymbol{u}'$, but not on $\bar{\boldsymbol{B}}$; they are large-scale, but the overbar is omitted.

Since the explicit evaluation of a_{ij} and b_{ijk} is clearly, even in the simplest cases, an intricate and laborious matter, it is worth noting that alternative general arguments are possible which are both simple and powerful. The situation is, in some sense, parallel to that arising in the statistical mechanics of nonuniform gases. Strictly, the motion of each molecule depends on that of every other, no matter how distant, but in fact, since these motions are correlated only over distances of order of the mean free path, only neighbouring molecules exert a decisive influence. This leads to the idea that a process, such as viscous shear, should in a fluid depend only on the contemporaneous values of local flow properties, such as rate of strain. Having made this postulate, it is possible, by the powerful methods of continuum mechanics, to make quite general statements about the relationship between stress and rate of strain. In any particular case, it is still necessary to undertake a difficult approximation procedure before the magnitude of constants, such as the coefficients of viscosity, can be estimated accurately; but, fortunately for the development of fluid mechanics, general consequences of such relationships between stress and rate of strain can be examined without such a calculation! In the same way, for many purposes a detailed knowledge of the turbulence is not required for an understanding of the relationship between equations governing the mean field and Maxwell's equations for the raw field. In any particular case, it is still necessary to undertake a difficult approximation procedure, of the type outlined above (see also §3.3), before the magnitude of constants, such as α, can be estimated accurately but, fortunately for the development of the subject (christened *mean field electrodynamics* by the East German School), general consequences can be examined without such a calculation!

In discussing the general properties of a_{ij} and b_{ijk}, it is convenient to introduce a distinction between scalars and pseudoscalars, and between polar vectors and axial (or skew) vectors. (It is then unnecessary to change Maxwell's equations under reflection of axes.) Scalars and axial vectors do not change their sign under axis reflection; polar vectors do. Fluid velocity, electric field and electric current are polar; vorticity and magnetic field are axial. The scalar product of two like vectors, i.e. two vectors of the same parity, is a scalar, and of two vectors of opposite parity (e.g. $\boldsymbol{u}' \cdot \boldsymbol{\omega}'$) is a pseudoscalar. In particular, the square k^2 of the length of a vector, $\boldsymbol{k}$, of either type is a scalar. Nevertheless, in taking the square root of this expression, we may, if we wish, define k to have one sign for right-handed coordinate frames and the opposite for left-handed frames. Then k, and any odd scalar function of k, will be a pseudoscalar. This remark is relevant to (3.3.19) below. The gradient operator is polar, while ε_{ijk} is skew. A tensor, composed of a product of an even (odd) number of ε symbols with a scalar or with a polar vector, is polar (skew); if instead it is with a pseudoscalar or with an axial vector, it is skew (polar). Thus, since $\bar{\boldsymbol{B}}$ is axial and $\mathscr{E}$ is polar, a_{ij} and b_{ijk} must both be skew. This itself is useful information: Suppose $\bar{\boldsymbol{u}} = 0$ and the turbulence is steady, homogeneous and isotropic, then a_{ij} and b_{ijk} must be constant. The only skew isotropic tensors of degree two and three are $\alpha\delta_{ij}$ and $\beta\varepsilon_{ijk}$, where α and β are a constant pseudoscalar and a constant scalar, respectively. Thus (20) becomes

$$\mathscr{E} = \alpha\bar{\boldsymbol{B}} - \beta \operatorname{curl} \bar{\boldsymbol{B}}, \tag{3.2.21}$$

and Ohm's law for the mean field becomes, averaging (1.2.5),

$$\bar{\boldsymbol{j}} = \sigma(\bar{\boldsymbol{E}} + \alpha\bar{\boldsymbol{B}} - \mu\beta\bar{\boldsymbol{j}}), \tag{3.2.22}$$

or

$$\bar{\boldsymbol{j}} = \sigma_T(\bar{\boldsymbol{E}} + \alpha\bar{\boldsymbol{B}}), \tag{3.2.23}$$

where

$$\sigma_T = \sigma/(1 + \mu\sigma\beta) \quad \text{and} \quad \eta_T = 1/\mu\sigma_T = \eta + \beta, \tag{3.2.24}$$

are respectively "the turbulent conductivity" and "the turbulent magnetic diffusivity." The latter is the analogue of the turbulent viscosity, ν_T, introduced earlier. Clearly, for strongly turbulent flows in which $\beta \gg \eta$, we have $\eta_T \simeq \beta$. In any case, as far as β is concerned, the effect of the turbulence is merely to amend the conductivity appearing in Ohm's law; in fact, since β is positive, the effective conductivity is reduced. An illuminating interpretation of (24) is given by Steenbeck and Krause [**87**, p. 52].

It is important to observe that, since all statistical moments in mirror-symmetric turbulence are, by definition, invariant under axis reflection, all associated pseudoscalars (e.g. the helicity $\overline{\boldsymbol{u}' \cdot \boldsymbol{\omega}'}$ and α) must vanish. Thus helicity and the α-effect can be present *only* in turbulence which is not mirror-symmetric.

What can cause lack of mirror-symmetry? The arguments of §3.1 suggested that rotation might be responsible, although some doubts were expressed at the end of that section whether rotation alone would suffice. Moreover a formal difficulty arises: whether or not rotation creates helicity, it is certain to destroy the isotropy assumed above, making that discussion irrelevant. Instead the turbulence will be axisymmetric. Let $\boldsymbol{\lambda}$ define the direction of anisotropy. General invariance arguments show immediately that

(3.2.25) $$a_{ij} = \alpha\delta_{ij} + \alpha_0\varepsilon_{ijk}\lambda_k + \gamma\lambda_i\lambda_j,$$

(3.2.26) $$b_{ijk} = \beta\varepsilon_{ijk} - \beta_1\lambda_i\delta_{jk} - \beta_2\lambda_j\delta_{ki} - \beta_3\lambda_k\delta_{ij} + \mu_1\varepsilon_{jkl}\lambda_l\lambda_i + \mu_2\varepsilon_{kil}\lambda_l\lambda_j + \mu_3\varepsilon_{ijl}\lambda_l\lambda_k + \nu\lambda_i\lambda_j\lambda_k,$$

where the scalars $\alpha - \nu$ depend on the invariant λ^2, and the properties of the turbulence; β, μ_1, μ_2 and μ_3 are scalars, and α and γ are pseudoscalars. If $\boldsymbol{\lambda}$ is axial, $\beta_1, \beta_2, \beta_3$ and ν are scalars and α_0 is a pseudoscalar; if $\boldsymbol{\lambda}$ is polar, the reverse is true. It is difficult to make further general progress without a deeper knowledge of the structure of the turbulence. One may however argue as follows. As the discussion of §3.2 has indicated, decreasing length scales show an increasing tendency towards isotropy, and, even if the large eddies possess helicity induced by, say, rotation, this is likely to be nearly absent in the small eddies. As far as the small eddies are concerned it should be possible to suppose that, to leading order, the turbulence is homogeneous, isotropic *and* mirror-symmetric, and that the deviations from this state are small and can be represented by the terms linear in $\boldsymbol{\lambda}$ above; as a corollary the remaining constants $\alpha, \alpha_0, \beta, \beta_1, \beta_2$ and β_3 appearing in (25) and (26) will depend on the properties of the mirror-symmetric turbulence and will therefore all be true scalars; the pseudoscalars must vanish:

$$\alpha = 0, \quad \text{and} \quad \begin{cases} \alpha_0 = 0, & \text{if } \boldsymbol{\lambda} \text{ is axial,} \\ \beta_1 = \beta_2 = \beta_3 = 0, & \text{if } \boldsymbol{\lambda} \text{ is polar.} \end{cases} \tag{3.2.27}$$

Suppose $\boldsymbol{\lambda}$ corresponds to a mean local angular velocity vector $\bar{\boldsymbol{\omega}}$ which is necessarily axial, then equations (25) to (27) give, by equation (20), the Ohm's law

$$\bar{\boldsymbol{j}} = \sigma_T[\bar{\boldsymbol{E}} + \bar{\boldsymbol{u}} \times \bar{\boldsymbol{B}} + \mu\beta_3\bar{\boldsymbol{\omega}} \times \bar{\boldsymbol{j}} - (\beta_2 + \beta_3)(\nabla\bar{\boldsymbol{B}}) \cdot \bar{\boldsymbol{\omega}}], \tag{3.2.28}$$

where σ_T is given by (24). Confirming the doubts raised at the end of §3.1, the first of (27) indicates that weakly axisymmetric turbulence is not associated with an α-effect. Nevertheless, it would be a mistake to suppose that the remaining terms in equation (28) are ineffective. The $\bar{\boldsymbol{\omega}} \times \bar{\boldsymbol{j}}$ term in (28) can, in a spherical V, create toroidal current from poloidal, i.e. can produce poloidal field from toroidal, and so complete the process discussed in §1.4; this has been confirmed by Rädler [**68**]–[**72**] through explicit examples. If $\boldsymbol{\lambda}$ is polar (e.g. the gradient, $\boldsymbol{g}$, of turbulent intensity or a gradient of density), (25) to (27) give, by (20), the Ohm's law

$$\bar{\boldsymbol{j}} = \sigma(\bar{\boldsymbol{E}} + \bar{\boldsymbol{u}}_e \times \bar{\boldsymbol{B}}), \tag{3.2.29}$$

where $\bar{\boldsymbol{u}}_e = \bar{\boldsymbol{u}} - \alpha_0\boldsymbol{g}$ is a new "effective" mean velocity; there is little help in dynamo maintenance here.

It is of some interest to examine the effects associated with the two preferred directions $\bar{\boldsymbol{\omega}}$ and $\boldsymbol{g}$. Restricting attention to terms linear in both $\bar{\boldsymbol{\omega}}$ and $\boldsymbol{g}$ separately, similar invariance arguments show that $\bar{\boldsymbol{j}}$ is not just the obvious combination of (28) and (29); instead

$$\begin{aligned} \bar{\boldsymbol{j}} = \sigma_T[\bar{\boldsymbol{E}} + \bar{\boldsymbol{u}}_e \times \bar{\boldsymbol{B}} + \mu\beta_3\bar{\boldsymbol{\omega}} \times \bar{\boldsymbol{j}} - (\beta_2 + \beta_3)(\nabla\bar{\boldsymbol{B}}) \cdot \bar{\boldsymbol{\omega}} \\ + \alpha_1(\bar{\boldsymbol{\omega}} \cdot \boldsymbol{g})\bar{\boldsymbol{B}} + \alpha_2(\boldsymbol{g} \cdot \bar{\boldsymbol{B}})\bar{\boldsymbol{\omega}} + \alpha_3(\bar{\boldsymbol{\omega}} \cdot \bar{\boldsymbol{B}})\boldsymbol{g}], \end{aligned} \tag{3.2.30}$$

which contains three new terms whose coefficients, α_i, are true scalars. The influence of the last two has not been examined; but the first (α_1) clearly represents an α-effect, with $\alpha = \alpha_1(\bar{\boldsymbol{\omega}} \cdot \boldsymbol{g})$. This fact was first recognized by Steenbeck, Krause and Rädler [**89**] who also estimated α_i (although for some reason they did not include the β_i terms). The form of the α_1 term in equation (30) makes the earlier (constant α) theory seem unrealistic, for in a spherical body we may expect $\boldsymbol{g}$ to be radial and therefore $\alpha_1\bar{\boldsymbol{\omega}} \cdot \boldsymbol{g}$ to be proportional to the sine of latitude, with opposite sign in opposite hemispheres. The constant α theory is, however, much simpler, and in fact models of both types have been examined (§3.4).

The method just given for including $\bar{\boldsymbol{\omega}}$ is a special case of a more general theory developed by Krause [**40**], [**41**] in which it is supposed that isotropic mirror-symmetric turbulence is perturbed by an arbitrary rate of strain $\partial \bar{u}_i/\partial x_j$. Concerning the general philosophy of regarding deviations from isotropy created by the mean flow as perturbations, one must be encouraged by the success of Krause [**43**] in obtaining in this way a convincing theoretical basis for the idea of turbulent viscosity, ν_T.

3.3. *Explicit calculation of "transport coefficients" in special cases.* Even in the case of steady homogeneous turbulence, the explicit evaluation of the a_{ij} and b_{ijk} of (3.2.20) in terms of the statistical properties of the turbulence is intricate. The calculation is, in fact, simple only in the two special cases (Krause and Rädler [**45**, §3.5, §3.6]) of high conductivity ($\lambda^2 \gg \eta\tau$) and low conductivity ($\lambda^2 \ll \eta\tau$). (The latter implies that the microscale Reynolds number, $R_m = \lambda v/\eta$, is small.)

Since $\bar{\boldsymbol{u}} = 0$, (3.2.16) reduces, in the high conductivity limit, to

$$\partial \boldsymbol{B}'/\partial t = \operatorname{curl}(\boldsymbol{u}' \times \bar{\boldsymbol{B}}), \tag{3.3.1}$$

whence, integrating over t,

$$\mathscr{E}(\boldsymbol{x}, t) = \overline{\boldsymbol{u}'(\boldsymbol{x}, t) \times \boldsymbol{B}'(\boldsymbol{x}, t_0)} + \int_{t_0}^{t} \overline{\boldsymbol{u}'(\boldsymbol{x}, t) \times \operatorname{curl}[\boldsymbol{u}'(\boldsymbol{x}, t') \times \bar{\boldsymbol{B}}(\boldsymbol{x}, t')]}\, dt'. \tag{3.3.2}$$

For $t - t_0 \gg \tau$, there will be no correlation between $\boldsymbol{u}'(\boldsymbol{x}, t)$ and $\boldsymbol{B}'(\boldsymbol{x}, t_0)$; the first term on the right of (2) therefore vanishes. Also, t_0 may be replaced by $-\infty$ in the second. (In fact, all this was done without comment in §3.2.) By a change of variable, we now have

$$\mathscr{E}_i = \int_0^\infty \overline{\varepsilon_{ijk} u_j(\boldsymbol{x}, t) \left[\bar{B}_m \frac{\partial u_k'}{\partial x_m} - u_m' \frac{\partial \bar{B}_k}{\partial x_m} \right]_{(\boldsymbol{x}, t-T)}}\, dT. \tag{3.3.3}$$

Assuming that $\bar{\boldsymbol{B}}$ is constant over the correlation time τ, we may replace $\bar{\boldsymbol{B}}(\boldsymbol{x}, t - T)$ by $\bar{\boldsymbol{B}}(\boldsymbol{x}, t)$, giving (3.2.20) with

$$a_{ij} = \int_0^\infty \overline{\varepsilon_{ikl} u_k'(\boldsymbol{x}, t) \frac{\partial u_l'(\boldsymbol{x}, t - T)}{\partial x_j}}\, dT, \tag{3.3.4}$$

$$b_{ijk} = \varepsilon_{ijl} \int_0^\infty \overline{u_l'(\boldsymbol{x}, t) u_k'(\boldsymbol{x}, t - T)}\, dT. \tag{3.3.5}$$

Now, if the turbulence is isotropic, the correlation tensors shown must also be isotropic in their two indices, i.e. proportional to the unit tensor; thus

(3.3.6) $$\overline{u'_l(\boldsymbol{x}, t)u'_k(\boldsymbol{x}, t - T)} = \tfrac{1}{3}\delta_{lk}\overline{\boldsymbol{u}'(\boldsymbol{x}, t)\cdot\boldsymbol{u}'(\boldsymbol{x}, t - T)}, \quad \text{etc.}$$

i.e. (3.2.21) is recovered with

(3.3.7) $$\alpha = -\frac{1}{3}\int_0^\infty \overline{\boldsymbol{u}'(\boldsymbol{x}, t)\cdot \operatorname{curl}\boldsymbol{u}'(\boldsymbol{x}, t - T)}\,dT,$$

(3.3.8) $$\beta = \frac{1}{3}\int_0^\infty \overline{\boldsymbol{u}'(\boldsymbol{x}, t)\cdot\boldsymbol{u}'(\boldsymbol{x}, t - T)}\,dT.$$

Parker [**61**], [**63**] provides an alternative approach to this high conductivity case.

In the low conductivity limit, the starting point replacing (1) is

(3.3.9) $$\eta\nabla^2\boldsymbol{B}' = \operatorname{curl}(\boldsymbol{u}' \times \bar{\boldsymbol{B}}),$$

leading by (3.2.19) to

(3.3.10)

$$\mathscr{E}(\boldsymbol{x}, t) = -\frac{1}{4\pi\eta}\iiint_{-\infty}^{\infty} \frac{\overline{\boldsymbol{u}'(\boldsymbol{x}, t)\times[(\boldsymbol{x} - \boldsymbol{x}')\times\{\boldsymbol{u}'(\boldsymbol{x}', t)\times\bar{\boldsymbol{B}}(\boldsymbol{x}', t)\}]}}{|\boldsymbol{x} - \boldsymbol{x}'|^3}\,d^3\boldsymbol{x}'.$$

As before, we may replace $\boldsymbol{B}(\boldsymbol{x}', t)$ by $\boldsymbol{B} - \boldsymbol{\xi}\cdot\nabla\boldsymbol{B}$ evaluated at $(\boldsymbol{x}, t)$ and obtain

(3.3.11) $$a_{ij} = \frac{1}{4\pi\eta}\varepsilon_{ikl}\varepsilon_{lmn}\varepsilon_{npj}\iiint_{-\infty}^{\infty} Q_{kp}(\boldsymbol{x}, t; \boldsymbol{x} + \boldsymbol{\xi}, t)\frac{\xi_m d^3\boldsymbol{\xi}}{\xi^3},$$

(3.3.12) $$b_{ijk} = \frac{1}{4\pi\eta}\varepsilon_{ilm}\varepsilon_{mnp}\varepsilon_{pqj}\iiint_{-\infty}^{\infty} Q_{lq}(\boldsymbol{x}, t; \boldsymbol{x} + \boldsymbol{\xi}, t)\frac{\xi_n\xi_k\, d^3\boldsymbol{\xi}}{\xi^3}.$$

Assuming homogeneity, these reduce to

(3.3.13) $$a_{ij} = -\frac{1}{8\pi\eta}(\delta_{im}\delta_{jn} + \delta_{jm}\delta_{in})\iiint_{-\infty}^{\infty} \varepsilon_{mpq}Q_{pq}(\boldsymbol{\xi})\frac{\xi_n\, d^3\boldsymbol{\xi}}{\xi^3},$$

(3.3.14) $$b_{ijk} = \frac{1}{4\pi\eta}\varepsilon_{ijl}\iiint_{-\infty}^{\infty} Q_{lm}(\boldsymbol{\xi})\frac{\xi_m\xi_k\, d^3\boldsymbol{\xi}}{\xi^3}.$$

Clearly a_{ij} is symmetric and b_{ijk} is antisymmetric in i and j.

In terms of the generalized Fourier transform of (3.2.6) viz.

(3.3.15) $$Q_{ij}(\boldsymbol{\xi}) = \iiint_{-\infty}^{\infty} \tilde{Q}_{ij}(\boldsymbol{k})\, e^{i\boldsymbol{k}\cdot\boldsymbol{\xi}}\, d^3\boldsymbol{k}.$$

Equations (13) and (14) become

$$(3.3.16)\qquad a_{ij} = \frac{i}{2\eta}(\delta_{im}\delta_{jn} + \delta_{jm}\delta_{in}) \int\int\int_{-\infty}^{\infty} \varepsilon_{mpq}\tilde{Q}_{pq}(\boldsymbol{k})\frac{k_n\, d^3\boldsymbol{k}}{k^2},$$

$$(3.3.17)\qquad b_{ijk} = \frac{1}{\eta}\varepsilon_{ijl} \int\int\int_{-\infty}^{\infty} \tilde{Q}_{lk}(\boldsymbol{k})\frac{d^3\boldsymbol{k}}{k^2}.$$

If the turbulence is isotropic, a_{ij} must be proportional to δ_{ij}, i.e., $a_{ij} = a_{kk}\delta_{ij}/3$. Treating the integral in (17) similarly, we may recover (3.2.21) with

(3.3.18)

$$\alpha = \frac{i}{3\eta}\int\int\int_{-\infty}^{\infty} \varepsilon_{pq,r}\tilde{Q}_{pq}(\boldsymbol{k})\frac{k_r\, d^3\boldsymbol{k}}{k^2}, \qquad \beta = \frac{1}{3\eta}\int\int\int_{-\infty}^{\infty} \tilde{Q}_{pp}(\boldsymbol{k})\frac{d^3\boldsymbol{k}}{k^2}.$$

In isotropic turbulence, $\tilde{Q}_{ij}$ must be an isotropic function of $\boldsymbol{k}$, the only possibility being (for incompressible fluids)

$$(3.3.19)\qquad \tilde{Q}_{ij}(\boldsymbol{k}) = \frac{E(k)}{4\pi k^4}(k^2\delta_{ij} - k_ik_j) + \frac{iF(k)}{8\pi k^4}\varepsilon_{ijk}k_k,$$

where $E(k)$ is a scalar, the energy spectrum function, and $F(k)$ is a pseudo-scalar which may be called "the helicity spectrum function":

$$(3.3.20)\qquad \overline{\boldsymbol{u}'\cdot\boldsymbol{\omega}'} = \int_0^{\infty} F(k)\, dk, \qquad \tfrac{1}{2}v^2 \equiv \tfrac{1}{2}\overline{\boldsymbol{u}'^2} = \int_0^{\infty} E(k)\, dk.$$

From (18) and (19), we obtain

$$(3.3.21)\qquad \alpha = -\frac{1}{3\eta}\int_0^{\infty} F(k)\frac{dk}{k^2}, \qquad \beta = \frac{2}{3\eta}\int_0^{\infty} E(k)\frac{dk}{k^2}.$$

The relationship between α-effect and helicity is clearly brought out by (20) and (21). Results (16) and the first of (21) have been obtained by Moffatt [**58**] in the case of constant $\overline{\boldsymbol{B}}$ by direct Fourier transformation of (3.2.16). This very simple method fails when $\overline{\boldsymbol{B}}$ is spatially varying. The expression (3.2.19) for $\mathscr{E}$ is, however, general, and subsequent expansion in $\lambda/\mathscr{L}$ [as described below (3.2.19)] yields, as we have seen, Moffatt's results to first order, and (17) and the second of (21) to second order, and so on, the coefficients a_{ij}, b_{ijk}, etc. being directly obtainable from the generalized Fourier transform of $\boldsymbol{u}'$.

It is common in turbulence theory to regard the expression of a field in terms of a generalized Fourier integral as the limit, for $L \to \infty$, of a Fourier sum representing a field periodic over a box of volume L^3. This point of view is particularly valuable here because it permits a rigorous treatment of the expansions in $\lambda/\mathscr{L}$ described above, a fact first recognized by Childress [**21**]. Suppose $\boldsymbol{u}'$ is steady and also cellular, i.e., satisfies

$$\boldsymbol{u}'(\boldsymbol{x} + \boldsymbol{l}_i) \equiv \boldsymbol{u}'(\boldsymbol{x}), \tag{3.3.22}$$

for three noncoplanar vectors $\boldsymbol{l}_i$. The microscale λ introduced earlier may now be thought of as a typical $|\boldsymbol{l}_i|$. It is readily found that the fields induced are of the form

$$\boldsymbol{B}(\boldsymbol{x}, t) = \boldsymbol{H}(\boldsymbol{x}) \exp(i\boldsymbol{n} \cdot \boldsymbol{x} + pt), \tag{3.3.23}$$

where $\boldsymbol{n}$ (real) and p are constants, and $\boldsymbol{H}(\boldsymbol{x}, t)$ is "$\boldsymbol{u}$-periodic," i.e. obeys the same periodicity condition (22) as $\boldsymbol{u}'$. The expression (23) may be displayed as a "mean field" and "fluctuating field" by dividing it into

$$\begin{aligned} \bar{\boldsymbol{B}} &\equiv \boldsymbol{H}_0(\boldsymbol{n}) \exp(i\boldsymbol{n} \cdot \boldsymbol{x} + pt), \qquad \text{and} \\ \boldsymbol{B}' &\equiv (\boldsymbol{H} - \boldsymbol{H}_0) \exp(i\boldsymbol{n} \cdot \boldsymbol{x} + pt), \end{aligned} \tag{3.3.24}$$

where $\boldsymbol{H}_0$ is the mean of $\boldsymbol{H}$ over the unit cell of periodicity. The induction equation provides a scalar condition of the form

$$\Gamma(p, \boldsymbol{n}) = 0, \tag{3.3.25}$$

which can be regarded, for $p = 0$, as a compatibility requirement on $\boldsymbol{n}$ for steady dynamo action; otherwise it determines whether, for the assumed $\boldsymbol{n}$, the field grows $[\mathscr{R}(p) > 0]$ or decays. The analogue of the expansion in $\lambda/\mathscr{L}$ introduced below (3.2.19) in the turbulent case is now an expansion of $\boldsymbol{H}_0(\boldsymbol{n})$ and $p(\boldsymbol{n})$ in powers of $n\lambda = \varepsilon$. In the time independent case ($p = 0$), Childress [**21**] showed that, for sufficiently small local magnetic Reynolds number ($\lambda v/\eta$) and sufficiently small ε, these expansions were convergent and could be evaluated explicitly. He found that, in this low microscale Reynolds number case, a_{ij} is [like (13) above] symmetric to leading order and that, provided at least two of its eigenvalues are of the same sign, condition (25) defines a surface of wave number vectors each of which generates a steady dynamo. On the basis of general arguments, G. O. Roberts [**74**] has noted that the corresponding results for unsteady fields ($p \neq 0$) have immediate implications at *all* resistivities (except possibly for a set of measure zero). In these cases, a_{ij} is not necessarily symmetric but, provided the condition given above is satisfied by the eigenvalues of the symmetric part of a_{ij}, amplified solutions could exist for all sufficiently small $|\boldsymbol{n}|$.

An example of a working periodic dynamo is [**21**]

$$\boldsymbol{u}' = \frac{v}{\sqrt{3}}\left[\sin\frac{y}{\lambda} \pm \cos\frac{z}{\lambda}, \sin\frac{z}{\lambda} \pm \cos\frac{x}{\lambda}; \sin\frac{x}{\lambda} \pm \cos\frac{y}{\lambda}\right], \tag{3.3.26}$$

where the $\pm$ signs are taken together. Since this is a Beltrami field ($\boldsymbol{u}' \times \operatorname{curl} \boldsymbol{u}' = 0$), it clearly possesses helicity; in fact

$$\overline{\boldsymbol{u}' \cdot \boldsymbol{\omega}'} = \pm \overline{\boldsymbol{u}'^2}/\lambda.$$

And consistent with this [cf. (20) and (21)] we obtain, using (15)–(17),

$$a_{ij} = \pm\alpha\delta_{ij}, \qquad b_{ijk} = \beta\varepsilon_{ijk}, \qquad \lambda\alpha = \beta = (\lambda v)^2/3\eta. \tag{3.3.27}$$

The two-dimensional examples of §1.3 are

$$\boldsymbol{u}' = \frac{v}{\sqrt{2}}\left[\cos\frac{y}{\lambda} \mp \cos\frac{z}{\lambda}, \sin\frac{z}{\lambda}, \sin\frac{y}{\lambda}\right], \tag{3.3.28}$$

the upper sign also giving a Beltrami field. We obtain, using (15)–(17),

$$\begin{aligned} a_{22} &= \pm a_{33} = -\lambda v^2/2\eta, \\ b_{123} &= -b_{213} = \tfrac{1}{2}b_{231} = -\tfrac{1}{2}b_{321} = b_{312} = -b_{132} = (\lambda v)^2/2\eta, \end{aligned} \tag{3.3.29}$$

other components of these tensors being zero. Referring to the results stated in the last paragraph, we see that the choice of the upper sign in (29) gives first dynamo action, while the fate of the other cannot be settled to that order. G. O. Roberts [**75**] was able to show, however, that it could regenerate for some k *larger* than a certain lower bound (k_0, say) provided the Reynolds number was sufficiently large.

The fact that the choice of lower sign in (28) leads to dynamo action for some $k > k_0$ even though that flow possesses no net helicity suggests that dynamo maintenance may be possible even in a mirror symmetric turbulence. Investigations of this type of magnetohydrodynamic turbulence have a long and, at times, inconclusive history. We may mention, however, advances made by Batchelor [**6**], Chandrasekhar [**19**], Moffatt [**56**], Saffman [**80**], and by Kraichnan and Nagarajan [**51**]. In particular, we should note recent reports by Lerche [**39**] who, on taking the Fourier transform of the full expression (3.2.19) for $\mathscr{E}$ without the subsequent expansion process (3.2.20), concludes that kinematic dynamo action will occur even in isotropic mirror symmetric turbulence provided the wave number, k, of the mean field exceeds some critical k_0.

3.4. *Illustrations.* We examine the simplest case first: the α^2-dynamo promised in §3.1 (Krause and Steenbeck [**47**]). Imagine a sphere, V, of radius a filled with homogeneous isotropic turbulence of constant helicity and surrounded by an insulator $\hat{V}$. The mean field induction equation (3.2.11) reduces to

$$\partial\boldsymbol{B}/\partial t = \operatorname{curl}(\alpha\boldsymbol{B}) + \eta\nabla^2\boldsymbol{B}, \tag{3.4.1}$$

where α is constant; the overbars on the average fields are here and henceforward omitted, as is the suffix T on η_T. Dividing $\boldsymbol{B}$ into its toroidal and poloidal parts, i.e. writing

(3.4.2) $$\boldsymbol{B} = \text{curl}(T\boldsymbol{r}) + \text{curl}^2(S\boldsymbol{r}),$$

we obtain

(3.4.3) $$\partial T/\partial t = \eta\nabla^2 T - \alpha\nabla^2 S, \qquad \partial S/\partial t = \eta\nabla^2 S + \alpha T,$$

where

(3.4.4) $$\langle T\rangle = \langle S\rangle = \langle \partial S/\partial r\rangle = 0, \text{ on } r = a,$$

and

(3.4.5) $$\hat{T} = \nabla^2\hat{S} = 0, \qquad \hat{S} \to 0, \quad \text{as } r \to \infty.$$

The solution to this problem is elementary, being essentially an extension of that determining the decay-modes of a stationary sphere (cf. e.g. P. H. Roberts [**76**, p. 97]).

The solution is first separated into spherical harmonic components by writing

(3.4.6) $$T = \sum_{n=1}^{\infty} T_n(r)Y_n(\theta, \phi)\exp(\sigma_n t), \quad \text{etc.}$$

Substitution into (3) then gives

(3.4.7) $$\eta\left[\frac{d^2}{dr^2} - \left(\frac{\sigma_n}{\eta} + \frac{n(n+1)}{r^2}\right)\right](rT_n) = \alpha\left[\frac{d^2}{dr^2} - \frac{n(n+1)}{r^2}\right](rS_n),$$

(3.4.8) $$\eta\left[\frac{d^2}{dr^2} - \left(\frac{\sigma_n}{\eta} + \frac{n(n+1)}{r^2}\right)\right](rS_n) = -\alpha(rT_n),$$

where, by (4) and (5),

(3.4.9) $$T_n = \partial S_n/\partial r + [(n+1)/r]S_n = 0, \quad \text{at } r = a.$$

Since singularities in V must be avoided, (7) and (8) solve as linear combinations of $J_{n+1/2}(k_i r)/(k_i r)^{1/2}$, where

(3.4.10) $$(k_1, k_2) = \alpha/2\eta \pm [(\alpha/2\eta)^2 - \sigma_n/\eta]^{1/2},$$

and the application of (9) yields the consistency condition

(3.4.11) $$J'_{n+1/2}(k_1 a)/J_{n+1/2}(k_1 a) = J'_{n+1/2}(k_2 a)/J_{n+1/2}(k_2 a),$$

determining the eigenvalues σ_n. For $\alpha \to 0$, the toroidal and poloidal decay-modes are recovered. For these, σ_n is negative, the largest being $-\eta(\pi/a)^2$ and corresponding to a poloidal $n = 1$ mode. As $|\alpha|$ increases, the eigenvalues remain real, and the largest approaches zero. In fact, this marginal case is reached when the turbulent magnetic Reynolds number,

$$R_{\alpha^2} \equiv a|\alpha|/\eta, \tag{3.4.12}$$

reaches the smallest positive root of $J_{n+1/2}(z)$. The least such root occurs for $n = 1$ and is 4.494..., i.e. the dynamo becomes self-excited once R_{α^2} reaches this value, and at larger values the fields will grow. (The critical value for the $n = 2$ mode is 5.763... .)

We have already remarked (below 3.2.30) that the assumption of constant α is artificial: in practice, we expect α to be proportional to $\cos\theta$. The resulting theory is still governed by (1), but the solutions no longer separate into spherical harmonics. Restricting attention to axisymmetric fields, as we shall do henceforward, we find that there are two families of solutions: (i) the dipole type, composed of S of odd n and T of even n, and (ii) the quadrupole type composed of S of even n and T of odd n. Defining R_{α^2} by the constant value of $\alpha \sec\theta$ in place of α in (12), Steenbeck and Krause [**85**], [**86**] obtained smallest critical values of 8.26 and 9.62, respectively, in these two cases. These values were derived by a drastic truncation of the harmonic expansion but are, nevertheless, reasonably accurate. By taking the harmonic series much further, the present author obtained the values 7.64 and 7.81, respectively. The corresponding fields are stationary in time. Steenbeck and Krause [**88**] attempt to account for the geomagnetic field by dynamo models of this type and have here examined the convergence of their solutions carefully.

Many astrophysical bodies possess not only a rotation which is responsible for an α-effect as already described, but also a *differential* rotation which creates toroidal field from poloidal (§1.4). If the Reynolds number defined by this shear is large compared with the α-effect Reynolds number (12), the production of toroidal field from poloidal by the α-effect may be neglected. Moreover, for the same reason, the toroidal fields produced will be large compared with the poloidal, roughly as the root of the ratio of the two Reynolds numbers just introduced. Dynamos of this type, which function on the "product" of α- and ω-effects might be called "$\alpha\omega$-dynamos." The models studied above would then be called "α^2-dynamos" since they function through a product of α-effects. Since differential rotation (rather than solid body rotation) is an essential ingredient of the $\alpha\omega$-dynamo, it might better be called an "$\alpha\omega'$-dynamo"! The α^2-dynamos, for which differential rotation is neglected and which rely on an α-effect associated with a solid body rotation, come close to reviving the old idea (Blackett [**9**]) that cosmic bodies might be intrinsically magnetized through their rotation.

Since the time of the early models of Parker [**60**], there have been many studies of $\alpha\omega$-dynamos: e.g. Krause and Steenbeck [**46**], Steenbeck and Krause [**85**]–[**87**], Parker [**62**], [**65**]–[**67**], and, in a different context (§4), Braginskiĭ [**11**, §4], [**12**]. The relationship between $\nabla\omega$ and the sign of α is crucial to the character of the solutions, as Parker [**60**], [**64**] first recognized. If the sense is as described in §3.1, i.e. positive helicity in the northern hemisphere and negative in the southern hemisphere, and if ω decreases outwards as ideas of conservation of angular momentum suggest, the dipole field can be regenerated, as Figure 14a indicates. Here the curves in $\hat{V}$ show the external dipole field. The toroidal lines in V give the direction in which this is sheared by a motion which is predominantly eastward (out of paper) in the interior and westward in the exterior [assuming $\omega = \omega(r)$ for definiteness; cf. Figure 3a]; the little loops on the toroidal field lines indicate the effect of the helicity. "Clearly" the sense is such as to regenerate the external field. If either the sign of ω' or the sign of the helicity had been different (but *not* both), this would not have been the case. Parker [**64**, p. 399] argues that in these cases steady dynamo solutions are not possible; Steenbeck and Krause [**85**], [**86**] however report numerical computations which indicate that steady dipole and quadrupole solutions may exist. [The situation is illustrated in Figure 14b. If the quadrupole is to regenerate, the net α-effect from low latitudes (between extreme dotted lines) must swamp that from high latitudes.] Whichever view is correct, there is little doubt that time-dependent solutions of the $\alpha\omega$-dynamo are the norm. They may be illustrated best by the migratory dynamo of Parker. In Figure 15, the $\pm$ signs represent flux tubes in which $\boldsymbol{B}$ has a component out (plus) or into (minus) the paper. If $\alpha > 0$, this component will generate through the α-effect a secondary field in the xy-plane given by the sense of the arrows. Now suppose that a z-component of velocity is present, which has a constant shear in the y-direction (i.e. $u_z = u_z' y$, where u_z' is a positive constant). We may picture this in Figure 15 as consisting of a large velocity out of the paper for $y > 0$ and into the paper for $y < 0$. Thus, in region F, the secondary field lines will be tipped into the paper, while at G they will be tipped out. In other words, the right side of the original flux tubes is regenerated while the left is depleted. Whether its strength suffers a net increase or decrease depends, of course, on the magnitude of $\alpha u_z'$, but it is clear that, if the correct value is chosen, the overall effect will be a motion of each flux tube to the right. If the sign of $\alpha u_z'$ is reversed, it will move to the left. In the case of the $\alpha\omega$-dynamos described below (and identifying z with ϕ, y with r, and x with latitude, $\frac{1}{2}\pi - \theta$), the waver will tend to move poleward if $\alpha\omega'(r)$ is predominately positive in the northern hemisphere, and equatorwards (like the Sun) otherwise. In both cases, the solution is

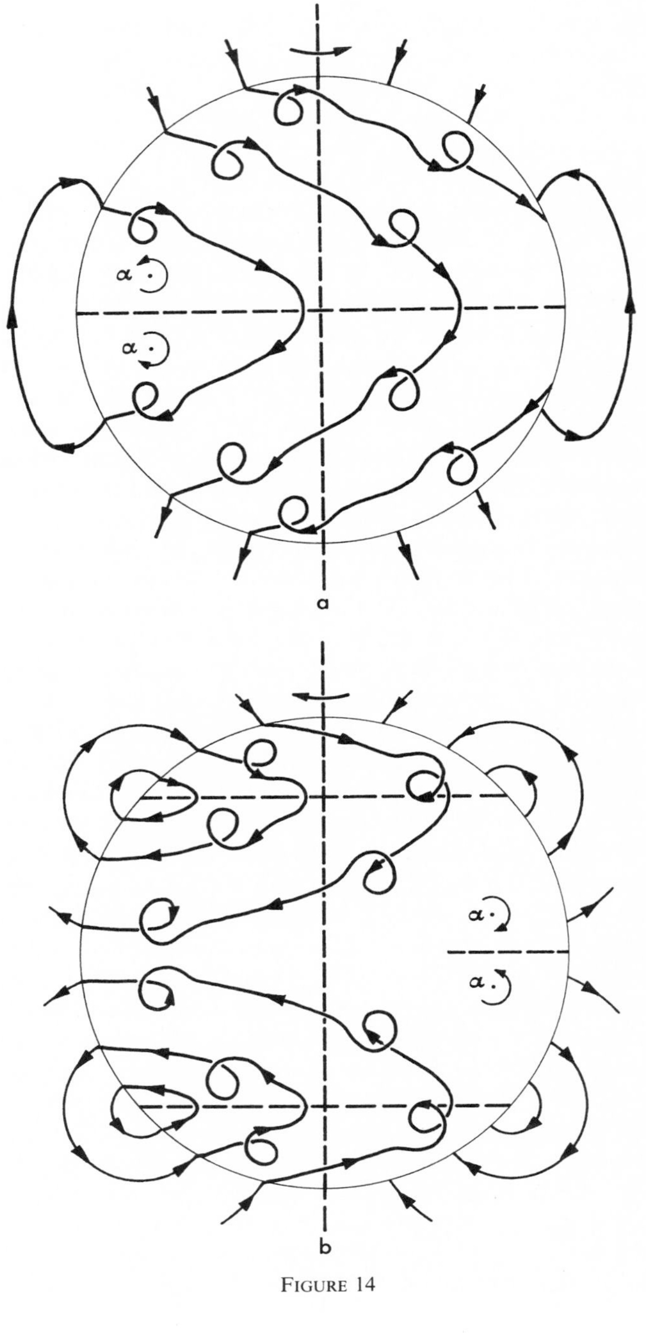

FIGURE 14

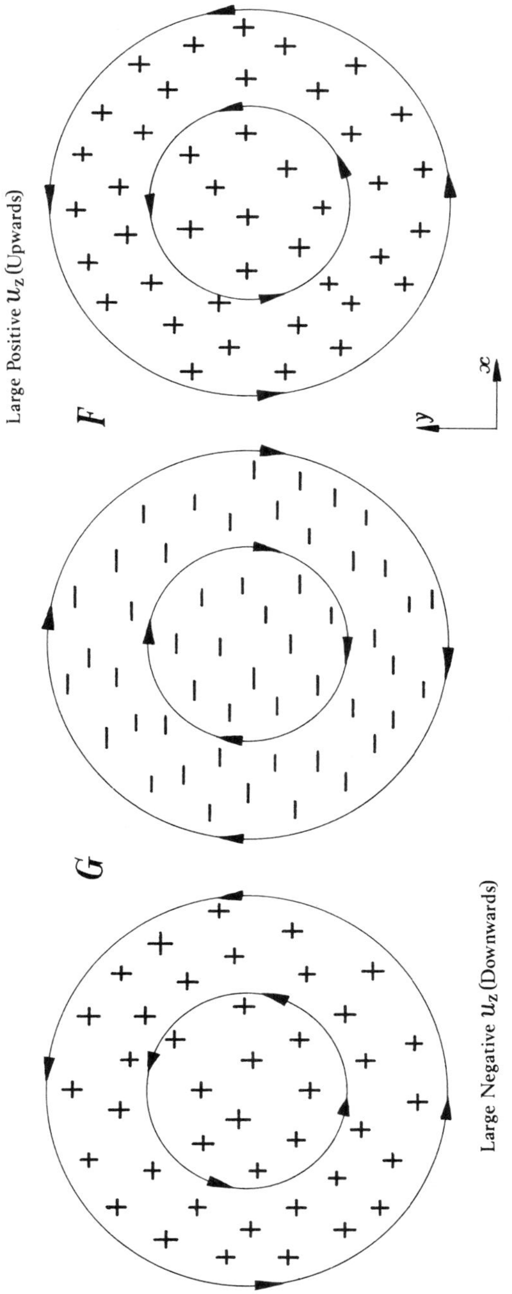
Large Positive u_z (Upwards)
F
y
x
G
Large Negative u_z (Downwards)

FIGURE 15

unsteady. Other types of dynamo waves have been examined by G. O. Roberts [**73**].

To examine the $\alpha\omega$-dynamos theoretically, we first write the mean field equation (3.2.11) as

$$\partial \boldsymbol{B}/\partial t = \operatorname{curl}(\boldsymbol{u} \times \boldsymbol{B} + \alpha \boldsymbol{B}_\phi) + \eta \nabla^2 \boldsymbol{B}. \tag{3.4.13}$$

In doing so, we have replaced $\alpha\boldsymbol{B}$ by $\alpha\boldsymbol{B}_\phi$, where $\boldsymbol{B}_\phi = B\boldsymbol{\phi}$ is, in the present case of axisymmetry, the toroidal component of $\boldsymbol{B}$, and have therefore recognized (as discussed above) that it is not essential to include the generation of poloidal current by the α-effect. In terms of the vector potential introduced in §1.4, for which (in cylindrical coordinates)

$$\boldsymbol{B} = \left[-\frac{\partial A}{\partial z}, B, \frac{1}{\varpi}\frac{\partial}{\partial \varpi}(\varpi A) \right], \tag{3.4.14}$$

Equation (13) may be divided into the Parker equations:

$$\frac{\partial A}{\partial t} + \frac{\boldsymbol{u}}{\varpi} \cdot \nabla(\varpi A) = \eta \Delta A + \alpha B, \tag{3.4.15}$$

$$\frac{\partial B}{\partial t} + \varpi \boldsymbol{u} \cdot \nabla\left(\frac{B}{\varpi}\right) = \eta \Delta B + \left[\nabla\left(\frac{u_\phi}{\varpi}\right) \times \nabla(\varpi A)\right]_\phi, \tag{3.4.16}$$

which differ from (1.4.15–16) primarily by the crucial αB source in (15), which can prevent the collapse of A. In the present case, in which $\boldsymbol{u}$ has only a ϕ-component

$$\boldsymbol{u} = \boldsymbol{\omega} \times \boldsymbol{r}, \tag{3.4.17}$$

(15) and (16) simplify to

$$\partial A/\partial t = \eta \Delta A + \alpha B, \tag{3.4.18}$$

$$\partial B/\partial t = \eta \Delta B + [\nabla \omega \times \nabla(\varpi A)]_\phi. \tag{3.4.19}$$

Assuming α is odd in $\frac{1}{2}\pi - \theta$ and ω is even, the solutions (as for the α^2-dynamo) separate into dipole and quadrupole families. Assuming

$$\alpha = \alpha_0 \cos\theta, \qquad \omega = \omega_0' r \tag{3.4.20}$$

where α_0 and ω_0' are constants, Steenbeck and Krause [**85**] obtained smallest eigenvalues for steady dynamo action of

$$R_{\alpha\omega} \equiv \left| \frac{\alpha_0 \omega_0' a^4}{\eta^2} \right|^{1/2} \doteqdot \begin{cases} 25.5, \text{ for the } \alpha_0\omega_0' > 0 \text{ dipole solution,} \\ 44.7, \text{ for the } \alpha_0\omega_0' < 0 \text{ quadrupole solution.} \end{cases} \tag{3.4.21}$$

It may be noted however that Braginskiĭ [**12**] integrated (15) and (16) for three models in which the second of (20) was assumed, while the first was replaced by $\alpha = \alpha_0 \sin^2\theta\cos\theta$; he did not obtain steady solutions in the absence of a meridional circulation additional to (17). And perhaps, on this occasion, the results (21) are misleading. As for the results quoted above for the α^2-dynamo, Steenbeck and Krause obtained (21) by a drastic truncation of the harmonic expansion of the field. On this occasion, the results so derived do not appear to resemble the actual solution sought. The present author repeated the calculations taking many more terms in the harmonic series and has found that the most easily excited modes are oscillatory, and that, despite the appearance of (21), the most easily excited dipole (quadrupole) type field is one in which $\alpha_0\omega_0'$ is negative (positive). The results obtained were:

(a) dipole $\alpha_0\omega_0' > 0$, $R_{\alpha\omega} = 87.13$ (68.75),
(b) dipole $\alpha_0\omega_0' < 0$, $R_{\alpha\omega} = 74.39$ (54.07),
(c) quadrupole $\alpha_0\omega_0' < 0$, $R_{\alpha\omega} = 85.36$ (67.44),
(d) quadrupole $\alpha_0\omega_0' > 0$, $R_{\alpha\omega} = 76.08$ (55.14).

The figures in parentheses in each case record the corresponding oscillation frequency, Ω, in units of a reciprocal diffusion time. Again, the dipole and quadrupole families are excited with almost equal ease.

Steenbeck and Krause [**87**] also examined spherical $\alpha\omega$-dynamos of more complex structure under less drastic truncation. They were generally able to obtain converged solutions which, they found, oscillated in time. Unlike Deinzer and Stix [**24**] and the present author, however, they discovered that the solutions of quadrupole type were considerably harder to excite than those of dipole type.

To examine layered dynamos, we specialize (18) and (19) to plane geometry, by supposing that $\boldsymbol{B} = \boldsymbol{B}(x, t)$ and $B_x = 0$. In this case, in which α and $\zeta = \partial u_z/\partial y$ are functions of x alone, we obtain

(3.4.22) $$\frac{\partial B_y}{\partial t} = \eta\frac{\partial^2 B_y}{\partial x^2} - \frac{\partial}{\partial x}[\alpha(x)B_z],$$

(3.4.23) $$\frac{\partial B_z}{\partial t} = \eta\frac{\partial^2 B_z}{\partial x^2} + \zeta(x)B_y.$$

To obtain Parker's migratory dynamo wave, we suppose α and ζ are constant and seek solutions of the form

(3.4.24) $$\boldsymbol{B} = \boldsymbol{B}_0 \exp[i(kx - \omega t)], \qquad (\boldsymbol{B}_0 = \text{constant}).$$

These exist provided

$$\eta k^2 - i\omega = \begin{matrix} (1 - i)(\frac{1}{2}k\alpha\zeta)^{1/2}, & \text{if } k\alpha\zeta > 0, \\ (1 + i)(-\frac{1}{2}k\alpha\zeta)^{1/2}, & \text{if } k\alpha\zeta < 0. \end{matrix} \tag{3.4.25}$$

Thus, if $\alpha\zeta > 0$, k and $\mathscr{R}(\omega)$ have the same sign, and the waves (24) move in the positive x-direction; if $\alpha\zeta < 0$ they move in the opposite direction. The condition, $\mathscr{I}(\omega) = 0$, for constant amplitude solutions is

$$R^2_{\alpha\omega} \equiv \alpha\zeta k^{-3}/\eta^2 = \pm 2. \tag{3.4.26}$$

The layer model of Steenbeck and Krause [**85**], [**86**] is also governed by (22) and (23), but different types of solutions are sought which do not progress in the x-direction. In their case, the shear is meant to model the variation in latitude of the solar differential rotation, so that now the solar (r, θ, ϕ)-coordinates correspond to (x, y, z) above, in that order. The solar dynamo is probably only associated with the outermost layers of the sun, so that it is reasonable to seek solutions which decay with depth, x, into the Sun, and those of Steenbeck and Krause are contained within the electromagnetic skin depth $(\eta/\Omega)^{1/2}$ of the surface. They found that such solutions cannot exist unless the regions of large ζ and large α in this layer are spatially separated, a conclusion supported by Deinzer and Stix [**24**]. This consideration affected their later choice of models of the solar dynamo (Steenbeck and Krause [**87**]);[2] the resulting similarity between the temporal behaviour of their models and the famous butterfly diagrams (obtained from observation) is very striking. Galactic $\alpha\omega$-dynamos are considered by Parker [**62**], [**66**].

It should be pointed out that even quite small poloidal flows have a profound effect on the behaviour of the $\alpha\omega$-dynamo, as the present author has shown in an unpublished study. So that the parity scheme used above should be preserved, it was supposed that the meridional circulation was composed of a linear combination of $\tilde{S}_{2n}$ flows; this is also dynamically plausible. Once this flow velocity is comparable to, or greater than, the wave speed implied by (25), its effects were felt, it being found that one sense of circulation helped dynamo action while the other hindered it. Confining attention to the former, it was noted that the critical Reynolds numbers for the two field types were separated, there being a tendency for steady dipole type fields to be preferred if $\alpha\omega_0'$ is positive in the $z > 0$ hemisphere, and for steady quadrupole type fields to be preferred if the opposite was true. The minimum Reynolds numbers, in each case, were comparable.

---

[2] See also Krause and Hiller [**44**].

4. Nearly symmetric dynamos.

4.1. *Solution by iteration.* The difficulties of the direct integration of the steady induction equation, viz. (in dimensionless form)

$$\nabla^2 \boldsymbol{B} + R\,\mathrm{curl}(\boldsymbol{u} \times \boldsymbol{B}) = 0, \tag{4.1.1}$$

in the sphere $r = 1$ containing a large-scale motion, $\boldsymbol{u}$, have already been touched on in §1.4. In fact, it was partly these difficulties which prompted the less direct approach of §3. The question arises whether (1) may be solved by iteration. The obvious small R expansion

$$\boldsymbol{B} = \sum_0^\infty R^m \boldsymbol{B}_m, \tag{4.1.2}$$

the $\boldsymbol{B}_m$ being determined successively from

$$\nabla^2 \boldsymbol{B}_m = -\mathrm{curl}(\boldsymbol{u} \times \boldsymbol{B}_{m-1}), \qquad (m \geqq 0,\ \boldsymbol{B}_{-1} \equiv 0), \tag{4.1.3}$$

is clearly useless, for, bearing in mind (1.2.8), the only possible starting point is $\boldsymbol{B}_0 = 0$! [Moreover, dynamo action can only occur if the magnetic Reynolds number, R, is at least $O(1)$; cf. §1.3 above; so, questions of convergence of (2) apart, the iteration (3) would have had to have been carried to impractical lengths.]

An alternative to (2) is a large R expansion

$$\boldsymbol{B} = \sum_0^\infty R^{-m} \boldsymbol{B}_m, \tag{4.1.4}$$

the $\boldsymbol{B}_m$ being determined successively from

$$\mathrm{curl}(\boldsymbol{u} \times \boldsymbol{B}_m) = -\nabla^2 \boldsymbol{B}_{m-1}, \qquad (m \geqq 0, \boldsymbol{B}_{-1} \equiv 0). \tag{4.1.5}$$

This is more promising. The starting point, $\boldsymbol{B}_0$, obeys the steady induction equation for a perfect conductor. This can certainly be satisfied non-trivially, e.g. by the aligned field

$$\boldsymbol{B}_0 = \chi_0 \boldsymbol{u}, \tag{4.1.6}$$

where χ_0 is constant on streamlines. Since $\boldsymbol{n} \cdot \boldsymbol{u}$ and $\langle \boldsymbol{n} \cdot \boldsymbol{B}_0 \rangle$ vanish on S, we must have

$$\hat{\boldsymbol{B}}_0 = 0. \tag{4.1.7}$$

It must be strongly emphasized that the expansion scheme (4)–(5) is dominated by the fact that the operator on the left of (5) is singular, and that in consequence the equation cannot be solved unless its right side satisfies a consistency condition at *every* stage of the iteration.

To demonstrate this, we uncurl (5) to give

$$(4.1.8) \quad \boldsymbol{u} \times \boldsymbol{B}_m - \operatorname{grad} \Phi_m = \operatorname{curl} \boldsymbol{B}_{m-1}, \quad \operatorname{div} \boldsymbol{B}_m = 0 \quad (m \geqq 0, \boldsymbol{B}_{-1} \equiv 0).$$

Now a streamline of $\boldsymbol{u}$ may either (i) be a closed curve (Γ), or (ii) trace out a complete closed surface (Σ) in a family filling V, or (iii) fill V completely. And (8) gives in these three cases

$$(4.1.9) \quad \text{(i)} \oint_\Gamma \boldsymbol{u} \cdot \operatorname{curl} \boldsymbol{B}_m \frac{ds}{u} = 0, \qquad \text{(ii)} \int_\Sigma \boldsymbol{u} \cdot \operatorname{curl} \boldsymbol{B}_m \, dS = 0,$$

$$\text{(iii)} \int_V \boldsymbol{u} \cdot \operatorname{curl} \boldsymbol{B}_m \, dV = 0, \quad (m \geqq 0),$$

conditions which, indeed, $\boldsymbol{B}$ itself satisfies, as (1) shows. In particular, by (6)

$$(4.1.10) \quad \text{(i)} \oint_\Gamma \boldsymbol{u} \cdot \boldsymbol{\omega} \frac{ds}{u} = 0, \qquad \text{(ii)} \int_\Sigma \boldsymbol{u} \cdot \boldsymbol{\omega} \, dS = 0,$$

$$\text{(iii)} \int_V \boldsymbol{u} \cdot \boldsymbol{\omega} \, dV = 0.$$

These necessary conditions [for dynamos based on (6)] are not particularly encouraging, bearing in mind the helicity arguments of §3. Even if (10) is met, there is no guarantee that other equally awkward demands will not arise at later stages of the iteration: the fact that solutions to (5) are arbitrary to an additive $\chi_m \boldsymbol{u}$ (where χ_m is constant on streamlines) is of no help in meeting (9) since, by (10), its contribution to the relevant integral is zero.

There are other less serious difficulties. In the parlance of the hydrodynamics of flows at large Reynolds numbers, (4) is actually an "outer solution" which is not valid at distances which are $O(R^{-1/2})$ from S, as $R \to \infty$. The differential equation (5) is of lower order than the original (1), and its solution consequently cannot meet all the continuity requirements of S, no matter what eligible $\hat{\boldsymbol{B}}_m$ is selected. In fact, after satisfying the primary requirement of continuity of $\boldsymbol{n} \cdot \boldsymbol{B}_m$, little further scope remains, and $\boldsymbol{n} \times \boldsymbol{B}_m$ will be discontinuous. To correct this, an "inner solution" must be developed by scaling distances from S in units of $R^{-1/2}$. Differential equations similar to (3) are obtained, which are of the same order as (1). The solutions selected can, and in fact must, match with both the outer and insulator fields, thereby eliminating the discontinuity in $\boldsymbol{n} \times \boldsymbol{B}_m$. In doing so, however, a fresh discontinuity, of order $R^{-1/2}|\boldsymbol{B}_m|$, is in general created in $\boldsymbol{n} \cdot \boldsymbol{B}$, and this will excite fields of like order throughout V and $\hat{V}$. Such terms lie outside the expansion scheme (4)–(5), which therefore must be amended to

$$\boldsymbol{B} = \sum_{0}^{\infty} R^{-m/2} \boldsymbol{B}_m, \tag{4.1.11}$$

where

$$\operatorname{curl}(\boldsymbol{u} \times \boldsymbol{B}_m) = -\nabla^2 \boldsymbol{B}_{m-2}, \qquad (m \geqq 0, \boldsymbol{B}_{-2} \equiv \boldsymbol{B}_{-1} \equiv 0). \tag{4.1.12}$$

According to (12), $\boldsymbol{B}_0$ and $\boldsymbol{B}_1$ are governed by the same equation. If we assume $\boldsymbol{B}_0$ is given by (6), we might still take the different solution

$$\boldsymbol{B}_1 = \chi_1 \boldsymbol{u} - (\boldsymbol{u} \times \nabla\Phi_1)/u^2, \tag{4.1.13}$$

for $\boldsymbol{B}_1$. Here Φ_1 is constant on streamlines, and (taking the divergence)

$$\boldsymbol{u} \cdot \nabla \chi_1 = \nabla\Phi_1 \cdot \operatorname{curl}(u^{-2}\boldsymbol{u}). \tag{4.1.14}$$

Solutions of this type do not seem to have been fully explored, although they appear to fail in the context of the Braginskiĭ expansion below, as he in fact states. If, instead, the special case of constant Φ_1 is assumed and $\chi_0 + R^{-1/2}\chi_1$ is replaced by a fresh χ_0, the $\boldsymbol{B}_1$ term in (11) is completely removed. This is at variance with the boundary layer argument unless we make $\langle \boldsymbol{B}_0 \rangle = 0$ on S, by choosing χ_0 to be zero on all streamlines in contact with it.

There is a practical difficulty also. Simple solutions to (12) cannot be expected from complicated $\boldsymbol{u}$; but, if we took $\boldsymbol{u}$ to be, for example, axisymmetric (with $u_\phi \neq 0$, for geophysical relevance), the iteration (12) based on (6) would also, in general, be axisymmetric, and we know that such solutions do not exist. An interesting and unusual way out is to expand $\boldsymbol{u}$ in a series similar to (11):

$$\boldsymbol{u} = \sum_{0}^{\infty} R^{-m/2} \boldsymbol{u}_m, \tag{4.1.15}$$

where $\boldsymbol{u}_0$ is simple, and $\boldsymbol{u}_1$ etc. contain the required asymmetric terms. This does not alter the form of the iteration (11), although (12) must be replaced by

$$\operatorname{curl}(\boldsymbol{u}_0 \times \boldsymbol{B}_m) = -\operatorname{curl} \sum_{p=1}^{m} (\boldsymbol{u}_p \times \boldsymbol{B}_{m-p}) - \nabla^2 \boldsymbol{B}_{m-2} \qquad (m \geqq 0, \boldsymbol{B}_{-2} \equiv \boldsymbol{B}_{-1} \equiv 0). \tag{4.1.16}$$

The first two of these are satisfied by

$$\boldsymbol{B}_0 = \chi_0 \boldsymbol{u}_0, \qquad \boldsymbol{B}_1 = \chi_0 \boldsymbol{u}_1 + \chi_1 \boldsymbol{u}_0, \tag{4.1.17}$$

provided (taking divergences) χ_0 is constant on streamlines of $\boldsymbol{u}_0$ and vanishes on S, and

$$\boldsymbol{u}_0 \cdot \nabla\chi_1 = -\boldsymbol{u}_1 \cdot \nabla\chi_0. \tag{4.1.18}$$

By uncurling (16), we may show that, at the mth level of the iteration, i.e. that at which all fields up to and including $\boldsymbol{B}_m$ are known, we must have

$$\text{(i)} \quad \int_{\Gamma_0} \boldsymbol{u}_0 \cdot \left[\operatorname{curl} \boldsymbol{B}_{m-1} - \sum_{p=1}^{m} (\boldsymbol{u}_p \times \boldsymbol{B}_{m+1-p})\right]\frac{ds}{u_0} = 0, \text{ etc.} \qquad (m \geqq 1), \tag{4.1.19}$$

[cf. (9)], the first of which implies (10) for $\boldsymbol{u} = \boldsymbol{u}_0$. It is clear from (18) that $\boldsymbol{B}_1$ is arbitrary to any multiple of $\boldsymbol{u}_0$ which is constant on streamlines, Γ_0, of $\boldsymbol{u}_0$. The same is true at all levels of iteration, but as before this does not help in meeting (19).

The consistency conditions may seem to be a nuisance, but they really contain the essential physics. To make this clearer, suppose we had assumed that $\boldsymbol{B}$ and $\boldsymbol{u}$ are time-varying on the scale of the free decay time [i.e. $\partial/\partial t = O(1)$ in these units]. We would then have needed to add $\partial\boldsymbol{B}_{m-2}/\partial t$ to the right of (16), and the consistency conditions would essentially have determined these time derivatives, i.e. the evolution of the field. For example, if we took the series only to the 0th level, the $m = 1$ conditions (19) (suitably modified to include time variation) would determine $\partial\chi_0/\partial t$. Without the consistency condition (19), χ_0 would have been arbitrary, as one would indeed expect if there were no ohmic dissipation.

Essential though consistency conditions are, it is desirable, if possible, to formulate the iteration in such a way that it is transparently obvious that they have been satisfied up to any level. It is the object of the next section to show how such a self-consistent scheme can be devised in a particular case.

4.2. *Braginskiĭ's self-consistent expansion.* The idea of expansions in large R is due to Braginskiĭ [**10**], who selected the simplest $\boldsymbol{u}_0$, and, at the same time, the one of probably greatest geophysical interest, viz.

$$\boldsymbol{u}_0 = u_0\mathbf{1}_\phi = \varpi\zeta\mathbf{1}_\phi \quad \text{(say)}, \qquad u_0 = u_0(\varpi, z), \tag{4.2.1}$$

(where, in this section, $\mathbf{1}_s$ is used to denote a unit vector in the s direction). Since χ_0 is constant on lines of $\boldsymbol{u}_0$, it is here axisymmetric. Thus, by (4.1.17), $\boldsymbol{B}_0 = \chi_0\boldsymbol{u}_0$ is also in the azimuthal direction and axisymmetric. Time dependence is also included, so that (4.1.16) becomes

$$\zeta\frac{\partial_1 \boldsymbol{B}_m}{\partial\phi} - \varpi(\boldsymbol{B}_m\cdot\nabla\zeta)\boldsymbol{1}_\phi = \operatorname{curl}\sum_{p=1}^{m}(\boldsymbol{u}_p\times\boldsymbol{B}_{m-p}) + \nabla^2\boldsymbol{B}_{m-2} - \frac{\partial\boldsymbol{B}_{m-2}}{\partial t} \qquad (m \geqq 0), \tag{4.2.2}$$

where $\partial_1/\partial\phi$ denotes differentiation with respect to ϕ holding the unit vectors constant. The consistency condition (4.1.19) is equivalent here to the statement that, since the meridional axisymmetric part of the left of (2) vanishes, the same must be true of the right. By (4.1.17) this is automatically true for $m = 0$ and 1.

To examine consistency questions further, it is useful to introduce the division of fields into axisymmetric and asymmetric parts, along the same lines as (3.2.1), where now, however, $\bar{F}$ denotes the ϕ-average,

$$\bar{F}(\varpi, z) = \frac{1}{2\pi}\int_0^{2\pi} F(\varpi, \phi, z)\,d\phi, \tag{4.2.3}$$

of a field F, and $F' \equiv F - \bar{F}$. Clearly (3.2.2–4) apply. For the asymmetric fields, the inverse of the $\partial_1/\partial\phi$ operation is useful: it will be denoted by $\hat{}$, where it will be understood that the arbitrary axisymmetric "field of integration" is zero, i.e. $\bar{\hat{F}} = 0$. Then equations (2) may be divided into symmetric and asymmetric parts (cf. (3.2.11–12)). The symmetric part gives

$$\frac{\partial\bar{\boldsymbol{B}}_{m,M}}{\partial t} = \operatorname{curl}\left[\sum_{p=1}^{m+2}(\bar{\boldsymbol{u}}_p\times\bar{\boldsymbol{B}}_{m+2-p})_\phi + \mathscr{E}_{m+2,\phi}\right] + \nabla^2\bar{\boldsymbol{B}}_{m,M}, \tag{4.2.4}$$

$$\frac{\partial\bar{\boldsymbol{B}}_{m-2,\phi}}{\partial t} = \operatorname{curl}\left[\sum_{p=1}^{m}(\bar{\boldsymbol{u}}_p\times\bar{\boldsymbol{B}}_{m-p})_M + \mathscr{E}_{m,M}\right] + \nabla^2\bar{\boldsymbol{B}}_{m-2,\phi} + \varpi(\bar{\boldsymbol{B}}_{m,M}\cdot\nabla\zeta)\boldsymbol{1}_\phi, \tag{4.2.5}$$

while the asymmetric part is (cf. (3.2.12))

$$\zeta\frac{\partial_1\boldsymbol{B}'_{m+1}}{\partial\phi} - \varpi(\boldsymbol{B}'_{m+1}\cdot\nabla\zeta)\boldsymbol{1}_\phi = \operatorname{curl}\left[\sum_{p=1}^{m+1}(\bar{\boldsymbol{u}}_p\times\boldsymbol{B}'_{m+1-p} + \boldsymbol{u}'_p\times\bar{\boldsymbol{B}}_{m+1-p}) + \boldsymbol{G}'_{m+1}\right] + \nabla^2\boldsymbol{B}'_{m-1} - \frac{\partial\boldsymbol{B}'_{m-1}}{\partial t}. \tag{4.2.6}$$

In (4)–(6) we have (cf. (3.2.10, 13))

$$\mathscr{E} = \overline{\boldsymbol{u}' \times \boldsymbol{B}'}, \qquad \boldsymbol{G}' = \boldsymbol{u}' \times \boldsymbol{B}' - \overline{\boldsymbol{u}' \times \boldsymbol{B}'}, \tag{4.2.7}$$

so that

$$\mathscr{E}_m = \sum_{p=1}^{m} \overline{\boldsymbol{u}'_p \times \boldsymbol{B}'_{m-p}}, \qquad \boldsymbol{G}'_{m+1} = \sum_{p=1}^{m+1} (\boldsymbol{u}'_p \times \boldsymbol{B}'_{m+1-p}) - \mathscr{E}_{m+1}. \tag{4.2.8}$$

The objective, it will be recalled, is to incorporate the consistency conditions, which in this case comprise (4) above, *within* the iteration scheme, and (5) now shows the desirability of redefining the mth level of iteration, as far as the symmetric field is concerned, to embrace all components up to and including $\overline{\boldsymbol{B}}_{m,M}$ and (*not* $\overline{B}_{m,\phi}$ but) $\overline{B}_{m-2,\phi}$. (To face up to the dissipative effects properly, m must be at least 2.) The hope, then, is to close the axisymmetric system completely at the level shown in (4) and (5). There are, however, several other terms in (4) and (5) which would appear to prevent this. The most obvious are the $p = 1$ terms, which seem to require, respectively, a knowledge of $\overline{\boldsymbol{B}}_{m+1,M}$ and $\overline{B}_{m-1,\phi}$, and these are inaccessible at the (newly defined) mth level. This objection is overcome by the simple expediency of setting $\overline{\boldsymbol{u}}_1$ zero, i.e.

$$\boldsymbol{u}_1 \equiv \boldsymbol{u}'_1. \tag{4.2.9}$$

There is clearly no point in including a symmetric part to χ_1 since the corresponding term in $\boldsymbol{B}_1$ could be absorbed into a fresh $\overline{\boldsymbol{B}}_0$. Thus $\boldsymbol{B}_1 \equiv \boldsymbol{B}'_1$, and (4.1.17–18) now give

$$\overline{\boldsymbol{B}}_0 = \chi_0 \overline{\boldsymbol{u}}_0, \qquad \boldsymbol{B}'_1 = \chi_0 \boldsymbol{u}'_1 - \varpi(\hat{\boldsymbol{u}}_1 \cdot \nabla \chi_0)\mathbf{1}_\phi. \tag{4.2.10}$$

The next problem is raised by (4): the $\mathscr{E}_{m+2,\phi}$ appears to require a knowledge of $\boldsymbol{B}'_{m+1}$. At first sight, it would seem sensible to define the mth level of iteration, as far as the asymmetric fields are concerned, to embrace all terms up to, and including, $\boldsymbol{B}'_{m+1}$. Now, it can be shown that, defining $\boldsymbol{Q}'_q$ by

$$\boldsymbol{Q}'_q = \zeta \boldsymbol{B}'_q - (\overline{B}_{q-1,\phi}/\varpi)\boldsymbol{u}'_1 - \varpi[\hat{\boldsymbol{B}}_q \cdot \nabla \zeta - \hat{\boldsymbol{u}}_1 \cdot \nabla(\overline{B}_{q-1,\phi}/\varpi)]\mathbf{1}_\phi \tag{4.2.11}$$

(so that $\operatorname{div} \boldsymbol{Q}' = 0$), an integration of the M component of (6) gives

$$\boldsymbol{Q}'_q = \operatorname{curl}\left[\sum_{p=2}^{q} (\overline{\boldsymbol{u}}_p \times \hat{\boldsymbol{B}}_{q-p} + \hat{\boldsymbol{u}}_p \times \overline{\boldsymbol{B}}_{q-p}) + \hat{\boldsymbol{G}}_q\right] + \nabla^2 \hat{\boldsymbol{B}}_{q-2} - \frac{\partial \hat{\boldsymbol{B}}_{q-2}}{\partial t}. \tag{4.2.12}$$

Even assuming that (12) determines $\boldsymbol{Q}'_q$ for all $q \leqq m+1$, it would still only be possible to determine $\boldsymbol{B}'_q$ fully from (11) for $q \leqq m-1$; for

$\bar{B}_{m,\phi}$ and $\bar{B}_{m-1,\phi}$ are unknown. Nevertheless, $\boldsymbol{B}'_{m+1}$ and $\boldsymbol{B}'_m$ are only required in order to evaluate the ϕ-components of $\mathscr{E}_{m+2}$ and $\mathscr{E}_{m+1}$, and for this purpose $\boldsymbol{Q}'_{m+1}$ and $\boldsymbol{Q}'_m$ are, by (8), sufficient *provided* we supplement (9) by $\boldsymbol{u}'_2 \equiv 0$, i.e.

(4.2.13) $$\boldsymbol{u}_2 \equiv \bar{\boldsymbol{u}}_2 .$$

We may note in passing that this assumption removes the problem of evaluating $\hat{\boldsymbol{u}}_2 \times \bar{\boldsymbol{B}}_{m-1}$ on the right of (12) for $q = m + 1$.

But do we really know the right of (12) for all $q \leqq m + 1$? The case $q = m + 1$ is the only possible exception. The arguments which gave us $\mathscr{E}_{m+1,\phi}$ also give us $\hat{G}_{m+1,\phi}$. And $\hat{\boldsymbol{G}}_{m+1,M}$ appears in (12) only as $\mathbf{1}_\phi \times \boldsymbol{G}'_{m+1,M}/\varpi$. The only dubious terms it contains are

$$\frac{1}{\varpi}\mathbf{1}_\phi \times [\boldsymbol{u}'_{1,\phi} \times \boldsymbol{B}'_{m,M} + \boldsymbol{u}'_{1,M} \times \boldsymbol{B}'_{m,\phi}] - \text{Average}$$

$$= \frac{1}{\varpi\zeta}\mathbf{1}_\phi \times [\boldsymbol{u}'_{1,\phi} \times \boldsymbol{Q}'_{m,M} + \boldsymbol{u}'_{1,M} \times \boldsymbol{Q}'_{m,\phi}$$

$$+ \varpi[\hat{\boldsymbol{B}}_m \cdot \nabla\zeta - \hat{\boldsymbol{u}}_1 \cdot \nabla(\bar{\boldsymbol{B}}_{m-1,\phi}/\varpi)](\boldsymbol{u}'_{1,M} \times \mathbf{1}_\phi)] - \text{Average},$$

in which the only unknown is

$$\frac{1}{\zeta}[\hat{\boldsymbol{B}}_m \cdot \nabla\zeta - \hat{\boldsymbol{u}}_1 \cdot \nabla(\bar{B}_{m-1,\phi}/\varpi)]\boldsymbol{u}'_{1,M} - \text{Average}.$$

But, as we can see from (8), this contribution to $\boldsymbol{B}'_{m+1}$ will not affect $\mathscr{E}_{m+2,\phi}$, which was the only motive for determining $\boldsymbol{G}'_{m+1}$.

Summarizing: provided $\boldsymbol{u}$ satisfies (1), (9) and (13), and provided the starting point (10) is chosen for $\boldsymbol{B}$, the iteration scheme, in which at the mth level

(i) $\bar{\boldsymbol{B}}_M$ is known as far as $\bar{\boldsymbol{B}}_{m,M}$,

(ii) $\bar{B}_\phi$ is known as far as $\bar{B}_{m-2,\phi}$,

(iii) $\boldsymbol{B}'$ is known completely up to $\boldsymbol{B}'_{m-1}$ and partially up to $\boldsymbol{B}'_{m+1}$ ($\boldsymbol{B}'_{m,M}$ and $\boldsymbol{B}'_{m+1,M}$ being undetermined to an arbitrary multiple of $\boldsymbol{u}'_{1,M}$), is self-consistent. In the developments which follow, it is assumed that $\boldsymbol{u}_m$ vanishes for $m \geqq 3$, and that $u_{2,\phi} = 0$.

4.3. *Results and discussion.* When the scheme outlined in the last section is put into its necessarily very cumbersome operation, some remarkable and beautiful simplifications finally occur both at the $m = 2$ level (Braginskiĭ [**10**]) and the $m = 3$ level (Tough [**91**], see also [**92**]). Briefly, provided $\bar{\boldsymbol{u}}_2$ and $\bar{\boldsymbol{B}}_2$ are replaced by suitably defined "effective" variables $\bar{\boldsymbol{u}}_{2e}$ and $\bar{\boldsymbol{B}}_{2e}$, Parker's equations (3.4.15–16) are found to govern the symmetric components, with an α which can be computed explicitly from u_0 and $\boldsymbol{u}_1'$. To give for definiteness the $m = 2$ results (henceforward omitting the overbar on all symmetric fields), we write $\boldsymbol{B}_e = \boldsymbol{B}_0 + \boldsymbol{B}_{2e}$ in the form (3.4.14)

$$\boldsymbol{B}_e \equiv B_0 \mathbf{1}_\phi + \boldsymbol{B}_{2e} = \left[-\frac{\partial A_e}{\partial z}, B_0, \frac{1}{\varpi}\frac{\partial}{\partial \varpi}(\varpi A_e) \right], \tag{4.3.1}$$

and obtain (in dimensional units)

$$\frac{\partial A_e}{\partial t} + \frac{\boldsymbol{u}_{2e}}{\varpi} \cdot \nabla(\varpi A_e) = \eta \Delta A_e + \alpha B_0, \tag{4.3.2}$$

$$\frac{\partial B_0}{\partial t} + \varpi \boldsymbol{u}_{2e} \cdot \nabla\left(\frac{B_0}{\varpi}\right) = \eta \Delta B_0 + \left[\nabla\left(\frac{u_0}{\varpi}\right) \times \nabla(\varpi A_e) \right]_\phi, \tag{4.3.3}$$

where

$$\boldsymbol{u}_{2e} = \boldsymbol{u}_2 + \operatorname{curl}(w \boldsymbol{u}_0), \qquad \boldsymbol{B}_{2e} = \boldsymbol{B}_2 + \operatorname{curl}(w \boldsymbol{B}_0), \tag{4.3.4}$$

and (generally omitting henceforward the suffix on $\boldsymbol{u}_1'$)

$$w = \frac{\varpi}{2u_0^2} \overline{(\boldsymbol{u}' \times \hat{\boldsymbol{u}})}_\phi, \tag{4.3.5}$$

$$\alpha = \frac{\eta}{\varpi u_0^2} \left[\overline{\boldsymbol{u}' \times \hat{\boldsymbol{u}}} + \overline{\boldsymbol{u}' \times \frac{\partial_1 \boldsymbol{u}'}{\partial \phi}} \right]_\phi + 2\eta \overline{\nabla_M \left(\frac{r u_r'}{u_0} \right) \cdot \nabla_M \left(\frac{\hat{u}_z}{u_0} \right)}.$$

Since w and B_0 vanish on S, A_e must satisfy thereon the same boundary conditions, as A did, viz.

$$\langle A_e \rangle = \left\langle \frac{\partial A_e}{\partial r} \right\rangle = B_0 = 0, \quad \text{on } r = a. \tag{4.3.6}$$

Once (2) and (3) have been solved subject to (6), the final term in (4) can be evaluated, and $\boldsymbol{B}_2$ found. The same general conclusions hold at the $m = 3$ level, although naturally (4) and (5) have to be refined. It is known, however, that equations (2)–(3) are not recovered for $m = 4$ (Soward [**82**]). The fact that α is here proportional to η is not surprising when it is recalled that the same is true of A/B_0.

An interesting interpretation of the effective variables has been given by Soward [**83**]. He demonstrates that an $O(R^{-1})$ axisymmetric velocity, $\boldsymbol{v}$, exists which, to order R^{-1}, coincides with $\boldsymbol{u}_2 - \boldsymbol{u}_{2e}$ and which is such that $\boldsymbol{U} = \boldsymbol{u}_0 + \boldsymbol{u}_1' + \boldsymbol{v}$ has closed streamlines. Corresponding to this velocity, an axisymmetric field $\boldsymbol{b}$ exists which, to order R^{-1}, coincides with $\boldsymbol{B}_2 - \boldsymbol{B}_{2e}$ and is such that $\boldsymbol{H} = \boldsymbol{B}_0 + \boldsymbol{B}' + \boldsymbol{b}$ is aligned to $\boldsymbol{U}$, i.e., $\boldsymbol{H} = \chi \boldsymbol{U}$ [cf. (4.2.10)]. The central mathematical idea is that of applying the expansion technique of §4.2 to the simpler "heat conduction" equation $\boldsymbol{U} \cdot \nabla \chi = 0$, which χ must obey in virtue of the solenoidal nature of $\boldsymbol{U}$ and $\boldsymbol{H}$. Soward (unpublished) has also exploited the fact that $\boldsymbol{U}$ possesses closed streamlines to obtain a simpler derivation of the results of Braginskiĭ [**11**] for the case in which $\boldsymbol{u}'_1$ is rapidly varying in time and its double ϕt average vanishes [cf. discussion after (15) below]. He has also succeeded in relating the α of (5) with the α-effect of §3, and has therefore forged a new and powerful link between the turbulent and the nearly symmetric dynamos.

It is easily seen that terms belonging to different m in the Fourier expansion of $\boldsymbol{u}'$, viz.

(4.3.7) $$\boldsymbol{u}'_M = \sum_1^\infty [\boldsymbol{u}'_{mcM} \cos m\phi + \boldsymbol{u}'_{msM} \sin m\phi],$$

do not interact to produce an α. Moreover, the self-interaction of an m term does not contribute to α when either $\boldsymbol{u}'_{mc}$ or $\boldsymbol{u}'_{ms}$ vanish (or are everywhere proportional). For example, the Bullard-Gellman motion (1.4.12) is not self-sustaining either on Braginskiĭ's large R theory or on Tough's refinement of it. Lilley's flow (1.4.13) does, however, possess a nonzero α. A jerky dynamo, of the Backus [**2**] variety, viz.

(4.3.8) $$\tilde{S} = f_1 P_1^1(\cos\theta)\sin\phi, \qquad \tilde{T} = f_2 P_2(\cos\theta) + f_3 P_1^1(\cos\theta)\sin\phi,$$

where $f_i = f_i(r, t)$, contains only the $m = 1$ asymmetries and gives by (1.4.3)

(4.3.9) $$u_0 = 3f_2\varpi z/r^2, \quad u'_{1cr} = 0, \quad ru'_{1sr} = 2\varpi f_1/r, \quad u'_{1cz} = -\varpi f_3/r.$$

This possesses the required interaction between sine and cosine terms, as may be seen by inspecting α in the form

(4.3.10) $$\alpha = \eta \sum_1^\infty \frac{1}{m}\left[\frac{(1-m^2)}{\varpi u_0^2}(\boldsymbol{u}'_{ms} \times \boldsymbol{u}'_{mc})_\phi - \nabla\left(\frac{ru'_{mcr}}{u_0}\right)\cdot\nabla\left(\frac{u'_{msz}}{u_0}\right) + \nabla\left(\frac{ru'_{msr}}{u_0}\right)\cdot\nabla\left(\frac{u'_{mcz}}{u_0}\right)\right],$$

though it does not seem to be true in this large R limit that almost any $\boldsymbol{u}'$ velocity having a poloidal part will be regenerative (Backus [**2**, p. 372]); the $\tilde{T}_1^1$ shear, is very necessary. If we substitute (9) directly into (10) only the last term contributes, giving

(4.3.11) $$\alpha = -\frac{2}{9}\eta\nabla\left(\frac{rf_3}{zf_2}\right)\cdot\nabla\left(\frac{rf_1}{zf_2}\right),$$

which clearly diverges on the equatorial plane $z = 0$. This may be attributed to the fact that u_0 vanishes there, although the parity of the solutions is such that B_0 does not. Thus, χ_0 becomes infinite, and the exterior solutions of §4.2 break down. This again calls attention to the need for examining diffusive layers. In the present case, an interior solution of boundary-layer type must be constructed to match (across the equatorial plane) the exterior solutions holding in each hemisphere separately. Braginskiĭ [**10**, §§4–5] looked at such possibilities and found that the integrated α-effect in such a "boundary layer" is not infinite, as (11) might suggest, but nevertheless might well be nonzero, a phenomenon he christened "concentrated generation." He also showed that boundary layers on S might be associated with a net α-effect.

With the geophysical application in mind, it is of interest to examine the form of $\hat{\boldsymbol{B}}$ and $\hat{\boldsymbol{B}}'$. [Here $\hat{}$ again refers to the insulator surrounding V; its previous use for ϕ integration is, except for (12) below, suspended until §5.] It is clear, by (10) and the vanishing of $\boldsymbol{n} \cdot \boldsymbol{u}$ and $\boldsymbol{n} \cdot \boldsymbol{u}'$ on S, that neither $\boldsymbol{B}_0$ nor $\boldsymbol{B}'_1$ pass through S; i.e. the larger the R, the less the field energy emerges into $\hat{V}$. It is also evident that the symmetric (poloidal) $\hat{\boldsymbol{B}}_2$ field will be present in general; for the case of the Earth, this will include the axial dipole (assuming that Coriolis forces decide the symmetry axis). The asymmetric terms require more care. It has just been noted that $\hat{\boldsymbol{B}}'_1$ vanishes. As far as the exterior solutions of §4.2 are concerned, it is readily shown that $\boldsymbol{n} \cdot \boldsymbol{B}'_2$ vanishes on S and (less readily) that $\boldsymbol{B}'_3$, in general, does not. One cannot, without examining the interior solution (i.e. the electromagnetic skin on S), conclude immediately that other fields, of boundary origin, do not arise. There are two main possibilities influencing the structure of these layers: (a) $\zeta \neq 0$, (b) $\zeta = 0$, on S.

In case (a), χ_0 is zero on S, and (4.2.10) shows that the exterior form of $\boldsymbol{B}'_1$ vanishes in *all* components on S; there is therefore no discontinuity of $\boldsymbol{n} \times \boldsymbol{B}'_1$ on S, and therefore no boundary layer which would generate a discontinuity in $\boldsymbol{n} \cdot \boldsymbol{B}'_2$ (cf. §4.1). The expression (Braginskiĭ [**10**, equations 3.13, 3.15]) for the exterior form of $\boldsymbol{B}'_2$ in V is

$$(4.3.12) \qquad \boldsymbol{B}'_{2M} = \frac{\varpi}{u_0} \nabla\chi_0 \cdot (\boldsymbol{u}'_M \hat{\boldsymbol{u}}_M - \overline{\boldsymbol{u}'_M \hat{\boldsymbol{u}}_M}).$$

Since $\nabla\chi_0$ is perpendicular to S, this shows that $B'_{2\theta}$ vanishes, and B'_{2r} vanishes quadratically, on S; therefore by continuity $B'_{2\phi}$ vanishes also. Again, therefore, no boundary layers will occur. Thus, the first asymmetric field to emerge is $\hat{\boldsymbol{B}}'_3$ (and this *does* excite a boundary layer); for the case of the Earth, $\hat{\boldsymbol{B}}'_3$ will include the equatorial dipole. Since both equatorial dipole and axial dipole decrease with distance from the geocentre in the same way, one might conclude that, on this theory (a), their ratio $|\hat{\boldsymbol{B}}_2|/|\hat{\boldsymbol{B}}'_3|$ (observed to be about 5 at the Earth's surface) provides an estimate of $R^{1/2}$, where R is the magnetic Reynolds number based on a typical toroidal velocity. The resulting value, $R \doteqdot 25$, is a little low compared with that inferred from the westward drift [see discussion beneath (1.4.19)]. One possibility is that the conductivity $3 \cdot 10^5$ mho/m assumed below (1.3.4) is too large (see also §5). Another (Braginskiĭ [**12**, §5]; Hide [**38**]) is that the westward drift is a wave phenomenon, and that the estimate 10^{-4} m/s made below (1.4.19) was really a wave velocity, greater than any toroidal motion. Finally, however, it should be remarked that the particular integrations by Braginskiĭ [**10**] gave a ratio of dipole moments of $kR^{1/2}$, where the constant k was of the order of 0.1. Whether this is a common feature of such models is not known, but, if so, R may be as large as 2500.

In case (b), we will suppose that the limit $\chi_0 = B_0/\varpi\zeta$ is finite and nonzero on S. It now can only be concluded from (4.2.10) that B'_{1r} vanishes on S. The boundary layer which reconciles the discontinuity in $\boldsymbol{n} \times \boldsymbol{B}'_1$ will introduce, as described in §4.1, a discontinuity on S between $\boldsymbol{n} \cdot \hat{\boldsymbol{B}}'$ and the exterior form of $\boldsymbol{n} \cdot \boldsymbol{B}'$ of order $R^{-1/2}B_0\delta$, setting up fields of like order throughout V and $\hat{V}$. Here, since ζ vanishes on S, the boundary layer thickness, δ, does not turn out to be $O(R^{-1/2})$; instead it is $O(R^{-1/3})$ in general. Thus, the boundary layer would break the strict ordering scheme of (4.1.11) by introducing $\boldsymbol{B}'$ fields of order $R^{-5/6}B_0$, i.e. of order $R^{1/6}\boldsymbol{B}_2$, and, in the geophysical application, the equatorial dipole would be large compared with the axial dipole (see Braginskiĭ [**10**, §5]). We may conclude that case (b) is not relevant to the Earth. This information can be used. We know that there can, in fact, be no relative slip between core and mantle, and the presumption $\zeta \neq 0$ of (a) cannot be literally correct. The postulates on ζ may however be modified to the statement that the thickness, δ_E, of the hydrodynamic boundary layer on S is (a) thin or (b) thick compared with that of the electromagnetic skin. If the flow in fact matches the no slip conditions through an Ekman layer (§5), we have

(4.3.13) $$\delta_E \doteqdot aE^{1/2} = (\nu/2\Omega)^{1/2},$$

where Ω is the angular velocity of the Earth ($7 \cdot 10^{-5}$ sec^{-1}) and E is the Ekman number

(4.3.14) $$E = \nu/2\Omega\mathscr{L}^2.$$

The geophysical relevance of (a) now suggests that $E \ll 1/R$, i.e.

(4.3.15) $$\nu \ll 2\Omega\eta\mathscr{L}/\mathscr{U} \doteqdot 10^7 \mathrm{m}^2/\mathrm{s},$$

for $\eta = 3\text{m}^2/\text{s}$, $\mathscr{U} = 10^{-4}$ m/s, and this is not very unlikely. For a recent discussion of boundary layers on the core surface, see Backus [**3**].

Braginskiĭ [**11**] extended his theory to include, in a geophysically interesting way, motions containing components (e.g. the secular variation) which fluctuate on a time-scale short compared with the free decay time. As far as $\boldsymbol{u}'$ is concerned, these consist of waves (e.g. westward drift) moving rapidly in the ϕ-direction. The analysis is similar to, but more intricate than, the case outlined above; and as before the final result is Parker's equations (1)–(3) for effective fields, but with a redefined w and α. The topic has obvious relevance to §3, as well as significance in §5 below.

5. **Hydromagnetic dynamos.**

5.1. *The zero order solution.* The most striking feature of the main field of the Earth is the near coincidence of its polar and magnetic axes; moreover, the palaeomagnetism of recent lavas shows strikingly that this is not an accident of the present. Clearly, the Coriolis force, $2\boldsymbol{\Omega} \times \boldsymbol{u}$, must exert a decisive effect on fluid motions in the core. If we take $\mathscr{U} \simeq 10^{-4}$ m/s [cf. discussion below (1.4.19)], we find that its magnitude is about 10^{-8} m/s^2. Now, for a toroidal field of 300 gauss ($3 \cdot 10^{-2}$ w/m^2), the Lorentz force per unit mass, $\mathscr{B}^2/\mathscr{L}\mu\rho$, for a fluid of density 10^4 kg/m^3 is also about 10^{-8} m/s^2 (corresponding to an Alfvén velocity of 0.3 m/s). This indicates not only that the Lorentz force $\boldsymbol{j} \times \boldsymbol{B}$ is important, but also that it adjusts its magnitude to that of the Coriolis force. And, of course, something must drive the motion against ohmic losses; we postulate a body force, $\boldsymbol{F}$, per unit mass lying in meridian planes

$$F_\phi \equiv 0. \tag{5.1.1}$$

We have primarily in mind a bouyancy force of thermal or chemical origin, ignoring the possibility of precessional excitation. The balance between Coriolis, Lorentz, bouyancy and, of course, pressure forces is summarized by the magneto-geostrophic equations

$$2\boldsymbol{\Omega} \times \boldsymbol{u} = -\nabla\widetilde{\Pi} + \frac{1}{\rho}\boldsymbol{j} \times \boldsymbol{B} + \boldsymbol{F}, \qquad \operatorname{div} \boldsymbol{u} = 0, \tag{5.1.2, 3}$$

where $\widetilde{\Pi} = p/\rho$, and ρ is density, assumed constant. It may be noted that this starting point is rather different from that adopted by Gilman for the solar dynamo, and in which Lorentz forces are not called into play at this level of the iteration (see Gilman [**33**], where earlier references will be found; see also Childress [**22**, p. 642]).

It will be seen that viscous forces have been omitted from (2). The geophysical evidence for this is plain. Even taking the value 10^7 m/s^2

of (4.3.15), $|\nu\nabla^2 \boldsymbol{u}| \simeq \nu\mathcal{U}/\mathcal{L}^2$ is only about 10^{-10} m/s^2. From the theoretical point of view, (2) governs the leading term in an "exterior solution" valid far from S and, since it is of lower order than the full Navier-Stokes equation, it will be impossible to satisfy the boundary conditions without introducing an "interior solution" of boundary-layer type on S. In invoking this to satisfy the no slip conditions on $\boldsymbol{n} \times \boldsymbol{u}$, a discontinuity in $\boldsymbol{n} \cdot \boldsymbol{u}$ of order $\partial_E|\boldsymbol{u}|$ will be introduced, where ∂_E is the boundary-layer thickness ($\delta_E \ll \mathcal{L}$). Thus, to leading order, solutions (2)–(3) must obey

$$\boldsymbol{n} \cdot \boldsymbol{u} = 0, \quad \text{on } S. \tag{5.1.4}$$

As in the case of the Braginskiĭ expansion, the "slow steady" approximation (2)–(3) can, for known $\boldsymbol{B}$, only be satisfied if a certain consistency condition is satisfied; this was first recognized by Taylor [**90**]. Let $\mathscr{C}$ be any cylinder (radius $\varpi_0 < a$; see Figure 16) coaxial with $\boldsymbol{\Omega}$. Let $\mathscr{S}$

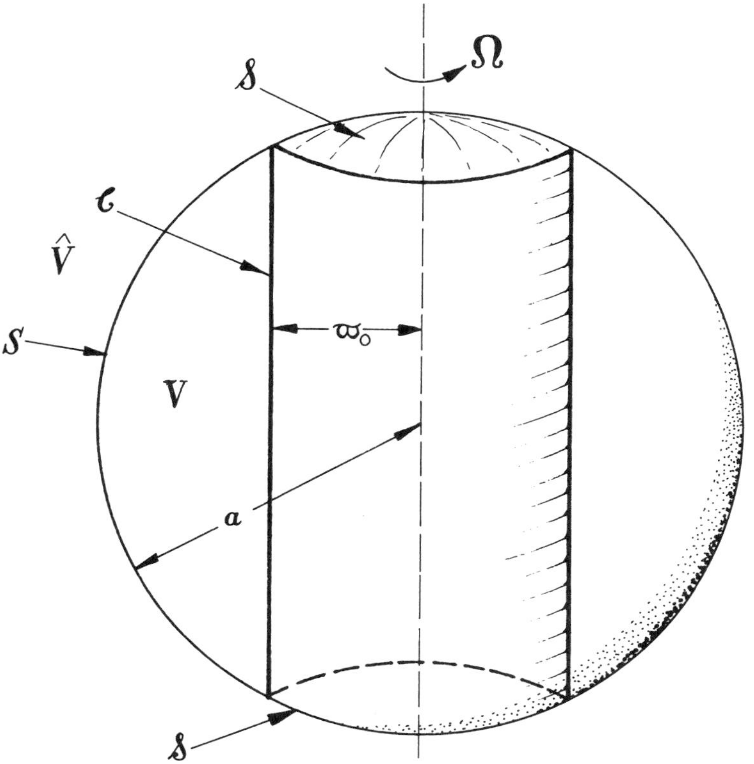

FIGURE 16

denote the spherical caps closing the cylinder at North and South Poles. On the assumption that the boundary layers on these caps do not generate significant radial flow, we have by (3) and (4)

$$\int_{\mathscr{C}} \boldsymbol{u} \cdot d\boldsymbol{S} = \int_{\mathscr{C}+\mathscr{S}} \boldsymbol{u} \cdot d\boldsymbol{S} = \int \operatorname{div} \boldsymbol{u} \, dV = 0. \tag{5.1.5}$$

Since the left-hand side of (5) may equally well be written as

$$\int_{\mathscr{C}} (2\boldsymbol{\Omega} \times \boldsymbol{u})_\phi \, dS,$$

and since, by supposition, $F_\phi \equiv 0$, (2) and (5) imply

$$\int_{\mathscr{C}} (\boldsymbol{j} \times \boldsymbol{B})_\phi \, dS = 0, \tag{5.1.6}$$

which is Taylor's result. It is equally valid for any axisymmetric container, and Soward (unpublished) has recently generalized it further, to any container possessing closed contours of constant height (cf. Greenspan [**34**, §2.6]).

Taylor's consistency condition (6) can be put to an interesting use, viz. that of obtaining $\boldsymbol{u}$ from an eligible $\boldsymbol{B}$; the curl of (2) gives

$$2\Omega \frac{\partial \boldsymbol{u}}{\partial z} = -\operatorname{curl}\left[\frac{1}{\rho}\boldsymbol{j} \times \boldsymbol{B} + \boldsymbol{F}\right]. \tag{5.1.7}$$

This determines $\boldsymbol{u}$ up to an arbitrary additive solenoidal $\boldsymbol{U}(\varpi, t)$. Condition (4), applied at the top and bottom of the cylinder of Figure 16, would appear to place two conflicting demands on $\bar{\boldsymbol{u}}_M$, but (6) is precisely the condition which makes them equivalent, and U_z is thereby fixed. No condition can be placed on U_ϕ in the framework of (2) above, but Taylor showed that (when the demand, that the time derivative of (6) vanishes, is appropriately transformed with the help of the induction equation) U_ϕ can be found. From the numerical point of view, (7) suggests the use of a ϖz-grid rather than an $r\theta$-grid, and this in fact dictated Thirlby's choice in his integrations of the induction equation reported in §1.4.

If $\boldsymbol{B}$ is axisymmetric and written in cylindrical coordinates as

$$\boldsymbol{B} = \left[-\frac{\partial A}{\partial z}, B, \frac{1}{\varpi}\frac{\partial}{\partial \varpi}(\varpi A)\right], \tag{5.1.8}$$

where

$$\langle A \rangle = \langle \partial A/\partial r \rangle = B = 0, \quad \text{on } S, \tag{5.1.9}$$

it is found that

$$\left(\frac{\partial}{\partial \varpi} + \frac{2}{\varpi}\right) \int_{\mathscr{C}} B \frac{\partial A}{\partial z}\, dS = -\mu \int_{\mathscr{C}} (\boldsymbol{j} \times \boldsymbol{B})_\phi \, dS. \tag{5.1.10}$$

Since A is bounded and B ($\propto \varpi$) is zero on $\varpi = 0$, the integral on the left of (10) is $o(\varpi^{-2})$ for $\varpi \to 0$. Hence Taylor's condition (6) may be written in the alternative form (Childress [**22**, equation 62])

$$\int_{\mathscr{C}} B \frac{\partial A}{\partial z}\, dS = 0. \tag{5.1.11}$$

Both the $\alpha\omega$-dynamo of §3 and the Braginskiĭ dynamo of §4 are dominated by the axisymmetric toroidal state

$$\boldsymbol{u} = u_0(\varpi, z, t)\mathbf{1}_\phi, \qquad \boldsymbol{B} = B_0(\varpi, z, t)\mathbf{1}_\phi. \tag{5.1.12}$$

In Braginskiĭ's case, it is also supposed that the axisymmetric poloidal deviations from (12) are $O(R^{-1}u_0)$ and $O(R^{-1}B_0)$, respectively. By (2) and (12) we have

$$2\Omega u_0 = \partial\Pi_0/\partial\varpi + B_0^2/\mu\rho\varpi - F_{0\varpi}, \qquad \partial\Pi_0/\partial z = F_{0z}, \tag{5.1.13, 14}$$

where $\Pi(\varpi, z, t) = P/\rho$, and $P = p + \boldsymbol{B}^2/2\mu$; also $\boldsymbol{F} = \boldsymbol{F}_{0M}(\varpi, z, t)$. (The arbitrariness $U_\phi(\varpi, t)$ in u_0 appears as an arbitrary addition $\xi(\varpi, t)$ to Π_0.) Clearly when (12) holds, none of the terms in (2) do work, and there is nothing to prevent B_0 from declining steadily to zero through ohmic decay.

We now add the $O(R^{-1})$ axisymmetric components. The ϕ-component of (2) gives directly

$$2\Omega u_\varpi = \frac{1}{\mu\rho\varpi}\left[\frac{\partial B}{\partial z}\frac{\partial}{\partial \varpi}(\varpi A) - \frac{\partial A}{\partial z}\frac{\partial}{\partial \varpi}(\varpi B)\right] = \frac{1}{\mu\rho\varpi}\boldsymbol{B}_M \cdot \boldsymbol{\nabla}(\varpi B). \tag{5.1.15}$$

From (3) and (15), u_z can be determined, (4) fixing U_z. The objection may be raised here (cf. P. H. Roberts [**77**, §7.1]) that this procedure could be invalid. We have already noted, in deriving (4), that u_0 will, through the Ekman layer on S, generate a component of flow (here clearly associated with the axisymmetric $\boldsymbol{u}_M$) which is $O(\delta_E u_0)$. If $\delta_E = O(a/R)$ or larger, (4) is invalid, and Taylor's condition (6) must be modified, as must (11). To remove this (harmless) complication, we will suppose from now onwards that $E \ll 1/R^2$, replacing (4.3.15) by the more restrictive demand

$$\nu \ll 2\Omega(\eta/\mathscr{U})^2 \simeq 10^5\ \mathrm{m^2/s}, \tag{5.1.16}$$

so that we can continue to apply (4), (6) and (11); geophysically, (16) is not implausible, but the possible importance of Ekman layers should, nevertheless, not be overlooked.

A dynamical proof of the impossibility of axisymmetric dynamos of zero $\boldsymbol{F}$ has been constructed by Childress [**22**, p. 645]. By "dynamical," we mean that it does not make full use of the information

$$(5.1.17)\qquad \frac{\partial A}{\partial t} + \frac{\boldsymbol{u}}{\varpi}\cdot\nabla(\varpi A) = \eta\Delta A,$$

$$(5.1.18)\qquad \frac{\partial B_0}{\partial t} + \varpi\boldsymbol{u}\cdot\nabla\left(\frac{B_0}{\varpi}\right) = \eta\Delta B_0 + \left[\nabla\left(\frac{u_0}{\varpi}\right)\times\nabla(\varpi A)\right]_\phi,$$

available from the induction equation: in fact, it does not use (17), nor (15) directly. If $\boldsymbol{F}$ is zero, (14) shows that $\Pi_0 = \Pi_0(\varpi)$, and (13) gives

$$(5.1.19)\qquad 2\Omega u_0 = \Pi_0' + B_0^2/\mu\rho\varpi.$$

Thus (18) may be written

$$(5.1.20)\qquad \frac{\partial B_0}{\partial t} + \varpi\boldsymbol{u}\cdot\nabla\left(\frac{B_0}{\varpi}\right) = \eta\Delta B_0 - \frac{\varpi}{2\Omega}\left(\frac{\Pi_0'}{\varpi}\right)'\frac{\partial A}{\partial z} + \frac{1}{2\Omega\mu\rho}\left[\nabla\left(\frac{B_0}{\varpi}\right)^2\times\nabla(\varpi A)\right]_\phi.$$

Multiplying this by B_0/ϖ^2, we obtain

$$(5.1.21)\qquad \frac{1}{2}\left(\frac{\partial}{\partial t} + \boldsymbol{u}\cdot\nabla\right)\left(\frac{B_0}{\varpi}\right)^2 = \frac{\eta B_0}{\varpi^2}\Delta B_0 - \left[\frac{1}{2\Omega\varpi}\left(\frac{\Pi_0'}{\varpi}\right)'\right]B_0\frac{\partial A}{\partial z} + \frac{1}{3\Omega\mu\rho\varpi}\left[\nabla\left(\frac{B_0}{\varpi}\right)^3\times\nabla(\varpi A)\right]_\phi.$$

Now integrating over V, using (4), the last of (9), and Taylor's condition in the form (11), we find

$$(5.1.22)\qquad \frac{\partial}{\partial t}\int_V \frac{1}{2}\left(\frac{B_0}{\varpi}\right)^2 dV = -\eta\int_V\left[\nabla\left(\frac{B_0}{\varpi}\right)\right]^2 dV,$$

so that B_0 disappears with increasing time, *irrespective* of any α-effect which might be added to (17). This last remark is important, since the interest in the analysis does not really lie in the direct disproof of regeneration at zero $\boldsymbol{F}$ (which could, in any case, have been concluded from elementary energy arguments) but rather in its extension to the mean field of a turbulent (§5.2) or of a nearly symmetric (§5.3) dynamo.

Here $\boldsymbol{F}$ is replaced by $\bar{\boldsymbol{F}}$ and must be taken to include the mean Lorentz force

$$\boldsymbol{S} = \overline{(\boldsymbol{j}' \times \boldsymbol{B}')}/\rho.$$

We then see that any turbulent dynamo, driven by a microscale force (§5.3) or any nearly symmetric dynamo driven by an asymmetric body force (§5.4), will fail unless $\boldsymbol{S}$ is nonzero. In fact, the obvious way of destroying the force of (22) is to suppose that S_ϕ is nonzero. Taylor's condition (11) does not then follow and the second term on the right of (21) does not vanish on integration. The possibility of nonzero S_ϕ arises naturally in §5.2 and §5.3.

5.2. *The development of a mean field magnetohydrodynamics.* The role of turbulence in the core is not clear (to me). The small-scale fields are shielded electromagnetically by the Earth's mantle because of their short-time scale, and are attenuated geometrically through their small-length scale; there is, then, no direct geomagnetic evidence for turbulence at all. The small Ekman number, E, of the core [$\simeq 10^{-2}$, even taking (4.3.15)] would suggest that rotation will tend to suppress the turbulent motions and may elongate the eddies along the direction of $\boldsymbol{\Omega}$ (see also Chandrasekhar [**20**, §28]). The large Hartmann number

$$M = \mathscr{B}\mathscr{L}(\sigma/\rho\nu)^{1/2} \tag{5.2.1}$$

of the core [$\simeq 150$ even taking (4.3.15)] also suggests suppression of the turbulent motions, but an elongation of the eddies in the direction of $\bar{\boldsymbol{B}}$ (see also Chandrasekhar [**20**, §44]). Indeed, in the absence of rotation, an attractive self-compensating mechanism may be possible: The turbulence generates an α-effect which creates $\bar{\boldsymbol{B}}$; the $\bar{\boldsymbol{B}}$ suppresses the small-scale motions until they are just sufficient to maintain the $\bar{\boldsymbol{B}}$ required; further decrease in $|\bar{\boldsymbol{B}}|$ would restore the α-effect leading again towards the equilibrium state. This argument also suggests that, for small $|\bar{\boldsymbol{B}}|$, $|\alpha|$ should be a nonlinear functional of $\bar{\boldsymbol{B}}$, decreasing with increasing $|\bar{\boldsymbol{B}}|$, see [**84**]. The dynamics of a system at small E and large M is quite different from that arising for either separately (it is in any case impossible for the eddies to align themselves simultaneously with field and rotation!). The primary balance on the large-scale is given by (5.1.2–3), with dissipation dominated by resistivity rather than viscosity (cf. Chandrasekhar [**20**, §45]). It is, however, no longer clear that a decrease in $|\boldsymbol{B}|$ will necessarily be accompanied by an enhancement of α by the turbulent motion (cf. §5.3 below), i.e. it is not obvious how the mechanism is self-compensating. It may be observed, however, that a suppression of the turbulence will, on the ideas of §3, create an increase in effective conductivity favourable for dynamo action. And there is also a completely new feature which must be examined, namely the large-scale component of the Lorentz force created by the small-scale fields [cf. (3.2.5)]. A reduction in this force might well create a poloidal flow which would crowd the toroidal field lines sufficiently together to restart the α-process once more. Much of this is (admittedly) speculative, but it does provide some crude objectives towards which to work. It should be noted that Braginskiĭ [**13**, §5] has discussed turbulence in the core. Should the turbulence in the core be truly two dimensional, the cascade of energy may be up the spectrum rather than down as in §3.2, and this may ultimately provide a superior way of treating the mean fields than that described below.

Consistent with the kinematic approach, it was supposed in §3 that $\bar{\boldsymbol{u}}$ and the statistical properties of $\boldsymbol{u}'$ were known; it is clear that here we should instead regard $\bar{\boldsymbol{F}}$ and the statistical properties of $\boldsymbol{F}'$ as being assigned. Now suppose, following the approach of §3.2, that the turbulence is basically homogeneous, isotropic *and* mirror-symmetric, and that small deviations from this state occur through quadratic terms in $\bar{\boldsymbol{B}}^2$, ignoring first the effect of $\bar{\boldsymbol{\varpi}}$ and $\boldsymbol{g}$. Then (cf. (3.2.25–26))

$$a_{ij} = \alpha\delta_{ij} + \gamma\bar{B}_i\bar{B}_j, \tag{5.2.2}$$

$$b_{ijk} = \beta\varepsilon_{ijk} + \mu_1\varepsilon_{jkl}\bar{B}_i\bar{B}_l + \mu_2\varepsilon_{kil}\bar{B}_j\bar{B}_l + \mu_3\varepsilon_{ijl}\bar{B}_k\bar{B}_l. \tag{5.2.3}$$

In general, $\alpha - \mu_3$ are functions of $\bar{\boldsymbol{B}}^2$ but, in the present perturbation theory, we have

$$\alpha = \alpha_0 + \alpha_1\bar{\boldsymbol{B}}^2, \qquad \beta = \beta_0 + \beta_1\bar{\boldsymbol{B}}^2, \tag{5.2.4}$$

where α_i, β_i, μ_i and γ depend only on the properties of the basic mirror-symmetric turbulence, and are therefore true scalars. Thus, $\alpha_0 = \alpha_1 = \gamma = 0$ and $a_{ij} = 0$. Redefining constants, we now find, from (3) and (3.2.20), that

$$\mathscr{E} = -[\beta - \gamma_2\bar{\boldsymbol{B}}^2]\bar{\mathbf{J}} - \alpha_2(\bar{\mathbf{J}}\cdot\bar{\boldsymbol{B}})\bar{\boldsymbol{B}} - \gamma_1\bar{\boldsymbol{B}}\times\nabla\bar{\boldsymbol{B}}^2. \tag{5.2.5}$$

On appropriately modifying σ_T [cf. (3.2.24)] by the new γ_2 term, Ohm's law for the mean field is found to be

$$\bar{\mathbf{J}} = \sigma_T[\bar{\boldsymbol{E}} + \bar{\boldsymbol{u}}_e \times \bar{\boldsymbol{B}} - \alpha_2(\bar{\mathbf{J}}\cdot\bar{\boldsymbol{B}})\bar{\boldsymbol{B}}]. \tag{5.2.6}$$

(Here we have included the $\alpha_1\nabla\bar{\boldsymbol{B}}^2$ term in $\bar{\boldsymbol{u}}_e$. Alternatively, following Rädler [**69**], we could absorb it in a turbulent permeability.) We may interpret (6) by saying that, in the presence of a large-scale field, α is changed from the $\alpha_1(\bar{\boldsymbol{\varpi}}\cdot\boldsymbol{g})$ value derived earlier to

$$\alpha = \alpha_1(\bar{\boldsymbol{\varpi}}\cdot\boldsymbol{g}) - \alpha_2(\bar{\mathbf{J}}\cdot\bar{\boldsymbol{B}}). \tag{5.2.7}$$

As before [cf. discussion below (3.2.28)] these are unlikely to be the only effects, but (7) has interesting physical content: In describing Parker's mechanisms in §3.1, it was clearly desirable to allow the cyclonic events to twist the basic loops through the optimum 90° angle (cf. Figure 10). If, through the success of the α-mechanism, $|\bar{\boldsymbol{B}}|$ grows, the driving forces provided will be less able to twist the loops through such a large average angle, and the current created parallel to $\bar{\boldsymbol{B}}$ will therefore be less. This reduction in α will make it impossible to maintain such a large $|\bar{\boldsymbol{B}}|$, and the mean field will decay until the indicated balance is attained. This argument also implies that, by Lenz's law, $\alpha_2 > 0$. If the presence of a field suppresses the turbulence and increases the effective conductivity, we may expect the γ_2 of (5) to be positive also.

Consider next the creation of a mean Lorentz force,

(5.2.8) $$\boldsymbol{S} \equiv \frac{1}{\rho} \overline{\boldsymbol{j}' \times \boldsymbol{B}'},$$

by the turbulence. Using the expression (3.2.18), we may express $\boldsymbol{S}$ as a quadratic functional of $\bar{\boldsymbol{B}}$ which vanishes for $\bar{\boldsymbol{B}} = 0$ and, following the procedure described below (3.2.19), we may approximate this by an expansion in the gradients of $\bar{\boldsymbol{B}}$, the first two terms having the form

(5.2.9) $$S_i = A_{ijk}\bar{B}_j\bar{B}_k + C_{ijkl}\bar{B}_j\frac{\partial \bar{B}_k}{\partial x_l},$$

[cf. (3.2.20)]. As explained below (3.2.19), the largest (A_{ijk}) term of such an expansion should be obtainable by assuming that $\bar{\boldsymbol{B}}$ is constant. In this case the turbulence will be homogeneous, and it is easily shown that, irrespective of isotropy or helicity, A_{ijk} must vanish. For, in homogeneous turbulence, the field correlation [cf. (3.2.6)],

(5.2.10) $$\overline{B_i'(\boldsymbol{x}, t)B_j'(\boldsymbol{x}', t')} \equiv F_{ij}(\boldsymbol{x}, t; \boldsymbol{x}', t'),$$

must, like any other correlation tensor, depend on $\boldsymbol{x}$ and $\boldsymbol{x}'$ only through $\boldsymbol{\xi} = \boldsymbol{x} - \boldsymbol{x}'$. Thus

(5.2.11) $$\partial F_{ji}/\partial x_j' = -\partial F_{ji}/\partial \xi_j = -\partial F_{ji}/\partial x_j,$$

or

(5.2.12) $$\overline{B_j'(\boldsymbol{x}, t)(\partial B_i'(\boldsymbol{x}', t')/\partial x_j')} = -\overline{(\partial B_j'(\boldsymbol{x}, t)/\partial x_j)B_i'(\boldsymbol{x}', t')} = 0,$$

by continuity. In particular

(5.2.13) $$\overline{B_j'(\boldsymbol{x}, t)(\partial B_i'(\boldsymbol{x}, t)/\partial x_j)} = 0,$$

and, since clearly

$$\overline{B'_j(\mathbf{x}, t)(\partial B'_j(\mathbf{x}, t)/\partial x_i)} = \tfrac{1}{2}\partial \overline{\mathbf{B}'^2}/\partial x_i = 0, \tag{5.2.14}$$

we have, by subtraction

$$\overline{\mathbf{j}' \times \mathbf{B}'} = 0. \tag{5.2.15}$$

This shows that $\mathbf{S}$ vanishes for vanishing grad $\bar{\mathbf{B}}$. The argument does not, however, show that C_{ijkl} also vanishes, for the field gradients create inhomogeneities in the turbulence which vitiates (15). Nevertheless, if the turbulence is (in the first approximation) isotropic, the C term appears to be of little interest since, in that case,

$$C_{ijkl} = C_1\delta_{ij}\delta_{kl} + C_2\delta_{ik}\delta_{lj} + C_3\delta_{il}\delta_{jk}, \tag{5.2.16}$$

where C_1 to C_3 are true scalars. This gives

$$\mathbf{S} = \mu C_2 \bar{\mathbf{j}} \times \bar{\mathbf{B}} + \tfrac{1}{2}(C_2 + C_3)\nabla\bar{\mathbf{B}}^2. \tag{5.2.17}$$

Clearly, the first term on the right of (17) may be incorporated into the Lorentz force of the mean field, perhaps by redefining the density, e.g., replacing ρ by $\rho_T = \rho/(1 + \mu\rho C_2)$; the second term represents a harmless addition to the mean total pressure. Equation (5.1.2) will still hold, Taylor's condition (5.1.6) will follow, and in the absence of a large-scale body force (or Ekman suction) the dynamo will run down by (5.1.22).

The initial discussion indicated that the local rotation $\bar{\boldsymbol{\omega}}$ should have a large effect on the character of the turbulence. The postulate (9) may therefore have been too limiting; let us instead examine deviations from mirror symmetry which depend linearly on $\bar{\boldsymbol{\omega}}$, $\mathbf{g}$ and $\bar{\mathbf{B}}^2$, with the proviso that, because of (15), we must use $\bar{\mathbf{j}}$ and $\bar{\mathbf{B}}$ together in place of $\bar{\mathbf{B}}^2$. The invariants $(\mathbf{g}\cdot\bar{\mathbf{j}})\bar{\mathbf{B}}$, $(\mathbf{g}\cdot\bar{\mathbf{B}})\bar{\mathbf{j}}$ and $(\bar{\mathbf{j}}\cdot\bar{\mathbf{B}})\mathbf{g}$ are skew and therefore ineligible. We therefore write

$$\mathbf{S} = \Lambda_1(\bar{\boldsymbol{\omega}}\cdot\bar{\mathbf{j}})\bar{\mathbf{B}} + \Lambda_2(\bar{\boldsymbol{\omega}}\cdot\bar{\mathbf{B}})\bar{\mathbf{j}} + \Lambda_3(\bar{\mathbf{j}}\cdot\bar{\mathbf{B}})\bar{\boldsymbol{\omega}}. \tag{5.2.18}$$

In application to dynamos of the $\alpha\omega$-type, the leading terms of $\bar{\boldsymbol{\omega}}\cdot\bar{\mathbf{B}}$ and $\bar{\mathbf{j}}\cdot\bar{\mathbf{B}}$ vanish, and (18) reduces to the single contribution of

$$\mathbf{S} = \frac{\Lambda B_0}{\mu\rho\varpi^2}[\nabla(\varpi u_0)\cdot\nabla(\varpi B_0)]\mathbf{1}_\phi. \tag{5.2.19}$$

Here Λ has the dimensions of time and, following the line of reasoning used by Steenbeck, Krause and Rädler [**89**], would appear to be of order $(\lambda v/\eta)^2\tau$, in the limit of small microscale Reynolds numbers, $R_m = v\lambda/\eta$.

The results of the foregoing theory seem encouraging. Not only has a very natural way of limiting the growth of magnetic energy been isolated, but it has also been shown that the mean Maxwell stress associated with the turbulence should have a nonzero azimuthal component. As yet no models have been integrated.

Attention should be drawn to an interesting study by Moffatt [**59**] of magnetohydrodynamic inertial waves [governed by (5.1.2) with the time derivative restored, but still without viscosity] in the low R_m limit. By introducing a preferred direction of energy transport, corresponding to $\boldsymbol{g}$, a net helicity and α-effect are found, and it is shown how these weaken as the resulting large-scale field grows. The system is supposed to be homogeneous, and some of the effects mentioned earlier in this section are therefore not present.

5.3. *Nearly symmetric self-consistent models.* The success of Braginskiĭ's kinematic theory, in creating an α-effect from a large-scale asymmetric component of flow, makes one wonder how a hydromagnetic dynamo, based on a large-scale asymmetric body force, would fare. Of course, Braginskiĭ's treatment of the induction equation would form an integral part of such an approach, and it is therefore necessary at the outset to find a plausible dynamical reason for the zero-order azimuthal flow (4.2.1). This theoretical consideration, quite apart from its obvious geophysical attractions, would probably suffice to suggest a model situation dominated by rotation, such as that described by (5.1.2). It is also immediately apparent that asymmetries create terms of the same order (R^{-1}) as the remaining terms in the expression (5.1.15) for u_ϖ. It is therefore possible to break the force of Childress's theorem (5.1.22) without introducing symmetric body forces or Ekman suction, i.e. a self-consistent theory can be constructed by keeping $\boldsymbol{F}'$ and discarding $\bar{\boldsymbol{F}}$ (but not, of course, by the reverse); we therefore assume

$$\boldsymbol{F} \equiv \boldsymbol{F}'. \tag{5.3.1}$$

Here, and later, the notation of §4 is revived.

It may be wondered why large-scale asymmetric bouyancy forces (1) should arise at all when the gravitational field, and other features of the Earth's interior, are so spherically symmetric. In this connection, it is worth noting that P. H. Roberts [**78**] has examined the stability of a rotating self-gravitating sphere containing a uniform distribution of heat sources and found, in the absence of magnetic fields, that, if the angular velocity is sufficiently large, the most unstable convective mode is *not* axially symmetric, and moreover the α of (4.3.5) is not zero. There is no reason to believe that the presence of a magnetic field will destroy this encouraging feature, though, of course, the structure of the motions would then be different.

This was brought out by Eltayeb and P. H. Roberts [**25**] who examined the stability of the Bénard layer in a number of special cases, one of which had also been studied much earlier by Chandrasekhar [**20**, Chapter 5]. Eltayeb and P. H. Roberts, however, confined their attention to large values of M and T, the Taylor number, where

$$(5.3.2) \qquad T \equiv (2\Omega\mathscr{L}^2/\nu)^2 = 1/E^2.$$

For simplicity they supposed that the magnetic Prandtl number,

$$(5.3.3) \qquad p_m = \nu/\eta,$$

was small. In the absence of field, it is well known that the critical Rayleigh number,

$$(5.3.4) \qquad R_a = \bar{g}\bar{\alpha}\bar{\beta}\mathscr{L}^4/\nu\kappa,$$

is $O(T^{2/3})$, for $T \to \infty$. (Here $\bar{g}$ is the acceleration due to gravity; $\bar{\alpha}$, the coefficient of volume expansion; $\bar{\beta}$, the adverse temperature gradient; and κ, the thermal diffusivity.) In a wide range of circumstances, including the interesting one in which $T = O(M^4)$ and in which (like the theory of §5.1) viscosity plays no essential role in the primary balance, Eltayeb and P. H. Roberts found that the critical R_a is only $O(T^{1/2})$, for $T \to \infty$. In other words if we consider Ω as essentially fixed at one value by the net angular momentum of the fluid, convection is easier to excite in the presence of a sufficiently strong magnetic field than without it. If one believes the heuristic principle that convection always acts to maximize the Nusselt number, he would, perhaps add as a corollary the belief that a convecting body of rotating fluid will generate a dynamo for purely thermodynamic reasons, viz. to maximize the heat transport through it. The link between this idea and dynamo theory is admittedly tenuous at the present time, but none the less intriguing.

The extension of the Braginskiĭ technique to the dynamical equations (5.1.2–3) is as cumbersome as for the induction equation, although Soward [**82**] has recently discovered a means of simplifying the original method of Tough and P. H. Roberts [**93**], and of extending it to the next order in R^{-1}. One of the most remarkable features of the final result is that the same effective fields that Braginskiĭ introduced to simply his mean field induction equations have an equal relevance to the dynamical mean field equations of Tough and P. H. Roberts. Moreover, the refinements to these fields introduced by Tough [**91**] for the induction equation have their counterpart by Soward [**82**] in the dynamical theory to give, for definiteness, the results of the simple theory, if we define $\boldsymbol{S}$ by [cf. (5.2.8)]

$$\boldsymbol{S} = (1/\rho)\overline{\boldsymbol{j}' \times \boldsymbol{B}'}, \tag{5.3.5}$$

the equation [cf. (5.1.15)]

$$2\Omega\bar{u}_{\varpi} = (1/\mu\rho\varpi)\bar{\boldsymbol{B}}_M \cdot \nabla(\varpi\bar{B}_0) + S_\phi, \tag{5.3.6}$$

[derived by taking the average of the ϕ-component of (5.1.2) and employing (1) above] can be simplified, using the leading-order asymmetric part

$$2\boldsymbol{\Omega} \times \boldsymbol{u}' = -\nabla\Pi' + \frac{1}{\rho}\boldsymbol{j}' \times \bar{\boldsymbol{B}} + \frac{1}{\rho}\bar{\boldsymbol{j}} \times \boldsymbol{B}' + \boldsymbol{F}' \tag{5.3.7}$$

of (5.1.2), to give

$$2\Omega\bar{u}_{e\varpi} = (1/\mu\rho\varpi)\bar{\boldsymbol{B}}_{eM} \cdot \nabla(\varpi\bar{B}_0) + \Lambda, \tag{5.3.8}$$

where

$$\Lambda = (\varpi/u_0)\overline{(\boldsymbol{u}' \times \operatorname{curl}\hat{\boldsymbol{F}})_\phi}. \tag{5.3.9}$$

The succinct manner in which the mean Lorentz force (5) is included in (8) is noteworthy.

Up to this stage, the application of the Braginskiĭ technique has run smoothly. Now, however, a difficulty arises. Neither the α of (4.3.5) obtained from Braginskiĭ's theory, nor the Λ of (8) can be evaluated until $\boldsymbol{u}'$ is known, and, to obtain this, (7) must be solved jointly with [cf. (4.2.10)].

$$\boldsymbol{B}_1' = (\bar{B}_0/\bar{u}_0)\boldsymbol{u}_1' - \varpi[\hat{\boldsymbol{u}}_1 \cdot \nabla(\bar{B}_0/\bar{u}_0)]\mathbf{1}_\phi, \tag{5.3.10}$$

for assigned $\bar{B}_0, \bar{u}_0$ and $\boldsymbol{F}'$. This problem can be "reduced" to that of solving (for each term in the Fourier decomposition of $\boldsymbol{F}'$ in ϕ) an inhomogeneous second-order partial differential equation, of hyperbolic or (if $|\bar{B}_0|$ is too large) mixed type, in a ϖz-section, $\mathscr{s}$, of V under assigned perimetric conditions. This is a computational problem about which little is known and which seldom arises naturally; indeed, Fox and Pucci [**26**], who examine a particular case of the theory, suggest that, when a physical situation seems to pose such a problem, too much of the physics has been lost in its formulation for reality. In this instance, the idea that diffusive effects can be confined to boundary layers on $\mathscr{s}$ may be in error.

The Tough-Roberts theory also suffers from quite another difficulty; assuming for clarity that $\boldsymbol{F}'$ has no ϕ-component, the Λ defined in (9) may be written as

$$\overline{\boldsymbol{u}' \cdot \boldsymbol{F}'}/u_0,$$

and is clearly related to the rate of working of the driving forces. If, the mth harmonic, say, of $\boldsymbol{F}'$ involves $\cos m\phi$ (or $\sin m\phi$) *alone*, the corresponding $\boldsymbol{u}'_M$ is proportional to $\sin m\phi$ (or $\cos m\phi$), and the harmonic will not contribute to the rate of working Λ; if all terms in the expansion are of this type, the dynamo must fail. It will be recalled that the kinematic dynamo suffered from a similar difficulty [see below (4.3.7)]. Tough and P. H. Roberts make no attempt to explain why both $\cos m\phi$ and $\sin m\phi$ terms should appear in $\boldsymbol{F}'$.

A more fundamental viewpoint is adopted by Braginskiĭ [**10**], [**13**]–[**15**] which holds some prospect of a consistent dynamical picture. It might be thought that a dynamo driven by an axisymmetric body force alone could not function. This, however, is not the case. Such a body force can increase the basic u_0 flow until an asymmetric Rossby type of instability appeared. This might, at the expense of the mean flow, amplify a stray magnetic field, and ultimately the dynamo might function through a similar type of instability of the basic aligned $(\bar{u}_0, \bar{B}_0)$ flow. This was appreciated early by Braginskiĭ who, in the papers cited above, initiated the study of Alfvén waves travelling along the curved $\bar{B}_0$ field lines, in the presence of the shear $\bar{u}_0$. He showed that, provided the time scale, τ_w, of the waves is short compared with the ohmic scale $\tau_\eta = \mathscr{L}^2/\eta \approx 3 \cdot 10^5$ years, the doubly averaged magnetic field (i.e. the field averaged over ϕ and the short-time scale, τ_w) still obeys Parker's equations, provided suitably amended effective fields are introduced. (Moreover, and very interestingly, he showed that a rapidly fluctuating axisymmetric field could be included in the ordering scheme, at the $R^{-1/2}$ level, without disrupting it.) The new α is the sum of contributions from each wave separately, and it is again necessary that waves of both $\sin[m(\phi - t/\tau_w)]$ and $\cos[m(\phi - t/\tau_w)]$ type should be present.

In order to examine the structure of the waves themselves, Braginskiĭ [**15**] made use of the very elegant formalism of Frieman and Rotenberg [**27**]. The wave speeds (complex in general) are revealed as the eigenvalues of a certain homogeneous second-order partial differential equation in ϑ. This equation possesses a number of symmetry properties which Braginskiĭ fully exploits (see also Barston [**5**]). In the stationary case it reduces to the zero $\boldsymbol{F}'$ form of the corresponding equation derived by Tough and P. H. Roberts, and its solution presents very similar difficulties. (That the equations are the same in this case is natural since Tough and P. H. Roberts are essentially studying the steady forcing of a system which Braginskiĭ subjects to a normal mode analysis.)

Braginskiĭ's linear wave model also presents another difficulty. At first sight, one might hope to construct a dynamo in which a stationary wave is superimposed on a steady $(\bar{u}_0, \bar{B}_0)$ state, the wave supplying the generation necessary to maintain the field. Since the amplitude of the wave is infinitesimal, viz. $O(R^{-1/2})$, compared with the basic state, one might hope that a linear theory of the waves, difficult though it is, would suffice. This is not the case. The generation coefficient, α, is only nonzero when the waves are "tilted," i.e. when, in the now familiar way, they can be expressed at all points, only by a linear combination of the $\sin m\phi$ and $\cos m\phi$ dependencies. And it may be shown easily from Braginskiĭ's theory that, when the basic state is marginally stable on the dynamic time scale, τ_w, the waves are, in fact, not tilted at all; thus the basic field would die away on the diffusive time scale, τ_η. But if we examined a basic state which is dynamically unstable, the waves would, on linear theory, grow in amplitude and destroy the basic state in a time of order τ_w. All this suggests strongly that the linear theory is inadequate, and that the basic state must be unstable on the linear theory, but that the nonlinear interactions between the waves themselves must prevent these amplitudes from increasing beyond the order $R^{-1/2}$ consistent with the Braginskiĭ ordering. The rapidly fluctuating axisymmetric $R^{-1/2}$ fields mentioned above contain the feed back, via the Maxwell stresses, of the wave on the basic state. Braginskiĭ [**16**], [**17**] has recently investigated these fields.

This picture, attractive though it is, presents formidable technical difficulties, but modifications of the Braginskiĭ formalism may lead to some simplification. For example, we might seek solutions in which the basic state lies "close" to the linear stability boundary. The angle of tilt would then be small and would have to be explicitly included in the ordering (see also Smagorinsky [**81**]). To obtain a sufficiently large generation coefficient, the amplitude of $\boldsymbol{u}'$ would have to be correspondingly increased (possibly to $R^{-1/4}$), but in fact the Braginskiĭ theory is sufficiently flexible to permit modifications of this type.

One may ask which of the two main themes of these lectures, the mean field electrodynamics and the nearly symmetric dynamo, should be held mainly responsible for (say) the geomagnetic field. The simple nature of the mean field theories will no doubt win many adherents, but its simplicity is to some extent deceptive. The absence of obstacles in the dynamical theory is an illusion; the difficulties are really more fundamental than those encountered by the Braginskiĭ approach and lie at the root of turbulence theory itself. Moreover one might wonder why, if turbulent generation were indeed the more potent effect, the geomagnetic field should show such large, consistent and characteristic deviations from axisymmetry. It is hard to escape the feeling that macroscopic regeneration is important. Nevertheless, it seems clear that some of the ideas of mean field electrodynamics should be incorporated in any macroscopic theory. In particular, turbulent conductivities and viscosities should be used [and for this reason estimates such as (4.3.15) or (5.1.16) should be applied cautiously].

ACKNOWLEDGEMENTS. I must thank Drs. P. A. Gilman and E. N. Parker for some helpful remarks, and Drs. S. Childress and G. O. Roberts for their advice in §3. Drs. Lerche, Parker, G. O. Roberts, Soward, Stix and Mr. Thirlby kindly communicated their recent work prior to its formal publication. Drs. Krause, Rädler and Steenbeck patiently explained their work to me during a recent visit to Potsdam, and §3 owes much to their influence; they must not, however, be blamed for the shortcomings of §5.2, in which they played no part.

REFERENCES[3]

1. G. E. Backus, *The axisymmetric self-excited fluid dynamo*, Astrophys. J. **125** (1957), 500–524.

2. ———, *A class of self-sustaining dissipative spherical dynamos*, Ann. Physics **4** (1958), 372–447. MR **20** #1512.

3. ———, *Kinematics of the geomagnetic secular variation in a perfectly conducting core*, Philos. Trans. Roy. Soc. London Ser. A **263** (1969), 239–266.

4. W. L. Bade, *Hydromagnetic effects of upwelling near a boundary*, University of Utah Report #1288(00) No. 10, 1954.

5. E. M. Barston, *Eigenvalue problem for Lagrangian systems*, J. Math. Phys. **8** (1967), 523–532; II, ibid, 1886–1892; III, ibid. **9** (1968), 2069–2075; see also: *Stability of dissipative systems*, Comm. Pure Appl. Math. **22** (1969), 627–637; *An energy principle for dissipative fluids*, J. Fluid Mech. **42** (1970), 97–109. MR **35** #5691; MR **36** #6746; MR **39** #4506.

6. G. K. Batchelor, *On the spontaneous magnetic field in a conducting liquid in turbulent motion*, Proc. Roy. Soc. London Ser. A **201** (1950), 405–416. MR **11**, 699.

7. ———, *The theory of homogeneous turbulence*, University Press, Cambridge, 1956. MR **14**, 597.

8. L. Biermann, *Bermerkungen über das Rotationsgesetz in irdischen und stellaren Instabilitätszonen*, Z. Astrophys. **28** (1951), 304–309.

9. P. M. Blackett, *A negative experiment relating to magnetism and the Earth's rotation*, Philos. Trans. Roy. Soc. London Ser. A **245** (1952), 309–370.

10. S. I. Braginskiĭ, *Self-excitation of a magnetic field during the motion of a highly conducting fluid*, Ž. Èksper. Teoret. Fiz. **47** (1964), 1084–1098 = Soviet Physics J.E.T.P. **20** (1965), 726–735. MR **32** #724.

11. ———, *Theory of the hydromagnetic dynamo*, Ž. Èksper. Teoret. Fiz. **47** (1964), 2178–2193 = Soviet Physics J.E.T.P. **20** (1965), 1462–1471. MR **32** # 1725.

12. ———, *Kinematic models of the Earth's hydromagnetic dynamo*, Geomag. Aeron. **4** (1964), 732–747; English, 572–583.

13. ———, *Magnetohydrodynamics of the Earth's core*, Geomag. Aeron. **4** (1964), 898–916; English, 698–712.

[3] Translations of papers **40**, **43**, **47**, **68**–**72**, **84**, **85**, and **87**–**89** have recently appeared in P. H. Roberts and M. Stix, *The turbulent dynamo: a translation of a series of papers by F. Krause, K.-H. Rädler and M. Steenbeck*, Tech. Note #60, National Center for Atmospheric Research, 1971.

14. ———, *Principles of the theory of the Earth's hydromagnetic dynamo*, Geomag. Aeron. **7** (1967), 401–410; English, 323–329.

15. ———, *Magnetic waves in the Earth's core*, Geomag. Aeron. **7** (1967), 1050–1060; English, 851–859.

16. ———, *Torsional magnetohydrodynamic vibrations in the Earth's core and variations of the day length*, Geomag. Aeron. **10** (1970), 3–12; English, 3–12.

17. ———, *Oscillation spectrum of the hydromagnetic dynamo of the Earth*, Geomag. Aeron. **10** (1970), 221–233; English, 172–181.

18. E. C. Bullard and H. Gellman, *Homogeneous dynamos and terrestrial magnetism*, Philos. Trans. Roy. Soc. London Ser. A **247** (1954), 213–278. MR **17**, 327.

19. S. Chandrasekhar, *The invariant theory of isotropic turbulence in magnetohydrodynamics*. II, Proc. Roy. Soc. London Ser. A **207** (1951), 301–306; *Hydromagnetic turbulence*. I. *A deductive theory*, ibid. **233** (1955), 322–330; II. *An elementary theory*, ibid. **233** (1955), 330–350.

20. ———, *Hydrodynamic and hydromagnetic stability*, Clarendon Press, Oxford, 1961. MR **23** #B1270.

21. S. Childress, *Construction of steady-state hydromagnetic dynamos* I. *Spatially periodic fields*, II. *The spherical conductor*. Courant Institute Reports AFOSR-67-0124 and 0976, New York; 1967; see also: *New solutions of the kinematic dynamo problem*, J. Math. Phys. **11** (1970), 3063–3076.

22. ———, *A class of solutions of the magnetohydrodynamic dynamo problem* in The application of modern physics to the Earth and planetary interiors, edited by S. K. Runcorn, Wiley, London, 1969, pp. 629–648.

23. ———, *Théorie magnétohydrodynamique de l'effet dynamo*, Report from Département Méchanique de la Faculté des Sciences, Paris, 1969.

24. W. Deinzer and M. Stix, *On the eigenvalues of Krause–Steenbeck's solar dynamo*, Astron. and Astrophys. (submitted).

25. I. A. Eltayeb and P. H. Roberts, *On the hydromagnetics of rotating fluids*, Astrophys. J. **162** (1970), 699–701.

26. D. W. Fox and C. Pucci, *The Dirichlet problem for the wave equation*, Ann. Mat. Pura Appl. (4) **46** (1958), 155–182. MR **21** #3653.

27. E. Frieman and M. Rotenberg, *On hydromagnetic stability of stationary equilibria*, Rev. Modern Phys. **32** (1960), 898–902. MR **23** #B2757.

28, A. Gailitis, *The self-excitation of a magnetic field in a pair of vortex rings*, Magnitnaja Gidrodinamika **6** (1970), 19–22.

29. R. D. Gibson, *The Herzenberg dynamo*. I, Quart. J. Mech. Appl. Math. **21** (1968), 243–255.

30. ———, *The Herzenberg dynamo*. II, Quart. J. Mech. Appl. Math. **21** (1968), 257–267.

31. ———, *The Herzenberg dynamo* in The application of modern physics to the Earth and planetary interiors, edited by S. K. Runcorn, Wiley, London, 1969, pp. 571–576.

32. R. D. Gibson and P. H. Roberts, *The Bullard–Gellman dynamo* in The application of modern physics to the Earth and planetary interiors, edited by S. K. Runcorn, Wiley, London, 1969, pp. 577–602.

33. P. A. Gilman, *A Rossby-wave dynamo for the Sun*. I, Solar Phys. **8** (1969), 316–330; II, ibid. **9** (1969), 3–18.

34. H. P. Greenspan, *The theory of rotating fluids*, University Press, Cambridge, 1968.

35. R. M. Halleen and J. P. Johnston, *The influence of rotation on flow in a long rectangular channel: an experimental study*, Stanford University Report MD-18, Stanford, Calif., 1967.

36. A. Herzenberg, *Geomagnetic dynamos*, Philos. Trans. Roy. Soc. London Ser. A **250** (1958), 543–585.

37. A Herzenberg and F. J. Lowes, *Electromagnetic induction in rotating conductors*, Philos. Trans. Roy. Soc. London Ser. A **249** (1957), 507–584.

38. R. Hide, *Free hydromagnetic oscillations of the Earth's core and the theory of geomagnetic secular variation*, Philos. Trans. Roy. Soc. London Ser. A **259** (1966), 615–650.

39. I. Lerche, *Kinematic dynamo theory*, Astrophys. J. (submitted).

40. F. Krause, *Eine Lösung des Dynamoproblemes auf der Grundlage einer linearen Theorie der magnetohydrodynamischen Turbulenz*, Habilitationsschrift, Jena, 1967.

41. ———, *Zum Anfangswertproblem der magnetohydrodynamischen Induktionsgleichung*, Z. Angew. Math. Mech. **48** (1968), 333–343.

42. ———, *Explanation of stellar and planetary magnetic fields by dynamo action of turbulent motion*, Proc. Fourth Consult. Solar Phys. and Hydromag. (Sopot, 1966), Acta Univ. Wratislavia No. 77 (1969), 157–170.

43. ———, *Zur Hydrodynamik des mittleren Geschwindigkeitsfeldes in einem turbulenten Medium*, Monatsb. Deutsch. Akad. Wiss. Berlin **11** (1969), 188–194.

44. F. Krause and H. Hiller, *Zur Dynamotheorie stellarer und planetarer Magnetfelder. III Über die Lösung der Eigenwertprobleme und die Berechnung der Feldgrössen*, Astronom. Nachr. **291** (1969), 287–294.

45. F. Krause and K.-H. Radler, In *Handbuch der Plasmaphysik und Gaselektronik*, edited by R. Rompe and M. Steenbeck, Akad. Verlag., E. Berlin, 1970.

46. F. Krause and M. Steenbeck, *Models of magnetohydrodynamic dynamos for alternating fields*, Proc. Third Consult. Solar. Phys. and Hydromag. (Tatranska Lomnica, 1964), Czech. Acad. Sci., Astron. Inst. Publ. No. 51 (1964), 36–38.

47. ———, *Untersuchung der Dynamowirkung einer nichtspiegelsymmetrischen Turbulenz an einfachen Modellen*, Z. Naturforsch. **22** (1967), 671–675.

48. E. P. Kropachev, *One mechanism of the excitation of a stationary magnetic field in a spherical conductor*, Geomag. Aeron. **4** (1964), 362–371; English, 281–288.

49. ———, *Excitation of a magnetic field in a spherical conductor*, Geomag. Aeron. **5** (1965), 945–947; English, 744–746.

50. ———, *Generation of a magnetic field near the boundary of a conductor*, Geomag. Aeron. **6** (1966), 548–555; English, 406–412.

51. R. H. Kraichnan and S. Nagarajan, *Growth of turbulent magnetic fields*, Phys. Fluids **10** (1967), 859–870.

52. F. E. M. Lilley, *On kinematic dynamos*, Proc. Roy. Soc. London Ser. A **316** (1970), 153–167.

53. D. Lortz, *Exact solution of the hydromagnetic dynamo problem*, Plasma Phys. **10** (1968), 967–972.

54. F. J. Lowes and I. Wilkinson, *Geomagnetic dynamo: A laboratory model*, Nature **198** (1963), 1158–1160.

55. ———, *Geomagnetic dynamo: An improved laboratory model*, Nature **219** (1968), 717–718.

56. H. K. Moffatt, *The amplification of a weak applied magnetic field by turbulence in fluids of moderate conductivity*, J. Fluid Mech. **11** (1961), 625–635.

57. ———, *The degree of knottedness of tangled vortex lines*, J. Fluid Mech. **35** (1968), 117–129.

58. ———, *Turbulent dynamo action at low magnetic Reynolds numbers*, J. Fluid Mech. **41** (1970), 435–452.

59. ———, *Dynamo action associated with random inertial waves in a rotating conducting fluid*, J. Fluid Mech. **44** (1970), 705–719.

60. E. N. Parker, *Hydromagnetic dynamo models*, Astrophys. J. **122** (1955), 293–314. MR **18**, 92; see also: *The solar hydromagnetic dynamo*, Proc. Nat. Acad. Sci. U.S.A. **43** (1957), 8–14.

61. ———, Lab. Astr. and Space Res. E. Fermi Inst., University of Chicago Report #70–17, Chicago, Ill., 1970; see also: *The generation of magnetic fields in astrophysical bodies*. I. *The dynamo equations*, Astrophys. J. **162** (1970), 665–673.

62. ———, Lab. Astr. and Space Res. E. Fermi Inst., University of Chicago Report #70-19, Chicago, Ill., 1970; see also: *The generation of magnetic fields in astrophysical bodies*. II *The Galactic field*, Astrophys. J. **163** (1971), 255–278.

63. ———, Lab. Astr. and Space Res. E. Fermi Inst., University of Chicago Report #70-25, Chicago, Ill., 1970; see also: *The generation of magnetic fields in astrophysical bodies*. III *Turbulent diffusion of fields and efficient dynamos*, Astrophys. J. **163** (1971), 279–285.

64. ———, *The origin of magnetic fields*, Astrophys. J. **160** (1970), 383–404.

65. ———, Lab. Astr. and Space Res. E. Fermi Inst., University of Chicago Report #70-52, Chicago, Ill., 1970; see also: *The generation of magnetic fields in astrophysical bodies*. IV *The solar and terrestrial dynamos*, Astrophys. J. **164** (1971), 491–509.

66. ———, *The generation of magnetic fields in astrophysical bodies*. V *Behavior at large dynamo numbers*, Astrophys. J. **165** (1971), 139–146.

67. ———, *The generation of magnetic fields in astrophysical bodies*. VI *Periodic modes of the Galactic field*, Astrophys. J. (submitted).

68. K.-H. Rädler, *Zur Elektrodynamik turbulent bewegter leitender Medien*. I. *Grundzüge der Elektrodynamik der mittleren Felder*, Z. Naturforsch. **23** (1968), 1841–1851.

69. ———, *Zur Elektrodynamik turbulent bewegter leitender Medien*. II. *Turbulenzbedingte Leitfähigkeits und Permeabilitätsänderungen*, Z. Naturforsch. **23** (1968), 1851–1860.

70. ———, *Zur Elektrodynamik in turbulenten Coriolis-Kräften unterworfenen leitenden Medien*, Monatsb. Deutsch. Akad. Wiss. Berlin **11** (1969), 194–201.

71. ———, *Über eine neue Möglichkeit eines Dynamomechanismus in turbulenten leitenden Medien*, Monatsb. Deutsch. Akad. Wiss. Berlin **11** (1969), 272–279.

72. ———, *Untersuchung eines Dynamomechanismus in turbulenten leitenden Medien*, Monatsb. Deutsch. Akad. Wiss. Berlin **12** (1970), 468–472.

73. G. O. Roberts, *Dynamo waves* in The application of modern physics to the Earth and planetary interiors, edited by S. K. Runcorn, Wiley, London, 1969, pp. 603–628.

74. ———, *Spatially periodic dynamos*, Philos. Trans. Roy. Soc. London Ser. A **266** (1970), 535–558.

75. ———, *Two dimensional spatially-periodic dynamos*, Philos. Trans. Roy. Soc. London Ser. A (to appear).

76. P. H. Roberts, *An introduction to magnetohydrodynamics*, Longmans, Green, London; American Elsevier, New York, 1967.

77. ———, Woods Hole Oceanographic Inst. Report #67-54, pp. 1, 51–165, 178–212, 1967.

78. ———, *On the thermal instability of a rotating-fluid sphere containing heat sources*, Philos. Trans. Roy. Soc. London Ser. A **263** (1968), 93–117.

79. ———, In *The world magnetic survey*, edited by A. J. Zmuda, IUGG Publ., 1971.

80. P. G. Saffman, *On the fine-scale structure of vector fields convected by a turbulent fluid*, J. Fluid Mech. **16** (1963), 545–572.

81. J. Smagorinsky, *Some aspects of the general circulation*, Quart. J. Roy. Meteorol. Soc. **90** (1964), 1–14.

82. A. M. Soward, *Nearly symmetric kinematic and hydromagnetic dynamos*, J. Math. Phys. (to appear).

83. ———, *Nearly symmetric advection*, J. Math. Phys. (to appear).

84. M. Steenbeck, et al., *Der experimentelle Nachweis einer elektromotorischen Kraft längs eines äusseren Magnetfeldes, induziert durch eine Strömung flüssigen Metalls* (α-*Effekt*), Monatsb. Deutsch. Akad. Wiss. Berlin **9** (1967), 714–719.

85. M. Steenbeck and F. Krause, *Erklärung stellarer und planetarer Magnetfelder durch einen turbulenzbedingten Dynamomechanismus*, Z. Naturforsch. **21** (1966), 1285–1296.

86. ———, *Die Entstehung stellares und planetarer Magnetfelder als Folge turbulenter Materiebewegung*, Gustav Hertz Festschrift, Akad. Verlag, E. Berlin, 1967, pp. 155–177, Russian transl., Magnetnaja Gidrodinamika **3** (1967), 19–44; English transl. **3** (1970), 8–22.

87. ———, *Zur Dynamotheorie stellarer und planetarer Magnetfelder*. I *Berechnung sonnenähnlicher Wechselfeldgeneratoren*, Astronom. Nachr. **291** (1969), 49–84.

88. ———, *Zur Dynamotheorie stellarer und planetarer Magnetfelder*. II *Berechnung planetähnlicher Gleichfeldgeneratoren*, Astronom. Nachr. **291** (1969), 271–286.

89. M. Steenbeck, F. Krause and K.-H. Rädler, *Berechnung der mittlerer Lorentz-Feldstärke* $\overline{\mathfrak{v} \times \mathfrak{B}}$ *für ein elektrisch leitendes Medium in turbulenter, durch Coriolis-Kräfte beeinflusster Bewegung*, Z. Naturforsch. **21** (1966), 369–376.

90. J. B. Taylor, *The magneto-hydrodynamics of a rotating fluid and the Earth's dynamo problem*, Proc. Roy. Soc. London Ser. A **274** (1963), 274–283.

91. J. G. Tough, *Nearly symmetric dynamos*, Geophys. J. Roy. Astron. Soc. **13** (1967), 393–396; corrigendum, ibid. **15** (1969), 343.

92. J. G. Tough and R. D. Gibson, *The Braginskiĭ dynamo* in The application of modern physics to the Earth and planetary interiors, edited by S. K. Runcorn, Wiley, London, 1969, pp. 555–569.

93. J. G. Tough and P. H. Roberts, *Nearly symmetric hydromagnetic dynamos*, Phys. Earth Planet. Int. **1** (1968), 288–296.

94. S. C. Traugott, *Influence of solid-body rotation on screen-produced turbulence*, Nat. Adv. Comm. Aeron. Technical Note #4135, 1958.

95. B. A. Tverskoy, *Theory of hydrodynamical self-excitation of regular magnetic fields*, Geomag. Aeron. **5** (1965), 11–18; English, 7–12.

THE UNIVERSITY OF NEWCASTLE UPON TYNE, ENGLAND

Do Precessional Torques Cause Geomagnetism?

W. V. R. Malkus

LECTURE I

Precessional torques and geomagnetism. The geomagnetic field is presumed to be due to motions in the earth's fluid core. Doubts concerning the possibility of dynamo action in a homogeneous electrically conducting fluid have been removed by the theoretical construction of velocity fields which exhibit magnetic instability. However, the origin and character of the motions responsible for the earth's dynamo have not been determined. That is to say, the elementary kinematic problem has been resolved while the numerous dynamic problems have barely been touched. It is generally believed that heat sources in the core produce convective motions which in turn drive the geodynamo. Convection is a theoretically attractive mechanism because of its possible applicability to stellar dynamos as well as to the earth, and there is some progress to report on the determination of interdependent velocity and magnetic fields due to buoyancy forces. However, it has been difficult to find a heat source sufficient to meet even the minimum energy needs of a geodynamo and compatible with surface observations of heat flux. Hence, after a preliminary discussion of kinematics, we will explore the possibility that flow due to the precession of the earth drives the geodynamo.

Figures 1 and 2 exhibit the complicated structure of that part of the magnetic field of internal origin which can be measured on the earth's surface. Several aspects of this data worthy of note are: the field is complicated; it changes with time (there is an apparent westward drift of the entire pattern of roughly 10^{-2} cm/sec); it exhibits no simple symmetries (e.g. symmetry or antisymmetry relative to the earth's rotation vector).

I will not make an attempt in this lecture to recall all the fascinating data that has been discovered regarding the history of the earth's field. Perhaps many of you are already familiar with the evidence that its dipole has

AMS 1970 *subject classifications.* Primary 86A25; Secondary 76W05.

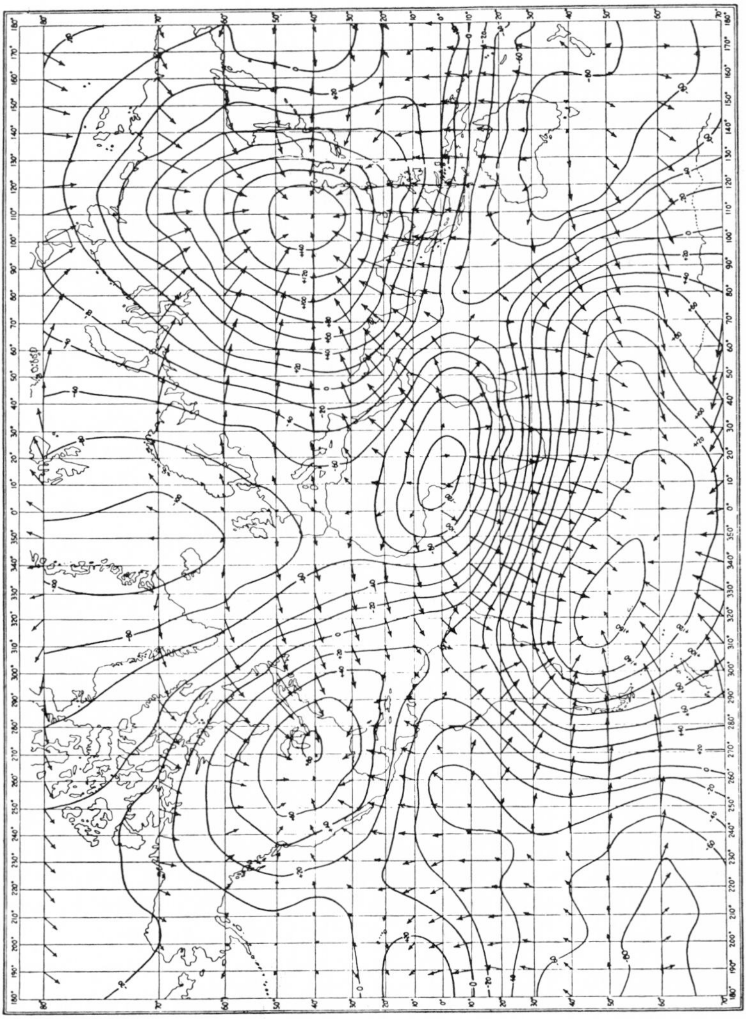

FIGURE 1

reversed sign frequently in the past and that the dipole axis tends to drift at a slow rate, remaining within a cone whose angle is 30° from the geographical pole.

The first and most obvious suggestion of the data is that we are looking at a process; not a static thing, not a remnant of the past. One infers the existence of a mechanism involved in transfering energy from fields of motion to magnetic fields to ohmic dissipation.

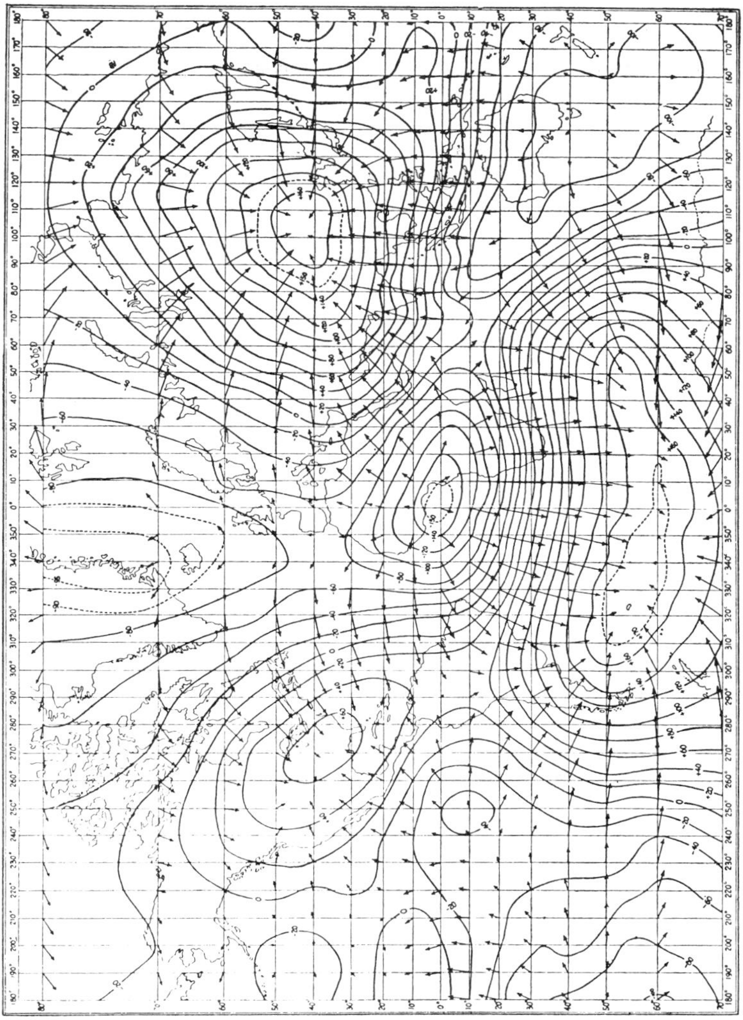

FIGURE 2

Let us first recall what a dynamo is. The simplest example is the "homopolar dynamo," seen in Figure 3. There, a simple disc of metal is rotated about its axis in the presence of a magnetic field, $\boldsymbol{B}$ imposed along that axis. As a consequence, a potential gradient $\boldsymbol{v} \times \boldsymbol{B}$ occurs in the plane of the disc, where $\boldsymbol{v}$ is the local velocity of the disc. If the charge accumulated at the periphery of the disc is permitted to flow through a coil back to the axis, then the resulting field can reinforce the original imposed field $\boldsymbol{B}$.

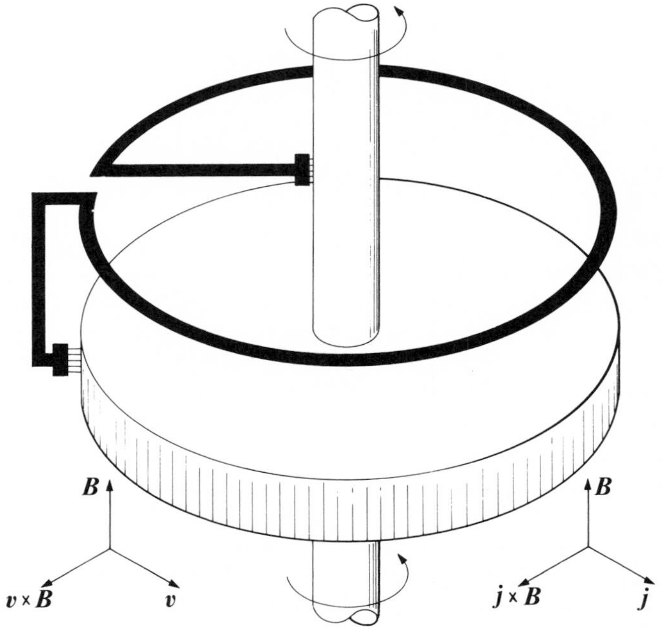

FIGURE 3

This system becomes a dynamo when this induced field becomes equal to the field required to produce it. The criterion of this critical behavior is determined by the nondimensional combination

$$R_m \equiv \mu\sigma|\boldsymbol{v}|L \gg 1 \tag{1}$$

where the magnetic Reynolds number R_m is defined as the product of the magnetic permeability μ, the conductivity σ, some measure of the velocity $|\boldsymbol{v}|$, some appropriate scale length L, of the dynamo system. When this number is large compared to 1, regeneration can occur. Note, however, that the homopolar machine is not a simply-connected dynamo. It requires complicated conduction paths and slipping brushes in order to operate. It was not clear that a homogeneous conducting fluid could be put into motion in such a way as to initiate this dynamo action.

We first address the question "Under what conditions can a homogeneous conducting fluid maintain a magnetic field?" The Maxwell equations, with the displacement current neglected, are written

$$\nabla \times \boldsymbol{H} = \boldsymbol{j}, \tag{2}$$

$$\nabla \times \boldsymbol{E} = -\partial \boldsymbol{B}/\partial t, \tag{3}$$

(4) $$\nabla \cdot \boldsymbol{B} = 0,$$

(5) $$\nabla \cdot \boldsymbol{D} = 0$$

where $\boldsymbol{B} = \mu \boldsymbol{H}$, $\boldsymbol{H}$ is the magnetic field strength, $\boldsymbol{D} = \varepsilon \boldsymbol{E}$, $\boldsymbol{E}$ is the electric field strength and ε is the dielectric constant. The current density is

(6) $$\boldsymbol{j} = \sigma(\boldsymbol{E} + \boldsymbol{v} \times \boldsymbol{B}) \qquad (\text{for } v^2/c^2 \ll 1).$$

Taking the curl of the equation (2) and using equations (3) and (6), one can write the magnetic diffusion equation

(7) $$\partial \boldsymbol{H}/\partial t + (1/\mu\sigma)\nabla^2 \boldsymbol{H} = \nabla \times \boldsymbol{v} \times \boldsymbol{H}.$$

If $\boldsymbol{v}$ is assumed to be a known function, equation (7) is a homogeneous equation of second order with nonconstant coefficients. Presumably, $\boldsymbol{H}$ can have an exponential instability for certain classes of these nonconstant coefficients. Among the first explorers of this problem was Cowling, who established that no axisymmetric $\boldsymbol{v}$ field could give rise to dynamo action, no field for which the radial component of velocity was zero could give rise to dynamo action, and no two-dimensional velocity fields could give rise to dynamo action. Hence, there was a period of time when it was very doubtful that homogeneous dynamos were possible.

This difficulty was resolved by Backus and Herzenberg. Herzenberg's model was particularly simple and is exhibited in Figure 4. Two spheres of conducting material are rotated in a surrounded medium of the same conductivity. One can see that the symmetric currents which would flow due to a field oriented along ω_1, will not sustain a dynamo action in the rotor ω_1. However, the magnetic field that would result can act to excite currents in the rotor ω_2. It is this coupling between the two asymmetric induction fields that permits dynamo action. The criterion for dynamo action to first order in the separation distance R is

(8) $$[(\tfrac{1}{5}R'_m \sin\theta_1 \sin\theta_2 \sin\phi)^2(\cos\theta_1\cos\theta_2 - \sin\theta_1\sin\theta_2\cos\phi) - 3]^2 = 0$$

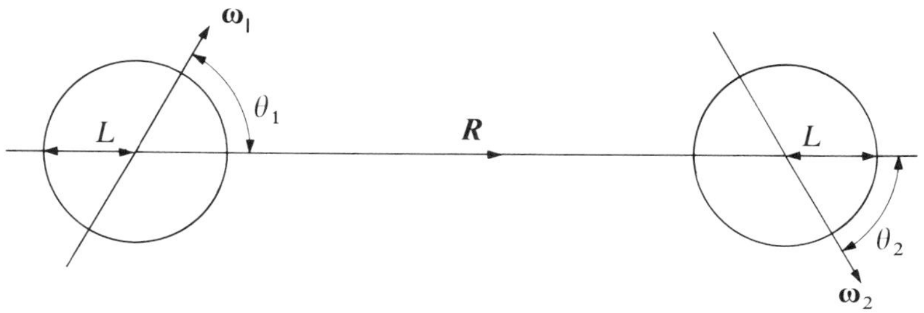

FIGURE 4

where $R'_m = R_m(L/R)^3$, $\theta_{1,2}$ are indicated on Figure 4 and θ is the angle between $\boldsymbol{\omega}_1$, $\boldsymbol{R}$ plane and the $\boldsymbol{\omega}_2$, $\boldsymbol{R}$ plane.

A laboratory model of this process constructed by Lowes and Wilkinson not only exhibited intense dynamo action above a critical angular velocity, but also had field reversals of long period. The field reversals were related to the strong back coupling of the magnetic and velocity fields arising from the Lorentz force $\boldsymbol{j} \times \boldsymbol{H}$. As the magnetic instability grew it slowed down the rotating spheres below the critical angular velocity. This caused the field to decay, the spheres to accelerate, and the process to repeat itself. A satisfactory theory for this dynamic aspect of a Herzenberg dynamo has yet to emerge.

With the question regarding kinematic homogeneous dynamo action behind us, the next problem is the determination of a "natural" flow process which can act as a dynamo. For example, can convection due to radial temperature contrasts in a rotating sphere cause a dynamo?—and what does the rotation have to do with it?

It is generally believed that the answer to the first part of this question is "yes," but it is not yet proved. An answer to the second part of the question may very well be the key to understanding the principal dynamic role of magnetic fields in nature. Many investigators have concluded that the magnitude of the Lorentz force is comparable to the Coriolis force in the earth and in the sun. Almost as many different reasons for this gross balance have been advanced. Since magnetic fields do not directly transfer heat, it is my view that the Lorentz force arises in conducting fluids to relax the gyroscopic constraints on motion due to the rotation, hence assisting in the release of available potential energy. This view is supported in a 1959 paper on initial finite-amplitude magnetoconvection. However, a definitive study has yet to be made.

Quite apart from the general problem of dynamo action due to convection is the problem of possible energy sources for the motion in the earth's core. If that energy source is thermal, then the total dissipation rate integral requires that

$$\Phi = (\alpha g/C_p)[WE], \tag{9}$$

where Φ is the total dissipation rate, both viscous and ohmic; α is the coefficient of thermal expansion of core material; g is the average gravitational acceleration; C_p is the specific heat at constant pressure; and $[WE]$ is the average convective heat flux due to W, the vertical velocity, and E, the thermal energy per unit volume. The minimum ohmic Φ necessary to support the largest scale components of the magnetic field has been estimated by various authors to be at least 10^{18} ergs/sec. Hence, equation 9 requires a minimum total heat flux of at least 6.10^{19} ergs/sec from the core. The observed surface heat flux is approximately 20.10^{19} ergs/sec and is mostly accounted for by radioactive decay in crustal material. Therefore, it is very difficult to picture a plausible distribution of radioactivity within the earth which both leads to the surface heat flux and provides the minimum required energy source for a convective geodynamo in the core.

Of the several mechanisms for inducing motion in the core that have been proposed, the precession of the earth is the only one whose magnitude and character are well known. The earth precesses with a period of 25,800 years, due to the gravity fields of the moon and the sun acting on the earth's equatorial bulge. Figure 5 illustrates the aspects of earth structure

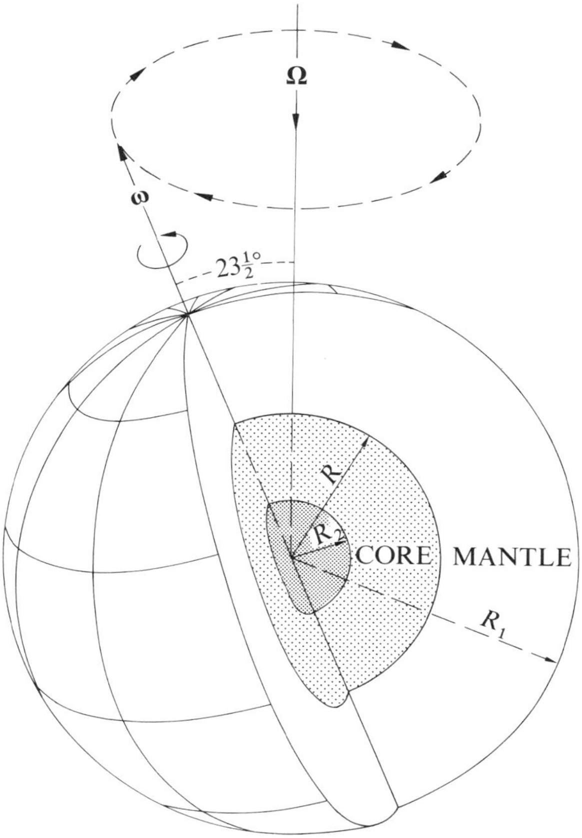

FIGURE 5

and average motion that are relevant to this study. The precession vector $\Omega = 7.71 \times 10^{-12}$ radian per second and is normal to the plane of the ecliptic. The earth's angular rotation vector $\omega = 7.29 \times 10^{-5}$ radian per second and is inclined at 23.5 degrees relative to $\mathbf{\Omega}$. The lightly shaded zone in Figure 5 is the earth's molten core, of mean radius $R = 3.47 \times 10^8$ centimeters. The mean radius of the earth as a whole, R_1, is roughly $2R$, while the radius, R_2, of the presumably solid inner core is roughly $0.4\,R$. The rotation of the earth causes an equatorial bulge, resulting in a difference between the moments of inertia about the polar and the equatorial axis. The ratio of this difference to the moment of inertia about the polar axis is known as the dynamic ellipticity. The dynamic ellipticity of the earth as a whole is 3.28×10^{-3}, and that of the core alone is $2.45 \times 10^{-3} \pm 2$ percent. The dynamic ellipticity of the core is three quarters that of the mantle, due to the core's greater density (approximately 10 grams per cubic centimeter).

The precession rate of a planetary body is directly proportional to its dynamic ellipticity, but independent of its radius or mass. Hence, if the core and mantle were not coupled, the core would precess at only three quarters the rate of the mantle. Thirty thousand years of such uncoupled precession would lead to relative velocities of 10^4 centimeters per second at the core-mantle boundary. Of course, core and mantle are coupled by torques resulting from their relative motion. These torques will increase until core and mantle precess at the same average rate.

The idealized flow induced by the relative precession of core and mantle was first investigated by Poincaré and Hough in the last century. They described the flow of the fluid inside a precessing spheroidal container of major radius R. The equations appropriate to these descriptions, first for the continuity of a mass,

$$\nabla \cdot \boldsymbol{v} = 0, \tag{10}$$

second for the continuity of momentum,

$$\partial \boldsymbol{v}/\partial t - E\nabla^2 \boldsymbol{v} + 2\mathbf{\Omega} \times \boldsymbol{v} + \boldsymbol{v} \cdot \nabla \boldsymbol{v} + \nabla P = 0, \tag{11}$$

are here scaled so that the velocity on the surface S of the container is written as

$$\boldsymbol{v} = \boldsymbol{k} \times \boldsymbol{r}, \tag{12}$$

where $\boldsymbol{k}$ is the unit normal vector along the axis of rotation and $\boldsymbol{r}$ is the position vector measured from the center of the system. The scaling is also such that the viscous term appears multiplied by the Ekman number,

$$E = \nu/\omega R^2, \tag{13}$$

where ν is the kinematic viscosity, ω the angular velocity of rotation, and R the principal radius of the container. The shape of the container is given by

$$r^2 - \eta(\boldsymbol{k} \cdot \boldsymbol{r})^2 = 1, \tag{14}$$

where η is the ellipticity of the oblate spheroid.

The classical problem treated only one aspect of the general boundary condition, equation (12); both Poincaré and Hough presumed that the boundary condition

$$\boldsymbol{v} \cdot \boldsymbol{n} = 0 \qquad \text{on } S, \tag{15}$$

where $\boldsymbol{n}$ is a unit vector normal to surface S, would be sufficient to describe all but subtle viscous aspects of the problem. A generalized solution to the problem with this simple boundary condition is easily shown to be one of constant vorticity, given by

$$\boldsymbol{v} = \boldsymbol{\omega}_f \times \boldsymbol{r} + \nabla A. \tag{16}$$

When this velocity field is substituted into the original equations, the solution for the vorticity of the fluid is given by

$$\boldsymbol{\omega}_f = \alpha\boldsymbol{k} - \frac{\alpha(2 + \eta)[(\boldsymbol{\Omega} \times \boldsymbol{k}) \times \boldsymbol{k}]}{\alpha\eta + 2(\boldsymbol{k} \cdot \boldsymbol{\Omega})(1 + \eta)}, \tag{17}$$

and for the potential term needed to satisfy the oblate spheroidal boundary conditions by

$$A = \frac{-\alpha\eta}{\alpha\eta + 2(\boldsymbol{k} \cdot \boldsymbol{\Omega})(1 + \eta)}(\boldsymbol{\Omega} \times \boldsymbol{k}) \cdot \boldsymbol{r}(\boldsymbol{k} \cdot \boldsymbol{r}). \tag{18}$$

As one might have anticipated, this theory of the flow is not unique and contains an arbitrary parameter α. When $\alpha = 1$, as was assumed by Poincaré and Hough, then $\boldsymbol{\omega}_f$ approaches the angular velocity of the container as $\boldsymbol{\Omega}$ approaches zero. What is fascinating about this solution is that it is an exact nonlinear solution of the entire problem, except for a small tangential component of the velocity at the boundary. If one defines a relative amplitude, ε, as

$$\varepsilon^2 = (\boldsymbol{\omega}_f - \boldsymbol{k})^2, \tag{19}$$

then in the linear theory, the order of magnitude of ε is given by

$$\varepsilon = O|\boldsymbol{\Omega} \times \boldsymbol{k}|/\eta. \tag{20}$$

The resulting flow is one of smooth elliptical stream lines around an axis, not quite in coincidence with the axis of rotation of the spheroidal shell.

It was believed by early investigators that this flow was too smooth and too uniform to provide any dynamo action.

A recent investigation of these Poincaré-like flows by Stewartson and Roberts included a viscous correction in the linear theory for the small inadequacy of this theory on the boundary surface. They concluded that the viscous correction would lead to a thin boundary layer, but that the principal flow would remain unaltered.

I constructed an experiment using a precessing oblate spheroidal cavity to test these conclusions in the laboratory. This experiment exhibited dramatic departures from the laminar Poincaré flow. Figure 6 illustrates

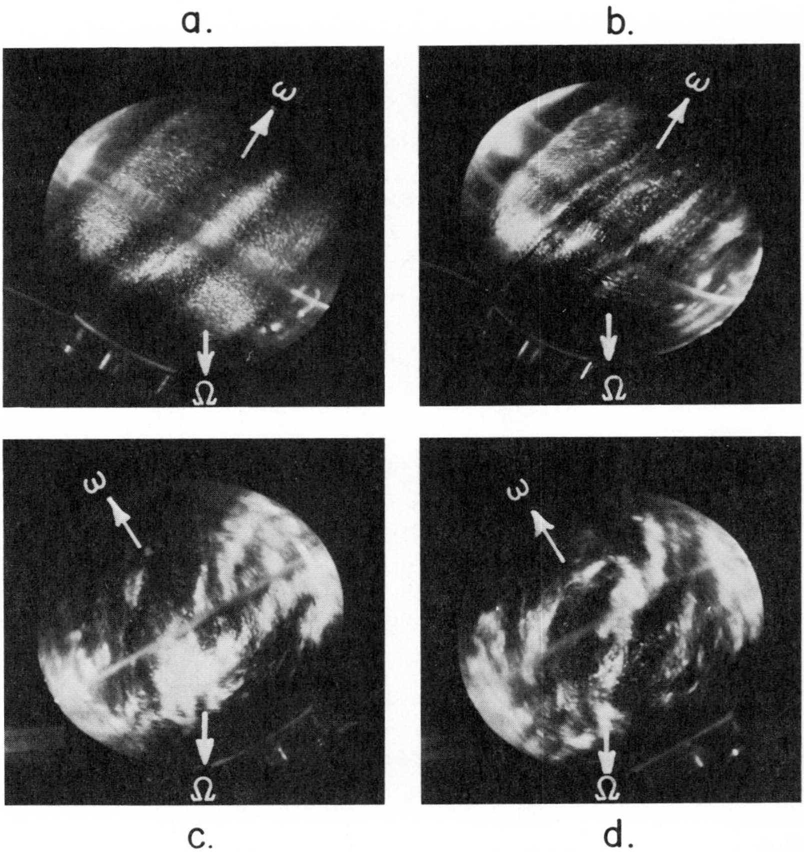

FIGURE 6

this study. A plastic spheroid of major axis 25 centimeters and minor axis 24 centimeters was filled with water and rotated at 60 revolutions per minute around its minor axis. The spheroid, its supports, and its motor were mounted on a horizontal, rotatable table, with the minor axis of the spheroid inclined at 30 degrees to the vertical. A pinch of aluminum powder dispersed in the water was illuminated by a beam of light normal to the rotation axis, permitting visualization of the process. Rotation of the table at $\frac{3}{4}$ revolution per minute caused the steady two-dimensional flow seen in Figure 6a. This steady flow already exhibits a departure from the general Poincaré solution, for the light and dark bands produced by the aluminum particles suggest shear zones of considerable magnitude in the toroidal flow in the container. Rotation of the table at 1 revolution per minute led to the wavelike instabilities of these toroidal flows seen in Figure 6b, while rotation at $\frac{4}{3}$ revolution per minute caused rather dramatic turbluent flows seen in Figures 6c and d.

A program of measurement was undertaken to determine the flow field prior to instability, the critical parameters determining instability, and the motor torques needed to sustain the flow. In a second lecture, the theoretical description of these various regimes of flow will be assessed. At this point I wish to explore the implications of these disordered flows as an energy source for a dynamo process.

Perhaps the most straightforward approach to the problem is to note that the transformation of coordinates from inertial space to a uniformly rotating and precessing space leads to the usual centrifugal and Coriolis forces plus an additional force. This unfamiliar precessional force per unit volume is written

$$[(\boldsymbol{\omega} \times \boldsymbol{\Omega}) \times \boldsymbol{r}]. \tag{21}$$

I will refer to this force as the Poincaré force, in honor of his early studies of precession. The Poincaré force is akin to the usual centrifugal force, but since it produces a net torque it cannot be balanced by pressure alone. Due to the difference in core and mantle ellipticity, one fourth of the Poincaré force remains unbalanced in the core and produces fluid motion.

A fluid velocity $\boldsymbol{V}$ in the core will give rise to the usual Coriolis force per unit mass, $2\boldsymbol{\omega} \times \boldsymbol{V}$. Such electric currents as may be associated with this flow will produce the Lorentz force per unit mass $(\mu/\rho)(\boldsymbol{j} \times \boldsymbol{H})$, where ρ is the density of the fluid, μ is the magnetic permeability, $\boldsymbol{H}$ is the magnetic field, and $\boldsymbol{j}$ is the electric current density. Here, and in what follows, we used unscaled variables and Gaussian electromagnetic units throughout.

I assume, tentatively, that the Poincaré force in the earth's core is large enough to cause the turbulent type of flow seen in the experimental situation, but presumably magnetic turbulence. I will define a fully magnetic turbulent flow as meaning that a gross balance has been struck between the fluctuating Coriolis and Lorentz forces in the interior of the flow and that each of these fluctuating responses is of the same magnitude as the steady Poincaré driving force. Forces due to pressure gradients can be as important as the Coriolis and Lorentz forces, but I will show that it is sufficient to consider only the two major inertial forces and the electromagnetic force in order to obtain an estimate of the velocities and magnetic fields which would be associated with such flow.

This gross balance is written

$$\tfrac{1}{4}|(\boldsymbol{\omega} \times \boldsymbol{\Omega}) \times \boldsymbol{r}| \cong |2\boldsymbol{\omega} \times \boldsymbol{V}| \cong |(1/4\pi\rho)(\nabla \times \boldsymbol{H}) \times \boldsymbol{H}|. \tag{22}$$

Hence the estimate for the magnitude of a characteristic velocity V is

$$V \cong \frac{|\boldsymbol{\omega} \times \boldsymbol{\Omega}|}{8|\boldsymbol{\omega}|} R \cong \tfrac{4}{3} \times 10^{-4}\ \text{cm/sec}. \tag{23}$$

For scales of motion comparable to the radius of the core, ∇ in equation (22) is replaced by $1/R$, and the estimate for the magnitude H of a characteristic magnetic field is

$$H \cong (\pi\rho|\boldsymbol{\omega} \times \boldsymbol{\Omega}|R^2)^{1/2} \cong 29\ \text{gauss}. \tag{24}$$

The vector character of the force balance (equation (22)) requires that the estimated magnetic field H^2 be interpreted as the product $H_p H_T$, where H_p is the magnitude of the poloidal components of the field and H_T is the magnitude of the toroidal component of the field. The poloidal components lie in meridian planes containing the axis of rotation. The toroidal component is at right angles to these planes. Similarly, the characteristic velocity V should be interpreted as representative of the geometric mean of poloidal and toroidal components. A balance of the average radial Coriolis force and the average radial Lorentz force is not included in the estimates given above, but is used in the following section. Neither V nor H are directly observable quantities since they are estimates for fields within the molten core material.

The magnitude of the earth's dipole field at the core-mantle boundary, as inferred from surface observations, is approximately 5 gauss. In order to determine the external dipole associated with the internal field H, one must establish how much of H can "leak out" of the electrically conducting mantle. A gross balance of the terms in the magnetic-diffusion equation provides such an estimate, but this estimate depends explicitly upon the value of core conductivity.

This balance between the loss of magnetic field by ohmic dissipation and its regeneration by the fluid motion is

$$|(1/4\pi\sigma)\nabla \times (\nabla \times \boldsymbol{H})| \cong |\nabla \times (\boldsymbol{V} \times \boldsymbol{H})|, \tag{25}$$

where σ is the electrical conductivity of the fluid. For scales of motion comparable to the radius of the core, the balance of equation (25) may be interpreted as the estimate

$$(26) \qquad H_0 \cong \frac{1}{4\pi\sigma R}\frac{H}{V} = \frac{2|\boldsymbol{\omega}|}{\sigma R}\left(\frac{\rho}{\pi|\boldsymbol{\omega} \times \boldsymbol{\Omega}|}\right)^{1/2},$$

where H_0 is defined as the magnitude of the magnetic field at right angles to the velocity V and is presumed to be small compared to H, and where the ratio $H:V$ is expressed in terms of the precession rate through the use of equations (22) and (23). Since the average fluid velocity must be parallel to the core-mantle boundary, H_0 is also an estimate for the magnetic field external to the core. A mean value for the electrical conductivity of core material, determined from the extrapolation of laboratory measurements, is 7×10^{-6} (abohm centimeter)$^{-1}$. However, an uncertainty of at least a factor of 3, in either direction, is given in the literature. With this choice for σ, the estimate (equation (26)) for the external magnetic field at the core-mantle boundary is

$$H_0 \cong 7 \text{ gauss} \pm 300\%.$$

If the estimate for H_0 had led to a value equal to or larger than H, the appropriate interpretation of the dissipation-regeneration balance equation would have been that the magnetic field could not be maintained by the velocity field and would vanish. This would occur if the precession $|\Omega|$ were $\frac{1}{16}$ its present value, or if the conductivity σ were $\frac{1}{4}$ the mean value used in arriving at the estimate given above.

Having found an approximate value for H_0, we can make a secondary estimate of the internal fields. Since the estimated value for H_0 is also an estimate for one component of the poloidal field just inside the core-mantle boundary, it sets a lower bound on the magnitude of the internal poloidal field H_p. Then $H^2 \equiv H_pH_T > H_0H_T$, so an upper bound for the internal toroidal field H_T can be found from the estimates for H^2 and H_0. From equations (24) and (26), we obtain the relation

$$(27) \quad H_T < H^2/H_0 = (\sigma/2|\omega|)\rho^{1/2}\pi|\boldsymbol{\omega} \times \boldsymbol{\Omega}|^{1/2}R^3 \cong 120 \text{ gauss} \pm 300\%.$$

Finally, a secondary estimate for a mean value of the toroidal velocity field, V_T, follows from the balance in equation (22) of the radial Coriolis force and the radial Lorentz force. This relation is written:

$$(28) \quad V_T = \frac{H_T^2}{8\pi\rho|\omega|R} < \frac{(\sigma\pi)^2R^5}{32}\left(\frac{|\boldsymbol{\omega} \times \boldsymbol{\Omega}|}{|\boldsymbol{\omega}|}\right)^3 \cong 2.3 \times 10^{-3} \text{ cm/sec} \pm 1000\%,$$

where equation (27) is used to express H_T^2 in terms of the precession rate and the conductivity. Order-of-magnitude accuracy is the most one might expect from a secondary estimate such as this. If V_T is interpreted as representative of the westward-drift component of core surface velocity, then the value in equation (28) is one-sixth the mean drift velocity inferred from recent geomagnetic data.

The energy consumption of the precession-driven geodynamo can be found directly from the earlier estimates. The characteristic stress acting on the core-mantle boundary must equal the momentum transport just inside the boundary. From equation (24) we find this momentum transport to be

$$H_p H_T/4\pi = (\rho/4)(|\boldsymbol{\omega} \times \boldsymbol{\Omega}| R^2) = 68 \text{ dyne/cm}^2. \tag{29}$$

An estimate of the total work W done by the mantle on the core is then, from equations (28) and (29).

$$\begin{aligned} W &= \left(\frac{H_p H_T}{4\pi}\right) V_T(4\pi R^2) < \frac{\rho\sigma^2\pi^3 R^9 |\boldsymbol{\omega} \times \boldsymbol{\Omega}|^4}{32|\boldsymbol{\omega}|^3} \\ &= 2.3 \times 10^{17} \text{ erg/sec} \pm 1000\%, \end{aligned} \tag{30}$$

and this would be consumed in ohmic heating primarily. For comparison, note that the estimated rate of dissipation due to tidal interaction of earth, moon, and sun is 3×10^{19} ergs per second. Like the tidal dissipation process, the energy for the precession-driven geodynamo must come from the kinetic energy stored in the earth's rotation. Unlike the tidal process, the response of the core fluid to precessional forces does not produce a reaction torque on the moon or sun. Hence, in order for the total angular momentum of the earth, moon, sun system to be conserved, only the rotational energy in the nonconserved component of the earth's rotation can supply the dynamo. This is the component of rotation in the plane of the ecliptic, the component at right angles to Ω in Figure 5. It is sufficient at present to maintain the geomagnetic field for many earth lifetimes. However, in an earlier eon when the moon was considerably closer to the earth than it is now, the dynamo dissipation might well have exceeded the total tidal dissipation and would have contributed significantly to internal heating in the earth.

It is rather fortunate that a scale analysis should lead to unique qualitative results as does the scale analysis starting with equation (22). It is then possible to test, certainly within an order of magnitude, the suggestion that precessional torques can be responsible for the dynamo process. One might have concluded from such an analysis that the physical parameters of the process could not possibly have led to a dynamo. One might have concluded that the physical parameters of the problem would have led to the dynamo many, many orders of magnitude greater than the one we presume exists in the earth's interior. That the analysis leads to values within the range of the real geodynamo strongly supports the suggestion that the geodynamo has this energy source.

LECTURE II

Flow in a precessing spheroid. The experimental study of fluid flow in a precessing spheroidal cavity has led to the isolation of four significantly different regimes. First, the tilt of the fluid's axis of rotation away from the axis of rotation of the container was indicative of the Poincaré type of solution. Second, a laminar zonal shearing flow appeared, whose intensity was a function of the Poincaré tilt. Thirdly, when this zonal flow became intense enough, a wavy instability was observed to occur. Lastly, with further precessional torque, the flow became quite turbulent.

The observed tilt in the laminar regime was in agreement with the classical theory described in the previous lecture. The emphasis in this lecture will be the laminar zonal flow. I will conclude the lecture with a discussion of progress towards an understanding of the wavy instability and some brief comments on the observations of the turbulent regime.

The origin of the zonal shearing flow must be the failure of the Poincaré solution to exactly satisfy the viscous boundary conditions. Stewartson and Roberts studied the linear boundary layer phenomenon. They concluded that the addition of a thin boundary layer flow met the viscosity requirements, and left Poincaré's solution for the fluid motion essentially unaltered. However, in contrast to their conclusion, the experiments suggest that the intensity of the observed zonal flow becomes greater as the viscosity of the fluid is decreased. It was clear that finite-amplitude precessional flow in the limit of vanishing viscosity would be very different from the classical flow found by assuming zero viscosity initially. I traced the origin of these zonal flows to the nonlinear momentum advection in the thin boundary layers. With the assistance of my colleague F. Busse, a first interpretation of the internal flow linked it with the resonant excitation of internal free modes driven by the boundary layer stresses. A detailed theoretical treatment of the problem in the limit of vanishingly small viscosity and small, but finite, amplitude has been prepared by Busse. It is this work which will be discussed in the following.

The basic equations appropriate for steady flows in the precessing system are: first, for the continuity of mass,

$$\nabla \cdot \boldsymbol{v} = 0, \tag{1}$$

second, for the continuity of momentum,

$$-E\nabla^2 \boldsymbol{v} + 2\boldsymbol{\Omega} \times \boldsymbol{v} + \boldsymbol{v} \cdot \nabla \boldsymbol{v} + \nabla P = 0, \tag{2}$$

here scaled so that the velocity on the surface S of the container is written as

$$v = k \times r, \tag{3}$$

where k is the unit normal vector along the axis of rotation, and r is the position vector measured from the center of the system. The scaling is also such that the viscous term appears multiplied by the Ekman number

$$E = \nu/\omega R^2, \tag{4}$$

where ν is the kinematic viscosity, ω the angular velocity of rotation and R the principal radius of the container. The shape of the container is given by

$$r^2 - \eta(k \cdot r)^2 = 1, \tag{5}$$

where η is the ellipticity of the oblate spheroid.

The boundary layer theory will be used to determine the flow to second order in amplitude. (A good reference for this use of singular perturbation theory is the book *The Theory of Rotating Fluids* by H. P. Greenspan.) We first express the velocity of the fluid as the sum of an interior inviscid velocity plus a boundary layer contribution as

$$v = u + \tilde{u}, \tag{6}$$

where u is the (inviscid) interior flow and $\tilde{u}$ is the boundary layer flow presumed to vanish in the interior. The sum of these two velocities is to satisfy the correct viscous boundary condition. The spatial variable normal to the boundary is scaled in terms of

$$\xi \equiv -[r(S) - r] \cdot n E^{-1/2}, \tag{7}$$

where n is the unit vector normal to the surface S.

From equations (7), (6), and (1), one may write

$$\nabla \cdot \tilde{u} = -E^{-1/2} \frac{\partial}{\partial \xi} \tilde{u} \cdot n + n \cdot \nabla \times (n \times \tilde{u}) + n \cdot \tilde{u} \nabla \cdot n = 0. \tag{8}$$

The departure of the total velocity from solid rotation divided by the velocity of solid rotation is defined as ε, the amplitude of the relative motion. The boundary layer and interior velocities are expanded in a double series in $E^{1/2}$ and the amplitude ε as

$$\begin{aligned} u &= \sum_{\mu=0;\nu=0}^{\infty} \varepsilon^{\mu} E^{\nu/2} u_{\nu}^{\mu}, \qquad u_0^0 \equiv k \times r, \\ \tilde{u} &= \sum_{\mu=0;\nu=0}^{\infty} \varepsilon^{\mu} E^{\nu/2} \tilde{u}_{\nu}^{\mu}, \qquad \tilde{u}_0^0 \equiv 0. \end{aligned} \tag{9}$$

The basic state is chosen as the one of solid rotation. The inhomogeneous forcing due to the precession will be introduced in the $\varepsilon E^{1/2}$ order to remove the indeterminancy of the problem. We treat η, the ellipticity, and $E^{1/2}$ as of the same order of magnitude to avoid a triple expansion.

The formal ordering of the sequence of problems is as follows:

$$\nabla \cdot \boldsymbol{u}_0^1 = 0, \qquad \boldsymbol{u}_0^1 \cdot \boldsymbol{n} = 0 \quad \text{on } S,$$
$$(10) \qquad \boldsymbol{u}_0^0 \cdot \nabla \boldsymbol{u}_0^1 + \boldsymbol{u}_0^1 \cdot \nabla \boldsymbol{u}_0^0 = -\nabla P_0^1,$$

where our purpose will be to find a class of $\boldsymbol{u}_0^1$ fields compatible with these equations. Secondly,

$$\tilde{\boldsymbol{u}}_0^1 + \boldsymbol{u}_0^1 = 0 \quad \text{on } S,$$
$$(11) \qquad E^{-1/2}\tilde{\boldsymbol{u}}_0^1 \cdot \boldsymbol{n} = 0 = \tilde{\boldsymbol{P}}_0^1,$$
$$2\boldsymbol{\Omega} \times \tilde{\boldsymbol{u}}_0^1 + \boldsymbol{u}_0^0 \cdot \nabla \tilde{\boldsymbol{u}}_0^1 + \tilde{\boldsymbol{u}}_0^1 \cdot \nabla \boldsymbol{u}_0^0 = \boldsymbol{n}(\partial \tilde{\boldsymbol{P}}_1^1/\partial \xi) + (\partial^2/\partial \xi^2)\tilde{\boldsymbol{u}}_0^1.$$

Our task here is to find $\tilde{\boldsymbol{u}}_0^1$ in terms of the $\boldsymbol{u}_0^1$ found in equation (10) above. Thirdly,

$$(12) \qquad -\partial \tilde{\boldsymbol{u}}_1^1 \cdot \boldsymbol{n}/\partial \xi + \boldsymbol{n} \cdot \nabla \times (\boldsymbol{n} \times \tilde{\boldsymbol{u}}_0^1) = 0.$$

Here we are to find $\tilde{\boldsymbol{u}}_1^1$ in terms of $\tilde{\boldsymbol{u}}_0^1$, and through the boundary condition to determine $\boldsymbol{u}_1^1 \cdot \boldsymbol{n}$ on the surface S. Fourthly,

$$\nabla \cdot \boldsymbol{u}_1^1 = 0,$$
$$(13) \qquad 2\boldsymbol{\Omega} \times \boldsymbol{u}_1^1 + \boldsymbol{u}_0^0 \cdot \nabla \boldsymbol{u}_1^1 + \boldsymbol{u}_1^1 \cdot \nabla \boldsymbol{u}_0^0 + \nabla P_1^1 = +\left(\frac{1}{\varepsilon E^{1/2}}\right)\boldsymbol{\Omega} \times \boldsymbol{u}_0^0.$$

It is not necessary to find $\boldsymbol{u}_1^1$ itself, however; the solvability condition for $\boldsymbol{u}_1^1$ determines the $\boldsymbol{u}_0^1$ which will be produced by the precessional forcing function.

We anticipate that $\varepsilon \boldsymbol{u}_0^1 = (\boldsymbol{\omega} - \boldsymbol{k}) \times \boldsymbol{r}$, where $\boldsymbol{\omega}$ is a constant vector, will fulfill the solvability condition. One determines from the above sequence of problems that

$$\omega^2 = \boldsymbol{k} \cdot \boldsymbol{\omega}, \qquad \varepsilon^2 = 1 - \omega^2$$

and

$$(14) \qquad \frac{\boldsymbol{\omega}}{\omega^2} = \boldsymbol{k} + \frac{A\boldsymbol{k} \times (\boldsymbol{\Omega} \times \boldsymbol{k}) + B(\boldsymbol{k} \times \boldsymbol{\Omega})}{A^2 + B^2},$$

where $A = (.259(E/\omega)^{1/2} + \eta\omega^2 + \boldsymbol{\Omega} \cdot \boldsymbol{k})$, $B = 2.62(E\omega)^{1/2}$.

Equation (14) is valid to second order in ε for the determination of the direction of $\boldsymbol{\omega}$ due to the precession. However, second order terms do cause a change in the zonal flow relative to the vector $\boldsymbol{\omega}$. Second order theory exactly parallels the linear theory above. One seeks that $\boldsymbol{u}_0^2$ which permits solvability of the equation for $\boldsymbol{u}_1^2$. The boundary conditions for $\boldsymbol{u}_1^2$ are dominated by the boundary layer efflux due to $\tilde{\boldsymbol{u}}_0^1 \cdot \nabla \tilde{\boldsymbol{u}}_0^1$. One may write

$$\boldsymbol{u}_0^2 = \boldsymbol{\omega} \times \boldsymbol{r} f(|\boldsymbol{\omega} \times \boldsymbol{r}|/\omega), \tag{15}$$

where with perseverance and skill Busse has found the f given in Figure 1.

The infinite shear layer in Figure 1 lies on the cylinder which intercepts the surface S at $\pm 30°$ latitude relative to the equator determined by $\boldsymbol{\omega}$. Although interior viscous effects certainly will smear this intense shear out, it is clear that inviscid theory differs dramatically from viscous theory in the limit of vanishing viscosity.

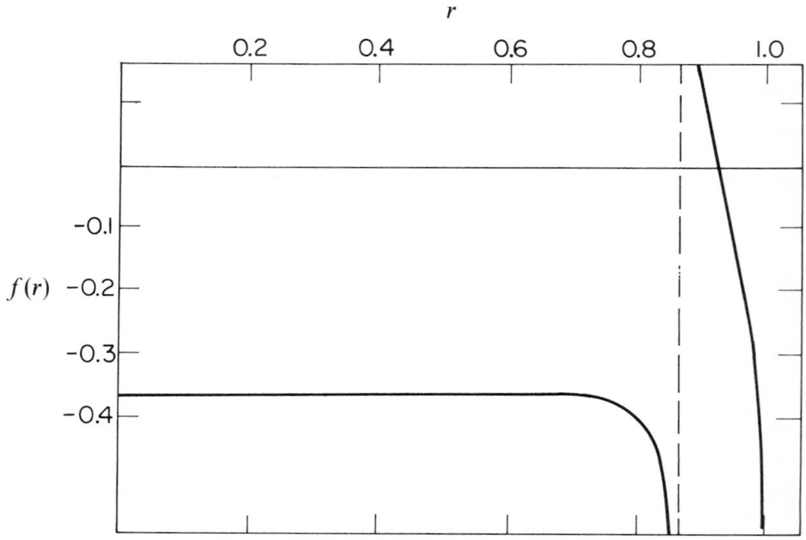

FIGURE 1

A work which parallels this study was done by doctoral candidate S. Suess on flow due to a tidal bulge. The off-equator component of the bulge gives rise to a $\boldsymbol{u}_0^2$ flow just as in Figure 1, while the equatorial component of the bulge gives rise to intense shear layers on the axis of the flow. This result is exhibited in Figure 2.

A representative sample of data for precessing flow in the laboratory exhibits the dramatic features of both Figures 1 and 2, but strongly smoothed by internal viscous effects. This is seen in Figure 3, which is a

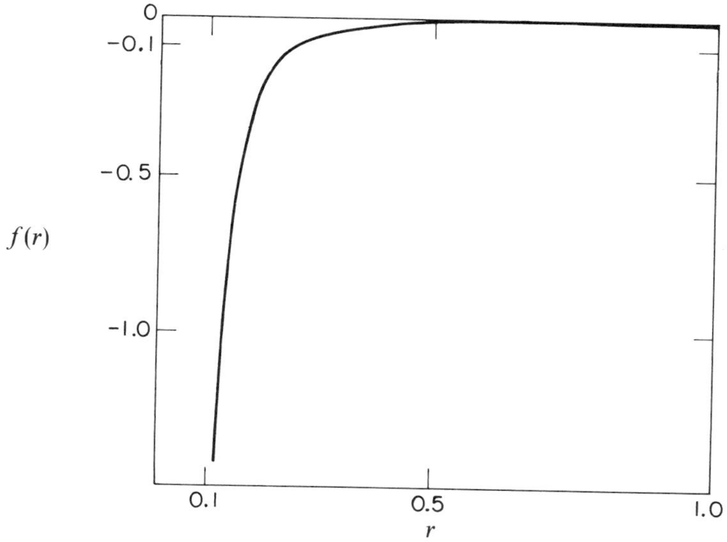

FIGURE 2

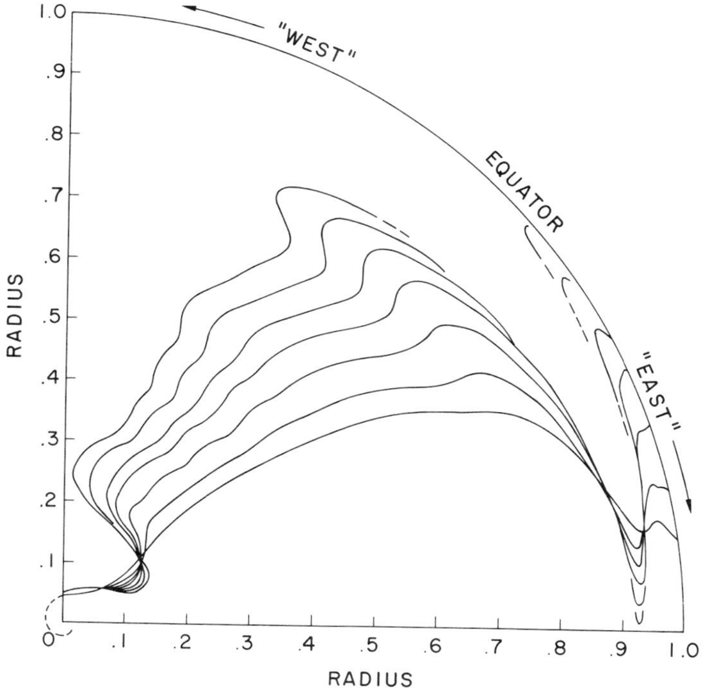

FIGURE 3

tracing from the photographic records of a dye streak introduced into a laminar flow such as we saw in the photographs in the previous lecture. The dye was produced electrically in a dilute solution of thymol blue by means of a straight wire probe extending from the equator of the spheroid to the center of the fluid. The probe was withdrawn and a very slow precession was started. Photographs of the developing flow were taken at one-minute intervals. A steady toroidal velocity relative to the container was reached after several minutes. The Ekman number for this flow is $E = 3.6 \times 10^{-6}$.

In Poincaré-like flows, and in the flows predicted by Stewartson and Roberts, the dye line of Figure 3 would not have moved at all. The interior flow in Poincaré core fluid is a smooth zero average diurnal oscillation. However, due to the second order features caused by the boundary layer efflux, one sees the beginnings of a sharp jet at 30° latitude and considerable shear along the axis of the flow. The jets became much sharper at smaller E. Data taken at Ekman numbers of 3×10^{-7} exhibit prograde jets five times as intense as shown in Figure 3.

These shearing flows are observed to become unstable at values of $\varepsilon^2 \gtrsim 5E^{1/2}F(E)$, where $F(E) \sim .4$ in these laboratory observations. The trend of the laboratory observations suggests that $F(E)$ may vary as $E^{-2/5}$

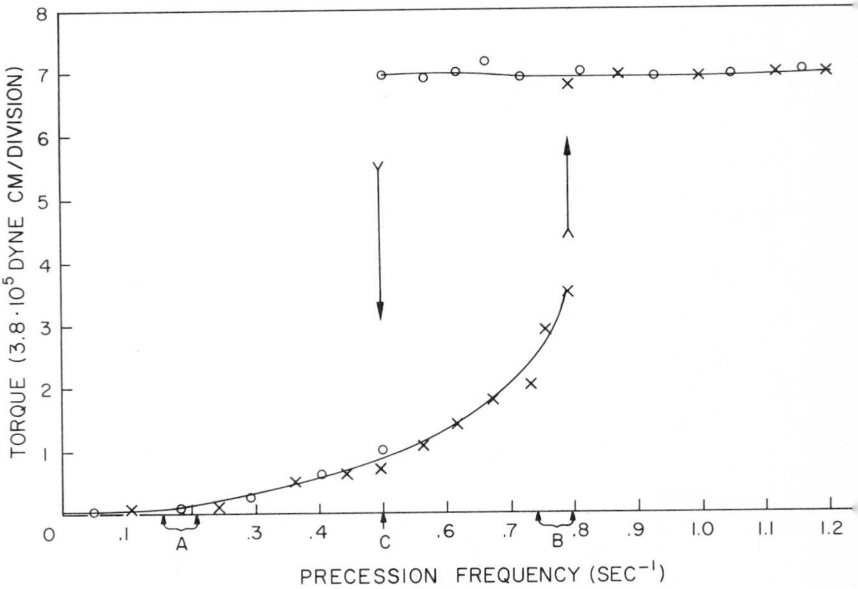

FIGURE 4

for extremely small E. However, this point has negligible theoretical foundation and only a few supporting experimental data. In any instance, if this scaling from the laboratory is appropriate to the earth, flow in the earth's core should be very unstable.

The character of unstable precessional flow has not been studied in any detail. Figure 4 exhibits the one kind of quantitative data yet obtained, namely the torque needed to maintain the flow. These particular data were taken from a spheroid rotating at 900 revolutions per minute. The angle between the rotation precession axis was 96°. The laminar regime of flow extends only to the range A, indicated in the figure. While between A and B, a wavy nonlinear flow persisted, becoming more and more intense. At B a dramatic transition occurred to a turbulence state which appears "saturated," since the torque required to maintain it was independent of precession frequency over a wide range. However, on reducing the precession frequency to the point labelled C in Figure 4, the flow relaxed back to the wavy regime.

Torque records containing hysteresis are commonplace in our studies of the flow due to precession. We anticipate that a transition to hydromagnetic turbulence will exhibit the same finite amplitude behavior.

References

Lecture I

1. G. E. Backus, *A class of self-sustaining dissipative spherical dynamos*, Ann. Physics **4** (1958), 372–447. MR **20**, #1512.

2. E. C. Bullard, *The secular change of the earth's magnetic field*, Monthly Notices Roy. Astron. Soc. Geophys. Suppl. **5** (1948), 248.

3. F. H. Busse, *Steady fluid flow in a processing spheroidal shell*, J. Fluid Mech. **33** (1968), 739.

4. T. G. Cowling, *The magnetic field of sunspots*, Monthly Notices Roy. Astron. Soc. **94** (1934), 39.

5. W. M. Elsasser, *On the origin of the earth's magnetic field*, Phys. Rev. **55** (1939), 489.

6. A. Herzenberg, *Geomagnetic dynamos*, Philos. Trans. Roy. Soc. London Ser. A. **250** (1958), 543.

7. R. Hide, *Motions of the earth's core and mantle, and variations of the main geomagnetic field*, Science **157** (1967), 55.

8. S. S. Hough, *The oscillations of a rotating ellipsoidal shell containing fluid*, Philos. Trans. Roy. Soc. London **186** (1895), 469.

9. W. V. R. Malkus, *Magnetoconvection in a viscous fluid of infinite electrical conductivity*, Astrophys. J. **130** (1959), 259–275. MR **21**, #6184.

10. ———, *Precession of the earth as the cause of geomagnetism*, Science **160** (1968), 259.

11. H. Poincaré, *Sur la précession des corps déformables*, Bull. Astron. **27** (1910), 321.

12. P. H. Roberts and K. Stewartson, *On the motion of a liquid in a spheroidal cavity of a precessing rigid body*, Astrophys. J. **137** (1963), 777.

LECTURE II

1. F. H. Busse, *Steady fluid flow in a precessing spheroidal shell*, J. Fluid Mech. **33** (1968), 739.

2. W. V. R. Malkus, *Precession of the earth as the cause of geomagnetism*, Science **160** (1968), 259.

3. S. Suess, *Effects of gravitational tides on a rotating fluid*, Doctoral Dissertation, Dept. Planetary and Space Science, U.C.L.A., 1969.

RENSSELAER POLYTECHNIC INSTITUTE

Planetary Waves

K. Stewartson

1. **The magnetic field of the earth.** It is well known that this magnetic field is mainly dipole, the contributions from the higher harmonics being smaller by a factor of at least 10, and has a total strength of about $\frac{1}{3}$ gauss. Further the field varies slowly with time (the geomagnetic secular variation or G.S.V.) in which two main features may be distinguished. First significant changes can occur on a continental scale over a period of a few decades. Second, harmonic analysis reveals a slow westward drift of the nondipole components of about $\frac{1}{4}°$ longitude per annum which seems to have been going on for about 500 years.

It now seems incontrovertible that the origin of the earth's magnetic field lies in the fluid core, whose outer radius is ~ 3500 km and inner radius ~ 1400 km, and arises from a dynamo action such as that considered by Lilley [1970]. A strongly toroidal field is produced, whose strength ~ 100 gauss in the core and much larger than the dipole field there (~ 1 gauss). This toroidal field is unobservable outside the core of the earth.

My own interest in planetary waves stems from the suggestion by Hide [1966] that the G.S.V. is associated with the free oscillation of the fluid in the core under magneto-hydrodynamic forces. On the time scale of the G.S.V. (~ 1000 years), we may assume that the lines of magnetic force move with the fluid and hence the observed variation in the observed field reflects the core motion. Hide used the β-plane approximation, first introduced by Rossby, for fluid in thin shells in which the variable term $f = 2\Omega \cos\theta$ (θ is co-latitude) is replaced by $f_0 + \beta y$, where y measures distance northward, f_0 is constant and so is $\beta = 2\Omega \sin\theta_0/R$ (Ω angular velocity, R the radius of the shell and θ_0 an appropriate and constant co-latitude). Clearly this idea has drawbacks when applied to planetary waves but may well be qualitatively correct. Hide extended the idea to thick shells and noted that whereas in thin shells moving a filament of fluid, lying parallel to the axis of rotation, polewards decreases its length and hence increases its vorticity, in thick shells the reverse happens. He accordingly suggested that the Rossby theory could be taken over to

---

AMS 1970 *subject classifications*. Primary 35L20, 35L05, Secondary 76C10, 76U05, 76W05.

thick shells such as the fluid part of the earth's core by taking $\beta < 0$. When in addition he added a toroidal magnetic field to Rossby's theory, he discovered a new class of (Hide) waves travelling westward with a wave speed $\sim H_0^2/\Omega R\rho$ where H_0 is the strength of the toroidal field, ρ the fluid density; for $H_0 \sim 100$ gauss this gives periods of oscillation ~ 1000 years.

Hide's argument was imprecise, and in order to quantify it Malkus [1967] considered a particular example of a complete fluid sphere with a magnetic field induced by a uniform current density. Specifically he supposed that

$$\boldsymbol{H}_0 = \boldsymbol{j} \times \boldsymbol{r}$$

where $\boldsymbol{j}$ is constant and parallel to the axis of rotation. The governing equations of the motion of the fluid may be taken as

$$\frac{\partial \boldsymbol{v}}{\partial t} + 2\boldsymbol{\Omega} \times \boldsymbol{v} - \boldsymbol{v} \times \operatorname{curl} \boldsymbol{v} = -\frac{1}{\rho}\operatorname{grad}(p + \tfrac{1}{2}\rho v^2) + \frac{1}{4\pi\rho}(\boldsymbol{H} \times \operatorname{curl} \boldsymbol{H}),$$

$$\operatorname{div} \boldsymbol{v} = \operatorname{div} \boldsymbol{H} = 0,$$

where $\boldsymbol{v}$ is the fluid velocity relative to axes rotating with the mantle and p is the pressure. The boundary condition is that $\boldsymbol{v} \cdot \boldsymbol{n} = 0$ at the mantle boundary, and it immediately follows that for small oscillations $\boldsymbol{H} \cdot \boldsymbol{n} = 0$ at the mantle boundary too so that the oscillations are directly unobservable. It is likely that oscillations would act as prime movers of the tertiary motions associated with the dipole-field and *these* would be observable.

On linearizing the equations, assuming all dependent variables are of the form $Q(r, z)\, e^{i(m\phi + \sigma\Omega t)}$, where (r, ϕ, z) are cylindrical polar coordinates, m is an integer and σ a constant to be found, and writing

$$p + \tfrac{1}{2}\rho v^2 + \tfrac{1}{2}\rho \boldsymbol{H}^2 = -\frac{i}{\sigma}\Phi(r, z) \exp[i(m\phi + \sigma\Omega t)],$$

we find that

$$\left(\nabla^2 - \frac{4}{\lambda^2}\frac{\partial^2}{\partial z^2}\right)\Phi = 0, \qquad r^2 + z^2 < 1,$$

and

$$\left(r\frac{\partial}{\partial r} + z\frac{\partial}{\partial z} + \frac{2m}{\lambda} - \frac{4}{\lambda^2}z\frac{\partial}{\partial z}\right)\Phi = 0, \qquad r^2 + z^2 = 1$$

where

$$\lambda = (m^2 j^2 - 4\pi\rho\Omega^2\sigma^2)/(mj^2 - 4\pi\rho\Omega^2\sigma).$$

This is Poincaré's classic problem for inertial oscillations of a rotating mass of fluid. Greenspan [1968] has shown that $|\lambda| \leqq 2$ (but see below) and solutions for Φ can be found by separation of variables using associated

Legendre polynomials. We also note that $j/\Omega\sqrt{\rho}$ is small for the earth and hence

$$\sigma \approx -\frac{mj^2}{4\pi\rho\Omega^2}\frac{m-\lambda}{\lambda}.$$

Thus for westward drifting waves $\lambda > m$ or $\lambda < 0$ and for eastward drifting waves $0 < \lambda < m$. Taking $H_m = jR$ as a representative toroidal field, we see that the speed of these waves $\sim H_m^2/\rho R\Omega$ in agreement with Hide's estimate.

Unfortunately the simplest solutions have $\Phi = r^m z$ and give $\lambda = 2/(m+1)$ so that for $m \geqq 1$ all waves drift eastward; the next have

$$\Phi = r^m \text{ [linear function of } r^2 \text{ and } z^2\text{] giving}$$

$$\lambda = \frac{2}{m+2}\left[1 \pm \left(\frac{(m+1)(m+2)}{2m+3}\right)^{1/2}\right].$$

Although for $m = 1$, $\lambda = -.18$ or 1.51 so that both waves drift westward, for $m = 2$ and above one drifts east and one west. In general it appears that roughly half the waves are eastward drifting and half westward. No preference has been established for the one rather than the other. There are actually ∞^3 possible values of λ, for there are three degrees of freedom (in r, ϕ, z).

It is generally believed that there is a central body roughly of radius 1400 km. In order to begin a study of its effect, the author [1967] considered the free oscillation of fluid in a thin shell, taking, like Hide, the toroidal field to be uniform and supposing that radial derivatives are finite. It turns out that all waves drift eastward although, to be sure, these are now only ∞^2 eigenvalues. Further it was shown that for $m \geqq 2$ it is impossible for the phase velocity to change sign by σ vanishing at some finite value of the thickness of the shell and very likely the same is true for $m = 1$. The inference seems to be that the westward drifting waves develop large radial derivatives as the shell becomes thin and therefore are more likely to be damped by viscosity. Must we infer then that the Hide mechanism is irrelevant and at best can be used to put an upper bound on the toroidal field? A deeper study of waves in a thick shell is needed before this question can be answered conclusively.

2. **Inertial waves in a rotating fluid.** The obvious starting point is the model adopted by Malkus which implies that we should first probe the properties of inertial waves in a rotating fluid. The corresponding results for Malkus' model then follow at once. We have to determine the properties of $\boldsymbol{v}$ given that

(2.1) $$\partial \boldsymbol{v}/\partial t + 2\Omega \times \boldsymbol{v} = -\nabla P, \qquad \nabla \cdot \boldsymbol{v} = 0$$

subject to $\boldsymbol{n} \cdot \boldsymbol{v} = 0$ on the bounding surfaces. Here $\boldsymbol{v}$ is the perturbed fluid velocity and P the reduced pressure. On supposing that the boundaries have axial symmetry and that

$$P = \Phi(r, z) \exp(i\lambda\Omega t + im\phi),$$

we have

$$\left[\frac{\partial^2}{\partial r^2} + \frac{1}{r}\frac{\partial}{\partial r} - \frac{m^2}{r^2} + \left(1 - \frac{4}{\lambda^2}\right)\frac{\partial^2}{\partial z^2}\right]\Phi = 0$$

with

$$-\lambda^2(\hat{\boldsymbol{n}} \cdot \nabla)\Phi + 4(\hat{\boldsymbol{n}} \cdot \boldsymbol{k})(\boldsymbol{k} \cdot \nabla\Phi) - (2\lambda m/r)\Phi = 0$$

on the boundaries where (r, ϕ, z) are cylindrical polar coordinates and $\boldsymbol{k}$, $\hat{\boldsymbol{n}}$ are unit vectors parallel to the axis of rotation and the normal to the surface respectively. It is very likely that for nontrivial solutions of these equations λ must be real and $|\lambda| < 2$ [Greenspan, 1968], but then we have to solve a hyperbolic equation subject to Dirichlet boundary conditions —an ill-posed problem in Hadamard's sense.

A simple example of such problems is to solve

$$\partial^2\phi/\partial x^2 - \alpha^2(\partial^2\phi/\partial y^2) = 0 \quad \text{in} \quad C \tag{2.2}$$

subject to $\phi = 0$ on C. Are there any values of α such that $\phi \not\equiv 0$? The earliest study I know of is discussed by Hardy and Wright [1960] when C is a square of unit side. Then if α is a rational number, ϕ need not vanish everywhere. It is interesting to note that if α is rational and $\phi \neq 0$ on C then in general there is no solution. If C is a circle, solutions can be found if $\alpha = \tan\beta$ where $\cos n\beta = 0$ or $\sin n\beta = 0$, n being an integer. Barcilon [1969] has solved (2.2) for a circle with similar results. Other solutions of (2.2) have been found for a circular cylinder [Stewartson, 1959] and the wedge has been discussed by Greenspan [1969]. This problem is also of interest because the apex accumulates energy and the spectrum of eigenvalues is continuous. All these cases, however, concern cavities without internal boundaries, which is the case of special interest here.

3. **Waves in a shell.** If the shell is thin, it is natural to assume that the prime motion takes place parallel to the bounding surfaces so that if u, v are the velocity components in the meridian and azimuthal directions respectively and θ denotes the co-latitude,

$$\frac{\partial v}{\partial t} + 2\Omega u \cos\theta = -\frac{1}{R\sin\theta}\frac{\partial p}{\rho\partial\phi}, \qquad \frac{\partial u}{\partial t} - 2\Omega v\cos\theta = -\frac{1}{\rho R}\frac{\partial p}{\partial\theta},$$

$$\frac{\partial}{\partial\theta}(u\sin\theta) + \frac{\partial v}{\partial\phi} = 0, \tag{3.1}$$

and so the stream function ψ satisfies

$$\frac{\partial}{\partial t}(\nabla^2\psi) + 2\Omega\frac{\partial\psi}{\partial\phi} = 0. \tag{3.2}$$

Hence, on assuming that $\psi \propto \exp(im\phi + i\omega t)$,

$$\omega = 2m/n(n+1), \tag{3.3}$$

m and n being integers. The stream function ψ is an associated Legendre polynomial, and we note that the waves are eastward moving. The governing equations are closely related to Laplace's tidal equations for oceans of depth h, viz.

$$\begin{gathered}\frac{\partial u}{\partial t} - 2\Omega v\cos\theta = -\frac{\partial}{\partial\theta}(g\zeta), \qquad \frac{\partial v}{\partial t} + 2\Omega u\cos\theta = -\frac{1}{\sin\theta}\frac{\partial}{\partial\phi}(g\zeta),\\ \frac{\partial\zeta}{\partial t} + \frac{1}{\sin\theta}\left[\frac{\partial}{\partial\theta}(hu\sin\theta) + \frac{\partial}{\partial\phi}(hv)\right] = 0,\end{gathered} \tag{3.4}$$

ζ being the elevation of the free surface and g the acceleration due to gravity. The tidal equations have been extensively studied (e.g. Longuet–Higgins and Pond [1970]) and numerical evaluation of the eigenvalues and eigenfunctions made for complete and hemi-spherical oceans. The values of ω depend significantly on $\bar{\varepsilon} = 4\Omega^2R^2/gh$, (3.3) being obtained in the limit $\bar{\varepsilon} \to 0$. The same is true for nonspherical rigid shell boundaries [Rickard, 1970]. In all this work, however, the thickness of the fluid layer is assumed small in comparison with the radius of the earth, which is not true for our core problem, and for the rest of these lectures we will consider the effect of h/R being small but not zero.

An early study of this question was made by Stern [1963] who was interested in the possibility of trapping modes of low frequency near the equator. Assuming the frequency of the oscillations to be $2\Omega\omega(h/R)^{1/2}$ where R is the radius of the outer sphere and $\omega \sim 1$, he obtained an expression for the stream function ψ describing axially symmetric motions in the form

$$\begin{aligned}\psi &= \operatorname{Re}\sum_{n=1}^{\infty} b_n \exp(-2ni\pi z)\\ &\quad \times \{\exp(n\pi i(y-\omega)^2) - \exp(-n\pi i(y-\omega)^2)\},\end{aligned} \tag{3.5}$$

where b_n are constants, hz measures distance from the outer sphere, $y(h/R)^{1/2}$ is the latitude, and $\omega = \frac{1}{2}\sqrt{m}$, m being an integer. He was able to choose b_n so that $|\psi| < B/|y|$ for some constant B and deduced that trapped modes could exist. Bretherton (1964) gave a geometrical argument which led to the same conclusion, and further support may be drawn from the solution (3.2)(3.3) when m, n are large. Stern's argument is, however, incomplete, for even if $\psi \to 0$ as $|y| \to \infty$, the velocity $(\psi_y, -\psi_z)$ need not do so as is clear from (3.5). We remark that although there are axially symmetric solutions for the complete sphere, there are no such smooth solutions in the limit $h \to 0$ from (3.2).

4. **Formal expansion in powers of** h/R. What is needed is some precise information about the general structure of the eigensolutions when h/R is small but not zero. It is clear that in some sense disturbances propagate along the characteristics and are therefore continually being reflected from one boundary to another, and yet in (3.2) the notion of a characteristic does not appear to be relevant. Let us, therefore, consider how (3.2)(3.3) are modified when h/R is small but not zero [Stewartson and Rickard, 1969]. Without loss of generality we shall suppose that all dependent variables are of the form

$$\operatorname{Re} q(r, \theta) \exp(i\phi + i\Omega\omega t),$$

and now r denotes distance from the centre of the shell. Further the exponential factors are omitted and we write

$$\begin{aligned} &\varepsilon = (a-b)/(a+b), \qquad r = \tfrac{1}{2}(a+b)(1+\varepsilon\xi), \qquad \mu = \cos\theta, \\ (4.1)\quad &p = i\Omega r \sin\theta P(r, \mu), \qquad u_\theta = U, \qquad u_\phi = iV \quad \text{and} \\ &u_r = \varepsilon^2 W/(1-\mu^2)^{1/2}. \end{aligned}$$

Here a, b are the outer and inner radii of the shell boundary. The governing equations satisfied by U, V, W, P then become

$$(4.2)\qquad \begin{aligned} \varepsilon^3\omega W/(1-\mu^2) - 2\varepsilon V &= -(1+\varepsilon\xi)\partial P/\partial\xi - \varepsilon P, \\ \omega U - 2\mu V &= (1-\mu^2)\partial P/\partial\mu - \mu P, \\ \omega V - 2\mu U &= -P + 2\varepsilon^2 W, \\ \varepsilon(1+\varepsilon\xi)\partial W/\partial\xi + 2\varepsilon^2 W - (1-\mu^2)&\partial U/\partial\mu - V = 0 \end{aligned}$$

and the boundary conditions are $W = 0$, $\xi = \pm 1$. Similar equations hold if the ϕ dependence is $e^{im\phi}$.

The first approximation is obtained on setting $\varepsilon = 0$ and is

$$(4.3)\qquad \omega = 2/n(n+1), \qquad U = P_n'(\mu), \qquad V = n(n+1)P_n(\mu) - P_n'(\mu),$$

where n is an integer and P_n is the Legendre polynomial. If, however, we attempt to set up the solution in the form of a series in integer powers of

ε, we find that the coefficient of ε^s in the series for U and V has a singularity when $\omega^2 = 4\mu^2$ being of the form

$$B_s(\xi, \mu)/(\omega^2 - 4\mu^2)^{2s-1} \tag{4.4}$$

where B_s is bounded. Thus there is a nonuniformity, in the expansion, in the neighborhood of $\omega^2 = 4\mu^2$ which needs separate treatment. The characteristics of the governing equation touch the spheres at $\mu = \pm\frac{1}{2}\omega_1$, to order ε. It follows from (3.7) that the singular region extends a distance $O(\varepsilon^{1/2})$ on either side of $\mu = \frac{1}{2}\omega_1$ (concentrating attention on the positive sign from now on) and we are led to write

$$\begin{aligned} U &= A_u + \varepsilon^{1/2}\bar{u}(\xi, x) + \cdots, \\ V &= A_v + \varepsilon^{1/2}\bar{v}(\xi, x) + \cdots, \\ W &= \varepsilon^{-1}\bar{w}(\xi, x) + \cdots, \\ P &= A_p + \varepsilon^{1/2}xB_p + \varepsilon(x^2C_p + D_p) \\ &\quad + \varepsilon^{3/2}[E_px^3 + x\xi F_p + G_p\Phi(\xi, x)] + \cdots \end{aligned} \tag{4.5}$$

when $x \sim 1$ where $\mu = \frac{1}{2}\omega_1 + (1 - \frac{1}{4}\omega_1^2)^{1/2}x$. A, B, C, E, F, G are constants known in terms of the first approximation (4.3) and Φ satisfies the differential equation

$$x(\partial^2\Phi/\partial\xi^2) + \partial^2\Phi/\partial\xi\partial x = 1$$

with boundary conditions

$$2x(\partial\Phi/\partial\xi) + \partial\Phi/\partial x = 2\xi \quad \text{on } |\xi| = 1. \tag{4.6}$$

In addition, in virtue of the expansion procedure adopted, we would like Φ to remain bounded as $|x| \to \infty$. In order to match with the regular solution already found when $|x|$ is large,

$$\Phi \sim (3\xi^2 + 1)/6x \quad \text{as } |x| \to \infty, \tag{4.7}$$

and so to begin with we need a solution of (4.6) which is an odd function of x. A general solution is

$$\Phi = x\xi + F(\tfrac{1}{2}x^2 - \xi) + G(x)$$

where

$$G(x) = x - 2x^3/3 + F(\tfrac{1}{2}x^2 - 1)$$

and

$$F(1 + \tfrac{1}{2}x^2) - F(\tfrac{1}{2}x^2 - 1) = 2x. \tag{4.8}$$

If we restrict attention to solutions odd in x we find that

$$\text{(4.9)}\qquad F(y) = 2^{3/2} \sum_{n=0}^{N-1} (y - 2n - 1)^{1/2} = \frac{1}{2i(2\pi)^{1/2}} \int_{c-i\infty}^{c+i\infty} \frac{e^{sy}\, ds}{s^{3/2} \sinh s}$$

where $|y - 2N| < 1$ and $c > 0$. Hence G may be found from (3.12) and finally

$$\text{(4.10)}\qquad \Phi(x, \xi) = x - \frac{2x^3}{3} + x\xi + \frac{1}{2i(2\pi)^{1/2}} \int_{c-i\infty}^{c+i\infty} \frac{e^{sx^2/2}\{e^{-s} + e^{-s\xi}\}\, ds}{s^{3/2} \sinh s}.$$

In order to evaluate the contour integral, the path of integration may be deformed onto the two sides of the negative real axis of s, and the resulting real integral then has an asymptotic form in agreement with (4.7) as $|x| \to \infty$. However, such a deformation of the contour of integration involves passing over the infinite set of poles of the integrand at the zeros of $\sinh s$, $s \neq 0$. The residues at these points give a total contribution to Φ of

$$K(\tfrac{1}{2}x^2 - \xi) + K(\tfrac{1}{2}x^2 - 1)$$

where

$$\text{(4.11)}\qquad K(y) = \frac{\sqrt{2}}{\pi} \sum_{n=1}^{\infty} \frac{(-)^n \cos(\lambda\pi y - 3\pi/4)}{n^{3/2}}.$$

As can be seen from (4.9) and (4.11),

$$K(y) \sim 2^{3/2}(y - 1)^{1/2} \quad \text{as } y \to 1+, \quad \text{and} \quad K(y) \quad \text{is smooth} \quad \text{as } y \to 1-.$$

Thus the nonuniformity is not confined to the neighborhood of the critical circles but extends over the whole shell of fluid. In physical terms the characteristic which touches the inner boundary of the shell has associated with it a singularity (like $x^{-1/2}$) of the velocity on the side nearer the inner boundary. This property of the solution is carried by the characteristic through its reflections from the two boundaries of the fluid and hence extends right through the fluid.

Many questions remain unanswered, for example: What happens when the reflecting wave enters the neighborhood of the circle where a member of the other family of characteristics touches the inner sphere ($\mu = -\tfrac{1}{2}\omega$) and near the poles? How does the value of ω depend on ε? More fundamentally the present analysis is inconsistent in that we have assumed that the solution (for p) is smooth almost everywhere as $\varepsilon \to 0$ and end up by deducing that it is nondifferentiable almost everywhere. Can it be put on a completely rational basis?

It seems to me that the main deduction we can make from the above argument is that when the fluid is confined in a shell whose inner and outer boundaries are smooth but otherwise unrestricted, then the velocities in any eigensolution can be expected to have singularities on one side of the characteristics touching the inner surface together with all their reflections. It follows that the velocities are no longer integrable in square and theorems which depend on this property, common in the literature, need re-examining. From a practical standpoint the determination of the eigenvalue in problems of geophysical interest has not been made easier by this analysis, and progress towards an understanding of the wave motions in the core of the earth is likely to be slow.

REFERENCES

F. E. M. Lilley, *On kinematic dynamos*, Proc. Roy. Soc. Ser. A **316** (1970), 153.

R. Hide, *Free hydromagnetic oscillation of the earth's core and the theory of the geomagnetic secular variation*, Philos. Trans. Roy. Soc. London Ser. A **259** (1966), 615.

W. V. R. Malkus, *Hydromagnetic planetary waves*, J. Fluid Mech. **28** (1967), 793–802. MR **36** #4869.

H. P. Greenspan, *Theory of rotating fluids*, Cambridge Univ. Press, New York, 1968.

K. Stewartson, *Slow oscillations of fluid in a rotating cavity in the presence of a toroidal magnetic fluid*, Proc. Roy. Soc. Ser. A **299** (1967), 173.

G. H. Hardy and E. M. Wright, *Theory of numbers*, 4th ed., Oxford Univ. Press, London, 1960.

V. Barcilon, *Axi-symmetric inertial oscillations of a rotating ring of fluid*, Mathematika **15** (1968), 93–102. MR **37**#7135.

K. Stewartson, *On the stability of a spinning top containing liquid*, J. Fluid Mech. **5** (1959), 577–592. MR **21**#4694.

H. P. Greenspan, *On the inviscid theory of rotating fluids*, Studies in Appl. Math. **48** (1969), 19.

M. S. Longuet-Higgins and G. S. Pond, *The free oscillations of fluid on a hemisphere bounded by meridians of longitude*, Philos, Trans. Roy. Soc. London Ser. A **266** (1970), 193.

J. A. Rickard, Ph.D. Thesis, London University, 1970.

M. E. Stern, *Trapping of low frequency oscillations in an equatorial "boundary layer"*, Tellus **15** (1963), 246.

F. P. Bretherton, *Low frequency oscillations trapped near the equator*, Tellus **16** (1964), 181.

K. Stewartson and J. A. Rickard, *Pathological oscillations of a rotating fluid*, J. Fluid Mech. **35** (1969), 759.

UNIVERSITY COLLEGE, LONDON

Ocean Tides and Related Waves[1]

George W. Platzman

1.1. **The ocean basins.** The factor most responsible for the complexity of ocean tides is the irregular configuration of the ocean bottom and lateral boundaries. Of the area covered by ocean, about three-fourths has depths between 3 and 6 km and slightly less than one-fourth is shallower than 3 km (Table 1.1.1). The most conspicuous features of the ocean bottom are numerous deep-sea basins the depths of which range between 3 and 6 km. Each of the three main oceans contains several of these basins.

TABLE 1.1.1. Distribution of ocean depths (from [Dietrich, 1963, p. 7]).

Depths (km)	Percent of Earth's surface	Percent of ocean surface
Land	29.2	—
0–3	16.2	22.9
3–6	53.6	75.7
>6	1.0	1.4

Also conspicuous are the mid-ocean ridges, notably in the South and North Atlantic, but prominent as well in the western Indian and eastern Pacific Oceans.

AMS 1970 *subject classifications.* Primary 86–02, 86A25, 76W05, 76E25.

[1] In these lectures I assumed the audience unacquainted with the tidal problem, and so began by formulating the tidal potential and the hydrodynamical equations for ocean tides. My aim then was to give an elementary account of some basic wave types in the range of tidal frequencies.

The deepest sounding is about 11 km but less than 2 percent of the ocean area is deeper than 6 km. Far more important for tides are the continental shelves, which range in depth from 0 to 200 m and account for about 8 percent of the total ocean area. It is in these shallow boundary regions that most of the existing observations of tidal phenomena have been made. Knowledge of the global tide has been highly inferential and speculative.

Of great interest from the standpoint of tides (as well as for general oceanography) are the marginal and mediterranean seas. Most of these are so small in lateral extent that the direct action of tidal forces does not produce as large a response as does the forcing of the tide in the ocean to which they are adjacent. (Obvious exceptions are the Arctic "Ocean" and Mediterranean Sea.) The mean depths and areas of the main oceans are listed in Table 1.1.2.

TABLE 1.1.2. Depths and areas of the main oceans (from [Dietrich, 1963, pp. 3, 7]).

Ocean	Mean depth* (km)	Area* (10^6 km^2)
Atlantic	3.87	82.22
Indian	3.96	73.44
Pacific	4.28	163.25
all three	4.10	320.91

* Excluding marginal seas.

1.2. **Geographical representation of tides.** The traditional way of showing the spatial configuration of the tide is by means of *cotidal lines*, which are simply isochrones of high water. Suppose that at each of a number of points x, y we make a harmonic analysis of the tide and thereby represent the constituent of a particular frequency σ as

$$A(x, y) \cos \sigma t + B(x, y) \sin \sigma t = C(x, y) \cos[\sigma t - \theta(x, y)],$$

$$(C, \theta) \equiv (\text{mod}, \arg)(A + iB).$$

It is conventional to take $0 \leqq \theta < 360°$, and to divide the full period (called a "constituent day") into 24 "constituent hours." Then at the constituent hours $0, 1, 2, \ldots, 23$ the isochrones of high water are respectively the lines $\theta = 0°, 30°, 60°, \ldots, 330°$. The amplitude of the wave usually is represented by *corange* lines $2 \cdot C(x, y) =$ constant.

Simple illustrations of patterns of cotidal and corange lines are provided by the plane progressive wave $\cos(\sigma t - kx)$, the plane standing wave $\sin kx \cos \sigma t$, and the oblique superposition of two plane progressive waves, $\sin ly \cos(kx - \sigma t)$. These patterns are shown respectively in Figure 1.2.1 upper left, upper right, lower left.

More relevant to tides is the superposition

$$\sin kx \cos \sigma t + \sin ly \sin \sigma t$$

of two standing waves that are in quadrature and have perpendicular nodal lines. In this case we get the picture shown in Figure 1.2.1 lower right. The cotidal lines, $\arg(\sin kx + i \sin ly) =$ constant, radiate from an *amphidromic point* ($x = 0$, $y = 0$ in this example) where the range is zero. In the example shown, the isochrone of high water rotates counterclockwise through 360 degrees in a constituent day.[2]

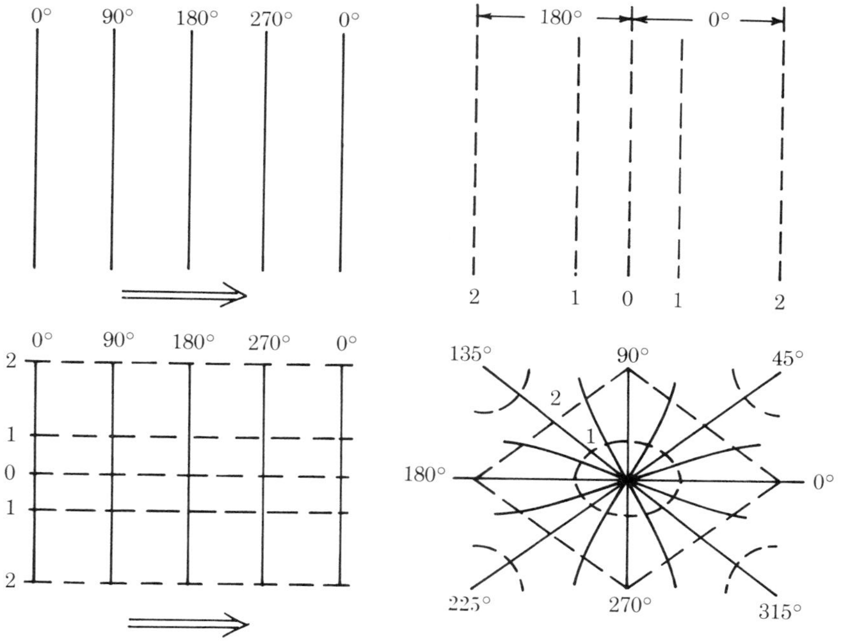

FIGURE 1.2.1. Cotidal (————) and corange (-----) lines for plane progressive wave (upper left), plane standing wave (upper right), oblique superposition of two plane progressive waves (lower left), and superposition of two standing waves in quadrature (lower right).

[2]In the lecture, I briefly described some features of cotidal charts of several marginal seas and of the global tide. This material is not included in the published version.

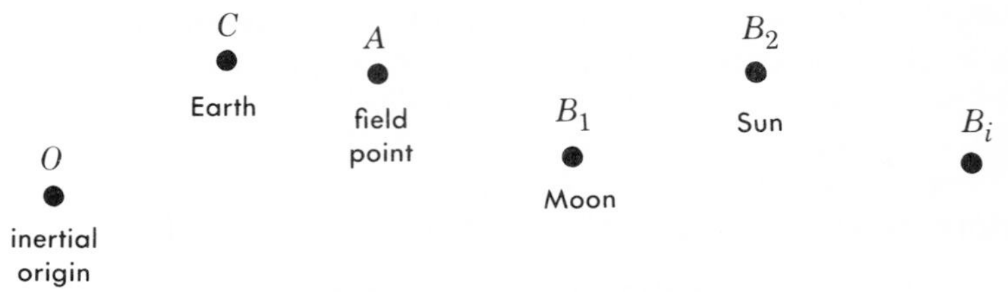

FIGURE 2.1.1. Identity of reference points.

2.1. **The tidal force.** Let C be the location of the center of the Earth and B_i ($i = 1, 2, 3, \ldots$) locations of other bodies in the solar system; also, let A be the location of an element of mass of the ocean, the motion of which we wish to investigate (Figure 2.1.1). If O is an origin with respect to which the position vector $\boldsymbol{OA}$ obeys Newton's second law of motion, then for a unit mass at A,

$$D_a^2 \boldsymbol{OA} = \boldsymbol{P}(A) + \boldsymbol{G}(A) + \sum_i \boldsymbol{F}_i(A). \tag{2.1.1}$$

Here D_a denotes "absolute" differentiation with respect to time (that is, in a reference system with inertial axes), $\boldsymbol{P}(A)$ is the nongravitational force (such as fluid or elastic stress), $\boldsymbol{G}(A)$ the Earth's gravitational attraction at A, and $\boldsymbol{F}_i(A)$ the gravitational attraction at A exerted by the body at B_i.

Let us estimate the magnitude of $\boldsymbol{F}(A)$ relative to $\boldsymbol{G}(A)$ when A is on or near the Earth's surface. For the Moon or Sun, $r \ll R$ (see Figure 2.1.2) so $|\boldsymbol{F}(A)| \approx |\boldsymbol{F}(C)| = \gamma M/R^2$ where γ is the universal constant of gravitation and M the mass of the body at B. Since $|\boldsymbol{G}(A)| \approx \gamma M_{\oplus}/r^2$, we have

$$|\boldsymbol{F}(A)|/|\boldsymbol{G}(A)| \approx (M/M_{\oplus})(r/R)^2.$$

The mass ratios Moon/Earth and Sun/Earth are

$$M_{☾}/M_{\oplus} = 1/81.53, \qquad M_{\odot}/M_{\oplus} = 333420. \tag{2.1.2}$$

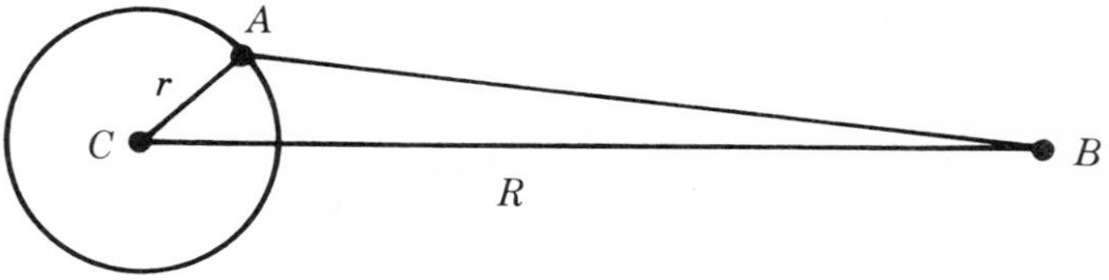

FIGURE 2.1.2. The circle represents the Earth and B the Moon or Sun.

If r is taken to be the Earth's equatorial radius a, the ratio r/R is the sine of the equatorial horizontal parallax of B, mean values of which for Moon and Sun are

$$a/c_{☾} = 0.016593, \qquad a/c_{\odot} = 4.2615 \times 10^{-5} \tag{2.1.3}$$

where $1/c$ is the orbital-mean value of $1/R$. Hence the ratios in question are

$$\frac{|\boldsymbol{F}(A)|}{|\boldsymbol{G}(A)|} \approx \begin{Bmatrix} 3.4 \times 10^{-6} & \text{for Moon} \\ 6.0 \times 10^{-4} & \text{for Sun} \end{Bmatrix}.$$

In spite of its greater distance from the Earth, the Sun exerts a stronger attraction than the Moon because of its preponderant mass. For the relative *tidal* forces, however, the opposite is true, as we shall soon see.

Since A is in the geosphere, we prefer to describe its motion by means of $\boldsymbol{CA}$ rather than $\boldsymbol{OA}$. To do so, we need an equation for the motion of C—that is, for the orbital motion of the Earth. As in (2.1.1), this is provided by Newton's second law of motion:

$$D_a^2 \boldsymbol{OC} = \sum_i \boldsymbol{F}_i(C).$$

(Strictly, there is a term corresponding to nongravitational forces on the Earth as a whole, such as radiation pressure, but such forces are negligible in this context.) Subtract the preceding equation from (2.1.1):

$$\begin{aligned} D_a^2 \boldsymbol{CA} &= \boldsymbol{P}(A) + \boldsymbol{G}(A) + \sum_i \boldsymbol{T}_i(A), \\ \boldsymbol{T}_i(A) &\equiv \boldsymbol{F}_i(A) - \boldsymbol{F}_i(C). \end{aligned} \tag{2.1.4}$$

The *tidal force* $\boldsymbol{T}(A)$ is the attraction due to the body at B as measured at the field point A, minus the attraction due to the same body as measured at the Earth's center C. This residual is the aspect of B's gravitational field that accelerates A *relative* to C and thus tends to deform the geosphere. To put it another way, in (2.1.4) the acceleration of A is expressed relative to the noninertial origin C; consequently, the relative motion of A is affected by "apparent" forces $-\boldsymbol{F}_i(C)$, which are the negatives of those that determine the motion of C. Each of these combines with the corresponding direct attraction $\boldsymbol{F}_i(A)$ to produce the tidal force $\boldsymbol{T}_i(A)$.

The configuration of $\boldsymbol{T}(A)$ can be inferred directly from the fact that the direction of $\boldsymbol{F}(A)$ is that of $\boldsymbol{AB}$ and the magnitude is inversely proportional to $|\boldsymbol{AB}|^2$. In Figure 2.1.3 the tidal force is constructed at each of eight field points symmetrically disposed on the meridian section of a sphere centered at C. Roughly speaking, on the hemisphere that faces B the tidal force tends *toward* B; on the opposite hemisphere it tends *away*

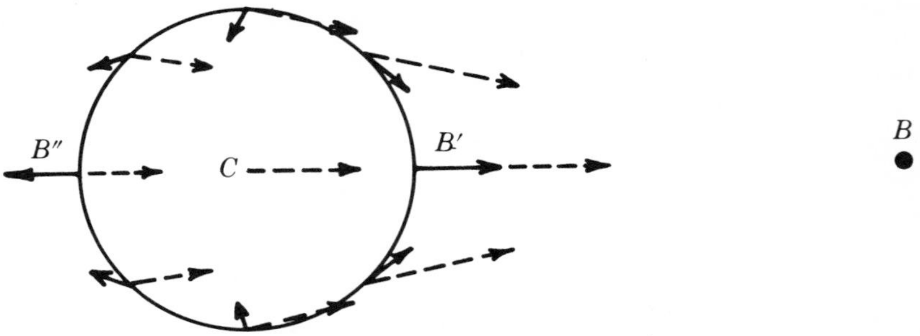

FIGURE 2.1.3. Tidal force (→) at eight points on a meridian section of a sphere. The broken arrows are $\boldsymbol{F}(A)$; at C it is $\boldsymbol{F}(C)$.

from B. When B is remote from C relative to the radius of the sphere (that is, for small parallaxes such as those of Moon and Sun), the configuration of $\boldsymbol{T}(A)$ on these two hemispheres will be virtually mirror images.

The order of magnitude of $\boldsymbol{T}(A)$ can be inferred by calculating $\boldsymbol{F}(A) - \boldsymbol{F}(C)$ at the sublunar (or subsolar) point B', where the directions of these vectors coincide. When $r\,(\equiv CB') \ll R\,(\equiv CB)$, we have

$$|\boldsymbol{T}(B')| = \gamma M/(R-r)^2 - \gamma M/R^2 \approx 2\gamma Mr/R^3.$$

Compared with the Earth's attraction $|\boldsymbol{G}(B')| \approx \gamma M_{\oplus}/r^2$,

$$|\boldsymbol{T}(B')|/|\boldsymbol{G}(B')| \approx 2\cdot(M/M_{\oplus})(r/R)^3.$$

Whereas the total gravitational force of Moon or Sun is proportional to the square of the parallax, the tidal force is proportional to the *cube*.

Using mass ratio from (2.1.2) and parallax from (2.1.3) we find

$$\text{(2.1.5)}\qquad \frac{M_{☾}}{M_{\oplus}}\left(\frac{a}{c_{☾}}\right)^3 = 0.5603\times 10^{-7},\qquad \frac{M_{\odot}}{M_{\oplus}}\left(\frac{a}{c_{\odot}}\right)^2 = 0.2580\times 10^{-7}.$$

Fortuitously, the Moon's tidal attraction is the same order of magnitude as (in fact, about twice) the Sun's.

We see from (2.1.5) that tidal forces are about 10^{-7} times the Earth's attraction. Why do we not ignore them entirely? There are two reasons. First, modern instrumental techniques make possible accurate measurement of tidal fluctuations of the intensity and direction of gravity; in other words, in spite of its smallness, the tidal force can be measured indirectly by detecting the response of the solid earth. Such measurements reveal much about the structure and elastic properties of the Earth.

More important for our present concern with ocean tides is the fact that throughout the geosphere (lithosphere, hydrosphere, atmosphere) the large downward force $\boldsymbol{G}$ is counteracted by an upward hydrostatic pressure force which is the main part of $\boldsymbol{P}$ in (2.1.4), so the sum $\boldsymbol{P} + \boldsymbol{G}$ can be extremely small in comparison with $\boldsymbol{G}$. In Figure 2.1.4 the broken line is normal to $\boldsymbol{G}$ and the solid, sloping line is a portion of a surface of hydrostatic pressure. The hydrostatic part of $\boldsymbol{P}$ is normal to this surface and the sum $\boldsymbol{P} + \boldsymbol{G}$ is a small residual nearly normal to $\boldsymbol{G}$. In fact, the response of the ocean to tidal excitation is such that the order of magnitude of this residual is the same as that of the tidal forces.

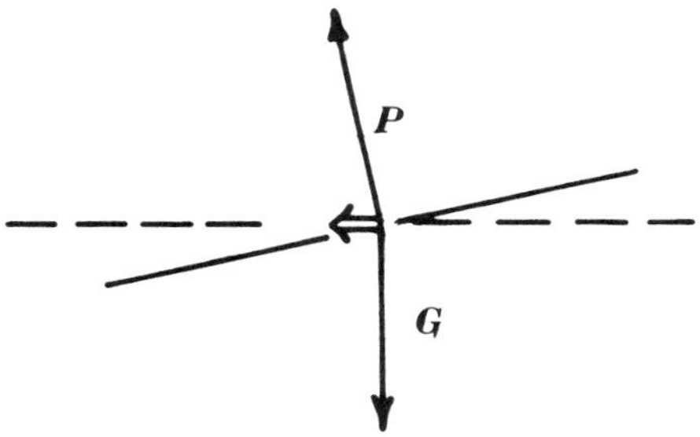

FIGURE 2.1.4. The small residual ($\Rightarrow$) of $\boldsymbol{P} + \boldsymbol{G}$.

Thus, the dynamically significant comparison to make between $\boldsymbol{T}$ and $\boldsymbol{G}$ is in the respective *directions*, rather than magnitudes: tidal displacements are produced by the component of tidal force at right angles to gravity and in that direction are unopposed except by pressure-gradient force. Referring to Figure 2.1.3 we can see that these "tractive" parts of the tidal force, which are tangent to the circle in that figure, tend to create *horizontal* displacements along the Earth's surface toward B' in one hemisphere and toward B'' in the other, and thus tend to cause mass to accumulate near B' and B''. This is the primary mechanism of the tides. The accompanying vertical (radial) displacements are a kinematic response to the horizontal displacements, dictated by conservation of mass. They are not caused by the radial component of tidal force.

2.2. **Rotational forces and gravity.** In (2.1.4) the acceleration of the vector $\boldsymbol{CA}$ is referred to axes whose directions are fixed in space. It is more convenient in tidal theory to use axes fixed in the Earth and thus rotating with the Earth's angular velocity $\boldsymbol{\Omega}$; so we must calculate the acceleration relative to such axes in terms of the absolute acceleration. If A were fixed to the Earth, the absolute velocity of $\boldsymbol{r} \equiv \boldsymbol{CA}$ would be $\boldsymbol{\Omega} \times \boldsymbol{r}$. Hence when A is free to move relative to the Earth, the relative velocity of $\boldsymbol{r}$ is $D_a\boldsymbol{r} - \boldsymbol{\Omega} \times \boldsymbol{r}$, which denote by $\boldsymbol{v}$. Similarly the relative acceleration is

$D_a \boldsymbol{v} - \boldsymbol{\Omega} \times \boldsymbol{v} \equiv \boldsymbol{a}$. It follows that

$$\boldsymbol{a} = D_a^2 \boldsymbol{r} - 2\boldsymbol{\Omega} \times \boldsymbol{v} - \boldsymbol{\Omega} \times (\boldsymbol{\Omega} \times \boldsymbol{r}) - (D_a \boldsymbol{\Omega}) \times \boldsymbol{r}.$$

Insert (2.1.4) for $D_a^2 \boldsymbol{r}$:

$$\boldsymbol{a} = \boldsymbol{P} + \boldsymbol{G} + \boldsymbol{T} - 2\boldsymbol{\Omega} \times \boldsymbol{v} - \boldsymbol{\Omega} \times (\boldsymbol{\Omega} \times \boldsymbol{r}). \tag{2.2.1}$$

The term $(D_a \boldsymbol{\Omega}) \times \boldsymbol{r}$ arising from changes in the direction or magnitude of $\boldsymbol{\Omega}$ can be safely neglected.

In (2.2.1) the two terms with $\boldsymbol{\Omega}$ are "apparent" forces that arise from adoption of rotating axes of reference. The *Coriolis force* $-2\boldsymbol{\Omega} \times \boldsymbol{v}$ (also called "deflecting force" and "geostrophic force") is perpendicular to both $\boldsymbol{\Omega}$ and $\boldsymbol{v}$; if viewed from the Northern Hemisphere, it is directed to the right of the projection of $\boldsymbol{v}$ into the equatorial plane. Its order of magnitude is $2\Omega|\boldsymbol{v}|$ where $\Omega = 7.292 \times 10^{-5}$ rad sec^{-1}, which gives about 1.5×10^{-3} cm sec^{-2} with $|\boldsymbol{v}| = 10$ cm sec^{-1}. Although this is little more than 10^{-6} times the magnitude of $\boldsymbol{G}$, we must recall that $\boldsymbol{G}$ is very nearly balanced hydrostatically, and that the Coriolis force, like the tidal force, can act at right angles to $\boldsymbol{G}$. In fact, the Coriolis force has a significant effect on virtually all tidal phenomena.

The *centrifugal force* $-\boldsymbol{\Omega} \times (\boldsymbol{\Omega} \times \boldsymbol{r})$ is directed outward from the axis of rotation and, in contrast to the Coriolis force, depends only on the position of the field point. Its magnitude is typified by $\Omega^2 r \approx 3.4$ cm sec^{-2} with $r = 6371$ km (the Earth's mean radius). In view of its direction, the centrifugal force can have a component at right angles to $\boldsymbol{G}$. Although this component evidently is much larger than the tidal or Coriolis force, it is not time-dependent and thus can be regarded merely as modifying $\boldsymbol{G}$. The resultant

$$\boldsymbol{g} \equiv \boldsymbol{G} - \boldsymbol{\Omega} \times (\boldsymbol{\Omega} \times \boldsymbol{r})$$

is called "gravity." Through the operation of centrifugal force over geologic time, the Earth has acquired a flattened figure (equatorial minus polar radius = 21.4 km) that corresponds very nearly to hydrostatic equilibrium with $\boldsymbol{g}$. Consequently, $\boldsymbol{g}$ is perpendicular to mean sea level (see Figure 2.2.1).

Previously I stated that the dynamically effective components of tidal and Coriolis force are those perpendicular to the Earth's attraction $\boldsymbol{G}$. It is now evident that to make this statement precise I must replace $\boldsymbol{G}$ by the gravity vector $\boldsymbol{g}$. Indeed, this distinction is essential because, as we have seen, the component of centrifugal force perpendicular to $\boldsymbol{G}$ is much larger than the tidal force. Having adopted the direction of $\boldsymbol{g}$ for reference, we need not deal further with $\boldsymbol{G}$ except for one additional complication that will now be mentioned.

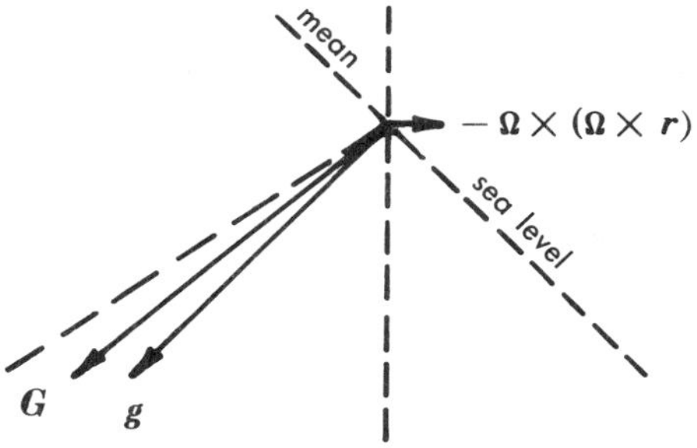

FIGURE 2.2.1. The vertical broken line is parallel to $\boldsymbol{\Omega}$, and mean sea level is perpendicular to $\boldsymbol{g}$. (The other broken line goes to the Earth's center.)

The displacements of the solid earth and ocean that occur in response to tidal forces are accompanied by corresponding changes in the Earth's gravitational field, owing to the tidally altered distribution of mass. Relative to the undisturbed $\boldsymbol{G}$ these changes are slight, but relative to $\boldsymbol{T}$ they are not, and we must take them into account. The details will be deferred, but for notation let $\boldsymbol{G}$ henceforth represent the *un*disturbed field and $\boldsymbol{G}'$, $\boldsymbol{G}''$ the disturbances (of the gravitational field) due to net mass displacements associated respectively with the ocean tide and the bodily tide. Then in place of $\boldsymbol{G}$ in (2.2.1) we must write $\boldsymbol{G} + \boldsymbol{G}' + \boldsymbol{G}''$, and thus obtain

$$\boldsymbol{a} = \boldsymbol{P} + \boldsymbol{g} + \boldsymbol{G}' + \boldsymbol{G}'' + \boldsymbol{T} - 2\boldsymbol{\Omega} \times \boldsymbol{v} \tag{2.2.2}$$

if at the same time we introduce $\boldsymbol{g}$ as previously defined.

To evaluate $\boldsymbol{G}'$ and $\boldsymbol{G}''$ we must know how the tide alters the configuration of the ocean and the solid earth. The bodily tide is very nearly a static response to tidal forces because the free periods of the Earth's elastic vibrations do not exceed about one hour, which is much less than the main tidal periods. In contrast, the free periods of the ocean span a range that completely contains the tidal periods, so the response of the ocean to tidal excitation is by no means statical. In principle, therefore, the disturbance $\boldsymbol{G}'$ due to the ocean tide cannot be determined independently of the solution of the ocean-tide problem. Although $\boldsymbol{G}'$ is somewhat smaller than $\boldsymbol{G}''$, it is not entirely negligible. I defer discussion of these terms to a later point.

Finally, we resolve (2.2.2) along specific axes. Generally, "geopotential axes" are convenient, the directions of which are

x-direction: horizontally eastward,
y-direction: horizontally northward,
z-direction: vertically upward.

The z-axis points in the direction of $-\boldsymbol{g}$ and thus is aimed at the astronomical zenith; the x, y-axes are in the plane of the horizon. Let $(0, 0, -g)$ denote the components of $\boldsymbol{g}$ on these axes and (u, v, w) the components of $\boldsymbol{v}$. Further, for brevity write $\boldsymbol{Q} \equiv \boldsymbol{G}' + \boldsymbol{G}'' + \boldsymbol{T}$. Then the components of (2.2.2) are

$$a_x = P_x + Q_x + 2\Omega_z v - 2\Omega_y w, \tag{2.2.3a}$$

$$a_y = P_y + Q_y - 2\Omega_z u, \tag{2.2.3b}$$

$$a_z = P_z + Q_z - g + 2\Omega_y u. \tag{2.2.3c}$$

These scalar equations make explicit the hydrostatic compensation previously noted. In the third equation P_z and g are individually at least five orders of magnitude larger than any of the other terms, so we have virtually $0 = P_z - g$. In the first and second equations all terms have roughly the same magnitude except $2\Omega_y w$ which is negligible. (In specific problems some terms may of course be more important than others.)

2.3 **The tidal potential.** Let $V(A)$ denote the gravitational potential at A due to the body at B—that is, $\nabla V(A) = \boldsymbol{F}(A)$—and let $W(A)$ be the tidal potential:

$$\nabla W(A) = \boldsymbol{T}(A) \equiv \boldsymbol{F}(A) - \boldsymbol{F}(C).$$

To express $W(A)$ in terms of $V(A)$, calculate the variation $\delta W = \nabla W \cdot \delta \boldsymbol{r}$ that accompanies displacement $\delta \boldsymbol{r}$ of A (Figure 2.3.1):

$$\delta W(A) = \boldsymbol{F}(A) \cdot \delta \boldsymbol{r} - \boldsymbol{F}(C) \cdot \delta \boldsymbol{r}.$$

But $\boldsymbol{F}(A) \cdot \delta \boldsymbol{r} = \delta V(A)$ and $\delta \boldsymbol{F}(C) = 0$ so

$$W(A) = V(A) + \boldsymbol{F}(C) \cdot \boldsymbol{r} + \text{constant}. \tag{2.3.1}$$

The constant is independent of A.

To make further progress we need an explicit formula for $V(A)$. The simplest assumption is that the body at B has the potential of a homogeneous sphere: $V(A) = \gamma M/d$. (This is reasonable not only because it is a

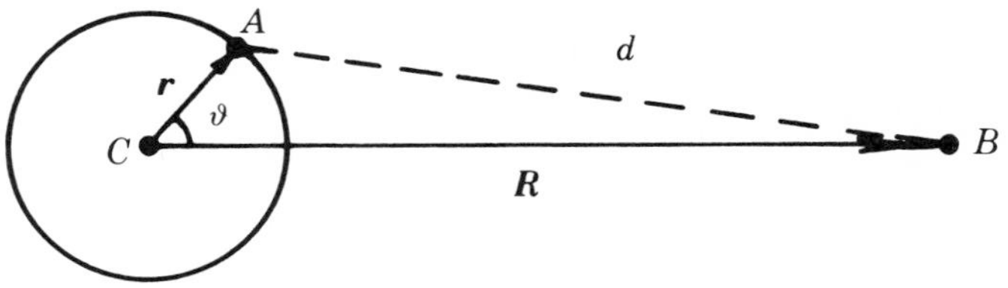

FIGURE 2.3.1. We seek the tidal potential at A in terms of the potential of the body at B whose geocentric zenith angle is ϑ.

good first approximation for any spheroidal body, but especially because we are looking at the far field.) Then $\boldsymbol{F}(C) = \gamma M\boldsymbol{R}/R^3$ and (2.3.1) becomes

$$W(A) = \frac{\gamma M}{R}\left(\frac{R}{d} - 1 - \frac{\boldsymbol{R}\cdot\boldsymbol{r}}{R^2}\right),$$

the additive constant having been chosen as $-\gamma M/R$. Now $r/R < 1$ (in fact $\ll 1$) so (Figure 2.3.1)

$$\frac{R}{d} = \left[1 - 2\frac{r}{R}\cos\vartheta + \left(\frac{r}{R}\right)^2\right]^{-1/2} = \sum_{n=0}^{\infty}\left(\frac{r}{R}\right)^n P_n(\cos\vartheta)$$

where $P_n(\cos\vartheta)$ are the Legendre polynomials

$$P_0(\cos\vartheta) = 1, \qquad P_2(\cos\vartheta) = \tfrac{1}{2}(3\cos^2\vartheta - 1),$$
$$P_1(\cos\vartheta) = \cos\vartheta, \qquad P_3(\cos\vartheta) = \tfrac{1}{2}(5\cos^3\vartheta - 3\cos\vartheta).$$
$$\cdots$$

Since $\boldsymbol{R}\cdot\boldsymbol{r}/R^2 = (r/R)\cos\vartheta$, the preceding expression for $W(A)$ reduces to

$$\begin{aligned} W(A) &= W_2(A) + W_3(A) + \cdots, \\ W_n(A) &\equiv (\gamma M/R)(r/R)^n P_n(\cos\vartheta). \end{aligned} \tag{2.3.2}$$

This expansion—fundamental for tidal theory—expresses the tidal potential at $A(r, \vartheta)$ as a sum of solid spherical zonal harmonics with respect to the axis CB.

The significant features of (2.3.2) are that the series begins with the harmonic of second degree, and that the convergence rate of the expansion

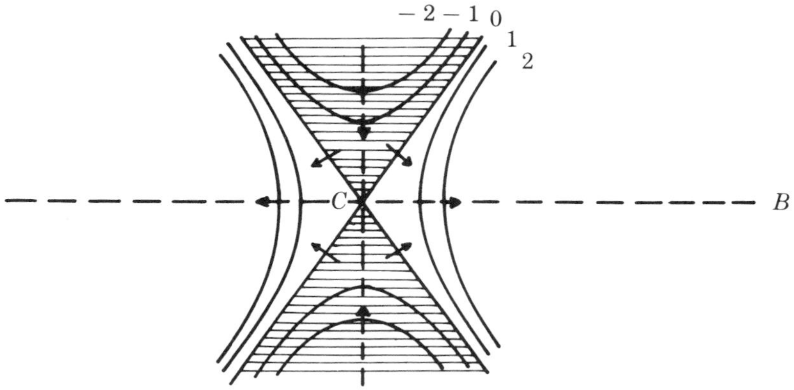

FIGURE 2.3.2. Equipotential surfaces of $(r/R)^2 P_2$ (cos ϑ). Arrows indicate directions of the corresponding tidal force. Shading indicates negative values of potential.

is r/R, the (sine of the) horizontal parallax of the body considered. The sine parallax of the Sun is less than 10^{-4} (see (2.1.3)) so $W_3(A)$ is negligible for the solar tide. The Moon's sine parallax varies between 0.0157 and 0.0180; hence, unless we require accuracy better than 2 percent, $W_3(A)$ can be neglected for the lunar tide as well.

Henceforth we consider only the leading term $W_2(A)$; in other words, we take the tidal potential to be a solid zonal harmonic of second degree. The equipotential surfaces of this term are hyperboloids of revolution about the axis CB. They have two sheets where $W_2 > 0$ (the sectors in syzygy with B) and one where $W_2 < 0$ (the sectors in quadrature with B). The potential W_2 is symmetric with respect to the plane perpendicular to CB at C.

For detailed calculation we must relate the zenith angle ϑ to coordinates in terms of which the positions of the field point and of the attracting body are conventionally expressed. From the definitions given in Figure 2.3.3, the law of cosines applied to the spherical triangle PAB is

$$\begin{aligned}\cos\vartheta &= \cos PA \cos PB + \sin PA \sin PB \cos(T+\lambda)\\ &= \sin\phi\sin\delta + \cos\phi\cos\delta\cos(T+\lambda).\end{aligned}$$

According to (2.3.2) we want to insert this expression into $P_2(\cos\vartheta)$. After some rearrangement,

$$\begin{aligned}P_2(\cos\vartheta) &= \tfrac{1}{4}(3\sin^2\phi - 1)(3\sin^2\delta - 1)\\ &\quad + \tfrac{3}{4}\sin 2\phi \sin 2\delta \cos(T+\lambda)\\ &\quad + \tfrac{3}{4}\cos^2\phi\cos^2\delta\cos 2(T+\lambda).\end{aligned} \tag{2.3.3}$$

When multiplied by $\gamma M r^2/R^3$ this gives the leading term $W_2(A)$ of the tidal potential as the sum of three parts, each of which has a factor dependent upon the latitude ϕ of the field point A, and a factor dependent in the same way upon the declination δ of the disturbing body B. The declination factor varies slowly through the orbital motion of B.

The principal distinction between the three parts of (2.3.3), however, is the way they involve the hour angle T, which increases by 360° in one day, owing to the Earth's rotation. The first part does not depend upon T and therefore is unaffected by the Earth's rotation. The tides to which this part gives rise are accordingly called *long-period* tides (or sometimes, tides of the "first species," following Laplace, to whom we owe (2.3.3) and its interpretation). The second part has the factor $\cos(T+\lambda)$, the period of which is one day (apart from slight deviations owing to orbital motion), and thus gives rise to *diurnal* tides (Laplace's "second species"). The third part has the factor $\cos 2(T+\lambda)$ with period virtually one-half day, and

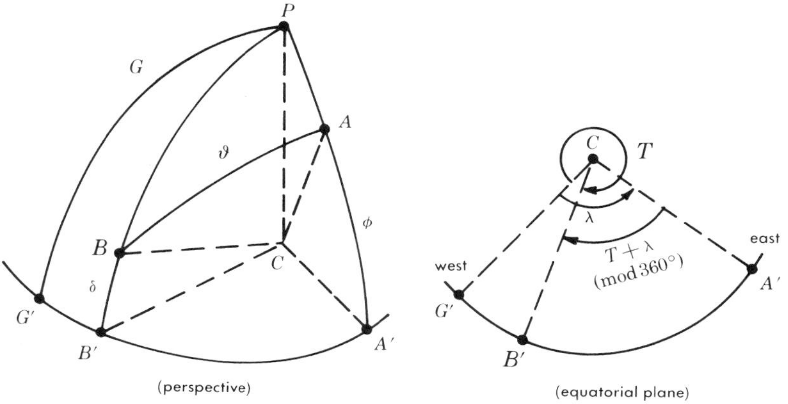

FIGURE 2.3.3. Left: perspective of the geocentric celestial sphere; right: plane of the celestial equator. Points and circles on the sphere are:

Points	*Circles*
A: field point	PAA': meridian of A
B: disturbing body	PBB': hour circle of B
G: Greenwich	PGG': meridian of G
P: North Pole	$G'B'A'$: celestial equator

Positions of A and B are fixed by the following angles:

Angles for A	*Angles for B*
$A'CA \equiv \phi$: latitude	$B'CB \equiv \delta$: declination
$G'CA' \equiv \lambda$: longitude	$G'CB' \equiv T$: hour angle

Angles for B relative to A

$ACB \equiv \vartheta$: zenith angle

$A'CB' \equiv T + \lambda \pmod{360°}$: hour angle

Longitude λ is measured eastwards from the meridian of Greenwich and hour angle T westwards.

generates the *semidiurnal* tides ("third species"). It is customary to write $\cos s(T + \lambda)$ for the hour-angle factors, thereby introducing a "species number" s equal to 0, 1, 2 for the first three species. As the partition into three species is applicable to each disturbing body, we have long-period, diurnal, and semidiurnal tides of solar as well as lunar origin.

The transformation (2.3.3) shifts the axis of reference for the zonal surface harmonic from CB to the Earth's axis CP (Figure 2.3.3). The three terms that arise in this way are respectively zonal, tesseral, and sectoral surface harmonics relative to the new axis. In latitude and longitude they have the configurations shown schematically in Figure 2.3.4. The nodal latitudes of the long-period term are at $\phi = \pm 35° \, 16'$ (where $\sin^2 \phi = \frac{1}{3}$). Poleward of these latitudes $W_2 < 0$ because the declinations of Sun and Moon do not attain values large enough to make $\sin^2 \delta > \frac{1}{3}$.

By means of (2.3.3) the tide potential of second degree in (2.3.2) can be written

$$
\begin{aligned}
W_2(A) &= W_2^0(A) + W_2^1(A) + W_2^2(A),\\
W_2^0(A) &\equiv D[(r/r_1)^2(\tfrac{1}{2} - \tfrac{3}{2}\sin^2\phi)][(c/R)^3(\tfrac{2}{3} - 2\sin^2\delta)],\\
W_2^1(A) &\equiv D[(r/r_1)^2 \sin 2\phi][(c/R)^3 \sin 2\delta \cos(T + \lambda)],\\
W_2^2(A) &\equiv D[(r/r_1)^2 \cos^2\phi][(c/R)^3 \cos^2\delta \cos 2(T + \lambda)],\\
D &\equiv 3\gamma M r_1^2/4c^3
\end{aligned}
\qquad (2.3.4)
$$

where r_1 and c are reference values of r and R, chosen to make r/r_1 and c/R of order unity. It is convenient to take r_1 equal to the areal mean value of r over the Earth's surface (strictly, over the reference ellipsoid), and c such that $1/c$ is the orbital-mean value of $1/R$ (strictly, so that a/c is the mean of the sine of horizontal equatorial parallax). Then the geographical mean of r/r_1 and the orbital mean of c/R are unity.

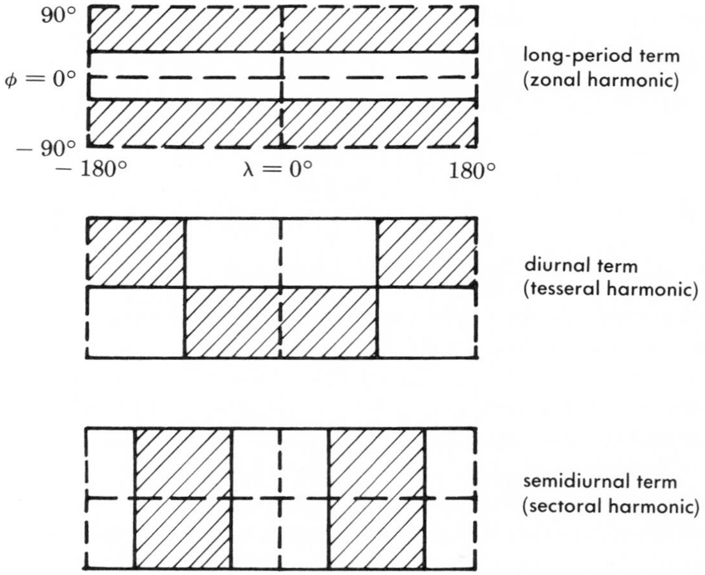

FIGURE 2.3.4. Schematic configuration of the three tidal species, in latitude and longitude. Solid lines are nodes; shading covers regions where the tidal potential is negative. The disturbing body is assumed to be on the Greenwich meridian ($\lambda = 0°$) and (for the diurnal term) to have positive declination.

In (2.3.4) the potential of each of the three species consists of three distinct factors. The first factor D, common to all species, is a dimensional constant (the "Doodson constant") that gives the order of magnitude of the tidal potential. The second (first bracket) is a "geodetic" factor which is independent of time and depends only upon location of the field point A. The third is an "astronomical" factor (second bracket) that depends upon orbital motion of the disturbing body as well as upon the Earth's rotation. The arrangement is such that when $r = r_1$ each of the geodetic factors, as a function of latitude, has maximum magnitude unity. Consequently the instantaneous magnitude of the astronomical factor for a particular species is a global measure of the intensity of the tidal potential of that species at that instant.

A numerical value for the Doodson constant can be assigned by writing first

$$D = \frac{3}{4} \cdot \frac{\gamma M_{\oplus}}{r_1} \cdot \left(\frac{r_1}{a}\right)^3 \cdot \frac{M}{M_{\oplus}} \left(\frac{a}{c}\right)^3$$

where a is the Earth's equatorial radius. From the theory of the Earth's figure [Jeffreys, 1970] it can be shown that

$$\gamma M_{\oplus}/r_1 = g_1 r_1/(1 - \tfrac{2}{3}m) \qquad \text{and} \qquad (r_1/a)^3 = 1 - e$$

where g_1 is gravity at the latitude of r_1 and

$$m = 0.003450, \qquad e = 0.003353$$

are respectively Clairaut's constant and the ellipticity. Consequently[3]

$$\frac{D}{g_1 r_1} = \frac{3}{4} \cdot \frac{1 - e}{1 - \frac{2}{3}m} \cdot \frac{M}{M_{\oplus}} \left(\frac{a}{c}\right)^3.$$

The final factor is given numerically in (2.1.5). From those values and e, m as above, we get

(2.3.5) $$D_{☾}/g_1 r_1 = 0.4198 \times 10^{-7}, \qquad D_{\odot}/D_{☾} = 0.4605.$$

The numerical values [Jeffreys, 1970]

$$g_1 = 979.76 \text{ cm sec}^{-2}, \qquad r_1 = 6371.27 \text{ km}$$

give further reductions in which r_1 and then g_1 are inserted:

(2.3.6) $$D_{☾}/g_1 = 26.75 \text{ cm},$$

[3] Bartels (1957) and others, following Doodson, write g_1 for what here is $g_1/(1 - \frac{2}{3}m) = 982.01$.

$$D_{\leftmoon} = 2.621 \times 10^4 \text{ cm}^2 \text{ sec}^{-2}. \tag{2.3.7}$$

The corresponding quantities involving $D_{\odot}$ can be found from $D_{\odot}/D_{\leftmoon} = 0.4605$.

It is often convenient to represent the tide potential W by an equivalent "equilibrium tide" or "equilibrium displacement," defined as

$$\bar{\zeta} \equiv W/g_1. \tag{2.3.8}$$

This quantity arises as follows. Let Ψ denote the geopotential (potential of gravity) and let Ψ = constant = c_1 be a potential surface in this field (such as mean sea level). Further, let $\Psi + W$ = constant = c_2 be a nearby surface in the tidally-modified field (see Figure 2.3.5), and let Δz denote the vertical displacement from A on c_1 to B on c_2. Then

$$\Psi(B) = \Psi(A) + (\partial\Psi/\partial z)_A \Delta z + \cdots,$$

$$W(B) = W(A) + (\partial W/\partial z)_A \Delta z + \cdots.$$

Add these equations and note that $\partial W/\partial z \ll \partial\Psi/\partial z = -g$:

$$c_2 = c_1 + W(A) - g(A)\,\Delta z + \cdots.$$

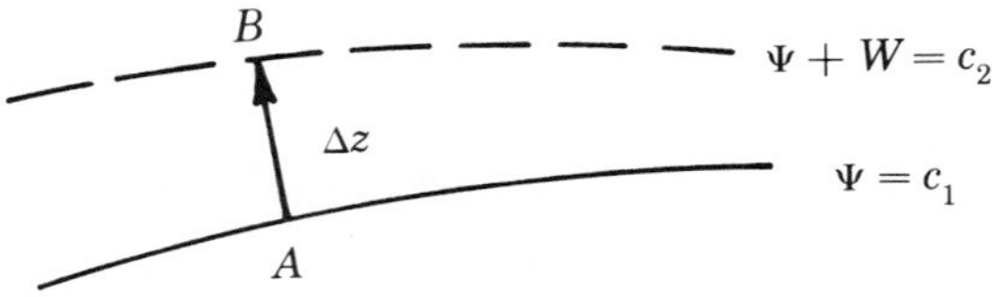

FIGURE 2.3.5. Displacement of the geopotential field by the tidal potential.

The constants c_1 and c_2 are arbitrary, but we can fix c_2 relative to c_1 by asking that the area average of $g\Delta z$ over the complete spheroid be zero (equivalent to saying that the volumes enclosed by c_1 and c_2 are equal). There will thus be compensating regions of negative and positive Δz. Since the average W is zero, this gives $c_2 = c_1$ and hence $\Delta z = W/g$. The quantity defined in (2.3.8) differs from Δz only in having the constant g_1 in place of the slightly variable g. (It is the "geopotential distance" between the two surfaces, whereas Δz is the geometric distance—a minor distinction in this context.)

Finally, I should mention that for the general term of (2.3.3) the shift of reference from axis CB to CP (Figure 2.3.3) is governed by the "addition theorem"

$$P_n(\cos \vartheta) = P_n(\sin \phi)P_n(\sin \delta) + 2 \sum_{s=1}^{n} \frac{(n-s)!}{(n+s)!} P_n^s(\sin \phi)P_n^s(\sin \delta) \cos s(T + \lambda)$$

of which (2.3.3) is the special case $n = 2$. From this we see that the nth harmonic term $W_n(A)$ contributes to tides of the first $n + 1$ species. For example, $W_3(A)$ contributes long-period, diurnal, and semidiurnal tides, and through the factor $\cos 3(T + \lambda)$ gives rise to the terdiurnal tide as well. In the theory of ocean tides, however, it suffices to confine attention to $W_2(A)$.

2.4. **Harmonic analysis of the tide potential.** The three astronomical factors of W_2 depend upon time in a way determined by the relative motion between the Earth and the body considered. By inspection it is evident that, although each factor contains periodic elements R, δ, and $\cos s(T + \lambda)$, its functional dependence on these elements is nonlinear, so its dependence upon time is not simple harmonic, or even strictly periodic.

We therefore seek to represent each astronomical factor as a trigonometrical series

$$\sum_j C_j \cos(\sigma_j t + s\lambda + \theta_j) \tag{2.4.1}$$

of simple-harmonic terms. The details of this "harmonic" analysis need not concern us here, but the nature of the process can be understood qualitatively by noting that the declination and parallax have periods of about one month (Moon) or one year (Sun), so these elements produce a modulation of the hour-angle factor (period one day). The latter may be regarded as the "carrier wave." We therefore expect a diurnal spectrum having a central frequency at one day and numerous side-band frequencies, and similarly a semidiurnal spectrum with a central frequency of one-half day and its side bands.

Each term of (2.4.1) is characterized by an amplitude coefficient C and a cosine argument $\sigma t + s\lambda + \theta$ (assumed to be augmented, if necessary, by 180° to make C positive). The argument depends linearly upon Greenwich mean solar time t and eastward longitude λ; its value at Greenwich when $t = 0$ is here denoted θ (a more usual symbol is V_0). The frequency σ —usually called the "speed" of the term and expressed in degrees per unit time—is a linear integer combination of the speed ω of the Earth's rotation and the angular speeds ω_k ($k = 1$ to 5) that correspond to the five fundamental astronomical periods listed in Table 2.4.1:

$$\sigma = s\omega + \sum_{k=1}^{5} m_k\omega_k.$$

TABLE 2.4.1. Fundamental periods in the orbital motion of Earth and Moon (from [Bartels, 1957])

Period (mean solar days, or years)	Description
$360°/\omega_1 = 27.321582$ days	Period of lunar declination
$360°/\omega_2 = 365.242199$ days	Period of solar declination
$360°/\omega_3 = 8.847$ years	Period of lunar perigee
$360°/\omega_4 = 18.613$ years	Period of lunar node
$360°/\omega_5 = 20{,}940$ years	Period of perihelion

Here $s = 0, 1, 2$ is the species number and $m_k = 0 \pm 1, \pm 2, \ldots$ is an integer "modulation" vector. The speed ω of the Earth's rotation can be taken as $\omega_☾ \equiv \Omega - \omega_1$ or $\omega_\odot \equiv \Omega - \omega_2$, where Ω is the sidereal speed of rotation and ω_1, ω_2 are the orbital speeds of the mean Moon and mean Sun (Table 2.4.1). These give respectively the *mean lunar day* and the *mean solar day* (Table 2.4.2).

TABLE 2.4.2. Fundamental periods in the rotation of the Earth

Period (mean solar days)	Description
$360°/\Omega = 0.997270$ days ($23^h56^m4^s$)	Sidereal day
$360°/\omega_\odot = 1$ day	Mean solar day
$360°/\omega_☾ = 1.035050$ days ($24^h50^m28^s$)	Mean lunar day

TABLE 2.4.3. The side bands for $C > 0.001$

$\|\Delta\sigma\|$ (deg hr^{-1})	$360°/\|\Delta\sigma\|$	Description
$\omega_4 = 0.0022064$	18.613 years	Nodal (lunar)
$\omega_2 - \omega_5 = 0.0410667$	365.260 days	Solar elliptical
$2\omega_2 = 0.0821372$	182.621 days	Solar declinational
$\omega_1 - 2\omega_2 + \omega_3 = 0.4715211$	31.812 days	Evectional (lunar)
$\omega_1 - \omega_3 = 0.5443747$	27.555 days	Lunar elliptical
$2(\omega_1 - \omega_2) = 1.0158958$	14.765 days	Variational (lunar)
$2\omega_1 = 1.0980330$	13.661 days	Lunar declinational

The form of the astronomical factors and the orbital equations is such that seven types of side bands account for all terms with $C > 0.001$ in (2.4.1) (there are 65 such terms in W_2), and thus for all cases of practical importance. These side bands are listed in Table 2.4.3 in order of increasing $|\Delta\sigma|$, the difference in speed between the center of the band and the line of first order.

The number of terms needed in the series is large if we require great accuracy (in principle, it is infinite); but if approximations of a few percent are tolerable, only a few terms suffice. In particular, if we exclude all terms with coefficient C less than 0.05, only seven long-period terms are needed, seven diurnal terms, and four semidiurnal terms. These 18 terms (13 of which are lunar) are listed in Table 2.4.4 in order of increasing speed. Only four of the seven side bands of Table 2.4.3 are needed to account for these terms, namely the two declinational bands and the nodal and lunar-elliptical bands. Except for the nodal terms, the band centers are the terms labeled M_0, S_0, O_1, P_1, M_2, S_2 in Table 2.4.4.

The symbols shown in Table 2.4.4 are those conventionally used to identify the main terms. M and S refer to Moon and Sun, and subscript 0, 1, 2 is the species number. The first two terms S_0 and M_0 of long-period species are independent of time and, owing to the form of the geodetic factor for this species, they have the effect of contributing very slightly (by about 0.002 percent) to the permanent flattening of the geoid. The third term, described as "nodal to M_0," has a period of 18.6 years, the time required for a complete revolution of the ascending node of the Moon's orbit. The fourth term *Ssa* is the principal solar (*S*) semi-annual (*sa*) tide, associated with the annual variation of the Sun's declination. The fifth, *Mm*, is the lunar (*M*) monthly (*m*) tide, associated with the monthly variation of the Moon's distance (hence designated an "elliptic" or "parallactic" tide). The sixth, *Mf*, is the lunar (*M*) fortnightly (*f*) tide, associated with the monthly variation of the Moon's declination. The seventh, described as "nodal to *Mf*," differs in speed from *Mf* by the speed of the Moon's node.

Descriptions of the diurnal and semidiurnal terms can be interpreted in an analogous way. It should be noted that there are no diurnal terms M_1 or S_1 corresponding to M_2 and S_2. The former, with periods exactly one lunar and one solar day, are absent because in the astronomical factor for the diurnal species the declination enters through $\sin 2\delta$ (see (2.3.4)), the mean value of which is zero. Declination modulation of the hour-angle factor therefore cannot produce the central frequency. (In the semidiurnal species we have $\cos^2 \delta$, with nonzero mean.) Terms M_1 and S_1 do, however, enter weakly from the tide potential of third degree.

TABLE 2.4.4. Tidal constituents with $C > 0.05$ (from [Bartels, 1957])

Coefficient C	Speed (deg hr^{-1}) σ	Period $360°/\sigma$	Symbol (and description)
	Long-period tides		
0.2341	0		S_0 (constant solar)
0.5046	0		M_0 (constant lunar)
0.0655	$\omega_4 = 0.00221$	18.613 years	—(nodal to M_0)
0.0729	$2\omega_2 = 0.08214$	182.621 days	Ssa (declinational to S_0)
0.0825	$\omega_1 - \omega_3 = 0.54437$	27.555	Mm (elliptical to M_0)
0.1564	$2\omega_1 = 1.09803$	13.661	Mf (declinational to M_0)
0.0648	$2\omega_1 + \omega_4 = 1.10024$	13.633	—(nodal to Mf)
	Diurnal tides		
0.0722	$(\omega_{\leftmoon} - \omega_1) - (\omega_1 - \omega_3) = 13.39866$	26.868 hours	Q_1 (elliptical to O_1)
0.0710	$(\omega_{\leftmoon} - \omega_1) - \omega_4 = 13.94083$	25.823	—(nodal to O_1)
0.3769	$(\omega_{\leftmoon} - \omega_1) = 13.94304$	25.819	O_1 (principal lunar)
0.1755	$(\omega_{\odot} - \omega_2) = 14.95893$	24.066	P_1 (principal solar)
0.1682	$(\omega_{\odot} - \omega_2) + 2\omega_2 (= \Omega) = 15.04107$	23.934	K_1^S (declinational to P_1)
0.3623	$(\omega_{\leftmoon} - \omega_1) + 2\omega_1 (= \Omega) = 15.04107$	23.934	K_1^M (declinational to O_1)
0.0718	$(\omega_{\leftmoon} + \omega_1) + \omega_4 = 15.04328$	23.931	—(nodal to K_1^M)
	Semidiurnal tides		
0.1739	$2\omega_{\leftmoon} - (\omega_1 - \omega_3) = 28.43973$	12.658 hours	N_2 (elliptical to M_2)
0.9081	$2\omega_{\leftmoon} = 28.98410$	12.421	M_2 (principal lunar)
0.4229	$2\omega_{\odot} = 30$	12	S_2 (principal solar)
0.0365	$2\omega_{\odot} + 2\omega_2 (= 2\Omega) = 30.08214$	11.967	K_2^S (declinational to S_2)
0.0786	$2\omega_{\leftmoon} + 2\omega_1 (= 2\Omega) = 30.08214$	11.967	K_2^M (declinational to M_2)
	Combination tides		
0.5305	$\Omega = 15.04107$	23.934 hours	K_1 (lunisolar declinational)
0.1151	$2\Omega = 30.08214$	11.967	K_2 (lunisolar declinational)

Owing to the fact that K_1^S and K_1^M have exactly the same period (one sidereal day), these two terms are normally combined and the resultant K_1 is called a "lunisolar" term. Similarly, K_2^S and K_2^M are combined to give the lunisolar K_2. These combination terms are listed at the end of Table 2.4.4. The reader may have noticed that K_2^S is the only term in the table for which $C < 0.05$ and will now understand that its presence is explained by the fact that it must be combined with K_2^M.

Inasmuch as all geophysical time series have a background spectrum produced by nontidal causes, upon which the response to tidal excitation is superimposed, we must bear in mind that in the spectral analysis of such data for tidal constituents, it is necessary to contend in some degree with a weak signal-to-noise ratio. Practically, this means that in order to resolve the very narrow nodal band (see Table 2.4.3), a record length of at least 19 years is needed. When records longer than a few years are not available (as is often the case), it is customary to use one year (actually 355 or 369 days) as the standard length, which moreover normally suffices for the main diurnal and semidiurnal constituents. In these circumstances an allowance must be made for the influence of nodal bands which, as is evident from Table 2.4.4, can contribute significantly to the over-all response and will be superimposed upon the central term if the nodal band is not resolved.

The conventional way of doing this is as follows. Let C and $E \equiv \sigma t + s\lambda + \theta$ denote coefficient and argument of the central term, and similarly for the side band write C_k and $E_k = (\sigma + k|\Delta\sigma|)t + s\lambda + \theta_k$ where $|\Delta\sigma|$ is the spacing of lines in the band and $k = \pm 1, \pm 2, \pm 3, \ldots$ is the order of a particular line. Then formally

$$C \cos E + \sum_k C_k \cos E_k = fC \cos(E + u)$$

where $f(t)$ and $u(t)$ are the amplitude and phase of the modulation

$$fe^{iu} \equiv 1 + C^{-1} \sum_k C_k \exp i(k|\Delta\sigma|t + \theta_k - \theta)$$

that results from superposition. The first-order terms $k = \pm 1$, which are dominant in all cases, have period $360°/|\Delta\sigma|$, equal to 18.6 years for the nodal bands (Table 2.4.3). Hence we may regard f and u as virtually constant over intervals of a year or less. The common practice is to assign values for the middle of the interval considered.

The nodal terms of (2.4.1) therefore are normally taken into account by writing the series in the form

$$\sum_j f_j C_j \cos(\sigma_j t + s\lambda + \theta_j + u_j). \tag{2.4.2}$$

Nodal terms are now excluded from the sum, but their effect is implicit in the "node factor" f_j and phase modulation u_j, numerical values of which are available in standard tide tables. This effectively reduces the number of terms in Table 2.4.4 from 19 to 15. If we also exclude the constant terms S_0 and M_0 and use the combinations K_1 and K_2 in place of their lunar and solar parts, there remain the following 11 main constituents:

$$\begin{aligned} \textit{long-period}&: \quad Ssa,\ Mm,\ Mf, \\ \textit{diurnal}&: \quad Q_1, O_1, P_1, K_1, \\ \textit{semidiurnal}&: \quad N_2, M_2, S_2, K_2. \end{aligned}$$

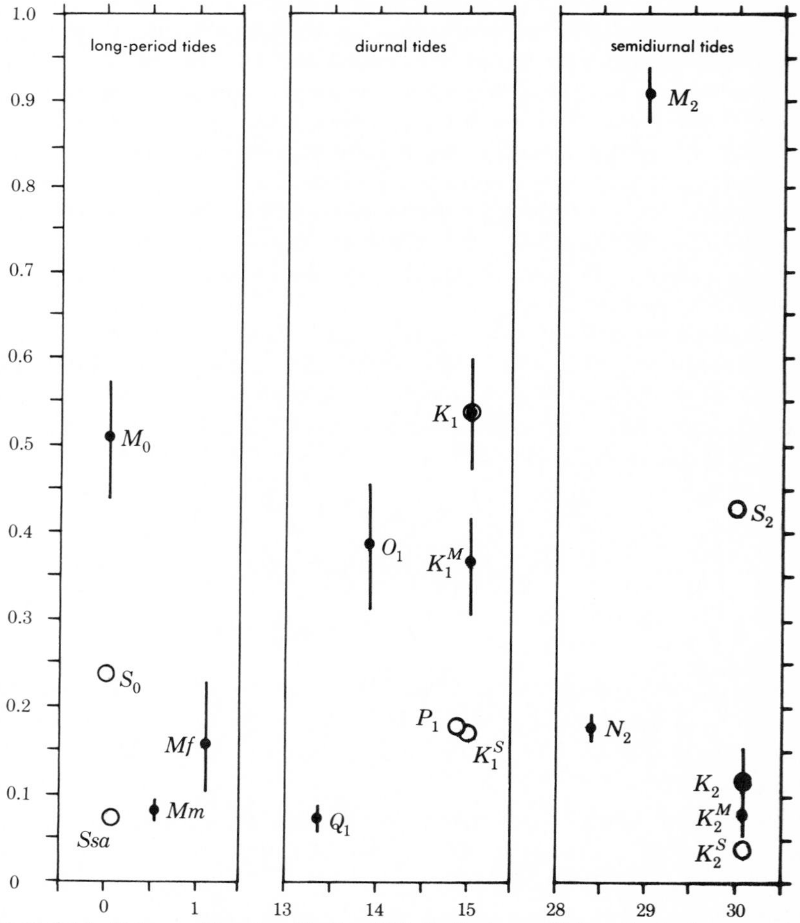

FIGURE 2.4.1. Tide constituents with $C > 0.05$. Ordinate: coefficient C; abscissa: speed σ in degrees per hour. The vertical bar on lunar (and lunisolar) terms shows the limits of nodal modulation.

The side-band relations among these terms are shown in Table 2.4.5.

Table 2.4.6 gives maximum and minimum values of f for the lunar terms in Table 2.4.4 (solar terms are unaffected). Also shown, for illustration, are values of f for midyear 1970 and $\theta + u$ for the same year, θ being the (unmodulated) argument at 00 GMT 1 January 1970 on the Greenwich meridian and u a value at midyear.

Figure 2.4.1 is a graphical display of the contents of Table 2.4.4.

TABLE 2.4.5. Side-band relations among the main tidal constituents

	Lunar terms			Solar terms	
Species	principal	elliptical	declinational	principal	declinational
0	M_0	Mm	Mf	S_0	Ssa
1	O_1	Q_1	K_1^M	P_1	K_1^S
2	M_2	N_2	K_2^M	S_2	K_2^S

TABLE 2.4.6. Node factors and arguments for the main constituents (from [Schureman, 1940])

Constituent	f (min)	f (max)	f (1970)	$\theta + u$ (1970)
		Long-period tides		
Ssa	1	1	1	200.5 deg
Mm	0.871	1.131	0.882	255.3
Mf	0.625	1.452	1.417	44.0
		Diurnal tides		
Q_1	0.805	1.183	1.170	255.4
O_1	0.805	1.183	1.170	150.7
P_1	1	1	1	349.8
K_1	0.882	1.113	1.105	13.4
		Semidiurnal tides		
N_2	0.963	1.038	0.966	270.1
M_2	0.963	1.038	0.966	165.4
S_2	1	1	1	0
K_2	0.748	1.317	1.289	207.2

2.5. **Hydrodynamical equations for ocean tides.** The driving frequencies that concern us are those of the main tidal constituents. These are contained in the range 0.6×10^{-7} to 3×10^{-5} cycles per second (cps). However, the ocean responds to excitation over a much broader range, notably at higher frequencies where we find wind waves (sea, swell, capillaries) which, literally, are in the "visible" part of the spectrum. We

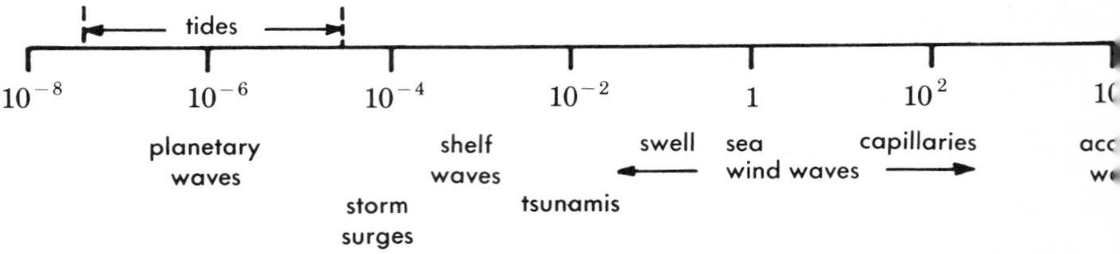

FIGURE 2.5.1. Tides are at the low-frequency end of a broad range that spans much diverse activity in the oceans.

should therefore simplify the hydrodynamical equations by means of low-pass filtering approximations that suppress resonances at frequencies higher than, say, 10^{-2} cps. The equations thus obtained are called the "long-wave" equations. (The word "long" usually qualifies wave length as well as period.) We adopt them not merely because it is logical to do so, but for the very pragmatic reason that they offer the only mathematically tractable basis for a theory of the tides.

Acoustic waves are eliminated from the theory if we regard the ocean as an incompressible fluid, and capillary waves are removed if we ignore surface (or interfacial) tension. We are now left with the dynamical effects of (a) gravity operating on the pressure field through changes in the mass distribution, and (b) the Earth's rotation. The mass distribution can be altered by displacement of the ocean surface and by rearrangement of the internal density field. Although "internal" tides are a worthwhile subject of inquiry, I shall in these lectures regard the ocean as a homogeneous (as well as an incompressible) fluid, and thereby limit the theory to what sometimes are called "external" waves. (This simplification is not, however, part of the long-wave approximation.) Resonances modified or controlled by dynamical effects of the Earth's rotation would naturally be expected to fall in the range of tidal frequencies; we must therefore incorporate these effects.

As a starting point in deriving the long-wave equations, refer to Figure 2.5.2, a vertical section of the ocean from surface to bottom. If the configuration of the ocean surface is averaged over a long time (many years), we presumably depict a portion of the geoid. This is indicated by the broken horizontal line $z = 0$. (To this point, z has designated only the vertical axis; henceforth, it will additionally be a local coordinate that measures distance along the local vertical from mean sea level.) Similarly, if the configuration of the ocean bottom is averaged over a long time, we obtain a function $h(x, y)$ which is simply the depth of the ocean. Further, let $\zeta_s(x, y, t)$ and $\zeta_b(x, y, t)$ be the instantaneous elevations of the ocean surface and ocean bottom above their respective levels of repose.

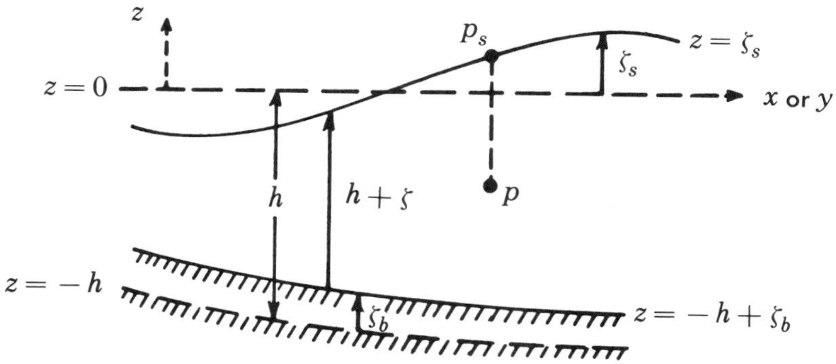

FIGURE 2.5.2. Definition sketch.

Conservation of mass in an incompressible fluid is expressed by asserting that the three-dimensional divergence of velocity is zero everywhere:

$$\text{div}(u, v) + \partial w/\partial z = 0. \tag{2.5.1}$$

Here u, v, w are components along geopotential axes, and $\text{div}(u, v)$ is the horizontal-surface divergence

$$\text{div}(u, v) = \partial u/\partial x + \partial v/\partial y - (r_1^{-1} \tan \phi) \cdot v$$

where r_1 is the Earth's mean radius and $\partial/\partial x$, $\partial/\partial y$ are eastward and northward directional derivatives. (In (2.5.1) a term is neglected that arises from the fact that neighboring verticals are not quite parallel. Relative to $\partial w/\partial z$ this term is of order $h/r_1 < 0.002$.) Integrate (2.5.1) from the bottom to the surface and use $w_s = d\zeta_s/dt$ and $w_b = d(-h + \zeta_b)/dt$ with $d/dt \equiv \partial/\partial t + (u, v)$ grad:

$$\frac{\partial(\zeta_s - \zeta_b)}{\partial t} = -\text{div} \int_{-h+\zeta_b}^{\zeta_s} (u, v)\, dz.$$

This evidently expresses the balance of mass in a fixed vertical column extending from ocean bottom to surface.

It will be shown below that, in the long-wave approximation, the horizontal velocity (u, v) is virtually independent of z. In these circumstances the preceding equation is approximately

$$\partial\zeta/\partial t = -\text{div}[(u, v)(h + \zeta)].$$

Here $\zeta \equiv \zeta_s - \zeta_b$ is the upward displacement of the surface relative to the bottom, which is in fact the quantity measured by a tide gauge. Formally, the ζ-term on the right side makes a second-order contribution, since it is multiplied by (u, v); in a perturbation analysis we may therefore neglect it. Moreover, practically, except in very shallow regions, $|\zeta| \ll h$. Hence for most purposes in tidal theory it is customary to use

$$\partial\zeta/\partial t = -\operatorname{div}[(u, v)h] \tag{2.5.2}$$

as the "continuity equation."

As a corollary to the horizontal velocity being independent of z, the vertical velocity is a linear function of z. Thus, if (2.5.1) is integrated from an arbitrary level z to the surface $z = \zeta_s$, we find the vertical velocity at z to be

$$w = d\zeta_s/dt + (\zeta_s - z)\operatorname{div}(u, v)$$

or, if second-order terms are ignored,

$$w = \partial\zeta_s/\partial t - z\operatorname{div}(u, v). \tag{2.5.3}$$

This is a "diagnostic" equation, which enables us to infer w from other dependent variables.

To incorporate the long-wave approximations explicitly in the dynamical equations (2.2.3), refer again to Figure 2.5.2 and consider two points on the same vertical: one interior to the fluid where the pressure is p, the other on the surface where the (atmospheric) pressure is p_s. Since the density is assumed uniform, and the length of the column between the two points is $\zeta_s - z$, the purely hydrostatic part of p is $p_s + \rho g(\zeta_s - z)$. The residual or "dynamic" pressure at the interior point therefore is

$$p' \equiv p - (p_s + \rho g\zeta_s) + \rho g z.$$

It follows that the horizontal components of pressure-gradient force per unit mass can be written

$$-\frac{1}{\rho}\left(\frac{\partial}{\partial x}, \frac{\partial}{\partial y}\right)p = -g\left(\frac{\partial}{\partial x}, \frac{\partial}{\partial y}\right)\left(\zeta_s + \frac{p_s}{\rho g}\right) - \frac{1}{\rho}\left(\frac{\partial}{\partial x}, \frac{\partial}{\partial y}\right)p'. \tag{2.5.4}$$

The essence of the long-wave approximation is that the contribution of p' to this force is negligible. Assuming this justifiable (the justification is given below), insert (2.5.4) for the stress components P_x, P_y in the dynamical equations (2.2.3a, b), and give the acceleration components a_x, a_y their linearized form $\partial(u, v)/\partial t$. Further, omit the term $2\Omega_y w$ from (2.2.3a), because except near the equator (where $\Omega_z = 0$) its ratio to $2\Omega_z v$ is roughly the ratio of w to u and this, by (2.5.1), is not greater than the vertical aspect ratio which for long waves is small. Hence (2.2.3a, b) become

$$\frac{\partial(u, v)}{\partial t} = -g\left(\frac{\partial}{\partial x}, \frac{\partial}{\partial y}\right)\left(\zeta + \zeta_b + \frac{p_s}{\rho g}\right) + (Q_x, Q_y) + 2\Omega_z(v, -u) \tag{2.5.5}$$

with the substitution $\zeta_s = \zeta + \zeta_b$. To these we join the continuity equation (2.5.2). The system of three equations thus obtained is closed on the dependent variables ζ, u, v. The z-component equation is superfluous: it is needed only to estimate the size of p' (as is done below). The vertical velocity can be calculated in terms of ζ, u, v from (2.5.3).

In deriving the continuity equation it was assumed that (u, v) is independent of z. The justification can now be given, for if (u, v) is initially independent of z, it will remain so because $\partial(u, v)/\partial t$ from (2.5.5) is independent of z when (u, v) is. (The tidally-generated forces Q_x, Q_y vary with z through the factor $(r/r_1)^2 \approx 1 + 2z/r_1$; hence we make an error only of order h/r_1 by assuming them independent of z.) This characteristic feature of long-wave motion has come about through omission of the p'-terms from the horizontal pressure-gradient force; the remaining, hydrostatic effects are inherently independent of z. Thus we see that in the long-wave approximation for a homogeneous fluid the main variables ζ, u, v are independent of z (ζ inherently, u and v approximately), while the vertical velocity is a linear function of z. In other words, the independent variable z has been "eliminated."

Equations (2.5.5) are fundamental for tidal theory. They explicitly show the form in which the atmospheric tide (through p_s) and bodily tide (through ζ_b) are coupled to the ocean tide. The pressure amplitude of the atmospheric tide at sea level (mainly S_2) is not much larger than 1 mb ($= 10^3$ dynes cm^{-2}), which gives $p_s/\rho g$ an amplitude of only about 1 cm. For this reason the effect of the atmospheric tide on the ocean tide usually is neglected. The displacement amplitude of the bodily tide is about 0.6 times the equilibrium displacement, which gives as much as 14 cm for M_2; this should not be neglected. Moreover, it will be recalled that an indirect effect of the bodily tide is contained in $\boldsymbol{Q} \equiv \boldsymbol{T} + \boldsymbol{G}' + \boldsymbol{G}''$, where $\boldsymbol{T}$ is the direct tidal force and $\boldsymbol{G}'$, $\boldsymbol{G}''$ are changes of the Earth's gravitational field brought about by tidal displacement of, respectively, the ocean and the solid Earth. Owing to the fact that the solid Earth's free periods of elastic vibration do not exceed about one hour, displacements associated with the bodily tide are very nearly in the form of an equilibrium response, with the result that

$$(2.5.6) \qquad g\left(\frac{\partial}{\partial x}, \frac{\partial}{\partial y}\right)\zeta_b = h_L(T_x, T_y), \qquad (G''_x, G''_y) = k_L(T_x, T_y)$$

where the "Love numbers" h_L and k_L are constants, the numerical values of which are respectively, about 0.6 and 0.3. (These numbers indirectly express the degree of elastic yielding of the solid Earth to tidal forces.)

Consequently, (2.5.5) can be written

$$\frac{\partial(u, v)}{\partial t} = -g\left(\frac{\partial}{\partial x}, \frac{\partial}{\partial y}\right)\left(\zeta + \frac{p_s}{\rho g}\right) + 2\Omega_z(v, -u) \tag{2.5.7}$$
$$+ (1 + k_L - h_L)(T_x, T_y) + (G'_x, G'_y).$$

Here the effects of the bodily tide are incorporated simply through multiplication of the tidal force by the factor $1 + k_L - h_L \approx 0.7$.

Two difficulties remain in this formulation. First, the formula for ζ_b in (2.5.6) expresses only that part of ζ_b directly produced by the bodily tide. There is, however, a contribution to ζ_b from variations of crustal loading that accompany tidal displacement of the ocean. This is not included in (2.5.7). It is, hopefully, of lesser importance than that expressed through h_L. The other difficulty stems from G'_x, G'_y: we can in principle express this effect only as an area integral involving the distribution of ζ over the ocean surface. (The loading effect first mentioned is similar in this respect.) The incorporation of G'_x, G'_y therefore would oblige us to deal with an integro-differential equation.

To make this point explicit, let V' be the potential associated with $\boldsymbol{G}'$. Since the ocean is a thin, spheroidal shell, we can regard V' as produced by a spheroidal surface layer of mass $\rho\zeta$ per unit area (where ρ is the mean density of the ocean). If we take the shell to be a sphere of radius r_1, its potential evaluated at A on the surface is

$$V'(A) = \frac{4\pi\gamma\rho r_1}{2^{1/2}} \int \frac{\zeta(B)\, d\omega_B}{\sqrt{1 - \cos\theta(A, B)}}$$

where B is a point of integration over the sphere, $d\omega_B$ the element of solid angle at B, and $\theta(A, B)$ the angular distance between A and B. From this expression we can calculate $(G'_x, G'_y) = (\partial/\partial x, \partial/\partial y)V'$ for insertion in (2.5.7). (The above formula for V' does not permit calculation of G'_z, but that component is not needed.)

In writing the working form of the tidal equations, it is customary to use the equilibrium displacement $\bar{\zeta}$ (defined in (2.3.8)) in place of W, so that $(T_x, T_y) = g_1(\partial/\partial x, \partial/\partial y)\bar{\zeta}$. The potential V' of the self-attraction can be expressed in a similar way: $V' = g_1\zeta'$ where

$$\zeta'(A) \equiv \frac{3}{2^{1/2}} \frac{\rho}{\rho_\oplus} \int \frac{\zeta(B)\, d\omega_B}{\sqrt{1 - \cos\theta(A, B)}} \tag{2.5.8}$$

if we use $g_1 \approx \gamma M_\oplus/r_1^2 \approx \gamma \cdot \frac{4}{3}\pi\rho_\oplus r_1$ where $\rho_\oplus$ is the mean density of the Earth. (The ratio $\rho/\rho_\oplus \approx 0.18$.) Omitting the effect of atmospheric pressure, we find then that (2.5.7) can be written

(2.5.9) $$\frac{\partial(u, v)}{\partial t} = -g\left(\frac{\partial}{\partial x}, \frac{\partial}{\partial y}\right)(\zeta + \zeta' - \gamma_L\bar{\zeta}) + f(v, -u)$$

where $\gamma_L \equiv 1 + k_L - h_L$ is the Love "reduction factor" and $f \equiv 2\Omega_z = 2\Omega \sin \phi$ is the "Coriolis parameter."

Finally, I must establish the conditions under which the long-wave approximation is valid—that is, under which the p'-terms in (2.5.4) are negligible. The test of whether this is so is

$$\rho^{-1}[p']/g[\zeta_s] \ll 1$$

where [ ] means the "scale" value (order of magnitude) of a quantity. The estimate of $[p']$ must come from the z-component equation (2.2.3c), where for P_z we can insert $-\rho^{-1}\,\partial p/\partial z = -g - \rho^{-1}\,\partial p'/\partial z$ (by definition of p'):

$$\rho^{-1}\,\partial p'/\partial z = Q_z + 2\Omega_y u - a_z.$$

The result of integrating this from the surface (where $p' = 0$) to an arbitrary depth is

$$\rho^{-1}[p'] = H \cdot [Q_z + 2\Omega_y u - a_z]$$

where H is a scale value of h. For tidally-generated motions we expect Q_z (which is of the same order as Q_x and Q_y) to be comparable in size to $2\Omega_y u$, so the preceding statement is not weakened by omitting it:

$$\rho^{-1}[p'] = H \cdot [2\Omega_y u - a_z] < (2|\Omega_y|\,[u] + [a_z]).$$

Here $[a_z] = [\partial w/\partial t] = \sigma[w]$ where σ is a characteristic frequency of the motion. According to (2.5.3), $[w] = [\partial\zeta_s/\partial t] = \sigma[\zeta_s]$ so $[a_z] = \sigma^2[\zeta_s]$ and we have

$$\rho^{-1}[p']/g[\zeta_s] < (2|\Omega_y| \cdot H[u])/g[\zeta_s] + \sigma^2 H/g.$$

The ratio $[u]/[\zeta_s]$ can be estimated from the x, y-component equations (2.2.3a, b), where P_x, P_y are given by (2.5.4). On the a posteriori condition that the p'-terms in (2.5.4) are negligible (and ignoring p_s), we have

$$g[\zeta_s] = L \cdot [(a_x, a_y) - 2\Omega_z(v, -u) + 2\Omega_y(w, 0)]$$

where L is the horizontal scale of the motion. (The Q-terms have been suppressed for the reason already given.) I shall assume that the size of ζ_s is fixed by that of (a_x, a_y), so $g[\zeta_s] = L \cdot [(a_x, a_y)] = L \cdot \sigma \cdot [u]$. (If this is not the case, the argument is a little different.) Then

$$\rho^{-1}[p']/g[\zeta_s] < (2|\Omega_y|/\sigma) \cdot (H/L) + \sigma^2 H/g.$$

At tidal frequencies σ is comparable to $|\Omega_y|$ so the first term is of order H/L, the vertical aspect ratio, which is $\ll 1$ for long waves (only 2 percent of the world ocean is deeper than 6 km). At higher frequencies this term is even smaller; but we must look at $\sigma^2 H/g$. With $H = 6$ km, we can write this

$$[\sigma/(6 \times 10^{-3}\text{ cps})]^2$$

which is small unless $\sigma > 10^{-3}$ cps. The general conclusion is that when $\sigma < 10^{-3}$ cps (about 90 cpd) the horizontal components of pressure-gradient force are predominantly hydrostatic.

2.6. **Potential vorticity.** The motion governed by the long-wave equations (2.5.9) and (2.5.2) is inviscid and is generated by conservative forces. We should therefore expect to find a first integral analogous to the Helmholtz Theorem that the strength of a vortex tube is conserved. In a long-wave motion the horizontal velocities are virtually independent of z and are much larger than the vertical velocity, so the vortex lines can be expected to be nearly vertical and the vorticity to consist primarily of the z-independent vertical component. The strength of a vortex tube is the product of this vorticity and the cross-section area of the tube. If we consider a tube that extends from the bottom $z = -h + \zeta_b$ to the surface $z = \zeta_s$ of the ocean, mass conservation requires that cross section to be inversely proportional to the length of the tube, $h + \zeta$. Hence the strength of the tube is proportional to the *potential vorticity* $\equiv$ (vorticity)/$(h + \zeta)$, which should be conserved.

A formal proof is easily made from (2.5.9) and (2.5.2), but before doing so I shall comment about the directions of the horizontal axes x, y. It will be recalled that we arrived at (2.5.9) by starting from (2.2.3) in which the axes are in the "geopotential" configuration: x east, y north, z up. However, it is clear by inspection that the long-wave equations (2.5.9) and (2.5.2) are invariant under a rotation of x, y into an arbitrary azimuth in the horizontal plane. Unless noted otherwise, I shall therefore not restrict x, y to the east, north directions, and assume only that the azimuth of these axes relative to east, north is the same at all points of the spherical surface.

In this connection it should be emphasized that x, y in (2.5.9) are "axes" but not "coordinates": no choice of horizontal coordinates has yet been made. Further, owing to the curvilinear properties of the spherical surface, the surface divergence is in general

(2.6.1) $$\operatorname{div}(u, v) = \partial u/\partial x + \partial v/\partial y - (r_1^{-1} \tan \phi) \cdot v_N$$

where v_N is the northward velocity and ϕ the latitude. Formally, $\operatorname{div}(u, v)$ can be defined as the scalar product of $(\partial/\partial x, \partial/\partial y)$ and (u, v); or as the limit $A^{-1}\oint(u, v)_n\,ds$ of the ratio of the outflow across a small horizontal circuit to the enclosed area A as $A \to 0$. We also want an analogous expression for the (vertical component of the) vector product of $(\partial/\partial x, \partial/\partial y)$ and (u, v), which is the same as the limit $A^{-1}\oint(u, v)_s\,ds$ of the ratio of the circulation in a small horizontal circuit to the enclosed area A as $A \to 0$. This is the z-component of $\operatorname{curl}(u, v)$. I shall denote it by $\operatorname{vor}(u, v)$; we find

$$\operatorname{vor}(u, v) = \partial v/\partial x - \partial u/\partial y + (r_1^{-1} \tan \phi) \cdot u_E \tag{2.6.2}$$

where u_E is the eastward velocity. This formula is equivalent to (2.6.1) in the sense that

$$\operatorname{vor}(u, v) = \operatorname{div}(v, -u), \qquad \operatorname{vor}(v, -u) = -\operatorname{div}(u, v).$$

Both (2.6.1) and (2.6.2) serve as reminders that on the sphere the directional derivatives $\partial/\partial x$ and $\partial/\partial y$ are not commutative.

Return to the question of potential vorticity and take the vorticity of (2.5.9): the gradient terms vanish and we get

$$(\partial/\partial t) \operatorname{vor}(u, v) = \operatorname{vor}[f(v, -u)].$$

Since $\operatorname{vor}(v, -u) = -\operatorname{div}(u, v)$, this can be written

$$(\partial/\partial t) \operatorname{vor}(u, v) = -f \operatorname{div}(u, v) - (u, v) \operatorname{grad} f.$$

From (2.5.2) we have a similar equation

$$\partial\zeta/\partial t = -h \operatorname{div}(u, v) - (u, v) \operatorname{grad} h$$

that can be used to eliminate $\operatorname{div}(u, v)$:

$$\frac{1}{h}\frac{\partial}{\partial t}\left(\operatorname{vor}(u, v) - \frac{f}{h}\zeta\right) = -(u, v) \operatorname{grad} \frac{f}{h} \tag{2.6.3}$$

both f and h being independent of t.

Now the "absolute" potential vorticity is $[f + \operatorname{vor}(u, v)]/(h + \zeta)$ where $f + \operatorname{vor}(u, v)$ is the absolute vorticity, consisting of the vorticity f of the Earth's absolute motion of rotation ($u_E = r_1\Omega \cos \phi$, $v_N = 0$) plus vorticity of the relative motion. When u, v and ζ are small,

$$\frac{f + \operatorname{vor}(u, v)}{h + \zeta} = \frac{f}{h} + \frac{1}{h}\left(\operatorname{vor}(u, v) - \frac{f}{h}\zeta\right) + \cdots.$$

This expresses total absolute potential vorticity as the absolute potential vorticity f/h of the undisturbed ocean plus the potential vorticity of the disturbance. It is clear then that (2.6.3) is the linearized form of the conservation theorem for potential vorticity

$$\frac{d}{dt}\frac{f + \mathrm{vor}(u, v)}{h + \zeta} = 0$$

where $d/dt \equiv \partial/\partial t + (u, v)\mathrm{grad}$ is the material derivative.

3.1. **Waves of the first class.** In the absence of tidal forces the long-wave equations (2.5.9), together with the continuity equation (2.5.2), are

$$\frac{\partial(u, v)}{\partial t} = -g\left(\frac{\partial}{\partial x}, \frac{\partial}{\partial y}\right)\zeta + f(v, -u), \tag{3.1.1}$$

$$\partial\zeta/\partial t = -\mathrm{div}[(u, v)h]. \tag{3.1.2}$$

These are linear homogeneous equations in three dependent variables u, v and ζ. The horizontal axes x, y are in an arbitrary azimuth, and the Coriolis parameter $f \equiv 2\Omega \sin\phi$ and depth h are spatially variable coefficients.

Let $(\zeta; u, v, w) = \mathscr{R}[(Z; U, V, W)e^{-i\sigma t}]$ be a harmonic solution, where $(Z; U, V, W)$ are space-dependent functions. Then (3.1.1) can be solved algebraically for U, V to give the *polarization equations*

$$U = \frac{g}{\sigma^2 - f^2}\left(-i\sigma\frac{\partial Z}{\partial x} + f\frac{\partial Z}{\partial y}\right), \tag{3.1.3a}$$

$$V = \frac{g}{\sigma^2 - f^2}\left(-i\sigma\frac{\partial Z}{\partial y} - f\frac{\partial Z}{\partial x}\right). \tag{3.1.3b}$$

The corresponding equation for W can be inferred from (2.5.3) (where ζ_s is now ζ); it is

$$W = -i\sigma(h + z)h^{-1}Z + h^{-1}z(U, V)\,\mathrm{grad}\,h \tag{3.1.3c}$$

with U, V as in (3.1.3a, b). Equations (3.1.3) determine the velocities—and hence the orbital motion—of fluid particles, when Z is known.

To find Z we can eliminate (u, v) from (3.1.2) by means of (3.1.3a, b). The outcome of this operation depends in a critical way upon the geographical variation of f and h. In particular, if the potential vorticity f/h of the undisturbed ocean is uniform, then by (2.6.3) the potential vorticity of the wave is zero. This is a significant constraint which in some respects is analogous to that of a velocity potential in a nonrotating fluid. To put it another way, uniform potential vorticity means uniform absolute angular momentum of individual fluid columns. If this state prevails in the undisturbed medium, no disturbance governed by conservative forces can alter it, so the perturbation must have zero angular momentum (zero potential vorticity).

It is precisely with this special circumstance that I am concerned in these lectures. Although it does not accord with the true geophysical setting for the oceans, it nevertheless enables us to examine conveniently some of the main features of the long-wave equations. The convenience stems from the fact that the constraint of uniform potential vorticity excludes a class of waves of very low frequency, sometimes called "rotational" waves or modes, or waves of the "second class." In this category are planetary waves, the existence of which depends upon the gradient of potential vorticity associated with the northward gradient of f; and topographic waves whose existence depends upon potential-vorticity gradients produced by variations of h (including, for example, a rotational type of edge wave associated with the continental shelf). What remains are long gravity waves, which are modified by rotation but would exist without it. These are the waves I shall consider.

We therefore now eliminate (u, v) from (3.1.2) by means of (3.1.3):

(3.1.4) $$\nabla^2 Z + [(\sigma^2 - f^2)/gh] \cdot Z = 0$$

assuming depth h and Coriolis parameter f each uniform. In one sense this assumption is less restrictive than it may seem, for if we deal with wave lengths small relative to the Earth's radius, the geographical variation of f can be ignored on the basis of scale considerations. A similar point of view can be adopted with respect to variations of h. The assumption of uniform h does, however, exclude "topographic" waves of the *first* class—such as gravitational edge waves—as well as those of the second class, so it should be recorded that (3.1.4) does not describe *all* rotationally-modified gravity waves.

Equation (3.1.4) is a classic starting point for vibration and wave problems. I shall freely quote some of its familiar solutions. It should be remarked that the effect of Coriolis force is not merely expressed through f^2 in (3.1.4): there are also significant effects owing to f-terms in the U, V-polarization equations (3.1.3a, b). Incidentally, those equations are not explicitly simplified by assuming f and h uniform, but (3.1.3c) reduces to

(3.1.5) $$W = -i\sigma(h + z)h^{-1}Z.$$

This expresses the fact that when the depth is uniform, the vertical velocity varies linearly from zero at the bottom ($z = -h$) to $\partial\zeta/\partial t$ at the free surface.

For later use it is convenient to note here that, since the hydrostatic pressure perturbation is $\rho g\zeta$, the energy flux through a vertical surface element, averaged over a wave period and integrated vertically, is

(3.1.6) $$\rho g h\overline{\zeta(u, v)} = \tfrac{1}{2}\rho g h\mathscr{R}[Z^*(U, V)]$$

per unit horizontal length. The kinetic and potential energy densities are, respectively,

$$\tag{3.1.7} \begin{aligned} \tfrac{1}{2}\rho g h\overline{(u^2 + v^2)} &= \tfrac{1}{4}\rho h(|U|^2 + |V|^2), \\ \tfrac{1}{2}\rho g\overline{\zeta^2} &= \tfrac{1}{4}\rho g|Z|^2 \end{aligned}$$

per unit horizontal area.

3.2. **The plane Sverdrup wave.** In view of the assumptions underlying (3.1.4), I shall confine attention to a geographically-limited region that can be approximated locally by the horizontal plane. In this region let x, y be Cartesian coordinates. Then for plane waves the solution of (3.1.4) is

$$\tag{3.2.1} \begin{aligned} Z &= h \exp[i(kx + ly)], \\ \sigma &= \pm[f^2 + (k^2 + l^2)gh]^{1/2} \end{aligned}$$

and the polarization equations give

$$\tag{3.2.2} \begin{aligned} (U, V, W) &= (A, B, C)\exp[i(kx + ly)], \\ A &\equiv (\sigma k + ifl)/(k^2 + l^2), \\ B &\equiv (\sigma l - ifk)/(k^2 + l^2), \\ C &\equiv -i\sigma(z + h). \end{aligned}$$

The complete plane-wave solution is

$$\tag{3.2.3} (\zeta; u, v, w) = \mathscr{R}\{(h; A, B, C)\exp[i(kx + ly - \sigma t)]\}.$$

For convenience, the amplitude of the wave (which of course is arbitrary) has been scaled to make the amplitude of ζ equal to the depth h.

When k and l are real—which assume until further notice—a wave of type (3.2.3) is sometimes called a "homogeneous" plane wave. In the present context I shall refer to it as a *Sverdrup wave*, which may be regarded as an elemental propagator of tidal signals.

According to (3.1.7) the densities of kinetic and potential energy of the Sverdrup wave (3.2.1, 2) are

$$\tfrac{1}{4}\rho g h^2 \cdot (\sigma^2 + f^2)/(\sigma^2 - f^2) \qquad \text{and} \qquad \tfrac{1}{4}\rho g h^2.$$

Thus we see that, whereas there is equipartition when $f = 0$, the operation of Coriolis force makes kinetic energy greater than potential in the ratio $(\sigma^2 + f^2)/(\sigma^2 - f^2)$. When σ is in the tidal range and therefore is close to f, this ratio is large.

The minimum frequency attainable for a Sverdrup wave is simply f ($\equiv 2\Omega \sin \phi$), which increases from zero at the equator to exactly 2 cycles per sidereal day at the pole. It follows that Sverdrup waves of given frequency $|\sigma|$ cannot exist poleward of the latitude (if any) where $|\sigma| = f$. These limiting latitudes for the main diurnal and semidiurnal constituents are:

Diurnal constituent	Limiting latitude	Semidiurnal constituent	Limiting latitude
Q_1	26.45°	N_2	70.98°
O_1	27.61	M_2	74.47
P_1	29.82	S_2	85.76
K_1	30	K_2	90

Although diurnal tides are in fact more conspicuous in low than in high latitudes, it should not be expected that the tide necessarily has the character of a Sverdrup wave.

To examine a single wave in detail it is convenient to orient axes with x in the direction of the wave-number vector k, l; then $l = 0$ (and $k > 0$) so after taking real parts,

$$\begin{aligned} (\zeta, u) &= (h, \sigma/k)\cos(kx - \sigma t), \\ (v, w) &= (f/k, \sigma(z + h))\sin(kx - \sigma t), \\ \sigma &= \pm(f^2 + k^2 gh)^{1/2}. \end{aligned} \tag{3.2.4}$$

It is clear that the Coriolis force (i) generates a particle motion horizontally transverse to the wave, (ii) increases the frequency of the wave for a given wave number, and (iii) makes the wave dispersive. We also see that, associated with a given wave number k, there is a backward Sverdrup wave ($\sigma < 0$) for each forward wave ($\sigma > 0$), as is characteristic of gravity waves. (By "forward" I mean a wave that travels in the direction of increasing values of phase angle $kx - \sigma t$.) The two waves have the same absolute frequency $|\sigma|$ and therefore the same speed $|\sigma|/k$.

Figure 3.2.1 shows the relation between $|\sigma|$ and k, as given in (3.2.4), for $f = 1$ cycle per day, the Coriolis parameter at latitude 29.91°. Figure 3.2.2 shows the phase and group speeds, respectively,

$$\frac{|\sigma|}{k} = \frac{|\sigma|}{(\sigma^2 - f^2)^{1/2}} \cdot (gh)^{1/2}, \qquad \left|\frac{d\sigma}{dk}\right| = \frac{(\sigma^2 - f^2)^{1/2}}{|\sigma|} \cdot (gh)^{1/2}$$

as functions of $|\sigma|$ for $f = 1$ cpd.

As can be seen by inspection of (3.2.4), the Sverdrup wave is polarized in a plane perpendicular to the wave fronts and inclined upward to the left of the direction of propagation (Figure 3.2.3). The orbit is an ellipse whose projections in the x, y and y, z-planes are ellipses with semi-axes

$$\alpha, \beta, \gamma = k^{-1}(1, |f/\sigma|, k(z + h))$$

and is traversed in the direction that gives clockwise rotation in the horizontal projection when $f > 0$ (Northern Hemisphere). The vertical aspect ratio $\gamma/\alpha = k(z + h)$ has its greatest value at the free surface ($z \approx 0$) where it is kh, which by the nature of the long-wave approximation is $\ll 1$. The horizontal aspect ratio $\beta/\alpha = |f/\sigma|$ is <1 and approaches zero in the high-frequency limit. Then the orbital plane is vertical and the wave is unaffected by Coriolis force. On the other hand, $|f/\sigma| \to 1$ in the low-frequency cut-off limit f. Near this limit the wave—strongly affected by Coriolis force—is very nearly circularly and horizontally polarized.

The profile of the Sverdrup wave is shown in Figure 3.2.4. Note that crests and troughs are horizontal. Streamlines are shown in Figure 3.2.5.

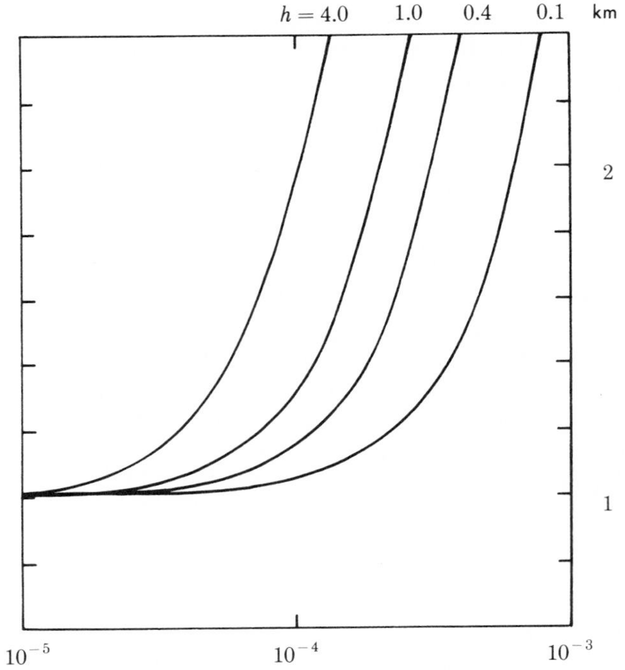

FIGURE 3.2.1. Ordinate: frequency $|\sigma|$ in cycles per day; abscissa: wave number of Sverdrup wave in cycles per kilometer.

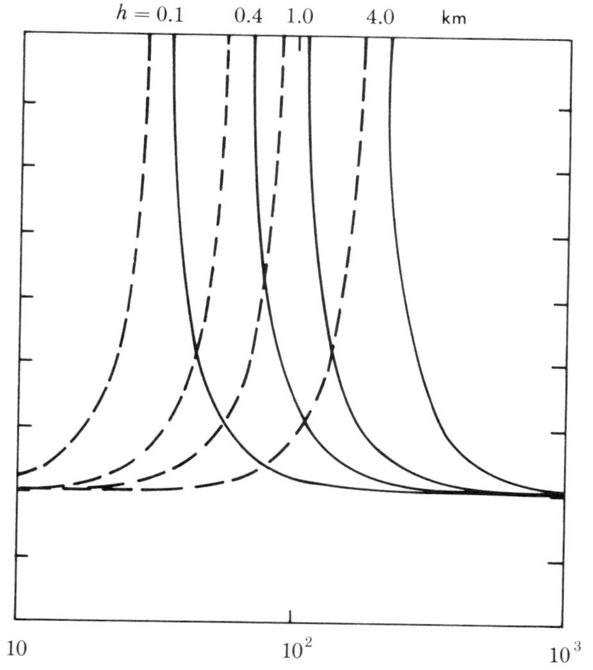

FIGURE 3.2.2. Ordinate: frequency $|\sigma|$ in cycles per day; abscissa: phase speed (————) and group speed (– – – – –) of Sverdrup wave in meters per second.

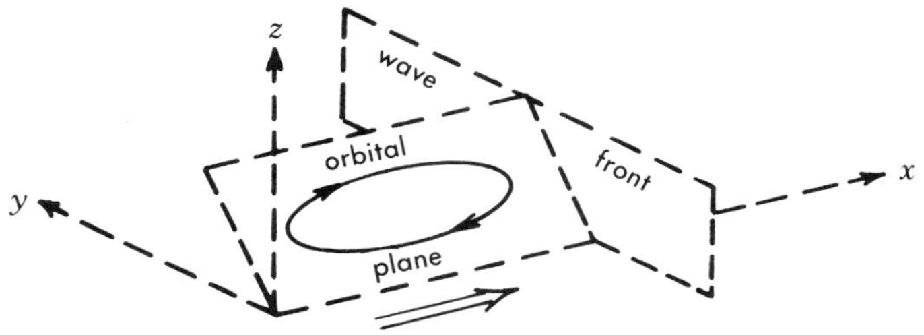

FIGURE 3.2.3. Orbital plane of a Sverdrup wave (⇒ shows direction of propagation of wave). Vertical scale is greatly magnified.

The energy flux of the Sverdrup wave, calculated from (3.1.6), is

$$\rho g h \overline{\zeta u} = \tfrac{1}{2}\rho g h^2 \cdot \sigma/k; \qquad \rho g h \overline{\zeta v} = 0$$

in the longitudinal and transverse directions. There is no flux transverse to the wave. By (3.1.7) the total energy density is $\frac{1}{2}\rho g h^2 \sigma^2/k^2 g h$ so the

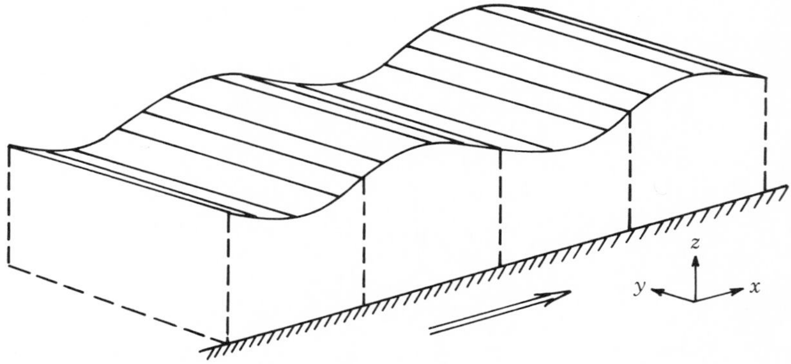

FIGURE 3.2.4. Contours of the Sverdrup wave profile.

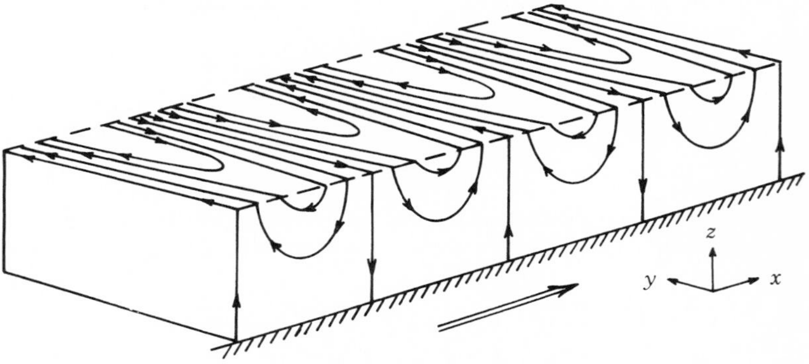

FIGURE 3.2.5. Streamlines of Sverdrup waves in x, y- and x, z-planes.

ratio of flux to density is kgh/σ. As is normally the case, this is equal to the group velocity $d\sigma/dk$.

In his study of tides on the North Siberian Shelf, as a member of the "Maud" expedition of 1922–24, Sverdrup constructed hodographs of tidal currents and cotidal lines of the surface profile for the "spring" tide (Moon and Sun is syzygy, so M_2 and S_2 in phase). His original diagrams are shown in Figure 3.2.6. The hodographs are clockwise, as in a Sverdrup wave, and some are nearly circular, as would be expected because of proximity to the critical latitudes for both M_2 (74.47°) and S_2 (85.76°). The cotidal lines (isochrones of wave crest) show a progressive wave advancing southward across the Siberian Shelf.

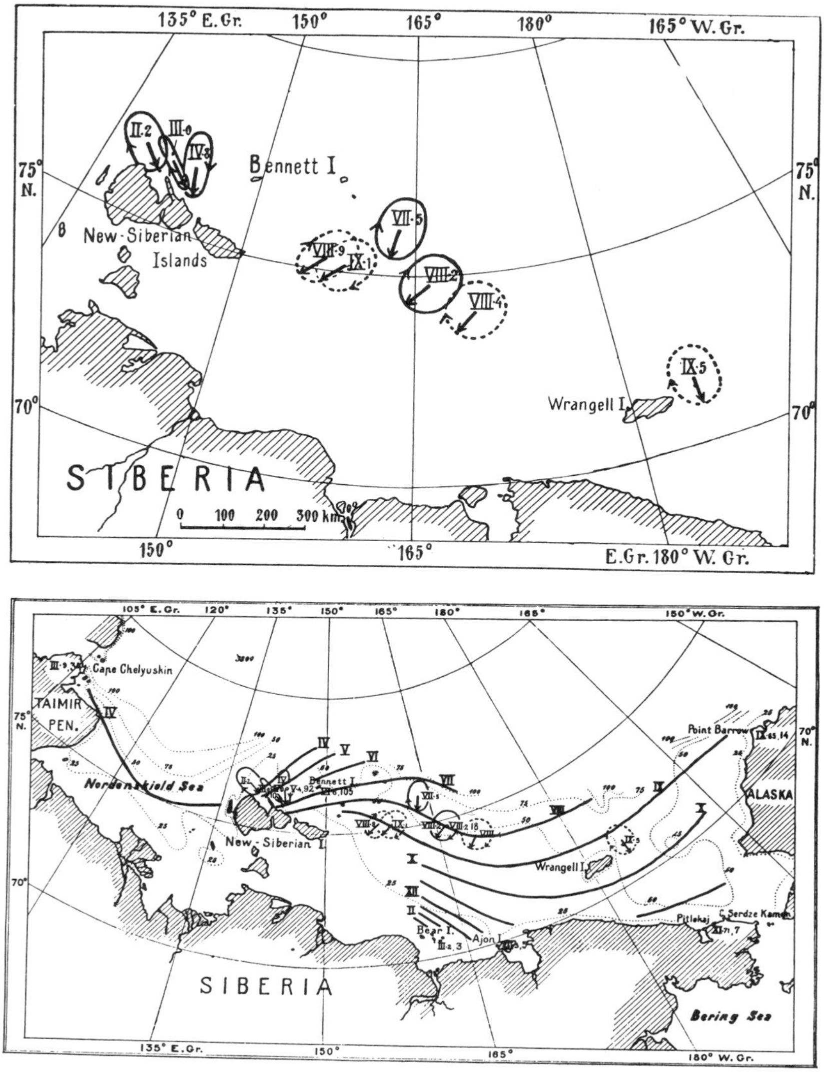

FIGURE 3.2.6. Current hodographs (upper diagram) and cotidal lines (lower diagram) of spring tide on the North Siberian shelf (from [Sverdrup, 1926, pp. 11, 13]).

3.3. **The plane Poincaré wave.** We consider now the reflection of a homogeneous plane wave (Sverdrup wave) from a straight coast. Take x-axis along the coast and y-axis seaward. To fix ideas, suppose the incident wave is moving forward along x, its trace velocity in that direction being

σ/k with $\sigma > 0$ and $k > 0$. The reflected wave must have the same trace velocity and frequency, so its wave-number vector is (k, l), that of the incident wave being $(k, -l)$, with $l > 0$. Hence according to (3.2.3) the incident and reflected waves are the real parts, respectively, of

$$\tfrac{1}{2}(h;A^*,-B^*,C)\exp[i(kx-ly-\sigma t)] \quad \text{and} \quad R\cdot\tfrac{1}{2}(h;A,B,C)\exp[i(kx+ly-\sigma t)] \tag{3.3.1}$$

where R is the reflection coefficient and A, B, C are the polarization coefficients of (3.2.2). (The factor $\frac{1}{2}$ is inserted for later convenience.)

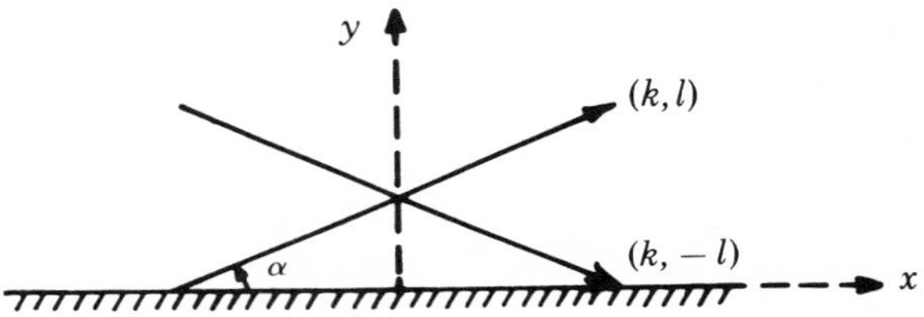

FIGURE 3.3.1. Wave-number vectors of incident and reflected waves.

Assuming perfect reflection, we require zero mass and energy flux in the y-direction at $y = 0$; that is, $\rho hv = 0$ and $\rho gh\zeta v = 0$ where v is the y-component velocity of the composite wave and ζ the composite disturbance. Both conditions are satisfied when $v = 0$ at $y = 0$, which by addition of the two appropriate parts of (3.3.1) means $-B^* + RB = 0$ so

$$R = B^*/B = \exp(-2i\arg B). \tag{3.3.2}$$

This confirms that in perfect reflection the amplitude does not change ($|R| = 1$), and shows that, owing to the Earth's rotation, the reflected wave receives a phase shift $-2\arg B = -2\arg(\sigma l - ifk)$.

After adding the two waves in (3.3.1) and taking real parts, we get

$$\begin{aligned} \zeta &= h\cos(ly-\arg B)\cos(kx-\sigma t), \\ u &= |A|\cos(ly-\arg B+\arg A)\cos(kx-\sigma t), \\ v &= -|B|\sin ly\sin(kx-\sigma t), \\ w &= \sigma(z+h)\cos(ly-\arg B)\sin(kx-\sigma t). \end{aligned} \tag{3.3.3}$$

(In writing these expressions the origin for x has been shifted so that what was $kx - \arg B$ has become kx.) This superposition of an incident and reflected Sverdrup wave is called a *Poincaré wave*. Its (composite) wave

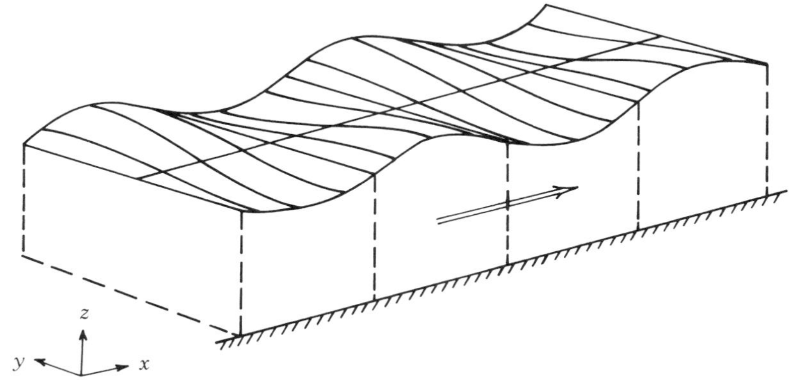

FIGURE 3.3.2. Profile of a Poincaré wave.

fronts are perpendicular to the coast, and its characteristic feature is the set of nodal lines normal to the wave fronts, which give a cellular pattern (Figure 3.3.2).

By changing the sign of σ in (3.3.3) we get a Poincaré wave moving in the opposite direction. It will be convenient to use the terms *right bounded* and *left bounded* in reference to the waves that have the coast respectively on the right or left of the direction of propagation. Thus, with axes as in Figure 3.3.1, we get a right-bounded wave by taking $\sigma > 0$ in (3.3.3) and a left-bounded wave with $\sigma < 0$.

There is a significant asymmetry between right-bounded and left-bounded Poincaré waves derived from the same wave numbers k, l. This can be demonstrated by locating the nodes of the wave profile (ζ) that are parallel to the coast. By (3.3.3) these are equally spaced a distance π/l apart, but their respective distances from the coast are different for the two waves. This can be inferred from the fact that $\arg B = \arg(\sigma l - ifk)$ is in the fourth quadrant for the right-bounded wave ($\sigma > 0$) and in the third quadrant for the left-bounded wave ($\sigma < 0$), which leads to the picture in Figure 3.3.3. In contrast to the picture at zero rotation, a "loop" (crest or trough) of the wave profile does not coincide with the coast. Specifically, the distance from the coast to the first node of the right-bounded wave is less than $\pi/2l$, one-quarter of the transverse wave length, whereas for the left-bounded wave it is greater than $\pi/2l$.

One important consequence of this unusual asymmetry is that the superposition of backward and forward Poincaré waves cannot produce a standing wave, and therefore cannot solve the problem of reflection of a Poincaré wave from a barrier perpendicular to the coast. This greatly

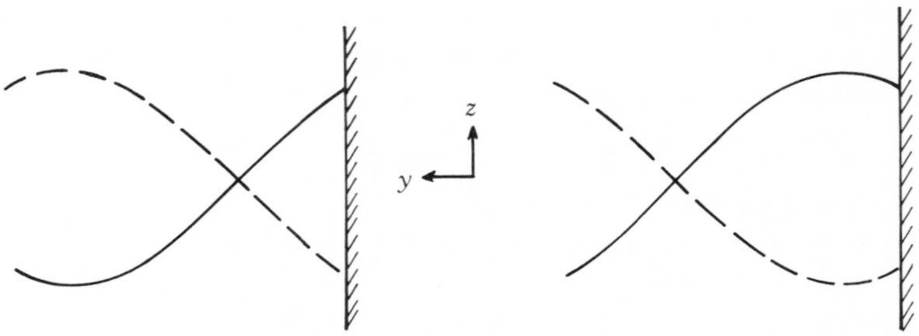

FIGURE 3.3.3. Left: transverse profiles of right-bounded Poincaré wave (moving away from reader) in two sections one-half wavelength apart; right: same for left-bounded wave (moving toward reader).

complicates the mathematical analysis of waves in rotating basins with sharp corners.

In terms of the longshore wave number k and direction α of the constituent Sverdrup waves (Figure 3.3.1), the frequency equation for Poincaré waves is, from (3.2.1):

$$\sigma = \pm(f^2 + k^2 gh \sec^2 \alpha)^{1/2}. \tag{3.3.4}$$

Figure 3.3.4 shows the relation between $|\sigma|$ and k, with α as a parameter. The spectrum is continuous in its dependence upon k and α and, for given k, has a cut-off at $(f^2 + k^2 gh)^{1/2}$, which corresponds to glancing incidence ($l = 0, \alpha = 0$). At normal incidence ($k = 0, \alpha = 90°$) the Poincaré wave is simply a standing Sverdrup wave.

The presence of nodal planes parallel to the direction of propagation makes it possible to use the Poincaré wave as a solution of the problem

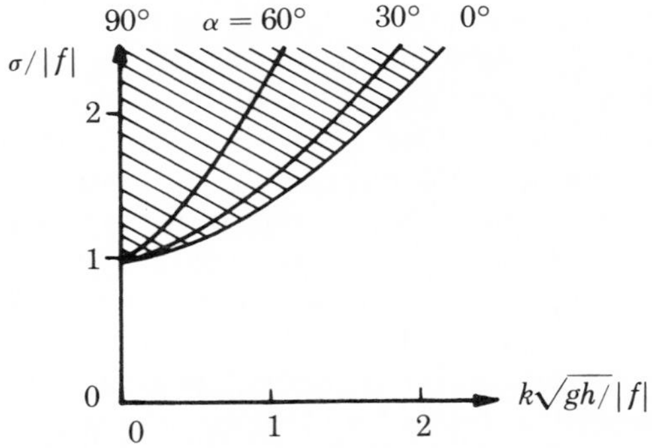

FIGURE 3.3.4. Relation between Poincaré-wave frequency, longshore wave-number k, and angle of incidence α. The spectrum is continuous and occupies the shaded region.

of Sverdrup waves guided through a channel with straight, parallel sides. For if Y is the width of the channel, the Poincaré wave (3.3.3) will satisfy the boundary condition $v = 0$ at $y = Y$ for any l in the sequence

$$l = n\pi/Y \qquad (n = 1, 2, 3, \ldots).$$

The corresponding "guided" modes have a frequency spectrum continuous in k but otherwise discrete (Figure 3.3.5). The cut-off frequency of the spectrum as a whole is that of the lowest mode, $(f^2 + \pi^2 gh/Y^2)^{1/2}$. For the English Channel, rough figures are $h = 50$ to 80 m and $Y = 100$ to 150 km, the latitude being about 50 degrees ($f \approx 1.5$ cpd). These give a cut-off between 8 and 10 cpd. The tide cannot therefore be expected to propagate through the English Channel as a Poincaré wave.

Most marginal seas are either too narrow or too deep to permit a Poincaré wave of tidal frequency to enter. (Some typical cases are listed in Table 3.3.1.) We must therefore inquire how the tide can propagate into these regions.

TABLE 3.3.1. Cut-off frequency of Poincaré wave in entrance channel of marginal seas (rough estimates)

	Latitude (deg)	Depth (m)	Width (km)	Cut-off frequency (cpd)
North Sea	60	180	500	4.0
Irish Sea	51	100	300	4.8
Gulf of Maine	42	200	400	5.0
Davis Strait	57	3000	1000	7.6
English Channel	50	80	150	8.2

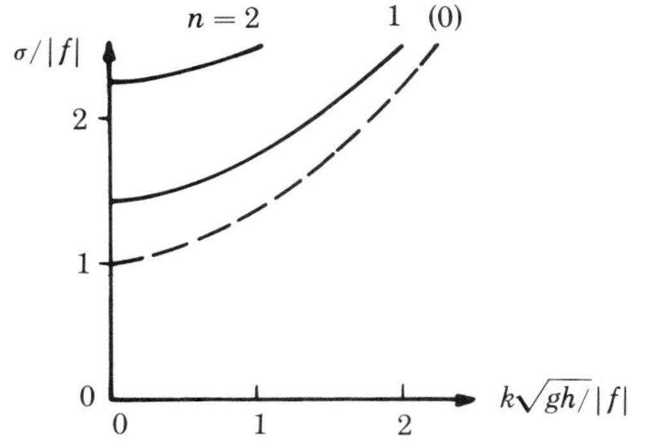

FIGURE 3.3.5. Semidiscrete spectrum of Poincaré waves guided along a channel with straight, parallel sides ($\pi(gh)^{1/2}/|f|Y = 1$).

3.4. **The plane Kelvin and Proudman waves.** The general plane-wave solution (3.2.1, 2, 3) is valid formally when k and l are complex. There is no loss of generality, however, in supposing k (at least) to be real, provided we continue to treat purely harmonic waves (σ real) and thus require $k^2 + l^2$ to be real. This can be seen by orienting the x-axis in the direction of $\mathscr{R}(k, l)$ so that $\mathscr{R}l = 0$ (and $\mathscr{R}k > 0$); then l^2 is real and so k^2 is real, since $k^2 + l^2$ is. Hence k is either real or imaginary. The latter choice would give a solution with no propagation in any direction. I exclude this in order to limit the discussion to propagating waves.

Suppose then that k is real and positive, and l purely imaginary. To make this explicit, replace l by $-il'$ and B by $-iB'$, where $l' \equiv -\mathscr{I}l$ and $B' \equiv -\mathscr{I}B$. Then after taking real parts,

$$\begin{aligned} (\zeta, u) &= (h, A)e^{l'y}\cos(kx - \sigma t), \\ (v, w) &= (B', C')e^{l'y}\sin(kx - \sigma t), \\ \sigma &= \pm[f^2 + (k^2 - l'^2)gh]^{1/2}, \\ A &\equiv (\sigma k + fl')/(k^2 - l'^2), \\ B' &\equiv (\sigma l' + fk)/(k^2 - l'^2), \\ C' &\equiv \sigma(z + h). \end{aligned} \tag{3.4.1}$$

This is an "inhomogeneous" plane wave (often called a "surface" or "boundary" wave); that is, its amplitude varies exponentially in the transverse direction.

As before, the sign alternative for σ gives forward- and backward-moving waves for each k and l'. However, inspection of (3.4.1) shows that these oppositely-moving waves are isomorphic, with signs of l' reversed. Hence it suffices to consider only forward waves ($\sigma > 0$). Figure 3.4.1 shows schematically the dependence of $|\sigma|$ on k for various l'^2. The shaded region corresponds to real values of l and hence to the Sverdrup wave, so we see that for given wave length $2\pi/k$, inhomogeneous plane waves exist only at frequencies below the Sverdrup-wave frequency $(f^2 + k^2gh)^{1/2}$.

Figure 3.4.1 does not reveal the fact that for each σ and k two waves are possible, according as

$$l' = \pm[k^2 + (f^2 - \sigma^2)/gh]^{1/2}.$$

The sign of l' determines whether the wave amplitude decreases ($l' < 0$) or increases ($l' > 0$) to the left of the direction of propagation (Figure 3.4.2)—assumed to be forward throughout this discussion. I shall refer to the former as a "left-decreasing" wave and to the latter as "right-decreasing."

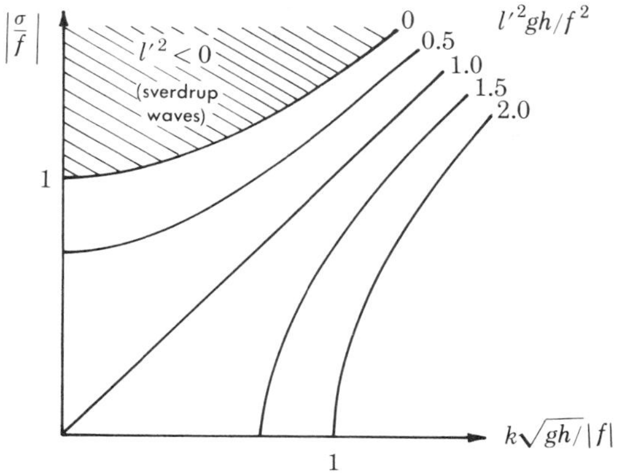

FIGURE 3.4.1. Frequency versus wave number of inhomogeneous plane waves.

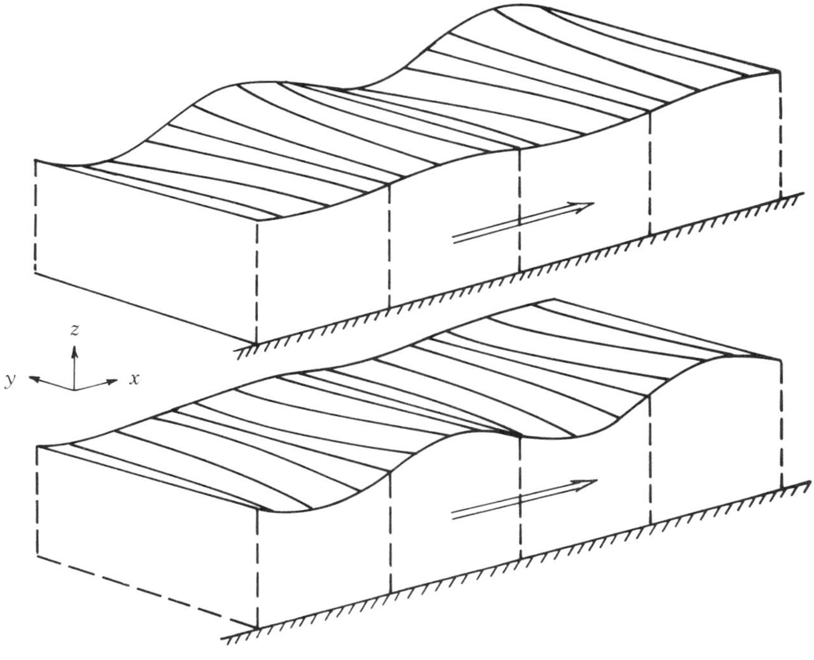

FIGURE 3.4.2. Profiles of right-decreasing (upper) and left-decreasing (lower) inhomogeneous plane waves.

These two wave types can be contrasted dynamically by examining the balance of forces in the transverse direction:

$$\partial v/\partial t = -g(\partial \zeta/\partial y) - fu.$$

In particular, from the ratio $(-g\,\partial\zeta/\partial y)/fu$ we can infer to what extent the transverse slope of the wave profile is in geostrophic balance with the longitudinal particle velocity. In the plane Sverdrup wave this ratio is zero at all frequencies because the wave crests are horizontal. However, for the inhomogeneous plane wave (3.4.1) we get the ratio

$$-\frac{g\,\partial\zeta/\partial y}{fu} = -\frac{gh}{f}\frac{l'}{A} \tag{3.4.2}$$

which is graphed in Figure 3.4.3. The upper branch in this figure corresponds to $l'f < 0$, the lower branch to $l'f > 0$. (Note that (3.4.2) is not altered by change in sign of both l' and f.) I shall refer to the corresponding waves, respectively, as the *Kelvin wave* and the *Proudman wave.*

Consider the Kelvin wave. Since $l'f < 0$, this is a left-decreasing wave in the Northern Hemisphere ($f > 0$, $l' < 0$), a right-decreasing wave in the Southern Hemisphere ($f < 0$, $l' > 0$). The transverse slope is quasi-geostrophic at all frequencies (ratio (3.4.2) positive). It becomes progressively more geostrophic (ratio → 1) as the frequency decreases from the

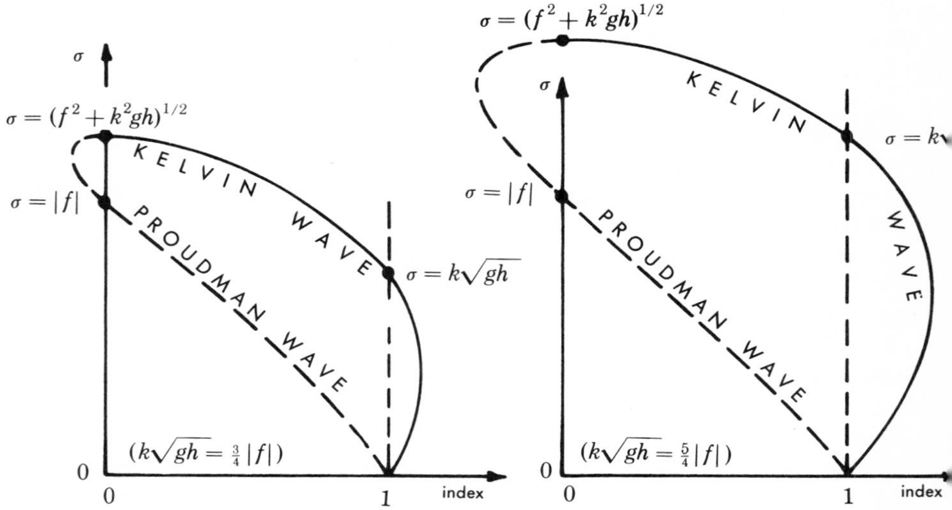

FIGURE 3.4.3. Index (3.4.2) of geostrophic balance in transverse slope of wave profile. Kelvin-wave branch: $l'f < 0$; Proudman-wave branch: $l'f > 0$.

cut-off $(f^2 + k^2gh)^{1/2}$, until at $\sigma = k(gh)^{1/2}$ it is *exactly* geostrophic. There $l' = -f/(gh)^{1/2}$ and the solution is

$$(3.4.3)\qquad \begin{aligned}(\zeta, u) &= (h, (gh)^{1/2}) \exp(-fy/(gh)^{1/2}) \cos(kx - \sigma t),\\ (v, w) &= (0, \sigma(z + h)) \exp(-fy/(gh)^{1/2}) \sin(kx - \sigma t)\end{aligned}$$

with $k = \sigma/(gh)^{1/2}$. Owing to its special role as a boundary wave, this important case usually is simply called "the" Kelvin wave. I shall refer to it as the "prototype" Kelvin wave. Its characteristic features are that it is vertically polarized and that its propagation speed is the zero-rotation, nondispersive $(gh)^{1/2}$.

Now consider the Proudman wave. As we move from the cut-off $(f^2 + k^2gh)^{1/2}$ to lower frequencies, we see that above $\sigma = |f|$ the transverse slope is *a*geostrophic (ratio (3.4.2) negative). In the limit $\sigma \to |f|$ the Proudman wave has infinite A and B' (see (3.4.1)), while $B'/A \to 1$. At this singular frequency, therefore, the wave is *horizontally* (and circularly) polarized and thus gives no disturbance to the free surface:

$$(3.4.4)\qquad \begin{aligned}(\zeta, u) &= (0, 1)e^{l'y} \cos(kx - |f|t),\\ (v, w) &= (1, 0)e^{l'y} \sin(kx - |f|t)\end{aligned}$$

where $l' = k \operatorname{sgn} f$ and $k\ (>0)$ is arbitrary. This is an *inertia motion* (particle acceleration controlled entirely by Coriolis force). In the range $0 < \sigma < |f|$ the transverse slope is quasigeostrophic. Here $A < 0$ and the wave has more of the character of a transverse than a longitudinal vibration, in the sense that the particles move backward rather than forward under a crest.

Thus we see that the Kelvin wave tends to be vertically polarized and is exactly so at $\sigma = k(gh)^{1/2}$, while the Proudman wave tends to be horizontally polarized and is exactly so at $\sigma = |f|$. Further, in the Northern Hemisphere the Kelvin wave is left-decreasing, the Proudman wave right-decreasing (vice versa in the Southern Hemisphere). In both wave types the sense of rotation of the hodograph is clockwise (N.H.) at all frequencies except in the range $0 < \sigma < |f|$, where the Kelvin-wave hodograph is counterclockwise (N.H.).

Figure 3.4.4 is a tidal chart for the M_2 tide in the English Channel. The cotidal lines indicate that the tide moves through the Channel from west to east as a progressive wave, and the corange lines show that on the English coast the range is significantly smaller than on the French coast. The tide therefore resembles a Kelvin wave which, in order to move through the Channel, must be nearly vertically polarized and therefore close to the prototype (3.4.3). To test this as to order of magnitude, place the x-axis in midchannel pointing eastward. If Y is the width of the Channel

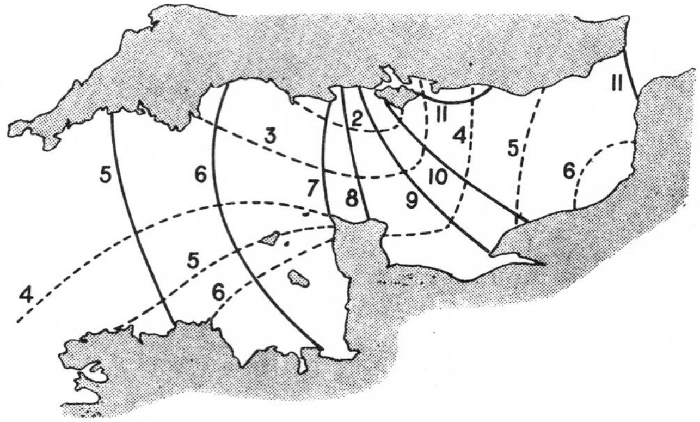

FIGURE 3.4.4. Cotidal lines (5–11 lunar hours) and corange lines (2–6 meters) for the M_2 tide in the English Channel.

Reprinted with permission of Methuen [Proudman, 1953, p. 262].

and $2\zeta_E$, $2\zeta_F$ the tidal ranges on the English and French ends of the transverse section, we should have

$$2\zeta_F/2\zeta_E = \exp(fY/\sqrt{gh})$$

according to (3.4.3). Rough values are $h = 50$ to 80 m and $Y = 100$ to 150 km, the latitude being about 50°N ($f = 1.12 \times 10^{-4}$ rad sec^{-1}). These give $fY/(gh)^{1/2}$ between about 0.5 and 0.6 so $2\zeta_F/2\zeta_E$ is between about 1.6 and 1.8. This certainly is in rough agreement with Figure 3.4.4. For example, where the range is 6 m on the French coast near the entrance to the Channel, it is 3 m on the English coast.

As a further comparison we can note from Figure 3.4.4 that the distance between cotidal lines 5^h and 11^h is roughly 460 km. Since $11^h - 5^h = 6$ hr is half the tidal period, the wave length should be $2 \times 460 = 920$ km. Now if the propagation speed is that of the Kelvin wave (3.4.3), we should have $\sigma = k(gh)^{1/2}$ so the wave length would be $2\pi/k = (2\pi/\sigma)(gh)^{1/2}$. For the M_2 tide $2\pi/\sigma = 12.4$ hr; hence, if we take a mean depth of 50 m, the computed wave length is about 950 km. These order-of-magnitude calculations tend to confirm that the tide moves through the English Channel as a Kelvin-like wave, but should by no means be construed as a model for what is in reality a more complicated process.

An M_2 tide in the South Atlantic Ocean was calculated by Proudman [1944] on the assumption that it consists partly of an "independent" tide (locally generated by tidal forces) and partly of free waves of tidal origin propagated westward from the Indian Ocean and then northward into the Atlantic. He placed the southern boundary of the region of calculation at 35°S, the latitude of the Cape of Good Hope, and assumed that a northbound Kelvin wave crosses this section exactly in the form of the "prototype." He calculated the further course of this wave as it moved northward, by stepwise integration across zonal sections of the ocean, up to 45°N. A southbound Kelvin wave was constructed in a similar way, so that it emerged across 35°S exactly in the form of the prototype. The two Kelvin waves that resulted from Proudman's calculation are shown in Figure 3.4.5. In the Southern Hemisphere the Kelvin wave is right-decreasing, so the northbound wave has its greatest amplitude on the South American coast and the southbound wave on the African coast. In the "prototype" wave the amplitude decreases by the factor e in a distance $(gh)^{1/2}/f$ which at 35°S and $h = 4.0$ km is about 2400 km. Thus, at latitude 35°S these Kelvin waves are not strongly bound to the coasts. Closer to the equator they are even less so.

At 35°S the width of the Atlantic ocean is about 6700 km and the mean depth roughly 4.0 km, from which we find a cut-off of about 1.7 cpd for the fundamental Poincaré mode and 2.8 cpd for the second mode. Thus a free wave at the M_2 frequency (1.93 cpd) can pass through this section as a Poincaré wave in the fundamental mode but not in any higher mode. The wave length (in direction of propagation) of such a wave is $2\pi[gh/(\sigma^2 - \sigma_c^2)]^{1/2}$ where σ_c is the cut-off frequency. With the values stated, we get about 20,000 km so the distance between nodes is 10,000 km which corresponds to about 90 degrees of latitude. Thus, when there is a node at 35°S there should be one at about 55°N. Proudman assumed that a southward-moving as well as a northward-moving fundamental-mode Poincaré wave crosses 35°S, and calculated the forms of these waves up to 45°N. His results are shown in Figure 3.4.6.

Superposition of the four free waves and the calculated independent tide (Figure 3.4.7, left) gave the synthesis shown in Figure 3.4.7, right. The amplitudes and phases of the free waves were chosen so that the synthesis gave best agreement with the observed M_2 tide at latitudes 32.5°S and 7.5°S on the South American coast and at the same latitudes on the African coast. The result was that the north-going Kelvin wave made the largest contribution to the synthesis and the south-going Poincaré wave the next largest.

Proudman's calculation of the forced tide (Figure 3.4.7, left) gave an amphidromic point at about 15°S, 15°W. In their recent numerical calculation of the global tide, Pekeris and Accad [1969] found an amphidromic point in virtually the same location. However, this point is not present in Proudman's synthesis of the whole tide, and in fact the existence of such a point may be difficult to reconcile with observations on the South American coast.

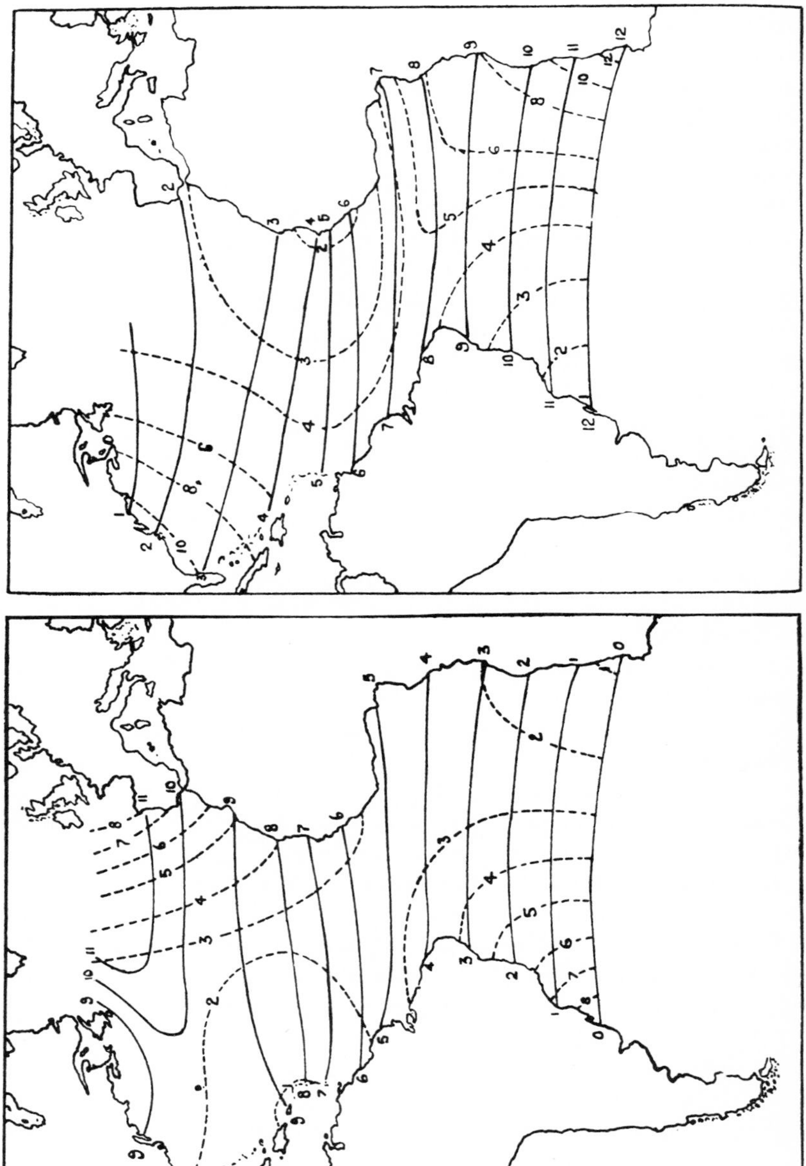

FIGURE 3.4.5. North-going (left) and south-going (right) Kelvin waves at M_2 frequency. Full lines: lunar hour of high water, time origin arbitrary;

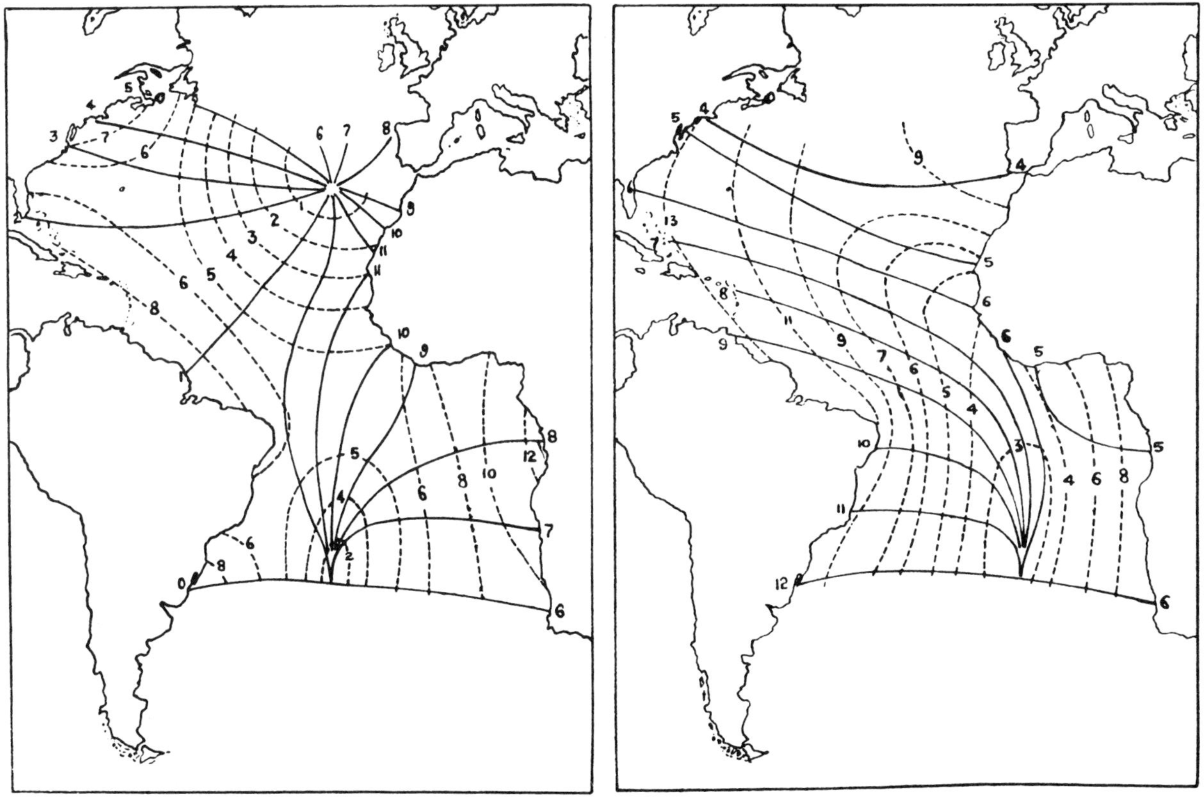

FIGURE 3.4.6. North-going (left) and south-going (right) Poincaré waves at M_2 frequency. Full lines: lunar hour of high water, time origin arbitrary; broken lines: amplitude, scale arbitrary.

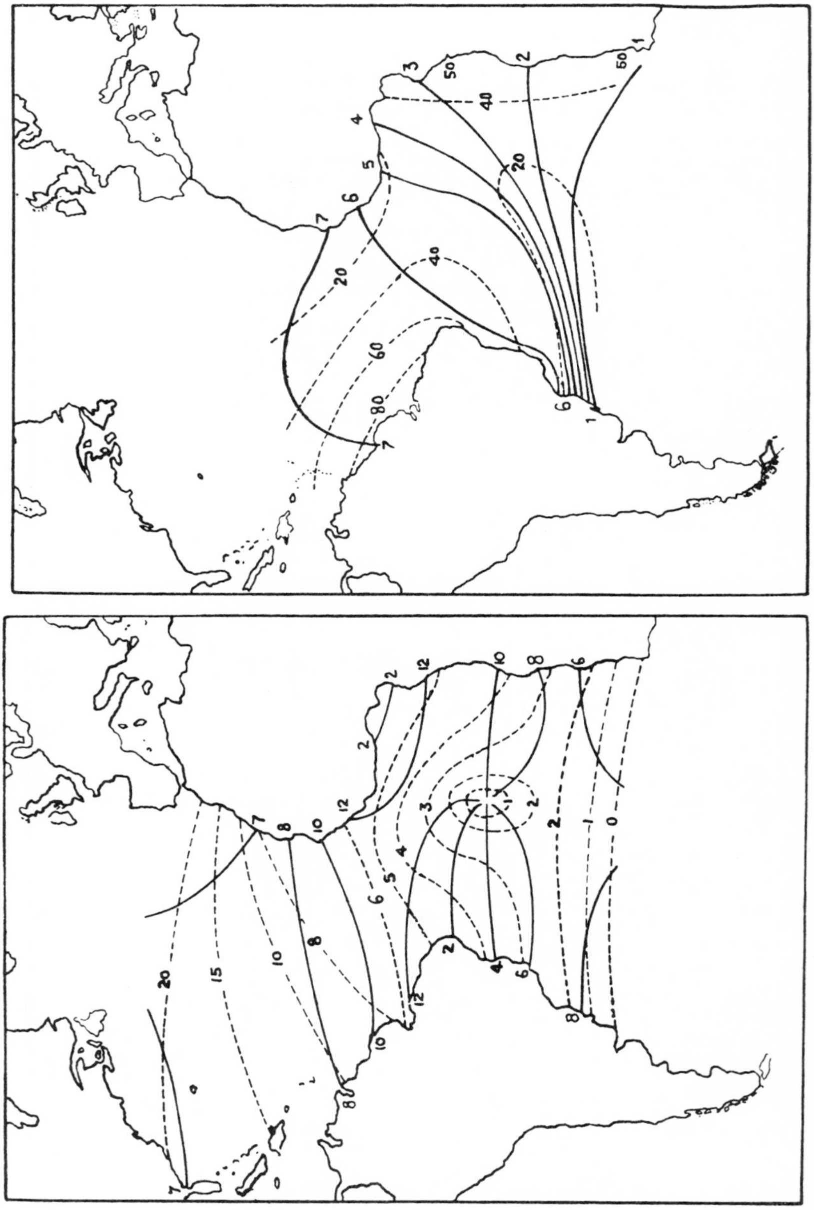

FIGURE 3.4.7. Left: independent (forced) M_2 tide; right, synthesis resulting from forced tide and free waves. Full lines: Greenwich lunar hour of high water. Broken lines: amplitude in centimeters.

REFERENCES

Julius Bartels 1957, *Gezeitenkräfte*, Handbuch der Physik, vol. 48, Springer-Verlag, Berlin, 1957, pp. 734–774.

Günter Dietrich 1963, *General oceanography*, Interscience, New York, 1963.

Harold Jeffreys 1970, *The Earth*, 5th ed., Cambridge Univ. Press, 1970.

C. L. Pekeris and Y. Accad 1969, *Solution of Laplace's equations for the M_2 tide in the world oceans*, Philos. Trans. Roy. Soc. London Ser. A **265** (1969), 413–436.

J. Proudman 1944, *The tides of the Atlantic Ocean*, Mon. Not. Roy. Astron. Soc. **104** (1944), 244–256.

——— 1953, *Dynamical oceanography*, Methuen, London, 1953.

Paul Schureman 1940, *Manual of harmonic analysis and prediction of tides*, U.S. Dept. of Commerce, Coast and Geodetic Survey, Special Publ., #98, U.S. Government Printing Office, Washington, D.C., 1940; reprint, 1958.

H. U. Sverdrup 1926, *Dynamics of tides on the North Siberian Shelf*, Geofys. Publ. **4** (1926), no. 5, 75 pp.

UNIVERSITY OF CHICAGO

Atmospheric Tides

R. Lindzen

I. Introduction, History, and Data

I.1. **Introduction.** Atmospheric tides refer to those oscillations in any atmospheric field whose periods are integral fractions of either a lunar or a solar day. The bulk of our concern, moreover, will be restricted to global scale, migrating tides. By migrating tides we mean tides which depend on local time; i.e., tides which follow the apparent motion of the sun or moon.

Atmospheric tides differ from sea tides in many important respects, most of which will become evident in the course of these lectures. Among the differences is the fact that atmospheric tides are excited not only by the tidal gravitational potential of the sun and moon but also (and as it turns out to a larger extent) by the daily variations in solar heating. Also, the atmosphere is approximately a spherical fluid shell; there are no coastal boundaries to worry about. Finally, the atmosphere responds to tidal or thermotidal excitation by means of internal gravity waves rather than by the barotropic surface waves of the sea.

There are a number of reasons for studying atmospheric tides. To begin with we know precisely their frequencies and zonal wavenumbers, and somewhat less precisely the magnitude and distribution of their excitation. We may, therefore, calculate the atmospheric response and isolate that response in the data. In other words they are a good vehicle for checking theories of internal waves in the atmosphere. Another reason for studying atmospheric tides is that they form a substantial part of the total meteorology of the upper atmosphere—at least above 50 km. However, the historical reasons for studying atmospheric tides were primarily intellectual. Indeed, both observations in the upper atmosphere and the recognition that atmospheric tides are internal gravity waves are products primarily of the last decade.

AMS 1970 *subject classifications.* Primary 86A10, 86A35, 76C15, 35A35, 86–02; Secondary 80A20.

I.2. **History.** Tidal oscillations in the atmosphere (apart from sea breezes which marginally fit our definition) were not recognized until the development of the barometer by Torricelli (ca. 1643). By contrast sea tides were first reported before 320 B.C. by Pytheas of Marseilles who in a voyage of exploration circumnavigated Britain. In the course of this voyage he noticed the approximately twice daily progression of high and low water whose phase appeared to parallel the time of the moon's transit. It was early suggested that the tides were due to some force of attraction emanating from the moon, but it awaited Newton to explain why the lunar sea tides were semidiurnal. Newton realized that the lunar tidal potential would excite oscillations in the atmosphere as well as in the sea. However, he thought the atmospheric oscillations would be too small to be detected. Given the amount of data available in Northern Europe in the seventeenth century he was certainly correct. However, the observation of atmospheric tides in the tropics proved to be an easy matter.

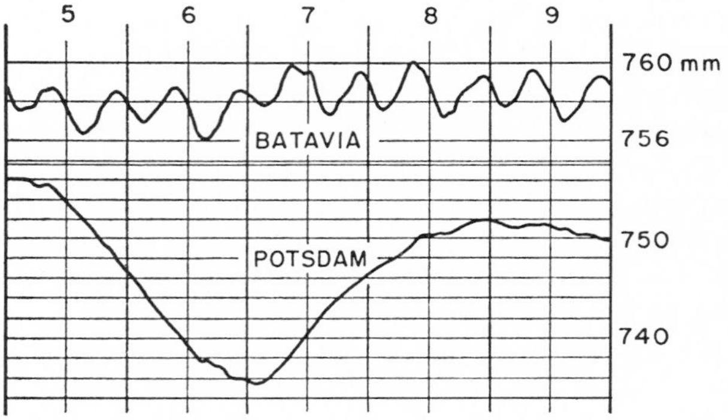

FIGURE 1. Barometric variations (on twofold different scales) at Batavia (6° S) and Potsdam (52° N) during November 1919. After Bartels [**29**].

In Figure 1 we see time records of surface pressure at two stations, Potsdam (52° N) and Batavia (now Jakarta, 6° S). The record at Potsdam is dominated by the moving highs and lows of midlatitude weather systems. At Batavia, however, an extraordinarily regular semidiurnal oscillation is evident. However, in contrast to sea tides, the period is one half of a solar day. Laplace was already aware of this observational fact. Since the lunar tidal potential is about twice as large as that due to the sun, Laplace quite properly concluded that the atmospheric tide was due to the thermal action of the sun, but he saw little hope of constructing a theory of thermally excited oscillations [**1**].

Lord Kelvin in the 1870's collected surface pressure data from thirty locations and harmonically analyzed this data for the diurnal (24 hr.), semidiurnal (12 hr.) and terdiurnal (8 hr.) components. In agreement with earlier findings he found that the oscillation was solar and that the semidiurnal component was substantially stronger than the diurnal component. In reporting these results Kelvin [**2**] concurred with Laplace that the atmospheric tides are thermally excited. He pointed out, however, that unlike gravitational excitation, thermal excitation will be primarily diurnal rather than semidiurnal. To explain the dominance of the semidiurnal surface pressure oscillation Kelvin suggested that the atmosphere had a suitable free oscillation with a period very near to 12 hrs. Kelvin's resonance hypothesis dominated thinking in this field for nearly sixty years. Rayleigh [**3**] and Margules [**4**] began the earliest searches for an atmospheric resonance. Margules showed that the atmosphere would be resonant to semidiurnal forcing if the atmosphere behaved like a homogeneous fluid with a depth of 7.85 km. This is known as the *atmosphere's equivalent depth*. Its physical and mathematical meaning will be explained when we turn to the theory of atmospheric tides. For the moment it suffices to say that the equivalent depth of an atmosphere depends on the thermodynamic state of the atmosphere and most significantly upon the mean thermal structure of the atmosphere. Both Rayleigh and Margules made rather crude and unrealistic assumptions concerning these matters which made their results uncertain. Nevertheless, they each concluded that resonance was a real possibility. Lamb [**5**] investigated the matter in greater detail. He found that for either isothermal atmospheres wherein density variations occur isothermally or for an atmosphere in adiabatic equilibrium, the equivalent depth of the atmosphere will be very nearly resonant. In addition, Lamb showed that in an atmosphere where the mean temperature varied linearly with height but not adiabatically there would be an infinite number of equivalent depths—thus greatly enhancing the possibilities for resonance. This particular finding was essentially forgotten for twenty years. Finally, Lamb suggested the possibility that the solar semidiurnal atmospheric tide might indeed be gravitationally excited since that would require a resonance magnification of about 70 which, in turn, would require that the resonant periods be within 2 minutes of 12 hrs. The lunar period would be 12 hr. 26 min. Arguing against this was, as Lamb noted, the intrinsic unlikelihood of such close agreement in period, and the fact that the observed surface pressure oscillations phase led rather than lagged the transits of the sun. Chapman [**6**] argued that the resonance theory might still be correct.

The existence of a thermal excitation comparable in magnitude to the gravitational excitation could explain the phase. In 1932 G. I. Taylor [**7**] put forth a rather significant criticism of the resonance theory. If the atmosphere had an equivalent depth h, then atmospheric disturbances excited by explosions, earthquakes, etc. should travel at a speed $(gh)^{1/2}$. Now data from the Krakatoa eruption of 1883 showed that the atmospheric pulse traveled at a speed of 319 m/s corresponding to $h = 10.4$ km —a value too far from 7.85 km to produce resonance. Later (1936), Taylor [**8**] reestablished Lamb's result (though more rigorously) that the atmosphere might have several equivalent depths—thus allowing some remaining hope for the, by now much modified, Kelvin resonance hypothesis. This hope received an immense boost from the work of Pekeris [**9**]. Pekeris investigated the equivalent depths of a variety of atmospheres with relatively complicated thermal structures and found that for an atmosphere whose temperature increased above the tropopause (ca. 12 km) to a high value (350°K) near 50 km and then decreased upwards to a low value, there existed a second equivalent depth close to 7.9 km. More convincing was the fact that the assumed temperature profile agreed excellently with a profile proposed by Martyn and Pulley [**10**] on the basis of then recent meteor and anomalous sound observations. It was almost as though Pekeris had adduced the atmosphere's thermal structure from tidal data at the earth's surface. His results, moreover, appeared to explain other observations of ionospheric and geomagnetic tidal variations. The vindication of the resonance theory seemed well nigh complete. Pekeris countered Taylor's early criticism by showing that a low level disturbance would primarily excite the faster mode associated with $h = 10.4$ km. A reexamination of the Krakatoa evidence by Pekeris even showed some evidence for the existence of the slower mode for which $h = 7.9$ km.

For the next fifteen years most research on this subject was devoted to the refinement and interpretation of Pekeris's work. This work is comprehensively reviewed by Wilkes [**11**]. Unfortunately, in the aftermath of World War II numerous rocket probings of atmospheric temperature were made and these showed a different temperature structure from that proposed by Martyn and Pulley. In particular the temperature maximum at 50 km was much cooler (about 280°K instead of 350°K). In addition the temperature decline above 50 km ended at 80 km above which the temperature again increases reaching very high values (600°K—1400°K) above 150 km [**12**]. For the new temperature profiles, the atmosphere no longer had a second equivalent depth and the magnification of the solar semidiurnal tide was no longer sufficient to account for the observed semidiurnal tide on the basis of any realistic combination of gravitational excitation, and excitation due to the upward diffusion of the daily variation of surface temperature (Jacchia and Kopal [**13**]).

With the demise of the resonance theory, the search began for additional thermal sources. Although most of the sun's radiation is absorbed by the earth's surface, about 10% is absorbed by the atmosphere directly and this appeared a likely source of excitation. Siebert [**14**] investigated the effectiveness of insolation absorption by water vapor in the troposphere and found that it could account for one third the observed semidiurnal surface pressure oscillation. This was far more than could be accounted for by gravitational excitation or surface heating. Siebert also investigated the effectiveness of insolation absorption by ozone in the mesosphere. He found its effect to be relatively small. As it turns out this last conclusion followed primarily from Siebert's use of an exceedingly unrealistic basic temperature distribution for the atmosphere above the lower troposphere. Butler and Small [**15**] corrected this error and showed that ozone absorption indeed accounted for the remaining two thirds of the surface semidiurnal oscillation.[1] At this point one was forced to return to Kelvin's original question: Why wasn't the diurnal oscillation stronger than the semidiurnal? The situation had, moreover, become more complicated. Data above the ground up to about 100 km showed that above the ground the diurnal oscillations were as strong and often stronger than semidiurnal oscillations. Lindzen [**16**] carried out theoretical calculations for the diurnal tide which provided satisfactory answers for the observational facts. Over half the globe (polewards of $\pm 30°$ latitude) 24 hours is longer than the local pendulum day and under these circumstances a twenty-four hour oscillation is incapable of propagating vertically. Because of this, it turns out that about 80% of the diurnal excitation goes into physically trapped modes which cannot propagate disturbances aloft to the ground. The atmospheric response to these modes in the neighborhood of the excitation is, however, substantial. (These trapped diurnal modes were discovered independently by Lindzen [**17**] and Kato [**18**].) In addition there exist (primarily within $\pm 30°$ latitude) diurnal modes which propagate vertically. However, as one could deduce from the dispersive properties of internal gravity waves, the long period and the restricted latitude scale of these waves causes them to have relatively short vertical wavelengths (25 km and less). They are, therefore, subject to some destructive interference effects. While Butler and Small suggested that this could explain the relatively small amplitude of the diurnal tide, subsequent calculations show that

[1] The problem of the semidiurnal oscillation at the ground is not yet completely settled. There remains a discrepancy of one half hour in phase between theory and observation.

this effect would be inadequate. What really proves to be important is that the propagating modes receive only 20% of the excitation.

After almost a century, Kelvin's question seems satisfactorily answered. However, as we shall see, numerous other problems remain in atmospheric tides.

I have devoted this much time to a somewhat sketchy history of the subject because I feel it is distinctly relevent to the subject of this seminar, "mathematical problems in the geophysical sciences." At each stage in the history of this subject there were equations to be solved and, indeed, they were tackled by some very eminent applied mathematicians. But rarely was the mere solution of an equation an important contribution in itself. Any geophysical system tends to be so complicated that the solution of equations is seldom a very sure approach and never a complete approach. Mathematical results must always be understood in terms of physical mechanism and related to observations. In dealing with atmospheric tides the physical problem of determining sources of excitation was clearly as important as the mathematical problem of calculating the atmosphere's response to the excitation. The mathematics was essential, but its role must be viewed in perspective.

I.3. **Data.** Before proceeding to the mathematical theory of atmospheric tides it is advisable to present a description of the phenomena about which we propose to theorize. In the short time allowed us we can hardly hope to do justice to the presentation of the data. Let it suffice to say that at many stages our observational picture is based on inadequate data, in almost all cases the analyses of data required the extrication of Fourier components from noisy data, and in some instances even the observational instruments have introduced uncertainties. Details of these matters may be found in Chapman and Lindzen [**28**].

Until recently almost all data analyses for atmospheric tides were based on surface pressure data. Although tidal oscillations in surface pressure are generally small, at quite a few stations we have as much as 50–100 years of hourly or bi-hourly data. As a result, even today, our best tidal data are for surface pressure.

Figure 2 shows the amplitude and phase of the solar semidiurnal oscillation over the globe; it was prepared by Haurwitz [**19**] on the basis of data from 296 stations. The phase over most of the globe is relatively constant, implying the dominance of the migrating semidiurnal tide, but other components are found as well (the most significant of which is the semidiurnal standing oscillation).

If we let $t =$ local time nondimensionalized by the solar day, then according to Haurwitz the solar semidiurnal tide is well represented by

FIGURE 2. World maps showing equilines of (below) the amplitude (s_2, unit 10^{-2} mb) and (above) the phase (σ_2) of $S_2(p)$, relative to local mean time. After Haurwitz [19].

$$S_2(p) =$$

$$1.16\ \text{mb} \sin^3\theta \sin(2t + 158°) + 0.085\ \text{mb}\ P_2(\theta) \sin(2t_u + 118°) \tag{1.1}$$

where θ = colatitude, t_u = Greenwich time and $P_2(\theta) = \frac{1}{2}(3\cos^2\theta - 1)$. One of the remarkable features of $S_2(p)$ is the fact that it hardly varies with season. This can be seen from Figure 4.

The situation is more difficult for $S_1(p)$. It varies with season, it is weak, and it is strongly polluted by nonmigrating diurnal oscillations (see Figure 5). There are values of s with amplitudes as large as $\frac{1}{4}$ of that pertaining to $s = 1$. For $S_2(p)$, $s = 2$ was twenty times as large as its competitor. Moreover large values of s, being associated with small scale (large gradients), produce larger winds for a given amplitude of pressure oscillation than $s = 1$. We will return to this later. According to Haurwitz [**20**], $S_1(p)$ is roughly representable as follows:

$$S_1^1(p) = 593\mu\text{b} \sin^3\theta \sin(t + 12°). \tag{1.2}$$

Data have also been analyzed for small terdiurnal and higher harmonics. Even $L_2(p)$ has been isolated.

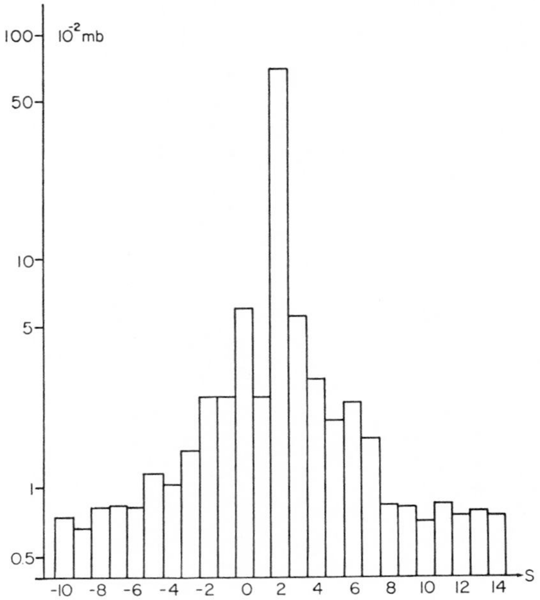

FIGURE 3. The amplitudes (on a logarithmic scale, and averaged over the latitudes 80° N to 70° S) of the semidiurnal pressure waves, parts of $S_2(p)$, of the type $\gamma_s \sin(2t_u + s\phi + \sigma_s)$, where t_u signifies universal mean solar time. After Kertz [**35**].

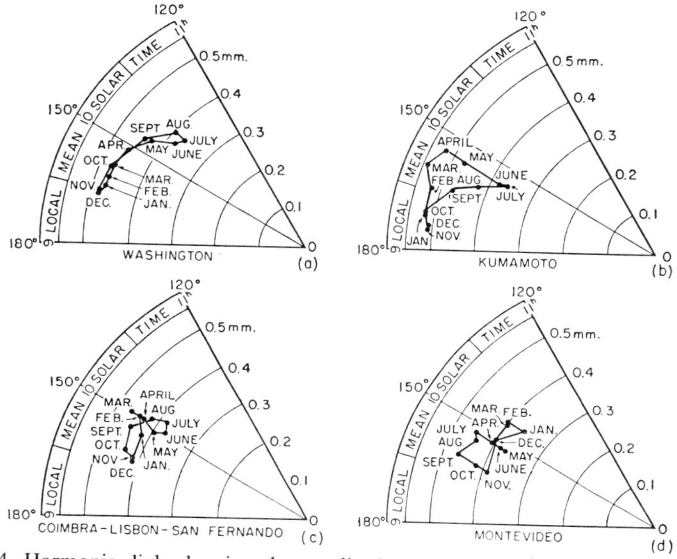

FIGURE 4. Harmonic dials showing the amplitude and phase of $S_2(p)$ for each calendar month for four widely spaced stations in middle latitudes, (a) Washington, D.C., (b) Kumamoto, (c) mean of Coimbra, Lisbon and San Fernando, (d) Montevideo (Uruguay). After Lindzen and Chapman [**28**].

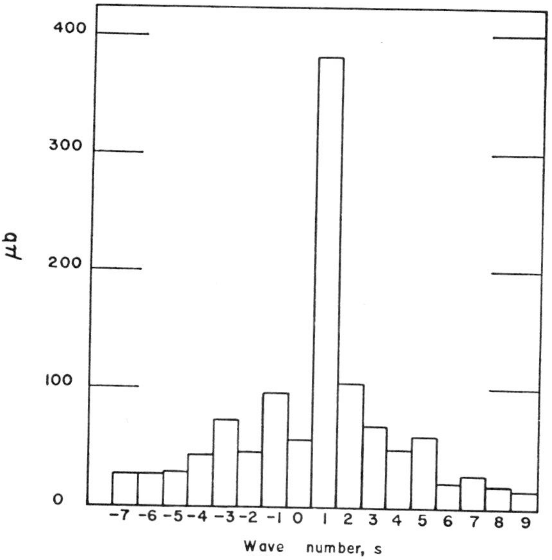

FIGURE 5. The amplitudes (on a logarithmic scale, and averaged over the latitudes from the north pole to 60° S) of the diurnal pressure waves, parts of $S_1(p)$, of the type $\gamma_s \sin(t_u + s\phi + \sigma_s)$. After Haurwitz [**20**].

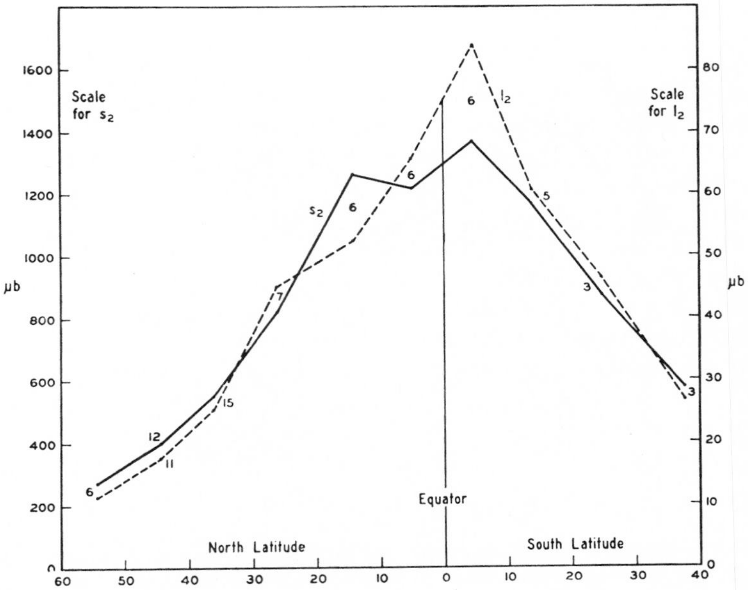

FIGURE 6. Mean values of the amplitudes s_2 (full line) and l_2 (broken line) of the annual mean solar and lunar semidiurnal air-tides in barometric pressure, $S_2(p)$ and $L_2(p)$, for 10° belts of latitude. The numbers beside each point show from how many stations that point was determined. After Chapman and Westfold [**31**].

As we see in Figure 6, the amplitude of $L_2(p)$ is about $\frac{1}{20}$ of $S_2(p)$. $L_2(p)$ also has a peculiar seasonal variation as can be seen in Figure 7. The seasonal variation occurs with both the northern and southern hemispheres in phase.

Data above the surface are rarer and less accurate. But, some are available. At a few stations there are sufficiently frequent balloon ascents to permit tidal analyses. The results of one such analysis are shown in Table I.

The results are sufficient to at least establish order of magnitude. In the troposphere horizontal wind oscillations have amplitudes ~ 10 cm/sec. In the stratosphere amplitudes are about 50 cm/sec. However, data at a single station does not permit separating migrating tides from other components. At many stations balloon soundings are made twice a day, 12 hours apart. The soundings are made, moreover, simultaneously at all stations (i.e. at the same universal time). Thus if one subtracts the average of measurements taken at 1200 GMT from the average of measurements

TABLE I.

Diurnal and semidiurnal variations of the eastward and northward components of the wind at Terceira, Azores. Annual mean values of the amplitude, s, in cm/sec, and phase, σ, in degrees. P.E. is the radius of the probable error circle of the annual means. Time of maximum wind is related to σ by the expression $t_{max} = (450° - \alpha)/15n$, where $n = 1$ for diurnal variations, and $n = 2$ for semidiurnal variations. After Harris, Finger and Teweles [**33**].

Mean pressure, mb	Variation of eastward wind component						Variation of northward wind component					
	Diurnal			Semidiurnal			Diurnal			Semidiurnal		
	s_1	σ_1	P.E.	s_2	σ_2	P.E.	s_1	σ_1	P.E.	s_2	σ_2	P.E.
Ground	2	75°	6	8	324°	7	7	341°	8	21	52°	6
1000	4	115	9	12	298	6	6	337	6	17	52	6
950	2	154	8	14	317	8	21	271	11	22	35	6
900	2	248	3	19	292	6	32	272	9	23	31	8
850	8	257	11	14	266	7	25	256	13	23	14	8
800	4	322	10	22	313	8	18	265	11	31	359	8
750	22	145	14	22	278	10	16	251	15	29	12	14
700	5	304	11	18	304	8	9	4	11	22	33	9
650	4	255	11	20	292	5	13	318	9	23	50	8
600	8	63	12	20	272	6	12	281	11	16	1	7
550	20	159	10	31	327	9	18	249	13	16	7	10
500	20	124	9	25	276	10	15	317	15	15	63	9
450	17	76	11	26	295	10	22	295	16	20	346	10
400	18	1	8	28	291	10	14	342	17	10	317	12
350	19	258	12	42	291	9	13	257	19	16	319	9
300	24	193	18	51	292	14	8	247	21	14	4	13
250	52	177	16	26	245	12	52	267	17	8	285	12
200	56	153	15	46	267	11	18	238	14	15	338	12
175	13	164	13	37	278	14	34	241	13	25	339	10
150	16	186	11	39	300	11	18	185	14	52	18	9
125	14	127	8	29	242	9	5	241	7	23	4	8
100	27	112	10	55	280	10	34	153	11	28	19	6
80	31	111	11	37	280	9	21	194	9	28	21	7
60	34	109	11	36	262	8	40	196	9	27	5	10
50	19	132	9	41	256	8	34	236	10	42	356	7
40	6	96	11	44	263	12	27	235	11	49	4	8
30	23	181	11	67	280	13	21	221	10	65	17	11
20	30	147	12	62	295	12	64	235	13	60	36	14
15	25	114	23	91	303	20	66	238	14	61	30	10
10	–	–	–	–	–	–	–	–	–	–	–	–

taken at 0000 GMT one should obtain a fair approximation to the diurnal component of the flow field at 0000 GMT. If the migrating tide dominates the wind oscillation then we should see a clear zonal wave number 1 pattern. The results are shown in Figures 8, 9 and 10 [**21**]. We see that the diurnal flow field at 700 mb is dominated by gyres associated with relatively small orographic features. At 60 mb orographic gyres remain, but they are associated with large features like the Pacific Ocean. By 15 mb we have a simple wavenumber 1 pattern. This suggests that orographic effects have died out by 30 km and that diurnal oscillations above this level will be representative of the migrating tide.

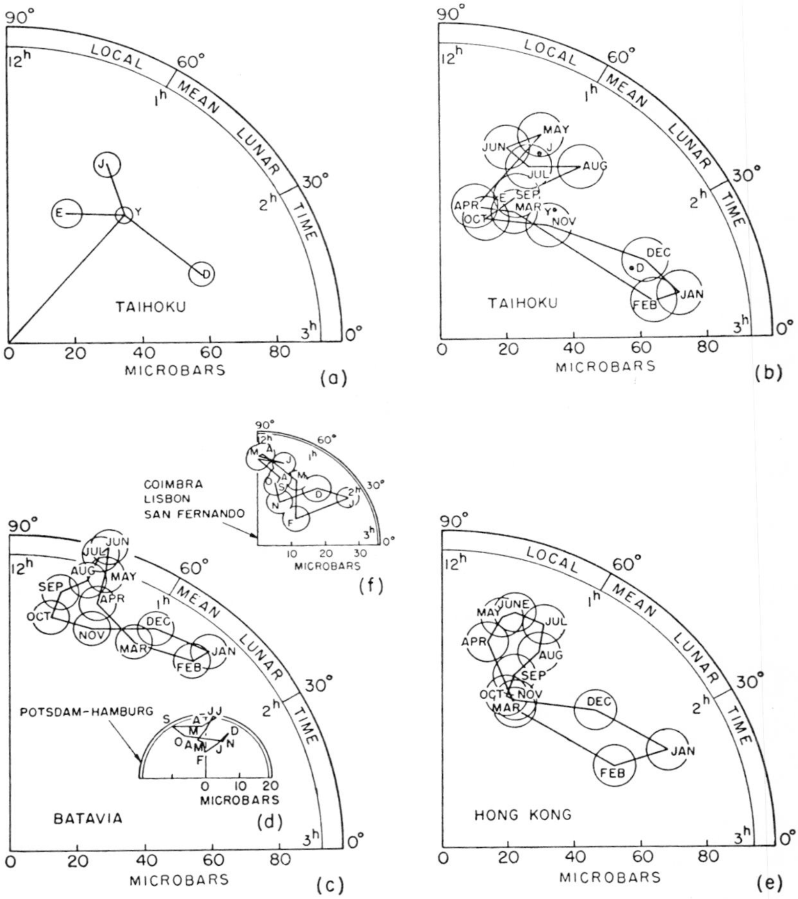

FIGURE 7. Harmonic dials, with probable error circles, indicating the changes of the lunar semidiurnal air-tide in barometric pressure in the course of a year, (a) Annual (y) and four-monthly seasonal (j, e, d) determinations for Tahoku, Formosa (now Taiwan, Taipei) (1897–1932). Also five sets of twelve monthly-mean dial points. See Table 2L.2 [**28**] for particulars of the seven stations. After Lindzen and Chapman [**28**].

In the region between 30 and 60 km most of our data come from meteorological rocket soundings. These are comparatively infrequent, and the method of analysis becomes, a priori, a serious problem. However, it turns out that results of different analyses appear to be compatible (at least for the diurnal component) because tidal winds at these heights

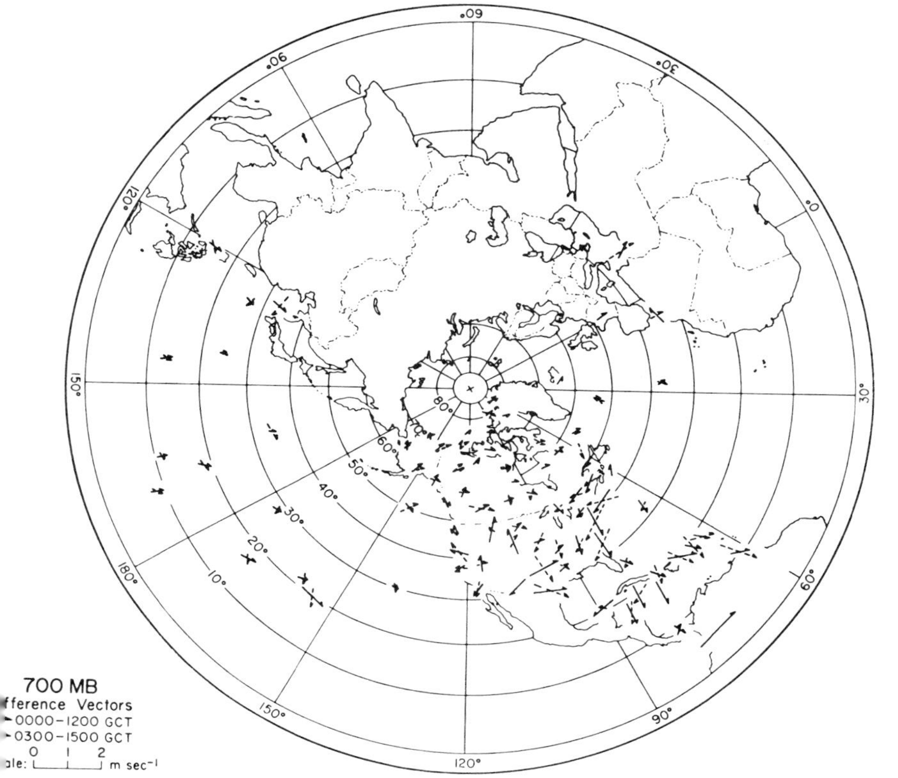

FIGURE 8. Annual average wind differences 0000–1200 GMT (solid) and 0300–1500 GMT (dashed) at 700 mb, plotted in vector form. The length scale is given in the figure. After Wallace and Hartranft [**21**].

are already a very significant part of the total wind (at least in the north-south direction). This is seen in Figure 11 where we show the southerly wind as measured over a period of 51 hours at White Sands, New Mexico. Analyses of tidal winds at various latitudes are now available. Figure 12 shows the phase and amplitude of the semidiurnal oscillation at about 30° N. Below 50 km, the results appear quite uncertain [**22**].

In Figures 13 and 14 we see the diurnal component at 61° N and at 20° N respectively. Amplitudes are on the order of 10 m/s at 60 km, but phase at 20° N is more variable than at 61° N [**24**].

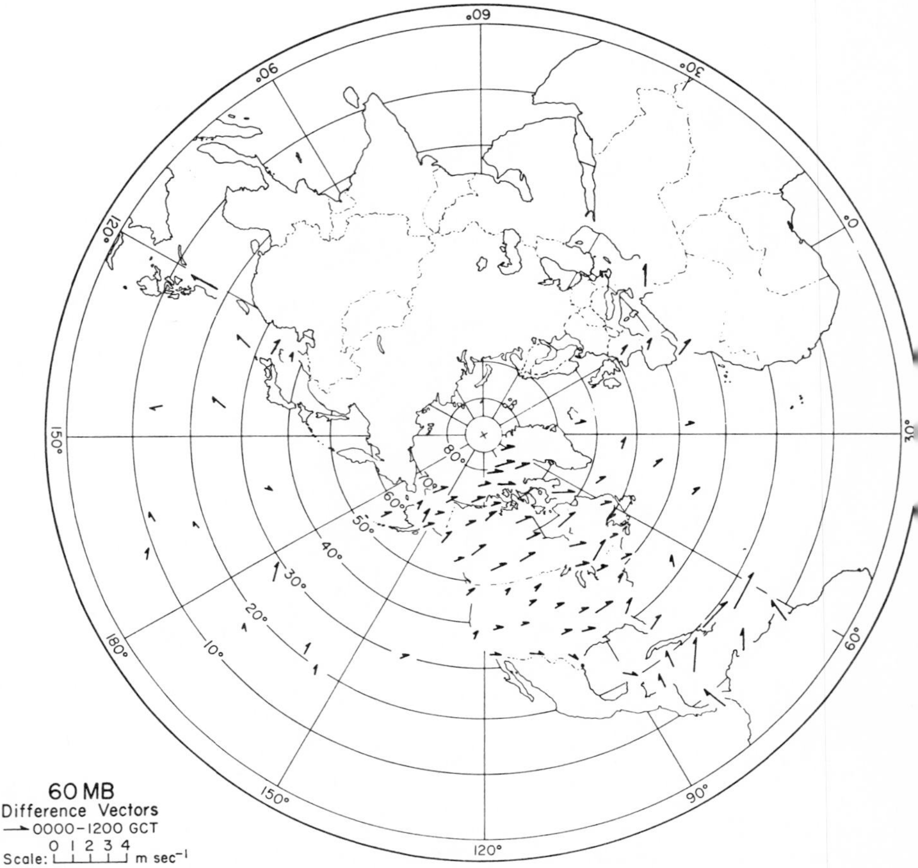

FIGURE 9. Annual average wind differences 0000–1200 GMT at 60 mb plotted in vector form. The length scale is given in the figure. After Wallace and Hartranft [**21**].

Between 60 km and 80 km there are too little data for tidal analyses. Between 80 and 105 km, there is a growing body of data from the observation of ionized meteor trails by doppler radar. Most of this data is for vertically averaged wind over the whole range 80–105 km. Some such data for Jodrell Bank [**24**] (53° N) and Adelaide [**25**] (35° S) are shown in Figures 15 and 16. Typical magnitudes are around 20 m/s. All quantities are subject to large seasonal fluctuations and error circles. At Adelaide diurnal oscillations predominate while at Jodrell Bank semidiurnal

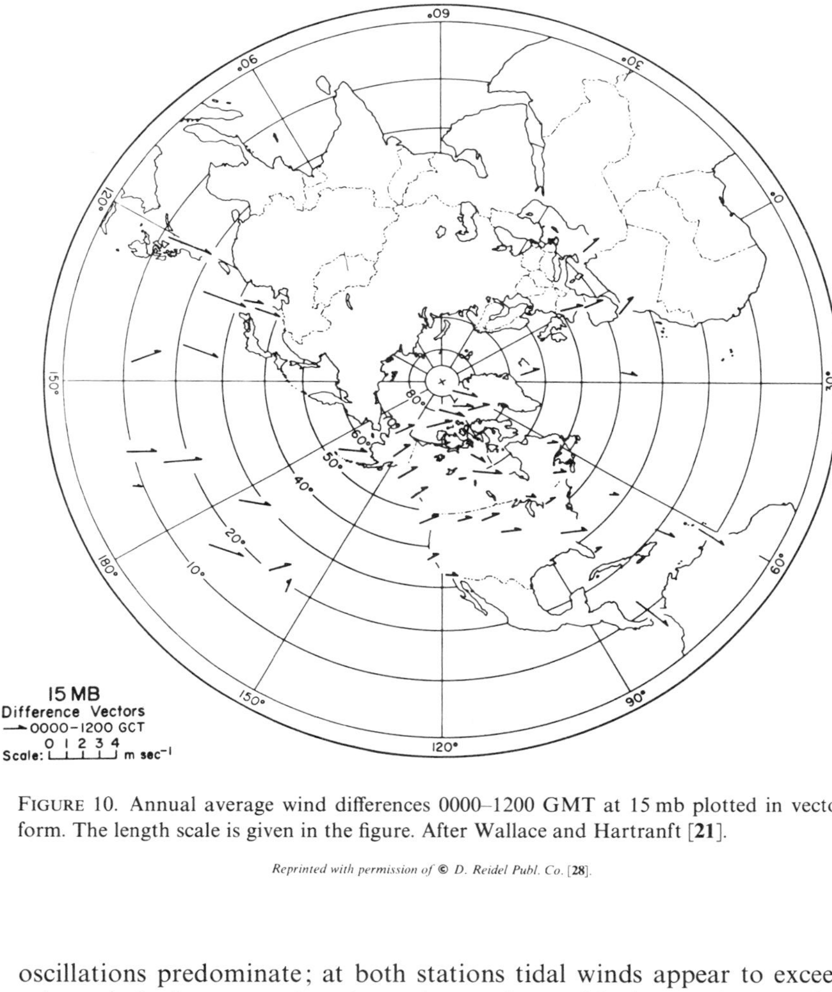

FIGURE 10. Annual average wind differences 0000–1200 GMT at 15 mb plotted in vector form. The length scale is given in the figure. After Wallace and Hartranft [**21**].

Reprinted with permission of © D. Reidel Publ. Co. [**28**].

oscillations predominate; at both stations tidal winds appear to exceed other winds. In comparing these observations with theory, due caution must be exercised in interpreting averages over such great depths.

Between 90 and 130 km (and higher) wind data can be obtained by visually tracking luminous vapor trails emitted from rockets. In most cases this is possible only in twilight at sunrise and sundown. Hines [**26**] used such data to form 12 hour wind differences which seemed likely to indicate the diurnal contribution to the total wind at dawn at Wallops Island (38° N). His result is shown in Figure 17.

There is an evident rotation of the wind vector with height characteristic of an internal wave with a vertical wavelength of about 20 km. Amplitude

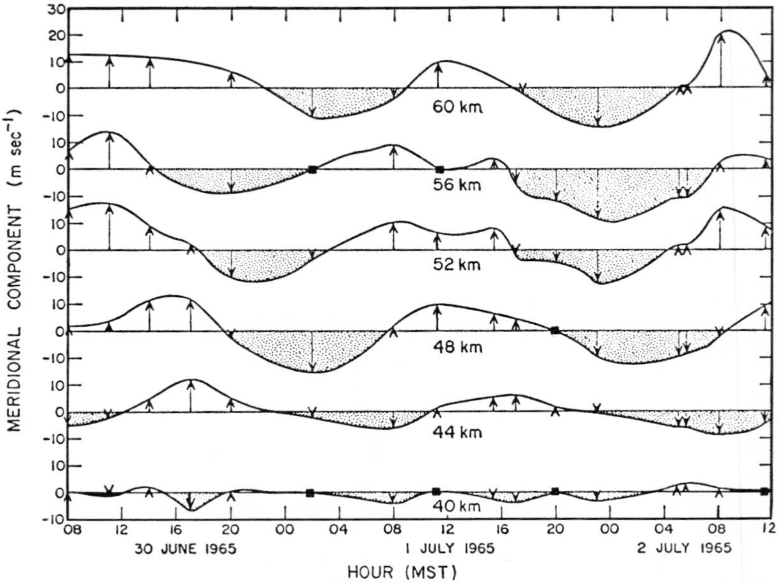

FIGURE 11. Meridional wind components u in m/sec. averaged over 4 km layers centered at 40, 44, 48, 52, 56 and 60 km. Positive values indicate a south to north flow. After Beyers, Miers and Reed [30].

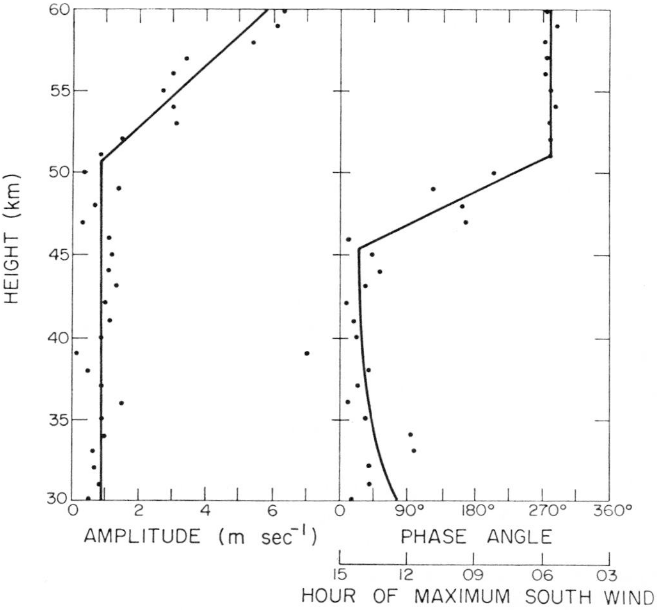

FIGURE 12. Phase and amplitude of the semidiurnal variation of the meridional wind component u at 30° N, based on data for White Sands (32.4 N) and Cape Kennedy (28.5 N). After Reed [22].

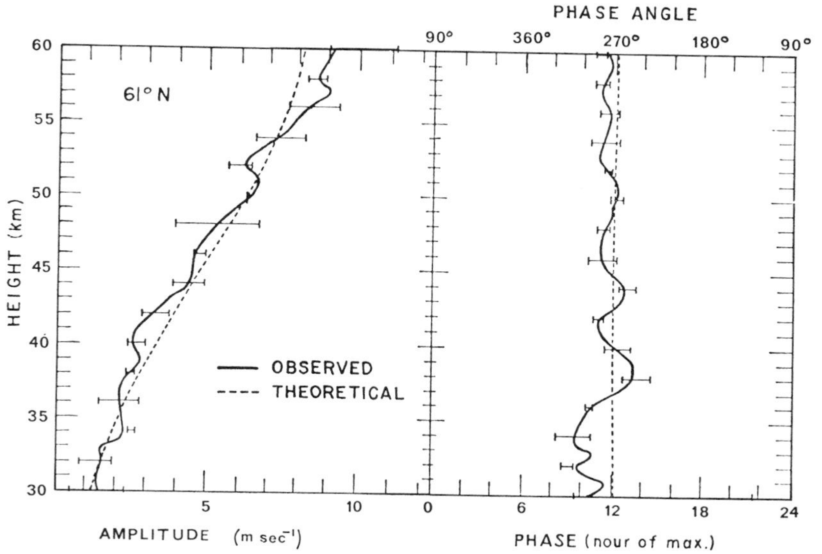

FIGURE 13. Amplitude and phase of the diurnal variation of the meridional wind component u at 61° N. Phase angle, in accordance with the usual convention, gives the degrees in advance of the origin (chosen as midnight) at which the sine curve crosses from − to +. After Reed, Oard and Sieminski [**23**].

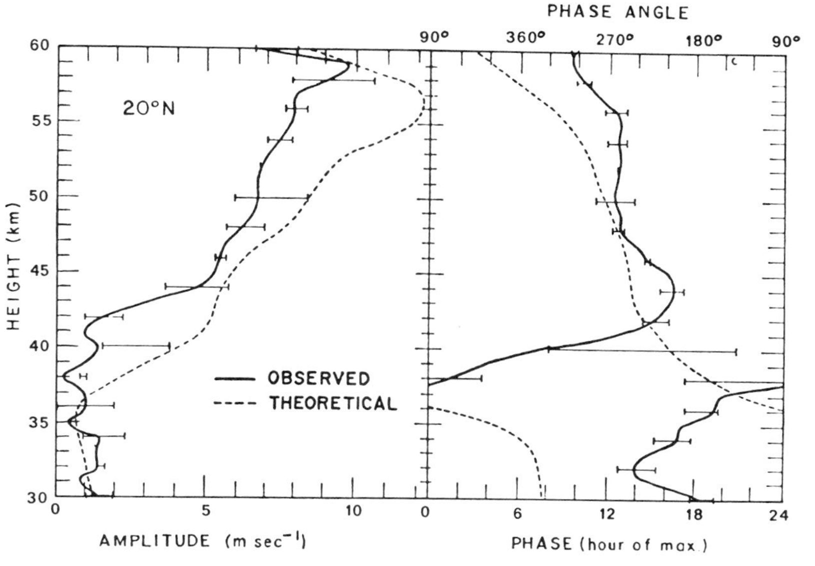

FIGURE 14. Amplitude and phase of the diurnal variation of the meridional wind component u at 20° N. After Reed, Oard and Sieminski [**23**].

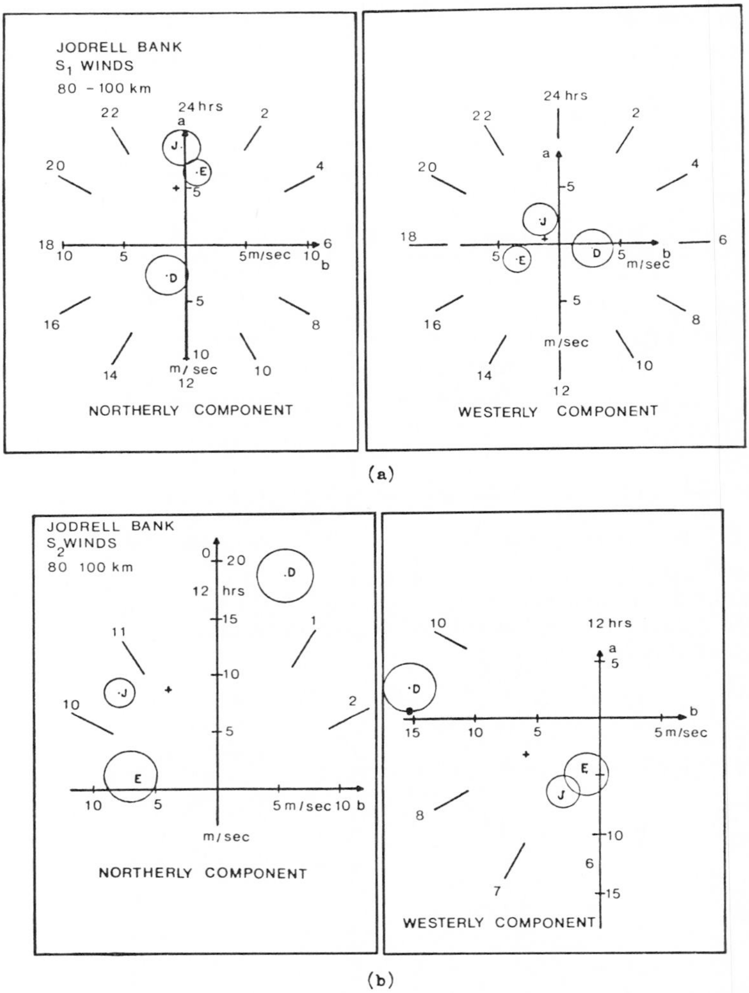

FIGURE 15. Harmonic dials for the mean northerly and westerly components of (a) the diurnal and (b) the semidiurnal wind variations at Jodrell Bank at 80–100 km. Crosses indicate annual mean values, dots seasonal values. Circles show probable errors of the seasonal means. From Haurwitz [**34**].

appears to grow with height up to 105 km and then decay. However, data from above 200 km (mostly from the analysis of satellite drag observations) shows an immense daily variation in the thermosphere [**27**]. This is seen in Figure 18 where the variation with time of the height of constant density surfaces is shown. Day-night variations in density of

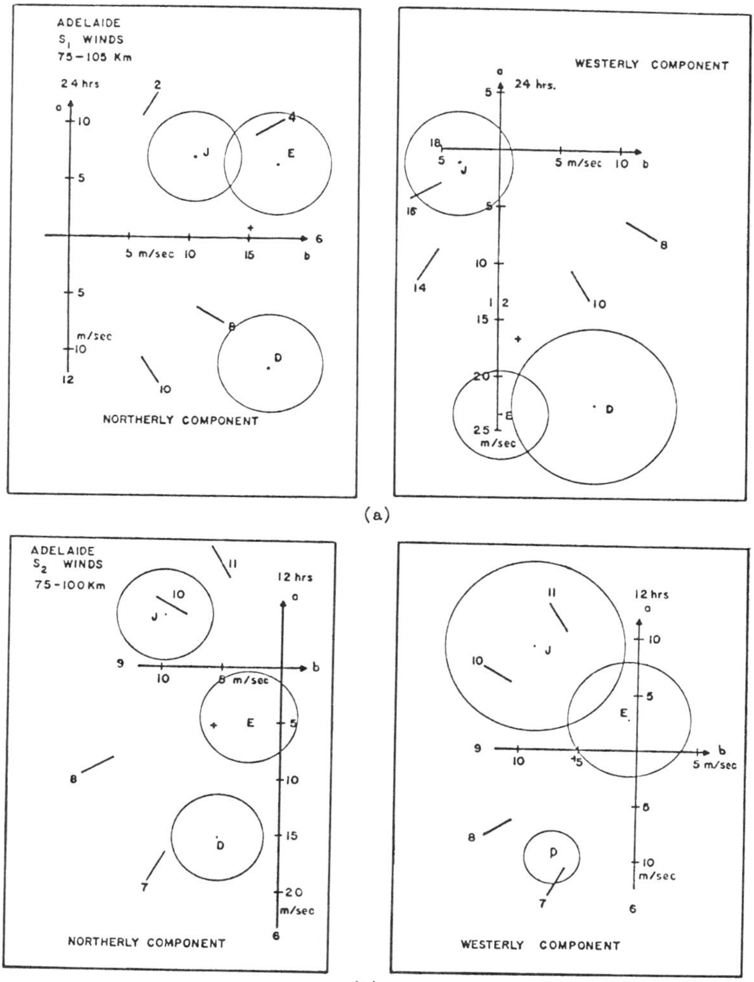

(b)

FIGURE 16. Harmonic dials for the mean northerly and westerly components of (a) the diurnal and (b) the semidiurnal wind variations at Adelaide, Australia, at 80–100 km. Crosses indicate annual mean values, dots seasonal values. Circles show probable errors of the seasonal means. From Haurwitz [**34**].

almost an order of magnitude are found at 600 km. There are problems with this observational technique and real variations could be much larger.

Having familiarized ourselves with atmospheric tides as they are actually observed we will proceed to their mathematical theory.

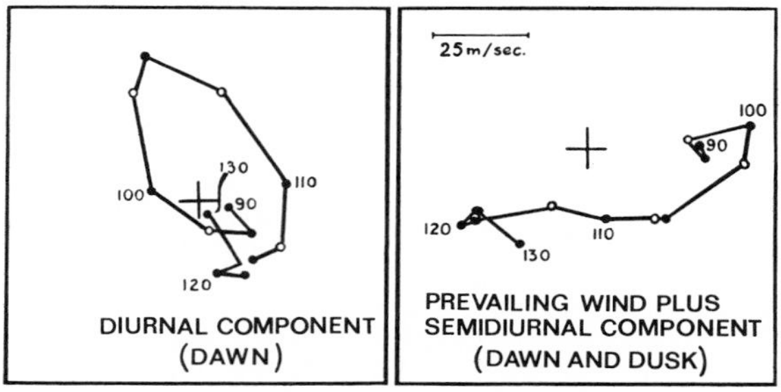

FIGURE 17. Vectograms showing the diurnal tide at dawn and the prevailing wind plus the semidiurnal tide at its dawndusk phase, as functions of height. After Hines [26].

Reprinted with permission of © D. Reidel Publ. Co. [28].

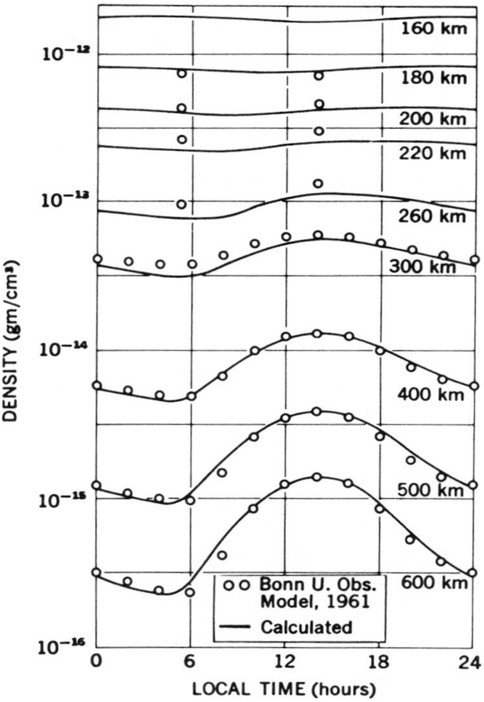

FIGURE 18. Diurnal variation of density for selected altitudes from 160 to 600 km. The solid curves give the values calculated with Table 1. The circles are densities taken from the Bonn University Observatory observational model of 1961.

Reprinted courtesy of NASA Goddard Space Flight Center [27].
Government endorsement not necessarily implied.

Appendix to Part I.

A. Representative thermal structure of the atmosphere and usual nomenclature.

B. Pressure vs. height for a typical atmosphere.

Height (km)	*Pressure* (mb)
0	1013.25
2	794.9
5	540.1
10	264.19
15	120.9
20	55.2
25	25.3
30	11.9
35	6.0

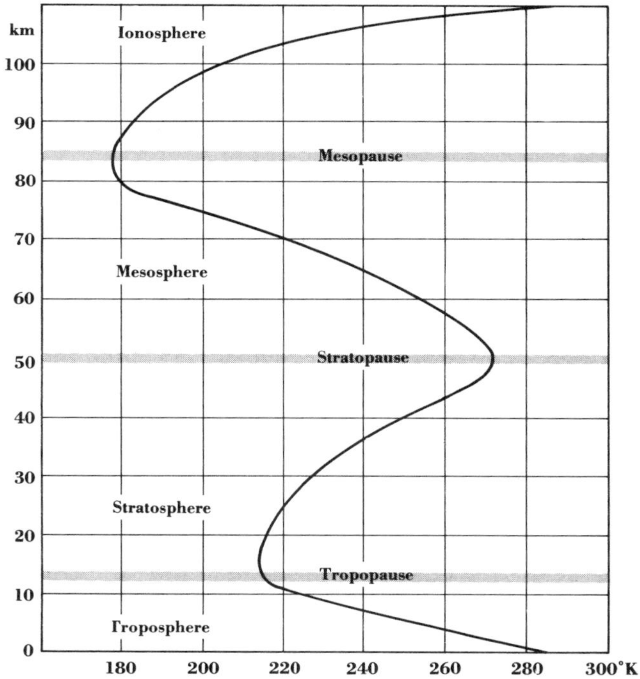

FIGURE A1. Principal features of the thermal structure of the atmosphere up to the lower part of the ionosphere.

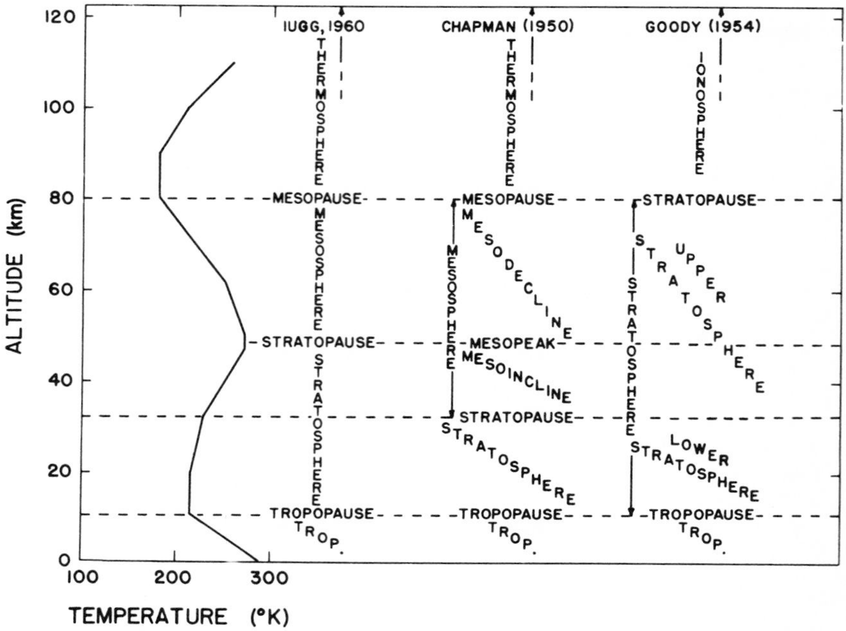

FIGURE A2. Vertical distribution of temperature up to 110 km according to the U.S. standard atmosphere, 1962, and common systems of nomenclature. The latter are discussed in the text.

Reprinted with permission of © Academic Press [Craig, 1965].

REFERENCES, PART I[2]

1. S. L. Laplace, *Mécanique céleste*, Duprat, Paris, 1799.

2. W. Thomson (Lord Kelvin), *On the thermodynamic acceleration of the Earth's rotation*, Proc. Roy. Soc. Edinburgh **11** (1882), 396–405.

3. Baron Rayleigh (J. W. Strutt), *On the vibrations of an atmosphere*, Philos. Mag. (5) **29** (1890), 173–180.

4. M. Margules, *Über die Schwingungen periodisch erwarmter Luft*, S.-B. Akad. Wiss. Wien, Abt. IIa **99** (1890), 204–227; *Luftbewegungen in einer rotierenden Sphäroidschale*, ibid. IIa **101** (1892), 597–626; **102** (1893), 11–56; **102** (1893), 1369–1421.

5. H. Lamb, *On atmospheric oscillations*, Proc. Roy. Soc. Ser. A **84** (1910), 551–572; *Hydrodynamics*, 4th ed., University Press, Cambridge, 1916.

6. S. Chapman, *The semi-diurnal oscillation of the atmosphere*, Quart. J. Roy. Meteorol. Soc. **50** (1924), 165–195.

7. G. I. Taylor, *The resonance theory of semidiurnal atmospheric oscillations*, Mem. Roy. Meteorol. Soc. **4** (1932), 41–52.

8. ———, *The oscillations of the atmosphere*, Proc. Roy. Soc. Ser. A **156** (1936), 318–326.

9. C. L. Pekeris, *Atmospheric oscillations*, Proc. Roy. Soc. Ser. A **158** (1937), 650–671.

10. D. F. Martyn and O. O. Pulley, *The temperatures and constituents of the upper atmosphere*, Proc. Roy. Soc. Ser. A **154** (1936), 455–486.

[2] A review of all subjects covered in this part may be found in [**28**]. Complete references may also be found in this monograph.

11. M. V. Wilkes, *Oscillations of the Earth's atmosphere*, Cambridge Univ. Press, New York, 1949. MR **11**, 756.

12. See Appendix A.

13. L. Jacchia and Z. Kopal, *Atmospheric oscillations and the temperature profile of the upper atmosphere*, J. Meteorol. **9** (1951), 13–23.

14. M. Siebert, *Zur theorie der thermischen Erregung gezeitenartiger Schwingungen der Erdatmosphäre*, Naturwissenschaften **41** (1954), 446; "Atmospheric tides" in *Advances in Geophysics*, vol. 7, Academic Press, New York, 1961, pp. 105–182.

15. S. T. Butler and K. A. Small, *The excitation of atmospheric oscillations*, Proc. Roy. Soc. Ser. A **274** (1963), 91–121.

16. R. S. Lindzen, *Thermally driven diurnal tide in the atmosphere*, Quart. J. Roy. Meteorol. Soc. **93** (1967), 18–42.

17. ———, *On the theory of the diurnal tide*, Mon. Wea. Rev. **94** (1966), 295–301.

18. S. Kato, *Diurnal atmospheric oscillation*, 1, *eigenvalues and Hough functions*, J. Geophys. Res. **71** (1966), 3201–3209.

19. B. Haurwitz, *The geographical distribution of the solar semidiurnal pressure oscillation*, Meteorol. Pap. **2(5)** (1956), New York University, New York.

20. ———, *The diurnal surface pressure oscillation*, Arch. Meteorol. Geophys. Bioklimatol. A **14** (1965), 361–369.

21. J. M. Wallace and F. R. Hartranft, *Diurnal wind variations; surface to* 30 km, Mon. Wea. Rev. **96** (1969), 446–455.

22. R. J. Reed, *Semidiurnal tidal motions between* 30 *and* 60 km, J. Atmospheric Sci. **24** (1967), 315–317.

23. R. J. Reed, M. J. Oard and Marya Sieminski, *A comparison of observed and theoretical diurnal tidal motions between* 30 *and* 60 km, Mon. Wea. Rev. **97** (1969), 456–459.

24. J. S. Greenhow and E. L. Neufeld, *Winds in the upper atmosphere*, Quart. J. Roy. Meteorol. Soc. **87** (1961), 472–489.

25. W. G. Elford, *A study of winds between* 80 *and* 100 km *in medium latitudes*, Planet. Sp. Sci. **1** (1959), 94–101.

26. C. O. Hines, *Diurnal tide in the upper atmosphere*, J. Geophys. Res. **71** (1966), 1453–1459.

27. I. Harris and W. Priester, *On the dynamical variation of the upper atmosphere*, J. Atmospheric Sci. **22** (1965), 3–10.

28. S. Chapman and R. S. Lindzen, *Atmospheric tides*, D. Reidel, Dordrecht, Holland; Gordon and Breach, New York, 1970.

29. J. Bartels, *Gezeitenschwingungen der Atmosphäre*, Handbuch der Experimentalphysik **25** (Geophysik **1**) (1928), 163–210.

30. N. J. Beyers, B. T. Miers and R. J. Reed, *Diurnal tidal motions near the stratopause during* 48 *hours at White Sands Missile Range*, J. Atmospheric Sci. **23** (1966), 325–333.

31. S. Chapman and K. C. Westfold, *A comparison of the annual mean solar and lunar atmospheric tides in barometric pressure as regards their world-wide distribution of amplitude and phase*, J. Atmos. Terr. Phys. **8** (1956), 1–23.

32. R. A. Craig, *The upper atmosphere*, Academic Press, New York, 1965.

33. M. F. Harris, F. G. Finger and S. Teweles, *Diurnal variations of wind, pressure, and temperature in the troposphere and stratosphere over the Azores*, J. Atmospheric Sci. **19** (1962), 136–149.

34. B. Haurwitz, *Tidal phenomena in the upper atmosphere*, W.M.O. Report #146 (1964), T. P. 69.

35. W. Kertz, *Components of the semidiurnal pressure oscillation*, Dept. of Meteor. and Ocean., Sci. Report #4, New York University, 1956.

II. Assumptions, Approximations and Equations

The theory of atmospheric tides consists both in the calculation of the atmospheric response to arbitrary gravitational and thermal forcing and in the specification of the sources of excitation. Historically, the framework for doing the former was developed first and we shall follow the same order in these lectures.

Our starting point in calculating the atmosphere's response to tidal (or thermotidal) excitation is the Navier-Stokes equations for a perfect gas in local thermodynamic equilibrium. This starting point already involves a substantial number of assumptions and approximations; but they are of the sort that experience has shown are almost completely untroublesome.

There are a number of assumptions which are also untroublesome, but which call for more justification:

(i) The gas constant for air is a constant. This is very nearly so up to 95 km, but by 200 km a 10% error is introduced.

(ii) The atmosphere is thin compared with the earth's radius. If we write the distance from the earth's center as $r = a + z$, where a = radius of solid earth, and z = distance from earth's surface, then we shall neglect terms in our equations of $O(z/a)$. It appears that for oscillations of planetary scales and tidal periods, this assumption also permits the neglect of the component of the earth's rotation vector parallel to the earth's surface (Phillips [**1**]).

(iii) Tidal oscillations are in hydrostatic equilibrium. This amounts to assuming vertical fluid accelerations will be much less than 10 m/sec^2 [**2**].

(iv) The ellipticity of the earth is ignored; the earth is assumed to be a sphere.

Although the above can be justified at some length, we will omit this in order to move on to those approximations which are most important in limiting the utility of conventional tidal theory. One of these is

(v) The earth's surface topography (land-sea distribution, mountain ranges, etc.) is ignored. The earth's surface is assumed to be a smooth rigid sphere where w (the vertical velocity) is zero. Related to this assumption is the assumption that the distribution of radiation absorbing gases is independent of longitude.

At this point our equations are the following:

$$\rho(\partial u/\partial t + \boldsymbol{u}\cdot\nabla u + wu/a - v^2 \cot\theta/a - 2\omega v\cos\theta) = -\frac{1}{a}\frac{\partial p}{\partial\theta} - \frac{\rho}{a}\frac{\partial\Omega}{\partial\theta} - F_\theta, \tag{1}$$

$$\rho(\partial v/\partial t + \boldsymbol{u}\cdot\nabla v + wv/a + uv\cot\theta/a + 2wu\cos\theta)$$

(2)
$$= -\frac{1}{a\sin\theta}\frac{\partial p}{\partial\varphi} - \frac{\rho}{a\sin\theta}\frac{\partial\Omega}{\partial\varphi} - F_\varphi,$$

(3)
$$\partial p/\partial z = -\rho(\partial\Omega/\partial z),$$

(4)
$$\partial\rho/\partial t + \boldsymbol{u}\cdot\nabla\rho + \rho\nabla\cdot\boldsymbol{u} = 0,$$

(5)
$$\rho c_v(dT/dt) = RT(d\rho/dt) + J - \mathscr{C},$$

(6)
$$p = \rho RT$$

where

$$\boldsymbol{u}\cdot\nabla = \frac{u}{a}\frac{\partial}{\partial\theta} + \frac{v}{a\sin\theta}\frac{\partial}{\partial\varphi} + w\frac{\partial}{\partial z},$$

$$\nabla\cdot\boldsymbol{u} = \frac{1}{a\sin\theta}\frac{\partial}{\partial\theta}(u\sin\theta) + \frac{1}{a\sin\theta}\frac{\partial v}{\partial\varphi} + \frac{\partial w}{\partial z},$$

$$d/dt = \partial/\partial t + \boldsymbol{u}\cdot\nabla,$$

and

θ = colatitude
φ = longitude
t = time
u = northerly velocity
v = westerly velocity
w = vertical velocity
T = temperature
p = pressure
ρ = density
Ω = gravitational potential
ω = earth's rotation rate
R = gas constant for air = 2.871×10^6 erg g^{-1} deg^{-1}
c_v = heat capacity at constant volume
F_θ = frictional force in θ direction
F_φ = frictional force in φ direction
J = thermotidal heating
$\mathscr{C}$ = heat diffusion and radiative cooling.

The form of F_θ, F_φ and $\mathscr{C}$ for molecular dissipation may be found in Goldstein [**3**]. Other dissipative processes (eddy diffusion, ion drag, infrared cooling) are described in Lindzen [**4**]. In general the expressions are extremely complicated and, in the cases of eddy diffusion and infrared cooling, quite uncertain. We next adopt the following:

(vi) F_θ, F_φ and $\mathscr{C}$ are neglected. It is a very complicated matter to systematically justify this assumption a priori, but we will check it a posteriori.

Two important simplifications remain.

(vii) Tidal oscillations are taken to be linearizable perturbations on a mean flow; i.e., if we write

$$\begin{Bmatrix} u \\ v \\ w \\ T \\ p \\ \rho \end{Bmatrix} = \underbrace{\begin{Bmatrix} u_0 \\ v_0 \\ w_0 \\ T_0 \\ p_0 \\ \rho_0 \end{Bmatrix}}_{\text{mean fields}} + \underbrace{\begin{Bmatrix} u' \\ v' \\ w' \\ \delta T \\ \delta p \\ \delta \rho \end{Bmatrix}}_{\text{tidal oscillations}} \tag{7}$$

then we shall neglect quadratic and higher order terms in the tidal perturbations. Let us explicitly consider Equation (1):

$$\begin{aligned}(\rho_0 + \delta\rho)\bigg(&\frac{\partial u'}{\partial t} + \boldsymbol{u}_0 \cdot \nabla u' + \boldsymbol{u}' \cdot \nabla u_0 + \boldsymbol{u}' \cdot \nabla u' + \boldsymbol{u}_0 \cdot \nabla u_0 \\ &+ \frac{w' u_0}{a} + \frac{w_0 u'}{a} + \frac{w_0 u_0}{a} + \frac{w' u'}{a} - (v_0^2 + 2 v_0 v' + v'^2)\frac{\cot\theta}{a} \\ &- 2\omega\cos\theta(v_0 + v')\bigg) \\ &= -\frac{1}{a}\frac{\partial p_0}{\partial \theta} - \frac{1}{a}\frac{\partial \delta p}{\partial \theta} - \frac{(\rho_0 + \delta\rho)}{a}\frac{\partial \Omega}{\partial \theta}.\end{aligned} \tag{1a}$$

Now Ω is considered to be of the form $\Omega = \Omega_0(z) + \Omega'(\theta, \varphi, z)$, where Ω' is a tidal forcing. If primed quantities are set to zero, mean terms must satisfy our equations exactly. Thus (1a) becomes

$$\begin{aligned}(\rho_0 + \delta\rho)\bigg(&\frac{\partial u'}{\partial t} + \boldsymbol{u}_0 \cdot \nabla u' + \boldsymbol{u}' \cdot \nabla u_0 + \boldsymbol{u}' \cdot \nabla u' \\ &+ \frac{w' u_0}{a} + \frac{w_0 u'}{a} + \frac{w' u'}{a} - (2 v_0 v' + v'^2)\frac{\cot\theta}{a} - 2\omega\cos\theta v'\bigg)\end{aligned} \tag{1b}$$

$$+\delta\rho\left(\boldsymbol{u}_0\cdot\nabla u_0+\frac{w_0u_0}{a}-v_0^2\frac{\cot\theta}{a}-2\omega\cos\theta v_0\right)$$

$$=-\frac{1}{a}\frac{\partial\delta p}{\partial\theta}-\frac{(\rho_0+\delta\rho)}{a}\frac{\partial\Omega'}{\partial\theta}$$

and linearization leads to

$$\rho_0\left(\frac{\partial u'}{\partial t}+\boldsymbol{u}_0\cdot\nabla u'+\boldsymbol{u}'\cdot\nabla u_0+\frac{w'u_0}{a}+\frac{w_0u'}{a}-2v_0v'\frac{\cot\theta}{a}-2\omega\cos\theta v'\right)$$

(1c)
$$+\delta\rho\left(\boldsymbol{u}_0\cdot\nabla u_0+\frac{w_0u_0}{a}-v_0^2\frac{\cot\theta}{a}-2\omega\cos\theta v_0\right)$$

$$=-\frac{1}{a}\frac{\partial\delta p}{\partial\theta}-\frac{\rho_0}{a}\frac{\partial\Omega}{\partial\theta}.$$

Similar results are obtained for Equations (2)–(6). Linearization clearly requires "small" tidal amplitudes, though the definition of small is not so clear, especially since it is not evident which of the neglected nonlinear terms will be most important. As with our other approximations we will check linearity a posteriori. However, at the least we expect that linearity requires that forcing be sufficiently small. Observations suggest, moreover, that linearity obtains in the troposphere where $|\delta p/p_0| \sim 10^{-3}$. We shall see, however, that neither of these considerations is sufficient to insure linearity at all levels—at least when dissipative processes are neglected. Finally, it is clear that for arbitrary choices for the basic state the solution of six coupled equations like (10) is likely to prove intractible. This leads us to our last major approximation.

(viii) $\boldsymbol{u}_0 = 0$ and p_0, ρ_0, and T_0 are independent of latitude and longitude. In a rotating fluid the assumptions for p_0, ρ_0, and T_0 imply $\boldsymbol{u}_0 = 0$. In general this approximation may seem initially poor since it is commonly supposed that linearization requires $\boldsymbol{u}' \ll \boldsymbol{u}_0$. This is simply untrue. For example, if in (1b) $|\boldsymbol{u}'\cdot\nabla u'| \ll |\partial u'/\partial t|$ then we drop $\boldsymbol{u}'\cdot\nabla u'$ even if $\boldsymbol{u}' > \boldsymbol{u}_0$. What does prove important is the ratio of $|\boldsymbol{u}_0|$ to the phase speed of the tide. For migrating tides this is the linear rotation speed of the earth's surface which at the equator is about 400 m/s. 400 m/s is certainly much greater than $|\boldsymbol{u}_0|$ and hence the neglect of $\boldsymbol{u}_0$ is at least plausible. The matter will be discussed in greater detail later.

Having made all the above assumptions and/or approximations we are left with a basic state where

(8)
$$\partial p_0/\partial z = -\rho_0 g$$

and

$$p_0 = \rho_0 R T_0. \tag{9}$$

$T_0(z)$ will be specified and (8) and (9) yield

$$p_0 = p_0(0)\, e^{-x}, \tag{10}$$

$$\rho_0 = p_0/RT_0 \tag{11}$$

where

$$x = \int_0^z \frac{dz}{H} \tag{12}$$

and

$$H = RT_0/g. \tag{13}$$

H is known as the *local scale height* while x is the height in scale heights. For our perturbation equations we have

$$\frac{\partial u'}{\partial t} - 2\omega \cos\theta v' = -\frac{1}{a}\frac{\partial}{\partial\theta}\left(\frac{\delta p}{\rho_0} + \Omega'\right), \tag{14}$$

$$\frac{\partial v'}{\partial t} + 2\omega \cos\theta u' = -\frac{1}{a\sin\theta}\frac{\partial}{\partial\varphi}\left(\frac{\delta p}{\rho_0} + \Omega'\right), \tag{15}$$

$$\partial\delta p/\partial z = -g\delta\rho - \rho_0(\partial\Omega'/\partial z), \tag{16}$$

$$d\rho/dt + \rho_0\chi = 0, \tag{17}$$

$$c_v(dT/dt) = (RT_0/\rho_0)(d\rho/dt) + J', \tag{18}$$

$$\delta p/p_0 = \delta\rho/\rho_0 + \delta T/T_0 \tag{19}$$

where

$$\chi \equiv \nabla\cdot \boldsymbol{u}' = \frac{1}{a\sin\theta}\frac{\partial}{\partial\theta}(u\sin\theta) + \frac{1}{a\sin\theta}\frac{\partial v}{\partial\varphi} + \frac{\partial w}{\partial z},$$

and

$$df/dt = \partial\delta f/\partial t + w'(df_0/dz).$$

Our aim is to reduce Equations (14)–(19) to a single equation in a single unknown. Several steps of this reduction are worth noting, but for the most part, the process is relegated to an appendix.

First we eliminate δT using (19). (18) becomes

$$dp/dt = \gamma g H(d\rho/dt) + \rho_0(\gamma - 1)J' \tag{20}$$

where

$$\gamma = c_p/c_v, \qquad c_p = R + c_v.$$

Next we assume solutions of the form

$$\exp(i(\sigma t + s\varphi)) \tag{21}$$

where

$$\sigma = 2\pi/\tau,$$
$$\tau = \text{solar or lunar day}/m,$$
$$m = \text{an integer},$$
$$s = 0, \pm 1, \pm 2, \cdots.$$

With solutions of this form

$$\partial/\partial t \to i\sigma, \qquad \partial/\partial\varphi \to is$$

and Equations (14) and (15) become

$$i\sigma u' - 2\omega\cos\theta v' = -\frac{1}{a}\frac{\partial}{\partial\theta}\left(\frac{\delta p}{\rho_0} + \Omega'\right), \tag{22}$$

$$i\sigma v' + 2\omega\cos\theta u' = -\frac{is}{a\sin\theta}\left(\frac{\delta p}{\rho_0} + \Omega'\right). \tag{23}$$

(22) and (23) are merely algebraic equations in u' and v' which may be solved in terms of δp and its θ-derivative yielding

$$u' = \frac{i\sigma}{4a^2\omega^2(f^2 - \cos^2\theta)}\left(\frac{\partial}{\partial\theta} + \frac{s\cot\theta}{f}\right)\left(\frac{\delta p}{\rho_0} + \Omega'\right), \tag{24}$$

$$v' = \frac{-\sigma}{4a^2\omega^2(f^2 - \cos^2\theta)}\left(\frac{\cos\theta}{f}\frac{\partial}{\partial\theta} + \frac{s}{\sin\theta}\right)\left(\frac{\delta p}{\rho_0} + \Omega'\right) \tag{25}$$

where $f \equiv \sigma/2\omega$. Now u' and v' enter the remaining equations only through the velocity divergence which can now be written, using (24) and (25)

$$\chi = \frac{i\sigma}{4a^2\omega^2}F\left[\frac{\delta p}{\rho_0} + \Omega'\right] + \frac{\partial w'}{\partial z} \tag{26}$$

where

$$F = \frac{1}{\sin\theta}\frac{\partial}{\partial\theta}\left(\frac{\sin\theta}{f^2 - \cos^2\theta}\frac{\partial}{\partial\theta}\right) - \frac{1}{f^2 - \cos^2\theta}\left(\frac{s}{f}\frac{f^2 + \cos^2\theta}{f^2 - \cos^2\theta} + \frac{s^2}{\sin^2\theta}\right). \tag{27}$$

Now Equations (26), (20), (17) and (16) form four equations in four unknowns, δp, χ, $\delta\rho$ and w. As it turns out the most convenient variable to solve for is

$$G = -(1/\gamma p_0)(dp/dt). \tag{28}$$

The manipulations are given in the appendix; one obtains

$$H\frac{\partial^2 G}{\partial z^2} + \left(\frac{dH}{dz} - 1\right)\frac{\partial G}{\partial z} = \frac{g}{4a^2\omega^2}F\left[\left(\frac{dH}{dz} + \kappa\right)G - \frac{\kappa J}{\gamma g H}\right] \tag{29}$$

where $\kappa = (\gamma - 1)/\gamma$. In terms of its mathematical symptoms (29) is a nearly fatal case. It is a mixed elliptic-hyperbolic equation with singularities at $\cos\theta = \pm f, \pm 1$. However, in practice its solution is straightforward if not simple. Recall that F is an operator involving only θ while the remainder of (29) depends only on z. We may therefore attempt to solve (29) by the method of separation of variables.

Let us assume G may be written

$$G = \sum_{\text{all } n} L_n(z)\Theta_n(\theta) \tag{30}$$

and let us assume further that $\{\Theta_n\}_{\text{all } n}$ is complete. Then we may write

$$J = \sum_{\text{all } n} J_n(z)\Theta_n(\theta). \tag{31}$$

Substituting (30) and (31) into (29) we get

$$F[\Theta_n] = -(4a^2\omega^2/gh_n)\Theta_n \tag{32}$$

and

$$H\frac{d^2 L_n}{dz^2} + \left(\frac{dH}{dz} - 1\right)\frac{dL_n}{dz} + \frac{1}{h_n}\left(\frac{dH}{dz} + \kappa\right)L_n = \frac{\kappa}{\gamma g H h_n}J_n. \tag{33}$$

The separation constant for the nth mode is $4a^2\omega^2/gh_n$ where h_n is called the *equivalent depth of the nth mode.* It is written in this fashion in order to make evident that (32) is the same as Laplace's Tidal Equation for a spherical fluid shell of depth h_n. We shall give a more meaningful interpretation of h_n shortly. If we require that Θ_n be bounded at the poles then (32) defines an eigenfunction-eigenvalue problem where the eigenfunctions are called *Hough Functions* and the eigenvalues are expressed in terms of h_n. The full solution associated with a given Hough Function is called a *Hough Mode.* The vertical structure of a given Hough Mode is given by the solution of Equation (33). (33) assumes a more easily interpreted form if we replace z by x as the independent variable and replace L_n by

$$y_n = e^{-x/2}L_n.$$

(33) becomes

$$\frac{d^2 y_n}{dx^2} + \left[\frac{1}{h_n}\left(\kappa H + \frac{dH}{dx}\right) - \frac{1}{4}\right]y_n = \frac{\kappa J_n}{\gamma g h_n} e^{-x/2}. \tag{34}$$

Once the Θ_n's and the associated y_n's are obtained we may easily solve for the other fields.

$$\begin{Bmatrix} \delta p \\ \delta \rho \\ \delta T \\ w' \end{Bmatrix} = \sum_n \begin{Bmatrix} \delta p_n(x) \\ \delta \rho_n(x) \\ \delta T_n(x) \\ w_n(x) \end{Bmatrix} \Theta_n(\theta), \tag{35}$$

$$u' = \sum_n u_n(x) U_n(\theta), \tag{36}$$

$$v' = \sum_n v_n(x) V_n(\theta) \tag{37}$$

where

$$\delta p_n = \frac{p_0(0)}{H(x)}\left[-\frac{\Omega_n(x)}{g} e^{-x} + \frac{\gamma h_n}{i\sigma} e^{-x/2}\left(\frac{dy_n}{dx} - \frac{1}{2}y_n\right)\right], \tag{38}$$

$$\begin{aligned} \delta \rho_n = \frac{p_0(0)}{(gH)^2}\Bigg\{ & -\Omega_n e^{-x}\left(1 + \frac{1}{H}\frac{dH}{dx}\right) \\ & + \frac{\gamma g h_n}{i\sigma} e^{-x/2}\left[\left(1 + \frac{1}{H}\frac{dH}{dx}\right)\left(\frac{dy_n}{dx} - \frac{y_n}{2}\right) + \frac{H}{h_n}\left(\kappa + \frac{1}{H}\frac{dH}{dx}\right)y_n\right] - \frac{\kappa J_n}{i\sigma}\Bigg\}, \end{aligned} \tag{39}$$

$$\begin{aligned} \delta T_n = \frac{1}{R}\Bigg\{ & \frac{\Omega_n}{H}\frac{dH}{dx} \\ & - \frac{\gamma g h_n}{i\sigma} e^{x/2}\left[\frac{\kappa H}{h_n} + \frac{1}{H}\frac{dH}{dx}\left(\frac{d}{dx} + \frac{H}{h_n} - \frac{1}{2}\right)\right]y_n + \frac{\kappa J_n}{i\sigma}\Bigg\}, \end{aligned} \tag{40}$$

$$w_n = -\frac{i\sigma}{g}\Omega_n + \gamma h_n e^{x/2}\left[\frac{dy_n}{dx} + \left(\frac{H}{h_n} - \frac{1}{2}\right)y_n\right], \tag{41}$$

$$u_n = \frac{\gamma g h_n e^{x/2}}{4a\omega^2}\left(\frac{dy_n}{dx} - \frac{1}{2}y_n\right), \tag{42}$$

$$v_n = \frac{i\gamma g h_n e^{x/2}}{4a\omega^2}\left(\frac{dy_n}{dx} - \frac{1}{2}y_n\right), \tag{43}$$

(44) $$U_n = \frac{1}{f^2 - \cos^2\theta}\left(\frac{d}{d\theta} + \frac{s\cot\theta}{f}\right)\Theta_n,^1$$

and

(45) $$V_n = \frac{1}{f^2 - \cos^2\theta}\left(\frac{\cos\theta}{f}\frac{d}{d\theta} + \frac{s}{\sin\theta}\right)\Theta_n.^1$$

An expansion for Ω' of the form

(46) $$\Omega' = \sum_n \Omega_n(x)\Theta_n$$

was assumed.

The solution of (34) requires two boundary conditions. One is obtained from (41) and the requirement that

$$w = 0 \quad \text{at} \quad z = x = 0.$$

Namely

(47) $$\frac{dy_n}{dx} + \left(\frac{H}{h_n} - \frac{1}{2}\right)y_n = \frac{i\sigma}{\gamma g h_n}\Omega_n \quad \text{at} \quad x = 0.^2$$

As a second boundary condition it often suffices to require boundedness as $x \to \infty$. However, if the top of our atmosphere is isothermal then (34) has solutions of the form

$$y_n \sim Ae^{+i\lambda x} + Be^{-i\lambda x},$$

where

$$\lambda = (\kappa H/h_n - 1/4)^{1/2}.$$

If λ is real then y_n is bounded for any choice of A and B. In this instance a radiation condition is imposed; i.e., it is required that there be no incoming energy from infinity. It can be shown that upward energy flux is associated with downward phase speed. The radiation condition therefore implies $B = 0$.

In passing it should be noted that if $J_n = \Omega_n = 0$, then the only solution to (34) is generally $y_n = 0$. There can exist values of h for which (34) can have nontrivial solutions (usually only one value). These are known as the *equivalent depths of the atmosphere.* For realistic atmospheres

---

[1] Equations (44) and (45) suggest infinite velocities when $f^2 = \cos^2\theta$. However, Brillouin [5] has shown that the behavior of Θ_n will be such that velocities are everywhere finite.

[2] Note that although gravitational tidal forcing acts throughout the atmosphere, mathematically it is equivalent to forcing at the lower boundary.

and hydrostatic waves, the atmosphere has only one equivalent depth ~ 10.4 km (see history). If an h_n is equal to the equivalent depth of the atmosphere, then we have a resonance.

Before going on to the methods of solving Equations (34) and (32), I would like to show the meaning of a mode's equivalent depth for some very simple cases. We must first note that all information about our geometric configuration is contained in Equation (32). Equation (34), apart from its dependence on h_n, depends only on the atmosphere's thermal and thermodynamic structure. Although the h_n's might be different, Equation (34) would be the same for waves on a nonrotating sphere, a rotating plane or even a nonrotating plane. It proves useful to study the counterparts of (32) for nonrotating and rotating planar atmospheres. For a nonrotating planar atmosphere Equations (14) and (15) are replaced by

$$\partial u'/\partial t = -(\partial/\partial x)(\delta p/\rho_0), \tag{48}$$

$$\partial v'/\partial t = -(\partial/\partial y)(\delta p/\rho_0). \tag{49}$$

We shall consider solutions of the form

$$\exp(i(\sigma t + kx + my)). \tag{50}$$

(48) and (49) become

$$u' = -(k/\sigma)(\delta p/\rho_0), \tag{51}$$

$$v' = -(m/\sigma)(\delta p/\rho_0) \tag{52}$$

and

$$\begin{aligned} \chi = \frac{\partial u'}{\partial x} + \frac{\partial v'}{\partial y} + \frac{\partial w'}{\partial z} &= \frac{i\sigma}{4a^2\omega^2} F\left(\frac{\delta p}{\rho_0}\right) + \frac{\partial w'}{\partial z} \\ &= -\frac{i\sigma}{gh}\frac{\delta p}{\rho_0} + \frac{\partial w'}{\partial z} = -\frac{i}{\sigma}(k^2 + m^2)\frac{\delta p}{\rho_0} + \frac{\partial w'}{\partial z}. \end{aligned} \tag{53}$$

Thus, by analogy with our earlier procedure we find

$$gh = \sigma^2/(k^2 + m^2). \tag{54}$$

h is thus a measure of the square of the wave's horizontal wavelength. Another interpretation (closely related to the name equivalent depth) is that h is the depth of a homogeneous fluid layer in which the phase speed of gravity waves is $\sigma/(k^2 + m^2)^{1/2}$. The identification of h with the thickness of a fluid can, however, be misleading. On a rotating plane

(14) and (15) are replaced by

$$\frac{\partial u'}{\partial t} - 2\omega v = -\frac{\partial}{\partial x}\left(\frac{\delta p}{\rho_0}\right), \tag{55}$$

$$\frac{\partial v'}{\partial t} + 2\omega u = -\frac{\partial}{\partial y}\left(\frac{\delta p}{\rho_0}\right). \tag{56}$$

Assuming solutions of the form of (50) again we find

$$(4\omega^2 - \sigma^2)u' = \sigma k(\delta p/\rho_0) - im2\omega(\delta p/\rho_0), \tag{57}$$

$$(4\omega^2 - \sigma^2)v' = \sigma m(\delta p/\rho_0) + ik2\omega(\delta p/\rho_0) \tag{58}$$

and

$$\chi - \frac{\partial w'}{\partial z} = (4\omega^2 - \sigma^2)^{-1}[i\sigma(k^2 + m^2)]\frac{\delta p}{\rho_0} = -\frac{i\sigma}{gh}\frac{\delta p}{\rho_0} \tag{59}$$

or

$$gh = (\sigma^2 - 4\omega^2)/(k^2 + m^2). \tag{60}$$

Once again h is a measure of the square of the horizontal wavelength. However, the interpretation of h as a depth runs into trouble since h can now be negative. (It is interesting to note that the name equivalent depth effectively prevented the discovery of negative equivalent depths until about 5 years ago.) This occurs whenever $\sigma^2 < 4\omega^2$. The depth interpretation breaks down because in a rotating fluid there is a minimum phase speed regardless of fluid depth. For atmospheric waves negative equivalent depths are perfectly meaningful. Consider Equation (34) (for simplicity consider an isothermal atmosphere away from excitation sources). When h_n is negative (34) will have exponential solutions of the form

$$y_n \sim e^{-\mu x} \tag{61}$$

where

$$\mu = (\tfrac{1}{4} - (1/h_n)(\kappa H))^{1/2} < \tfrac{1}{2}. \tag{62}$$

Thus the rate of decay of y_n will exceed the $e^{x/2}$ growth indicated in Equations (40)–(43), leading to actual decay of amplitude. For $0 < h_n < 4\kappa H$ solutions will be oscillatory in the vertical with amplitudes growing as $e^{x/2}$. For $h_n > 4\kappa H$ there is no phase change with height, but amplitudes will grow exponentially with height, though more slowly than $e^{x/2}$.

Incidentally, it is easily shown that the equivalent depth of an isothermal atmosphere is $h = \gamma H$ where $\gamma = 1.4$ and $H \sim 8$ km.

As in the above examples, h_n, on a rotating sphere, is a measure of the square of some characteristic horizontal scale. Also, there will be negative equivalent depths for any frequency for which $\sigma^2 < 4\omega^2 \cos^2 \theta$ for some θ.

Appendix to Part II.

We wish to reduce the following to a single equation:

$$G = -(1/\gamma p_0)(dp/dt), \tag{A1}$$

$$d\rho/dt = -\rho_0 \chi, \tag{A2}$$

$$dp/dt = \gamma g H(d\rho/dt) + (\gamma - 1)\rho_0 J, \tag{A3}$$

$$\chi - \frac{\partial w'}{\partial z} = \frac{i\sigma}{4a^2\omega^2} F\left(\frac{\delta p}{\rho_0} + \Omega'\right), \tag{A4}$$

$$\partial \delta p/\partial z = -g\delta\rho - \rho_0(\partial \Omega'/\partial z) \tag{A5}$$

where

$$df/dt = \partial f'/\partial t + w'(df_0/dz). \tag{A6}$$

From (A1)

$$\delta p = -\frac{p_0}{i\sigma}\left(\gamma G - \frac{1}{H} w'\right). \tag{A7}$$

From (A2) and (A3)

$$\chi = G + \kappa J/gH. \tag{A8}$$

From (A8) and (A2)

$$\delta \rho = -\frac{\rho_0}{i\sigma}\left(G + \frac{\kappa J}{gH}\right) - \frac{w'}{i\sigma}\frac{d\rho_0}{dz}. \tag{A9}$$

Differentiating (A7) we get

$$\frac{\partial}{\partial z}\delta p = -\frac{p_0}{i\sigma}\left(\gamma\frac{\partial G}{\partial z} - \frac{\gamma}{H}G - \frac{1}{H}\frac{\partial w'}{\partial z} + \frac{1}{H^2}\left(1 + \frac{dH}{dz}\right)w'\right) \tag{A10}$$

while from (A5) and (A9) we get

$$\frac{\partial}{\partial z}\delta p = \frac{g\rho_0}{i\sigma}\left(G + \frac{\kappa J}{gH}\right) + \frac{g}{i\sigma}\frac{d\rho_0}{dz}w' - \rho_0\frac{\partial \Omega'}{\partial z}. \tag{A11}$$

Equating (A10) and (A11) we get

$$\text{(A12)} \qquad \frac{\partial w'}{\partial z} = \gamma H \frac{\partial G}{\partial z} - (\gamma - 1)G - \frac{i\sigma}{g}\frac{\partial \Omega'}{\partial z} + \frac{\kappa J}{gH}.$$

Substituting (A12), (A8) and (A7) into (A4) we get

$$\text{(A13)} \qquad -\gamma H \frac{\partial G}{\partial z} + \gamma G + \frac{i\sigma}{g}\frac{\partial \Omega'}{\partial z} = \frac{i\sigma}{4a^2\omega^2} F\left(-\frac{gH}{i\sigma}\left(\gamma G - \frac{1}{H}w'\right) + \Omega'\right).$$

The quantity $(\sigma/g)(\partial\Omega'/\partial z)$ proves negligible compared with other terms and is dropped.

Finally we differentiate (A13) and use (A12) to get

$$\text{(A14)} \qquad H\frac{\partial^2 G}{\partial z^2} + \left(\frac{dH}{dz} - 1\right)\frac{\partial G}{\partial z} = \frac{g}{4a^2\omega^2} F\left(\left(\frac{dH}{dz} + \kappa\right)G - \frac{\kappa J}{\gamma g H}\right).$$

REFERENCES, PART II

1. N. A. Phillips, *The equations of motion for a shallow rotation atmosphere and the "traditional approximation"*, J. Atmospheric Sci. **23** (1966), 626–628; Reply to "comments on Phillips' simplification of the equations of motion", by G. Veronis, J. Atmospheric Sci. **25** (1968), 1155–1157.

2. M. Yanowitch, *A remark on the hydrostatic approximation*, Pure Appl. Geophys. **64** (1966), 169–172.

3. S. Goldstein, *Modern developments in fluid dynamics* (2 vols.), Oxford Univ. Press, Oxford, 1938.

4. R. S. Lindzen, *Internal gravity waves in atmospheres with realistic dissipation and temperature; Part* I, *Mathematical development and propagation of waves in the thermosphere*, Geophys. Fluid Dynamics **1** (1970), 303–355.

5. Marcel Brillouin, *Marées dynamiques—Les latitudes critiques*, C. R. Acad. Sci. Paris **194** (1932), 801–807.

III. MATHEMATICAL METHODS OF SOLUTION

III.1. **Laplace's Tidal Equation.** Laplace's Tidal Equation is given by Equations (II.27) and (II.32). It may be rewritten

$$\text{(1)} \qquad \frac{d}{d\mu}\left(\frac{1-\mu^2}{f^2-\mu^2}\frac{d\Theta_n}{d\mu}\right) - \frac{1}{(f^2-\mu^2)}\left[\frac{s}{f}\frac{f^2+\mu^2}{f^2-\mu^2} + \frac{s^2}{1-\mu^2}\right]\Theta_n + \frac{4a^2\omega^2}{gh_n}\Theta_n = 0,$$

where $\mu = \cos\theta$.

There are many theorems which can and have been proven about the solutions to (1). These can be found in Hough [**1**], Longuet-Higgins [**2**], and Flattery [**3**]. We shall state without proof the most important of these:

(i) All solutions are bounded at $f = \pm\mu$; the singularities at these points are in Ince's terminology "apparent."

(ii) The boundary condition that Θ_n be bounded at $\mu = \pm 1$ leads to an eigenfunction-eigenvalue problem (analogous to the equation for Associated Legendre Polynomials).

(iii) For a given choice of s, the eigenfunctions Θ_n are orthogonal.

(iv) The eigenvalues h_n are real.

(v) If $f \neq \pm 1$, then Θ_n behaves as $(1 - \mu^2)^{s/2}$ at the poles; if $f = \pm 1$ then θ_n behaves as $(1 - \mu^2)^{s/2+1}$ at the poles.

In general Equation (1) has no simple closed form solutions[1] and must be solved approximately. There are many approaches to this problem including the use of WKB methods [**4**]. Here, I shall present the most commonly used numerical approach.

We first assume that Θ_n may be expanded in a series of associated Legendre polynomials:

$$\Theta_n^{\sigma,s} = \sum_{m=s}^{\infty} C_{n,m}^{\sigma,s} P_m^s(\mu) \tag{2}$$

where $P_m^s(\mu)$ is defined as in Whittaker and Watson [**5**]. The advantage of (2) over a simple power series in μ is that the latter leads to a set of 5th order recursion relations in the coefficients while the (2) leads to more convenient 3rd order relations. As is shown in Hough [**1**] the substitution of (2) into (1) leads to the following relations for the coefficients

$$\begin{aligned} &\frac{(m-s)}{(2m-1)\{m(m-1)-s/f\}}\left[\frac{(m-s-1)}{(2m-3)}C_{n,m-2} + \frac{(m-1)^2(m+s)}{m^2(2m+1)}C_{n,m}\right] \\ &\quad - \left[f^2\frac{m(m+1)-s/f}{m^2(m+1)^2} - \frac{h_n g}{4\omega^2 a^2}\right]C_{n,m} \\ &\quad + \frac{(m+s+1)}{(2m+3)\{(m+1)(m-2)-s/f\}} \\ &\qquad \cdot\left[\frac{(m+2)^2(m-s+1)}{(m+1)^2(2m+1)}C_{n,m} + \frac{(m+s+2)}{2m+5}C_{n,m+2}\right] = 0. \end{aligned} \tag{3}$$

---

[1] There are three exceptions to this statement: (i) When $s = 0, f = 1$, there are trigonometric solutions; (ii) When rotation is neglected, solutions are spherical harmonics; (iii) When $h = \infty$, solutions are the sum of two spherical harmonics. Details and further references may be found in Flattery [**3**].

Equation (3) may be simplified by the introduction of auxiliary coefficients $D_{n,m}$ thus defined:

(4)
$$2\left(\frac{m+1}{m} - \frac{s}{fm^2}\right)D_{n,m} = \frac{2(m+1)^2(m-s)}{m^2(2m-1)}C_{n,m-1} + \frac{(m+s+1)}{(2m+3)}C_{n,m+1}.$$

Love [6] showed that the $D_{n,m}$'s are the expansion coefficients for the stream functions for horizontal flow. Substituting (4) into (3) one obtains

(5)
$$\frac{2(m+1)^2(m-s)}{m^2(2m-1)}D_{n,m-1} + \frac{(m+s+1)}{(2m+3)}D_{n,m+1} = 2\left[f^2\left\{\frac{m+1}{m} - \frac{s}{fm^2}\right\} - (m+1)^2\frac{h_n g}{4\omega^2 a^2}\right]C_{n,m}.$$

(4) and (5) may be written more concisely as follows:

(6)
$$K_m^s C_{n,m-1} - N_m^{\sigma,s} D_{n,m} + L_m^s C_{n,m+1} = 0,$$

(7)
$$K_m^s D_{n,m-1} - M_{n,m}^{\sigma,s} C_{n,m} + L_m^s D_{n,m+1} = 0$$

where

(8)
$$K_m^s = 2(m+1)^2(m-s)/(2m-1)m^2,$$

(9)
$$L_m^s = (m+s+1)/(2m+3),$$

(10)
$$N_m^{\sigma,s} = 2\left\{\frac{m+1}{m} - \frac{1}{m^2}\frac{s}{f}\right\},$$

(11)
$$M_{n,m}^{\sigma,s} = 2\left\{f^2\left[\frac{m+1}{m} - \frac{1}{m^2}\frac{s}{f}\right] - (m+1)^2\frac{h_n^{\sigma,s}g}{4\omega^2 a^2}\right\}.$$

We see from (6) and (7) that equations for $\{C_{n,m}\}$, $(m-s)$ odd, $\{D_{n,m}\}$, $(m-s)$ even (i.e., for Hough Functions antisymmetric about the equator) are decoupled from the equations for $\{C_{n,m}\}$, $(m-s)$ even, $\{D_{n,m}\}$, $(m-s)$ odd (i.e., for Hough Functions symmetric about the equator). Also (6) and (7) are homogeneous. Thus, nontrivial solutions will only exist for certain values of h_n (the eigenvalues). These eigenvalues are the roots of the infinite determinant of the coefficients of Equations (6) and (7), i.e.,

$$(12)\qquad \begin{vmatrix} -M_{n,s}^{\sigma,s} & L_s^s & 0 & 0 & 0 & \cdots \\ K_{s+1}^s & -N_{s+1}^{\sigma,s} & L_{s+1}^s & 0 & 0 & \cdots \\ 0 & K_{s+2}^s & -M_{n,s+2}^{\sigma,s} & L_{s+2}^s & 0 & \cdots \\ 0 & 0 & \ddots & \ddots & \ddots & \\ \vdots & \vdots & & \ddots & \ddots & \ddots \end{vmatrix} = 0$$

for symmetric eigenfunctions, and

$$(13)\qquad \begin{vmatrix} -N_s^{\sigma,s} & L_s^s & 0 & 0 & 0 & \cdots \\ K_{s+1}^s & -M_{n,s+1}^{\sigma,s} & L_{s+1}^s & 0 & 0 & \cdots \\ 0 & K_{s+2}^s & -N_{s+2}^{\sigma,s} & L_{s+2}^s & 0 & \cdots \\ 0 & 0 & \ddots & \ddots & \ddots & \\ \vdots & \vdots & & \ddots & \ddots & \ddots \end{vmatrix} = 0$$

for antisymmetric eigenfunctions. Insofar as (2) converges, the roots of sufficiently large truncations of (12) and (13) should be approximations to the actual eigenvalues. This is essentially Galerkin's method.[2] If one uses this method, it is by no means clear that the use of polynomial rather than associated Legendre polynomial expansions would be significantly less efficient. The virtue of the third order recursion relations (at least in the precomputer days of Hough) is that they permit the determinantal Equations (12) and (13) to be replaced by continued fractions. For example, from (7)

$$(14)\qquad M_{n,s}^{\sigma,s} = L_s^s \frac{1}{C_{n,s}/D_{n,s+1}}$$

and from (6)

$$(15)\qquad \frac{C_{n,s}}{D_{n,s+1}} = \frac{N_{s+1}^{\sigma,s}}{K_{s+1}^s} - \frac{L_{s+1}^s}{K_{s+1}^s}\frac{C_{n,s+2}}{D_{n,s+1}}.$$

(14) becomes

$$(16)\qquad M_{n,s}^{\sigma,s} = L_s^s\left(1 \Bigg/ \left(\frac{N_{s+1}^{\sigma,s}}{K_{s+1}^s} - \frac{L_{s+1}^s}{K_{s+1}^s}\frac{C_{n,s+2}}{D_{n,s+1}}\right)\right).$$

[2] For an explicit example see [7].

Alternately using (7) and (6) and letting $m \to \infty$ one gets the desired infinite continued fraction equation which in many cases is suitable for iterative solution procedures. Once an h_n has been determined (6) and (7) may be used to determine $\{C_{n,m}\}$ within some constant factor. The factor may be arbitrarily set by choosing $C_{n,s}$ (or $C_{n,s+1}$) to one. Then

$$\begin{aligned}\int_{-1}^{1} (\Theta_n(\mu))^2\, d\mu &= \sum_{m=s}^{\infty} (C_{n,m})^2 \int_{-1}^{1} (P_m^s(\mu))^2\, d\mu \\ &= \sum_{m=s}^{\infty} (C_{n,m})^2 \frac{2}{2m+1} \frac{(m+s)!}{(m-s)!} \\ &\equiv (F_n)^2.\end{aligned} \tag{17}$$

If one works with normalized associated Legendre Polynomials then

$$\Theta_n = \sum_{m=s}^{\infty} \hat{C}_{n,m}^{\sigma,s} P_{m,s}(\mu)$$

where

$$\hat{C}_{n,m}^{\sigma,s} = C_{n,m}\left(\frac{2}{2m+1} \frac{(m+s)!}{(m-s)!}\right)^{1/2}$$

and

$$(F_n)^2 = \sum_{m=s}^{\infty} (\hat{C}_{n,m}^{\sigma,s})^2.$$

It generally proves useful to use normalized Hough Functions

$$\overline{\Theta}_n(\mu) = \sum_{m=s}^{\infty} \overline{C}_{n,m} P_{m,s}(\mu)$$

where

$$\overline{C}_{n,m} = \hat{C}_{n,m}/F_n.$$

In general the normalization of Θ_n requires a knowledge of all $\hat{C}_{n,m}$'s. However, the use of truncated expansions generally leads to only small errors in $\overline{C}_{n,m}$ providing the truncation is sufficiently large. In what follows we will always use normalized Hough Functions and omit the overbar. If we wish to expand some function $g(\mu)$ in Hough Functions thus

$$g(\mu) = \sum_n g_n \Theta_n(\mu),$$

then

$$g_n = \int_{-1}^{1} g(\mu)\Theta_n(\mu)\, d\mu.$$

In particular if $g(\mu) = P_{m,s}(\mu)$, $g_n = \overline{C}_{n,m}$.

III.2. **Vertical structure equation.** The vertical structure depends on the distribution of the mean temperature and both the thermal and gravitational excitation. Apart from a resonant value of h_n, the vertical solution will be zero without excitation. For realistic mean distributions of temperature and excitation there are no simple closed form solutions.[3] However, extremely effective numerical approaches are available. In particular a finite difference scheme described by Bruce, et al. [**8**] has proven useful.

Let us repeat (II.34):

$$\frac{d^2 y_n}{dx^2} + \left[\frac{1}{h_n}\left(\kappa H + \frac{dH}{dx}\right) - \frac{1}{4}\right] y_n = \frac{\kappa J_n}{\gamma g h_n} e^{-x/2} \tag{18}$$

and (II.47):

$$\frac{dy_n}{dx} + \left(\frac{H}{h_n} - \frac{1}{2}\right) y_n = \frac{i\sigma}{\gamma g h_n} \Omega_n \qquad \text{at} \quad x = 0. \tag{19}$$

Let us now divide our x-domain into a number of discrete levels $x_0, x_1, x_2, \cdots$ where $x_0 = 0$ and the remaining levels are equally spaced with a separation δx. At $x = x_m$

$$\frac{d^2 y_n(x_m)}{dx^2} \approx \frac{y_n(x_{m+1}) - 2y_n(x_m) + y_n(x_{m-1})}{(\delta x)^2}. \tag{20}$$

Substituting (20) into (18) we get

$$A_m y_n(x_{m+1}) + B_m y_n(x_m) + C_m y_n(x_{m-1}) = D_m \tag{21}$$

where

$$A_m = 1,$$

$$B_m = -\left[2 + \frac{(\delta x)^2}{4}\left\{1 - \frac{4}{h_n}(\kappa H(x_m) + \left.\frac{dH}{dx}\right|_{x=x_m}\right\}\right],$$

$$C_m = 1$$

and

$$D_m = (\delta x)^2 (\kappa J_n(x_m)/\gamma g h_n).$$

[3] There are, of course, simplified atmospheres for which analytic solutions have been found. Some of these are cited in Part I. However, even for isothermal atmospheres one must use Green Function techniques to handle distributed heat sources.

(21) is solved as follows: let

$$y_n(x_m) = \alpha_m y_n(x_{m+1}) + \beta_m \tag{22}$$

where α_m and β_m are new variables. Similarly

$$y_n(x_{m-1}) = \alpha_{m-1} y_n(x_m) + \beta_{m-1}. \tag{23}$$

Substituting (23) into (21) we get

$$y_n(x_m) = -\frac{A_m y_n(x_{m+1})}{B_m + \alpha_{m-1}C_m} + \frac{D_m - \beta_{m-1}C_m}{B_m + \alpha_{m-1}C_m}, \tag{24}$$

and comparing (24) with (22) we get

$$\alpha_m = -A_m/(B_m + \alpha_{m-1}C_m) \tag{25}$$

and

$$\beta_m = (D_m - \beta_{m-1}C_m)/(B_m + \alpha_{m-1}C_m). \tag{26}$$

Now, once we know α_0 and β_0, we can solve for all α_m's and β_m's. α_0 and β_0 can be obtained from (19), the lower boundary condition. For convenience, let us approximate (19) by a one-sided difference:

$$\frac{y_n(x_1) - y_n(x_0)}{\delta x} + \left(\frac{H(x_0)}{h_n} - \frac{1}{2}\right) y_n(x_0) = \frac{i\sigma\Omega_n(x_0)}{\gamma g h_n}. \tag{27}$$

(27) may be rewritten

$$y_n(x_0) = \frac{1}{1 - (H(x_0)/h_n - \frac{1}{2})} y_n(x_1) - \frac{i\sigma\Omega_n(x_0)\delta x}{\gamma g h_n(1 - [H(x_0)/h_n - \frac{1}{2}]\delta x)}, \tag{28}$$

and comparing (28) with (22)

$$\alpha_0 = \frac{1}{1 - (H(x_0)/h_n - \frac{1}{2})\delta x}, \tag{29}$$

$$\beta_0 = \frac{-i\sigma\Omega_n(x_0)\delta x}{\gamma g h_n(1 - [H(x_0)/h_n - \frac{1}{2}]\delta x)}. \tag{30}$$

Finally, in order to solve for y_n we must know y_n at some high level, at which point (22) may be used to determine y_n at all lower levels. Sufficiently high in the atmosphere $H = T_0 =$ constant. As an example of the use of the upper boundary condition we will consider $h_n < 4\kappa H_\infty$, where H_∞ is the constant H in the high thermosphere (viz. the first appendix of Part I). Then from the radiation condition

$$y_n = A\, e^{i\lambda x} \tag{31}$$

in the high thermosphere where

(32) $$\lambda = (\kappa H_\infty/h_n - \tfrac{1}{4})^{1/2}.$$

Differentiating (31) we get

(33) $$dy_n/dx = i\lambda y_n,$$

and approximating (33) by finite differences we get

(34) $$(y_n(x_M) - y_n(x_{M-2}))/2\delta x = i\lambda y_n(x_{M-1})$$

where x_M is taken to be our top level. Now from (22) we have

(35) $$y_n(x_{M-2}) = \alpha_{M-2} y_n(x_{M-1}) + \beta_{M-2}$$

and

(36) $$y_n(x_{M-1}) = \alpha_{M-1} y_n(x_M) + \beta_{M-1}.$$

From (34)–(36) we get

(37) $$y_n(x_M) = (\beta_{M-2} + \beta_{M-1}(2i\lambda\delta x + \alpha_{M-2}))/(1 - \alpha_{M-1}[2i\lambda\delta x + \alpha_{M-2}]),$$

which is our desired result.

In practice, we have found that the above integration scheme works for all reasonable cases provided δx is sufficiently small. We have obtained better than 1% accuracy by choosing

(38) $$\delta x = \left(\text{minimum value of } 2\pi \Big/ \left(\frac{1}{h_n}\left(\kappa H + \frac{dH}{dx}\right) - \frac{1}{4}\right)^{1/2}\right) \times 10^{-2}.$$

However, to the best of my knowledge, the only rigorous proof of the validity of the above procedure is for the situation where

(39) $$[(1/h_n)(\kappa H + dH/dx) - \tfrac{1}{4}] < 0. \quad [\mathbf{8}]$$

Clearly this is a sufficient but not a necessary proof. The above procedure can also be extended to a variety of higher order ordinary differential equations and partial differential equations [**9**], but as with the above procedure, necessary conditions have not yet been proven.

REFERENCES, PART III

1. S. S. Hough, *The application of harmonic analysis to the dynamical theory of the tides*, Part II. *On the general integration of Laplace's dynamical equations*, Philos. Trans. Roy. Soc. London Ser. A **191** (1898), 139–185.

2. M. S. Longuet-Higgins, *The eigenfunctions of Laplace's tidal equations over a sphere*, Philos. Trans. Roy. Soc. London Ser. A **262** (1967/68), 511–607. MR **36** #4885.

3. T. W. Flattery, Technical Report #21, Dept. of the Geophysical Sciences, University of Chicago, Chicago, Ill., 1967.

4. G. S. Golitsyn and L. A. Dikiĭ, *Oscillations of planetary atmospheres as a function of the rotational speed of the planet*, Izv. Atmos. Ocean. Phys., English ed., **2** (1966), 137–142.

5. E. T. Whittaker and G. N. Watson, *A course of modern analysis*, 4th ed., Cambridge Univ. Press, London, 1927.

6. A. E. H. Love, *Notes on the dynamical theory of the tides*, Proc. London Math. Soc. **12** (1913), 309–314.

7. R. S. Lindzen, *On the theory of the diurnal tide*, Mon. Wea. Rev. **94** (1966), 295–301.

8. G. F. Carrier and C. E. Pearson, *Ordinary differential equations*, Blaisdell, Waltham, Mass., 1968.

9. R. S. Lindzen and H.-L. Kuo, *A reliable method for the numerical integration of a large class of ordinary and partial differential equations*, Mon. Wea. Rev. **97** (1969), 732–734.

IV. Explicit Calculation of Atmospheric Tides

IV.1. **Sources of excitation.** As mentioned in Part I, a large part of the development of atmospheric tidal theory has been concerned with the specification of the gravitational and thermal sources of excitation. However, time does not permit a detailed derivation of these sources here. The gravitational excitation has already been discussed in Platzman's lectures. A discussion of both gravitational and thermal excitations may also be found in Chapman and Lindzen [**1**] where additional references are included. Here, we will merely cite, in approximate form, the most important sources of excitation.

It has already been mentioned that thermal excitation is of dominant importance for the atmosphere and that thermal excitation is mainly due to absorption of solar radiation by water vapor and ozone. Thermal excitation is commonly expressed by a function τ rather than J, where

$$\tau = \kappa J / i\sigma R. \tag{1}$$

τ is, approximately, the amplitude of the temperature oscillation that would be produced by J in the absence of motion and dissipation. For a given absorber G, the excitation at frequency σ, and zonal wavenumber s can usually be written with sufficient accuracy as

$$\tau_G^{\sigma,s} = f_G^{\sigma,s}(z) g_G^{\sigma,s}(\theta). \tag{2}$$

For the diurnal excitation $\sigma = 2\pi/1$ solar day and $s = 1$; for the semidiurnal excitation $\sigma = 4\pi/1$ solar day and $s = 2$. The $f_G^{\sigma,s}$'s for both the diurnal and semidiurnal excitations can be taken to be the same:

$$f_{H_2O}(z) = \exp(-z/22.8 \text{ km}) \tag{3}$$

and

$$f_{O_3}(z) = \exp(0.0116\,(z - z_1))\sin(\pi(z - z_1)/60) \quad \text{for} \quad z_1 < z < z_2$$

(4) $$= 0 \quad \text{elsewhere},$$

$$z_1 = 18.\text{ km}, \qquad z_2 = 78.\text{ km}.$$

The distributions of $g_{H_2O}(\theta)$ and $g_{O_3}(\theta)$ are also taken to be the same for both the diurnal and semidiurnal excitations—with the exception of their overall amplitudes. Figure 1 shows both the vertical and latitudinal distributions of excitation.

For the diurnal excitation the phase is such that τ has a maximum at 1800 LT; for the semidiurnal excitation τ has maxima at 0300 and 1500 LT.[1]

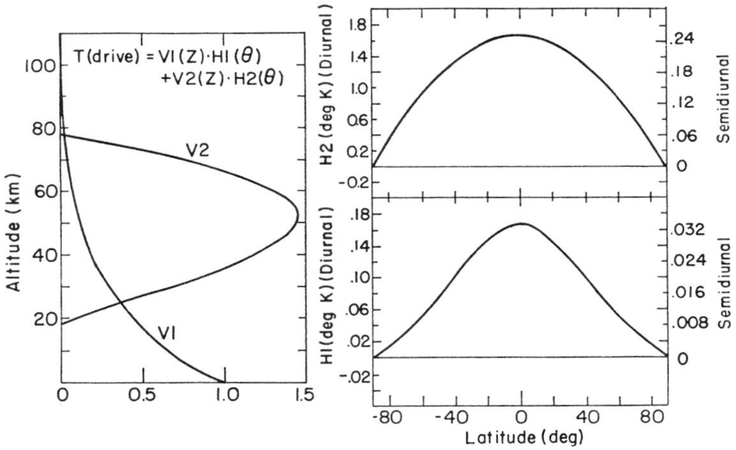

FIGURE 1. Vertical distributions of thermal excitation due to water vapor (V1) and ozone (V2); latitude distributions for water vapor (H1) and ozone (H2). After Lindzen [**5**].

IV.2. **Solar semidiurnal thermal tide.** The first three symmetric Hough Functions for this tide are shown in Figure 2.

Also shown are the corresponding expansion functions for the northerly and westerly velocity in Figures 3 and 4 respectively.

Associated with these three modes are the following equivalent depths: $h_2^{2\omega,2} = 7.8519$ km, $h_4^{2\omega,2} = 2.1098$ km and $h_6^{2\omega,2} = 0.9565$ km.[2]

---

[1] There is some reason to doubt the phase of the semidiurnal excitation; viz. [**2**].

[2] These numbers are taken from [**3**].

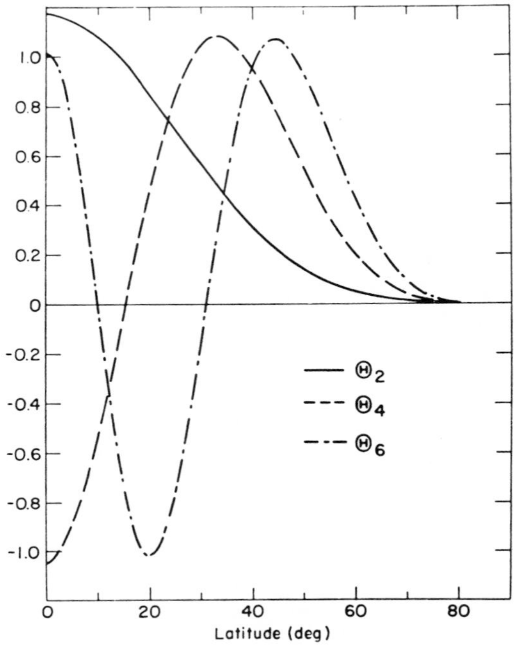

FIGURE 2. Latitude distribution for first three symmetric solar semidiurnal migrating Hough functions.

Expanding the latitude distributions of the excitation we get

(5) $g_{O_3}^{2\omega,2} = 0.249°\text{K}\,\Theta_2^{2\omega,2} + 0.0645°\text{K}\,\Theta_4^{2\omega,2} + 0.0365°\text{K}\,\Theta_6^{2\omega,2} + \cdots,$

(6) $g_{H_2O}^{2\omega,2} = 0.0307°\text{K}\,\Theta_2^{2\omega,2} + 0.00796°\text{K}\,\Theta_4^{2\omega,2} + 0.00447°\text{K}\,\Theta_6^{2\omega,2} + \cdots.$

We see from Figures 2 and 1 that the distributions of excitation are similar to $\Theta_2^{2\omega,2}$. This is reflected in Equations (5) and (6) by the fact that $\Theta_2^{2\omega,2}$ receives the bulk of the excitation. Moreover, the equivalent depth of this mode, 7.852 km, is such that $\lambda^2 = \{(1/h_2^{2\omega,2})(\kappa H + dH/dx) - \frac{1}{4}\}$ is everywhere close to zero; i.e., the vertical wavelength of this mode is very large (ca. 200 km). Thus, not only does most semidiurnal excitation go into this mode, but $\Theta_2^{2\omega,2}$ responds with particular efficiency since all excitation below 100 km acts "in phase." These facts explain the two most striking observational features of the solar semidiurnal surface pressure oscillation; namely, its strength and regularity. The latter results from the fact that the excitation going into $\Theta_2^{2\omega,2}$ is determined by the overall global distribution of excitation which does not change

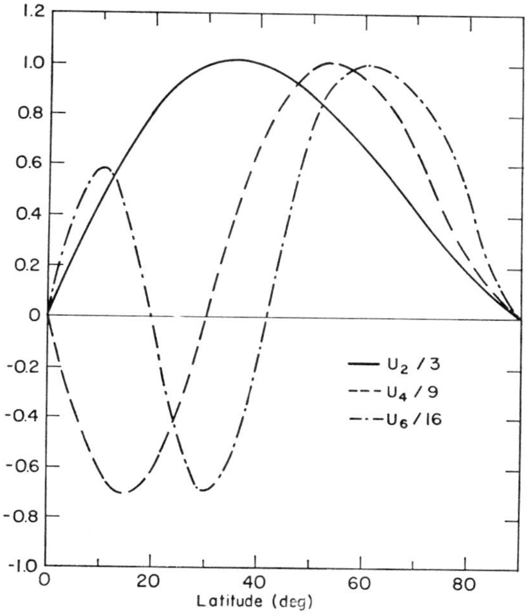

FIGURE 3. The expansion functions for the latitude dependence of the solar semidiurnal component of u, the northerly velocity. The functions have been divided by the amounts shown.

Reprinted with permission of © D. Reidel Publ. Co. [1].

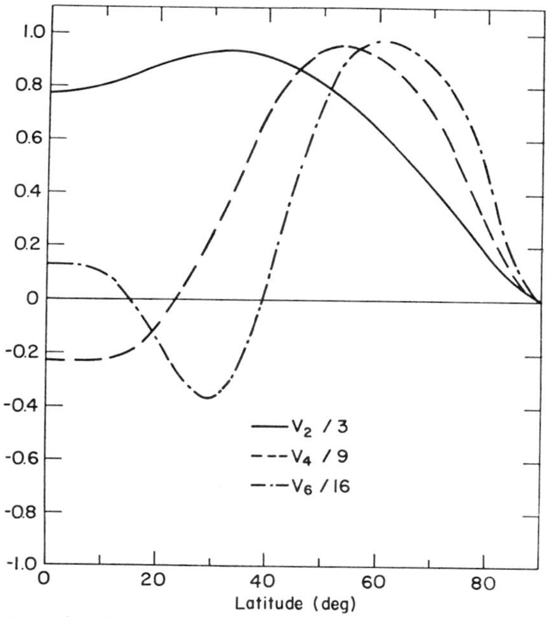

FIGURE 4. The expansion functions for the latitude dependence of the solar semidiurnal component of v, the westerly velocity. The functions have been divided by the amounts shown.

Reprinted with permission of © D. Reidel Publ. Co.

much; the larger local variations excite the higher order, less efficient Hough Modes.

The surface pressure oscillation resulting from the above excitation is relatively insensitive to the precise choice of basic temperature distribution. For the ARDC, Equatorial, and isothermal profiles shown in Figure 5, the amplitudes of the semidiurnal surface pressure oscillation at the equator are 1.18×10^{-3} mb, 1.27×10^{-3} mb, and 1.05×10^{-3} mb respectively; the phases correspond to maxima at 0862 LT, 0844 LT, and 0817 LT. The particular choice of temperature profile does, however, make a difference at higher altitudes. In Figure 6 we see the amplitude of the westerly velocity over the equator for the various choices to T_0. For realistic T_0's the $\Theta_2^{2\omega,2}$ mode is evanescent between about 50 and 80 km, and the lower the mesopause temperature the less the response above 80 km. The variation of semidiurnal tidal fields with latitude is indicated in Figures 7 and 8 where the amplitude and phase distributions of the northerly velocity at different latitudes are shown. Note the increasing importance of higher order modes at high altitudes and latitudes. These modes do not become evanescent in the mesosphere.

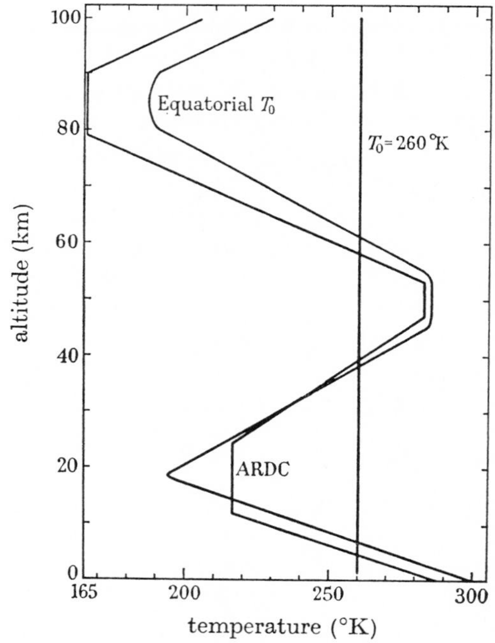

FIGURE 5. Different basic temperature profiles used in examining the semidiurnal thermal tide. After Lindzen [5]; Minzner, Champion and Pond [8].

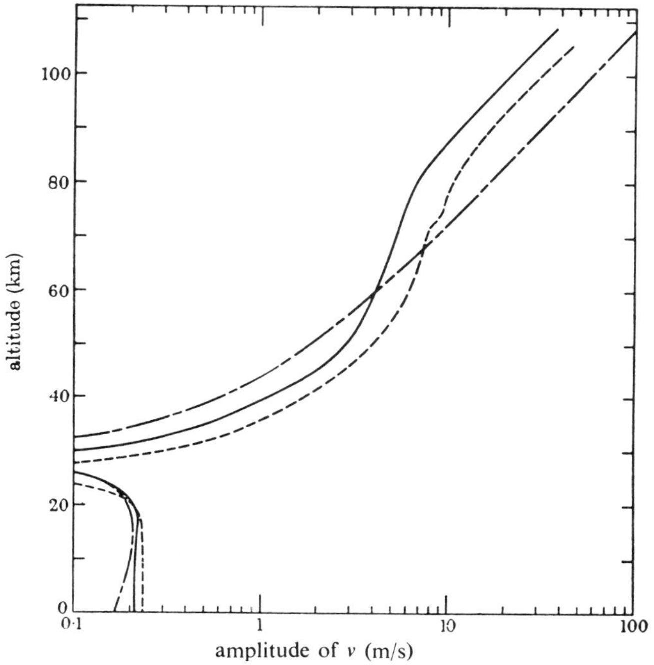

FIGURE 6. Amplitude of the solar semidiurnal component of v over the equator for different basic temperature profiles; ——, ARDC; ---, equatorial; —— – ——, isothermal ($T_0 = 260$ K). After Lindzen [5].

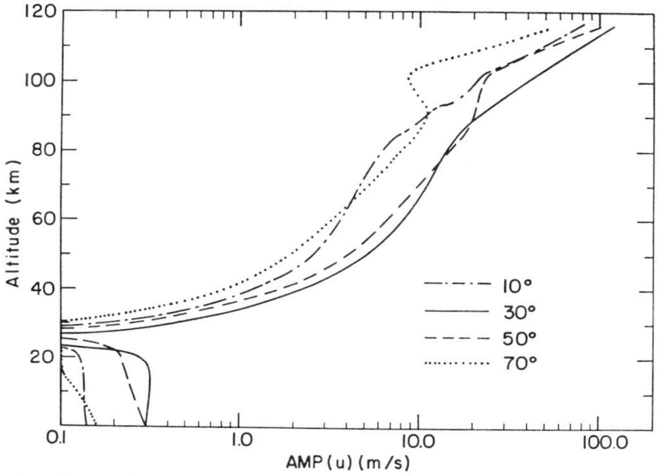

FIGURE 7. Amplitude of the solar semidiurnal component of u at various latitudes; equatorial $T_0(z)$ assumed.

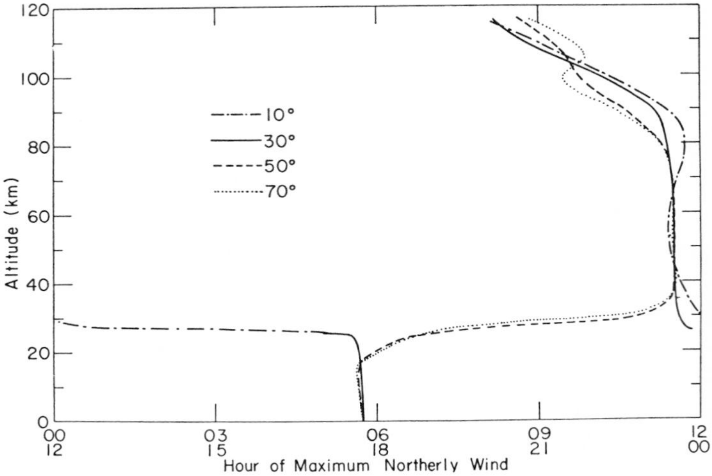

FIGURE 8. Phase (hour of maximum) of the solar semidiurnal component of u at various latitudes; equatorial $T_0(z)$ assumed.

Reprinted with permission of © D. Reidel Publ. Co. [1].

IV.3. **Solar diurnal thermal tide.** The first five symmetric Hough Functions (as well as the main antisymmetric Hough Function) for the migrating solar diurnal tide are shown in Figure 9. The corresponding expansion functions for the northerly and westerly velocity are shown in Figures 10 and 11. The equivalent depths are $h_1^{\omega,1} = 0.6909$ km, $h_3^{\omega,1} = 0.1203$ km, $h_5^{\omega,1} = 0.0484$ km, ..., $h_{-2}^{\omega,1} = -12.2703$ km, $h_{-4}^{\omega,1} = -1.7581$ km; for the main antisymmetric mode $h \approx \infty$.[3] The existence of negative equivalent depths has already been discussed in Part II. As we see in Figures 9–11, modes with negative equivalent depths have their amplitudes mostly polewards of $\theta = (60°, 120°)$ while modes with positive equivalent depths are confined primarily equatorwards of these colatitudes.

The expansions of the thermal excitations latitude dependences in terms of diurnal Hough Functions are

(7)
$$g_{O_3}^{\omega,1} = 1.6308°\text{K}\,\Theta_{-2}^{\omega,1} - 0.5128°\text{K}\,\Theta_{-4}^{\omega,1} + \cdots + 0.5447°\text{K}\,\Theta_1^{\omega,1} - 0.1411°\text{K}\,\Theta_3^{\omega,1} + 0.0723°\text{K}\,\Theta_5^{\omega,1}\cdots,$$

(8)
$$g_{H_2O}^{\omega,1} = 0.157°\text{K}\,\Theta_{-2}^{\omega,1} - 0.055°\text{K}\,\Theta_{-4}^{\omega,1} + \cdots + 0.062°\text{K}\,\Theta_1^{\omega,1} - 0.016°\text{K}\,\Theta_3^{\omega,1} + 0.008°\text{K}\,\Theta_5^{\omega,1}\cdots.$$

[3] These numbers are taken from [3].

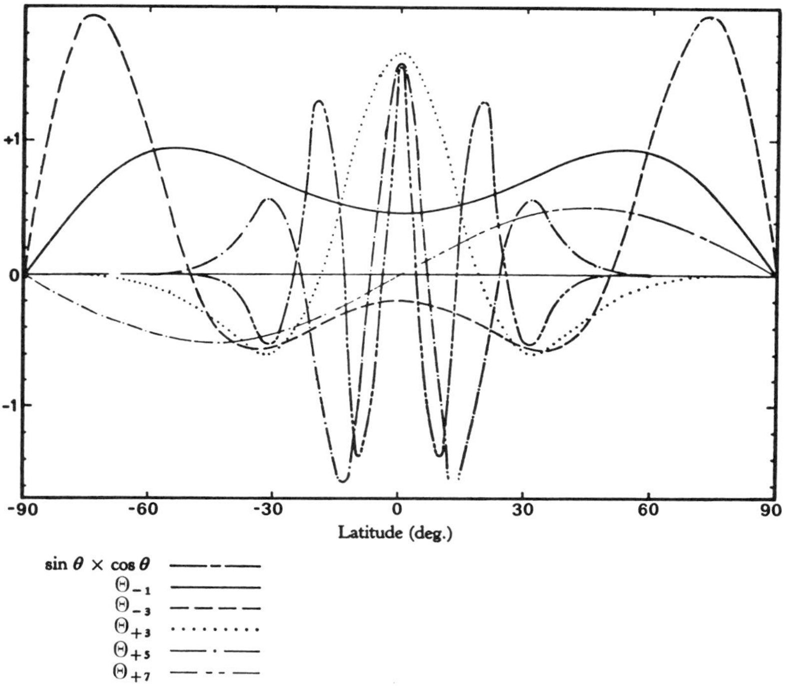

FIGURE 9. Symmetric Hough functions for the migrating solar diurnal thermal tide. Also shown is $\sin\theta\cos\theta$, the most important odd mode. After Lindzen [**4**].

Reprinted with permission of © D. Reidel Publ. Co. [**1**].

Comparing Figures 9 and 1, we see that no diurnal Hough Function is particularly like the latitude distribution of excitation—though $\Theta_{-2}^{\omega,1}$ comes the closest. Indeed, we see from Equations (7) and (8) that the $\Theta_{-2}^{\omega,1}$ mode does receive most of the excitation though not to so relatively great an extent as does the $\Theta_2^{2\omega,2}$ semidiurnal mode. Moreover, since the $\Theta_{-2}^{\omega,1}$ mode cannot propagate vertically, it is a relatively poor responder with respect to surface pressure. This is seen in Equations (9) and (10)

$$(9)\quad \begin{aligned}\delta p_{\mathrm{H_2O}}(0) = \{&137\,\Theta_{-2}^{\omega,1} - 68.2\,\Theta_{-4}^{\omega,1} + \cdots + 117\,e^{56^\circ i}\,\Theta_1^{\omega,1}\\ &- 13\,e^{73.3^\circ i}\,\Theta_3^{\omega,1} + 4.11\,e^{80.5^\circ i}\,\Theta_5^{\omega,1}\cdots\}\exp(i(\omega t + \varphi))\mu\mathrm{b},\end{aligned}$$

$$(10)\quad \begin{aligned}\delta p_{\mathrm{O_3}}(0) = \{&44.1\,\Theta_{-2}^{\omega,1} - 3.4\,\Theta_{-4}^{\omega,1} + \cdots + 94.1\,e^{12.75^\circ i}\,\Theta_1^{\omega,1}\\ &- 3.75\,e^{16.1^\circ i}\,\Theta_3^{\omega,1} + 0.754\,e^{-6.6^\circ i}\,\Theta_5^{\omega,1}\cdots\}\exp(i(\omega t + \varphi))\mu\mathrm{b},\end{aligned}$$

where we see the calculated surface pressure oscillations due to water vapor and ozone excitation [**4**]. We see that $\Theta_{-2}^{\omega,1}$ no longer dominates;

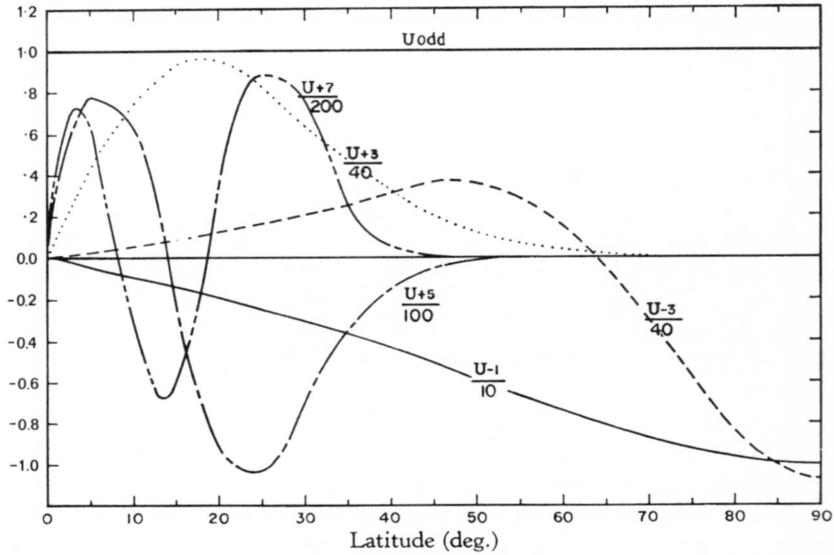

FIGURE 10. The expansion functions for the latitude dependence of the solar diurnal component of u, the northerly velocity. The functions have been divided by the amounts shown. After Lindzen [**4**].

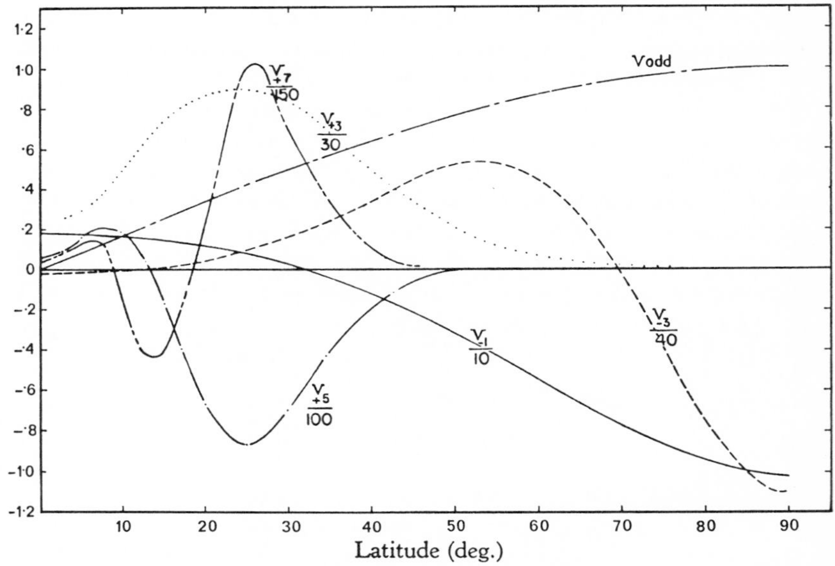

FIGURE 11. The expansion functions for the latitude dependence of the solar diurnal component of v, the westerly velocity. The functions have been divided by the amounts shown. After Lindzen [**4**].

$\Theta_1^{\omega,1}$ is just as important. It is interesting to note, however, that even for the propagating modes, H_2O is the largest contributor to the surface pressure oscillation. This is because the relatively thick O_3 source region together with the relatively short wavelengths of the propagating modes (approximately 25 km, 12 km and 7 km for the modes considered here) leads to some destructive interference.

The above discussion explains why the diurnal surface pressure oscillation is both weak and irregular. The former is due to inefficiency of the diurnal modes because of either trapping or interference; the latter is due to the fact that the response consists in several modes, each of which is relatively sensitive to local variations in excitation, temperature, etc.

The dependence of diurnal oscillations on latitude is particularly interesting. In Figures 12 and 13 we see calculated distributions for the amplitude and phase of the diurnal oscillation in northerly wind at different latitudes. Within regions of excitation, there are comparable amplitudes at all latitudes, but above the excitation there is continued $e^{x/2}$ growth at low latitudes and decay at high latitudes. Similarly, there is phase propagation at low latitudes, and almost none at high latitudes.

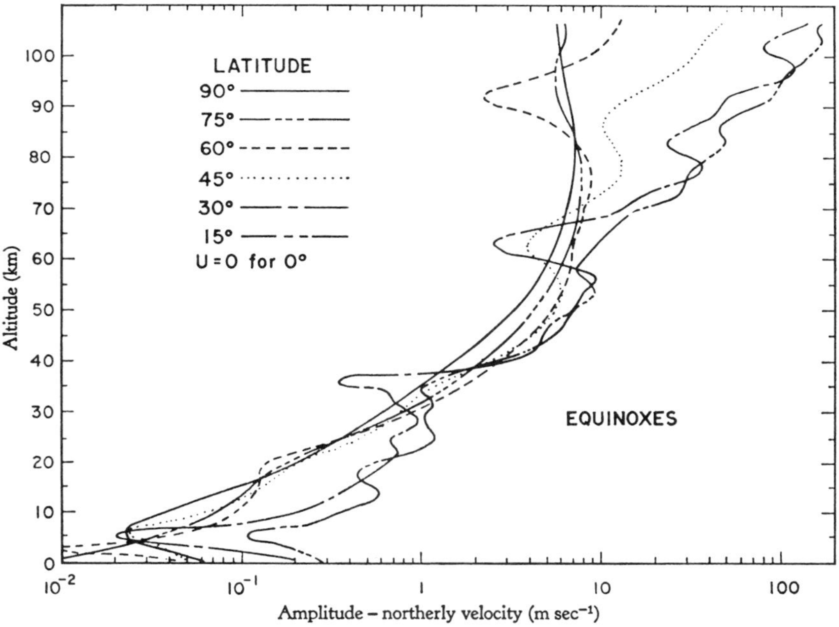

FIGURE 12. Altitude distribution of the amplitude of the solar diurnal component of u at 15° intervals of latitude; isothermal basic state assumed. After Lindzen [**4**].

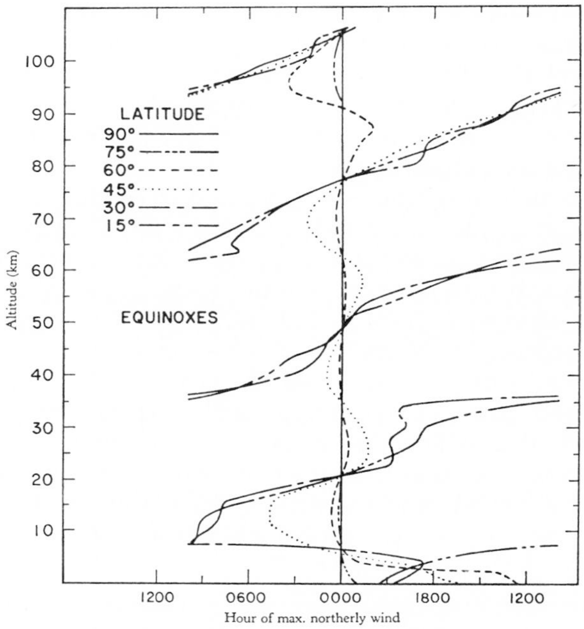

FIGURE 13. Altitude distribution of the phase of the solar diurnal component of u at 15° intervals of latitude; isothermal basic state assumed. After Lindzen [**4**].

Of particular importance to upper atmosphere dynamics at low latitudes is the fact that the diurnal propagating modes are not subject to trapping in the mesosphere.

The above results were calculated for an isothermal basic temperature, but they do not change in overall aspect for more realistic atmospheres [**5**].

IV.4. **Lunar semidiurnal tide.** The Hough Functions for the lunar semidiurnal tide are very similar to those for the solar semidiurnal tide; the equivalent depth of the main lunar mode is 7.07 km. Lunar excitation is gravitational; it is given by

$$\Omega = (-23{,}662\,\Theta_2 - 5{,}615\,\Theta_4 - 2{,}603\,\Theta_6 - \cdots) \cdot \cos(2(\sigma_2^L t + \varphi))\ \mathrm{cm}^2/\mathrm{sec}^2. \tag{11}$$

As we have already mentioned, lunar atmospheric tides are much smaller than solar tides. However, their theoretical behavior is, at least, pedagogically interesting. As may be recalled from Part I the amplitude of the lunar semidiurnal surface pressure oscillation is about 60–70μb with a maximum occurring about an hour after transit. If we calculate the response of an isothermal atmosphere to (11) we get

$$\delta p_2^{2\sigma_2^L,2}\,(\text{surface}) \approx \frac{34.17\ \mu\text{b}\exp(i(2(\sigma_2^L t + \varphi) + 90°))}{((H/h_2 - \tfrac{1}{2}) + i(\kappa H/h_2 - \tfrac{1}{4})^{1/2})}. \tag{12}$$

(It will prove a simple but useful exercise for the reader to derive (12).) For $H \sim 7$ km, (12) gives results close to the observed ones. However, what happens when one considers more realistic temperature profiles is somewhat surprising. In Figure 14 we see a number of temperature profiles differing only in their stratopause temperatures. Sawada [6] computed the lunar semidiurnal oscillation for each of these profiles. His results are shown in Figure 15.

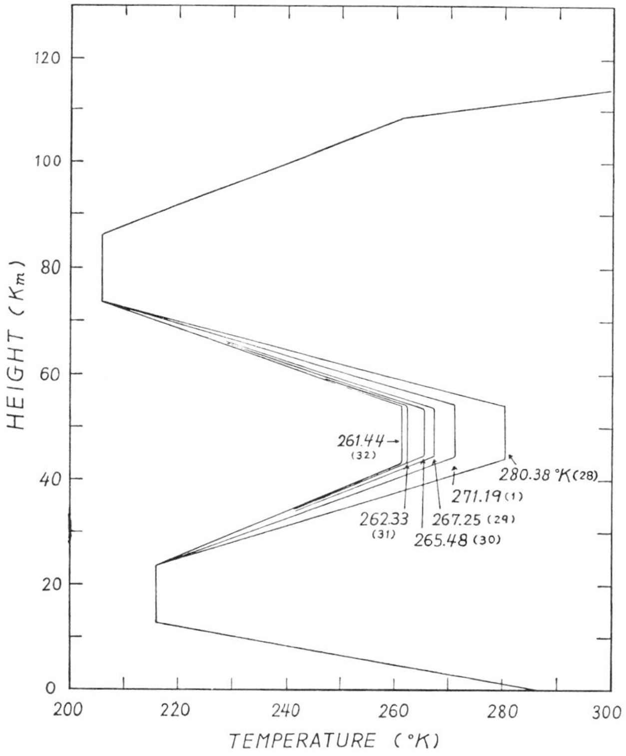

FIGURE 14. Various temperature profiles used in calculating the lunar semidiurnal surface pressure oscillation. The maximum temperature of the ozonosphere and a profile number are shown for each of the profiles. After Sawada [6].

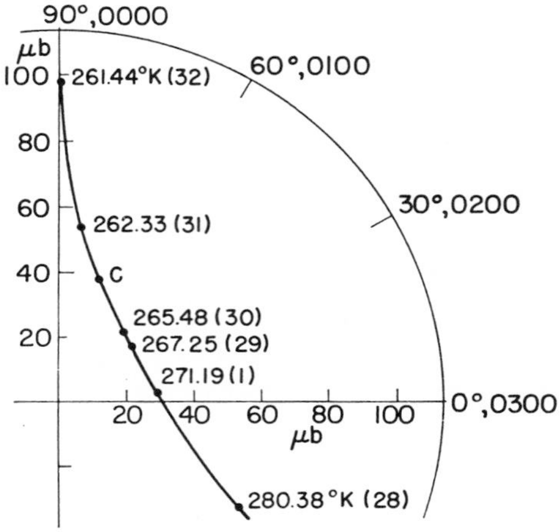

FIGURE 15. A harmonic dial for the lunar semidiurnal surface pressure oscillation. Amplitude and phase are shown as functions of the basic temperature profile. After Sawada [**6**].

Reprinted with permission of © D. Reidel Publ. Co. [1].

We see that the amplitude and phase vary significantly with small changes in the basic temperature—in distinct contrast to the solar semidiurnal oscillation. The Hough Functions for these two oscillations are so similar that the different behavior can only result from the different natures of the excitations. The gravitational excitation behaves as though it were a coherent source at the ground while the thermal excitation is distributed throughout the atmosphere. Apparently, small changes in the distribution of T_0 change the effective height of levels where semi-diurnal tides are partially reflected and the reflectivities. Repeated reflections produce significant interference for coherent gravitational excitation; for distributed sources these effects tend to cancel each other.

IV.5. **Comparisons with data.** Agreement between the above calculations and observations appear to be fairly good for the solar diurnal and semidiurnal surface pressure oscillations. The comparison for the former is shown in Figure 16. Regions of significant disagreement are also associated with low station density. For the semidiurnal surface pressure oscillation there is a consistent phase error of about one hour.

There is good agreement between observations of the lunar semidiurnal surface pressure oscillation and calculations for an isothermal atmosphere. However, the calculated variability of the surface oscillation with small changes in T_0 does not appear in the data.

It is difficult to compare theory and observations between the ground and 30 km because of orographic effects, but between 40 and 90 km

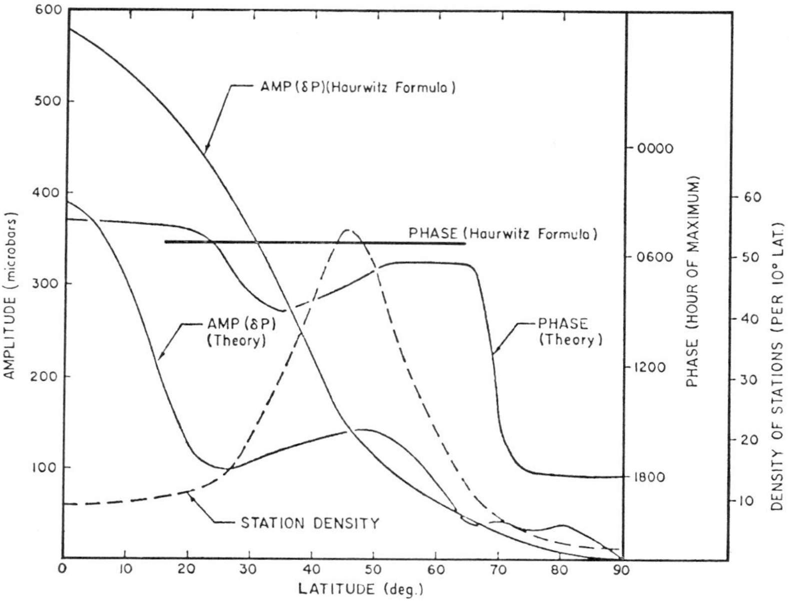

FIGURE 16. Calculated amplitude and phase of the solar diurnal surface pressure oscillation. These quantities are also shown as derived from the Haurwitz (1965) empirical formula. The distribution of stations on which his formula is based are shown. After Lindzen [**4**].

Reprinted with permission of © D. Reidel Publ. Co. [**1**].

there is again good agreement for solar oscillations. Lunar tides tend to be buried in noise.

Above 90 km what little data there is suggests significant disagreement with the above theory.

REFERENCES, PART IV

1. S. Chapman and R. S. Lindzen, *Atmospheric tides*, D. Reidel Publ., Dordrecht, Holland, and Gordon and Breach/Science Publ., New York, 1970.

2. R. S. Lindzen and Donna Blake, *Internal gravity waves in atmospheres with realistic dissipation and temperature*; *Part* II, *Thermal tides excited below the mesopause*, Geophys. Fluid Dyn. **2** (1971), 31–61.

3. T. W. Flattery, Technical Report #21, Dept. of Geophysical Sciences, University of Chicago, Chicago, Ill., 1967.

4. R. S. Lindzen, *Thermally driven diurnal tide in the atmosphere*, Quart. J. Roy. Meteorol. Soc. **93** (1967), 18–42.

5. ———, *The application of classical atmospheric tidal theory*, Proc. Roy. Soc. Ser. A **303** (1968), 299–316.

6. R. Sawada, *The atmospheric lunar tides and the temperature profile in the upper atmosphere*, Geophys. Mag. **27** (1956), 213–236.

7. B. Haurwitz, *The diurnal surface pressure oscillation*, Archiv. Meteorol. Geophys. Biokl. **A14** (1965), 361–379.

8. R. A. Minzer, K. S. W. Champion and H. L. Pond, *The ARDC Modal atmosphere*, Air Force Surveys in Geophysics, no. 115, 1959.

V. Further Developments

In the preceding parts we have described atmospheric tides as observed, as well as a mathematical model for atmospheric tides whose consequences have served to explain many aspects of the observed phenomena. The mathematical model incorporated a number of assumptions and approximations, whose consequences I would like to briefly explore. It might be mathematically soothing to have treated the assumptions and approximations as the asymptotic limits of expansions in specified parameters. However, such a procedure would be exceedingly complicated and, in all likelihood, illusory as far as rigour goes. As a geophysicist, it is my preference to reexamine my assumptions and approximations at least partly in the light of discrepancies between calculations and observations.

V.1. **Neglect of mean winds and horizontal temperature gradients.** As we saw in Part II our equations become immensely complicated when this approximation is dropped. Even if we were to restrict ourselves to steady zonal winds and north-south temperature gradients, we would already lose separability in z and θ dependence. In general, no comprehensive study of the effects of mean winds and temperature gradients has been made. As a rule, both must be studied together since mean winds and temperatures in the atmosphere are related (to a fair degree of approximation) by the thermal wind equation

$$(1) \qquad (\partial/\partial z + 1/H)(\rho u) = -(g\rho/afT)(\partial T/\partial \theta).$$

Limited aspects of the problem have been investigated, [**1**], [**2**], [**3**], but the only useful result appears to be that these approximations will not significantly change the resonance properties of the atmosphere [**4**]. The behavior of internal gravity waves in shear flows for both nonrotating [**5**] and rotating atmospheres [**6**], [**7**], [**8**] has been investigated in substantial detail. The latter results all show that mean flows affect internal waves to the degree that they doppler shift the wave's frequency. This must undoubtedly apply to tides as well. Indeed the recognition that tides are simply rotationally modified internal gravity waves permits us without further mathematical analysis to prescribe two necessary (though not necessarily sufficient) conditions for the neglect of mean winds and latitude variations of mean temperature:

(i) Latitude variations in temperature must be sufficiently small so that the variation of vertical wave number (i.e. $H^{-2}(h_n^{-1}(\kappa H + dH/dX) - \frac{1}{4})$; viz. Equation II. 34) be small compared to the vertical wavenumber itself for the mode in question.

(ii) Mean zonal winds must be small compared to the zonal phase speed of a tide. If U is a characteristic zonal speed, then

(2) $$U \ll c = \frac{2\pi \times \text{earth's radius}}{1 \text{ solar day}}$$

where $c \sim 450$ m/s.

Condition (ii) is certainly satisfied on earth. Condition (i) is also fairly well observed. That it is not perfectly satisfied is indicated by the observation of seasonal variations in tides (vis. Part I) which cannot be accounted for by seasonal variations in excitation alone.[1]

V.2. **Neglect of dissipation.** The consideration of this approximation is not only of intrinsic importance, but is also a necessary prerequisite to the consideration of nonlinearity. As we see from Equations (II. 34–43) the amplitudes of inviscid, adiabatic wave solutions grow as $e^{x/2}$. Thus we might expect eventual nonlinearity. However, as we shall show, the inclusion of viscosity and thermal conductivity in the linear problem suppresses this growth at sufficiently great heights. Hence, the consideration of nonlinear effects must be deferred until the linear dissipative solutions have been obtained.

It proves useful to begin with a qualitative discussion of an arbitrary dissipative mechanism whose time scale, τ_{diss}, may be specified. If a given tidal mode has a period, τ_{tide}, then two distinct situations may exist with respect to dissipation:

(i) If $\tau_{\text{diss}} < \tau_{\text{tide}}/2\pi$ then the tidal dynamics are fundamentally altered since dissipation is more important than inertia.

(ii) The presence of any dissipation will tend to reduce the $e^{x/2}$ growth of propagating modes. If, moreover,

(3) $$\frac{\tau_{\text{tide}}}{2\pi} < \tau_{\text{diss}} < \frac{\tau_{\text{tide}}}{(L/2H)}$$

(where L is the vertical wavelength of the tidal mode), then the vertical wavelength of the tide will be relatively unaffected, but the $e^{x/2}$ growth above the region of excitation will be replaced by decay of amplitude.

Condition (i) arises in the atmosphere from molecular viscosity and conductivity whose effectiveness increases as $1/\rho$ and possibly from hydromagnetic drag. For the most important tidal modes molecular effects assume importance above 100 km while hydromagnetic drag becomes important primarily above 200 km. Below 100 km infrared cooling is of moderate importance—leading to small reductions of amplitude ($\lesssim 20\,\%$) but not to decay with height.

[1] Calculation of the effects of seasonal variations in excitation may be found in [**9**], [**10**].

Having discussed dissipation qualitatively the problem remains as to how to deal with it quantitatively. If we return to Part II we will see that separation of variables resulted from the possibility of expressing the horizontal divergence of velocity as an operator on p/ρ_0 which depended only on latitude; and on the fact that the equations of state, hydrostatic pressure, continuity and energy depended on θ only through their dependence on velocity divergence. The inclusion of almost any meaningful model for friction will invalidate the first condition and eliminate the possibility of separating variables. On the other hand any model of thermal dissipation which may be expressed as an operator on temperature involving only altitude will simply complicate the vertical structure equation [**11**], [**12**]; Hough Functions will remain the appropriate latitude expansion functions. Basically, the presence of rotation leads to this state of affairs; friction leads to modified Coriolis torques with the consequence that the horizontal structure of modes changes with height. This does not necessarily happen with thermal dissipation. Thus, reasonable calculations of the effect of infrared cooling have been made [**13**]; however, the proper study of the effects of viscosity on tides remains to be made. In general, we are unable to analytically solve nonseparable equations, and the numerical solution would be exceptionally cumbersome (with the inclusion of conductivity, viscosity and anisotropic hydromagnetic drag we would have an eighth order equation, and we would need very high resolution).

Once again, however, we can begin to develop some intuition by studying internal gravity waves in a planar nonrotating fluid. Certainly, not all the effect of viscosity is to alter horizontal structure. Moreover, as we saw in Part II, there always exist internal gravity waves in planar nonrotating fluid whose frequency and inviscid vertical wavenumber are identical to those of tidal modes (negative equivalent depths are modelled by imaginary north-south wavenumbers). Before discussing studies of the effects of dissipation on internal gravity waves, I would like to point out that there are two important points at which inviscid, adiabatic wave theory may be affected by dissipation.

(i) In the upper atmosphere where molecular viscosity and conductivity assume dominant importance our solutions will be *locally* very different.

(ii) In adopting the radiation condition it was assumed that dissipation in the upper atmosphere would absorb all upcoming energy; it will be shown that for very large vertical wavelengths dissipation increasing in effectiveness as $1/\rho$ can actually cause partial reflection—thus affecting the solution at all levels.

Perturbation studies of the effects of dissipation had been made by Pitteway and Hines [**14**]. However, Yanowitch [**15**] was the first to make a full study of the effects of viscosity increasing as $1/\rho$ in a very difficult analytic study wherein thermal conductivity and compressibility (but not stratification) were neglected. Time would not permit reproducing Yanowitch's analysis. Fortunately, however, I was able to show that Yanowitch's most important results depended on the presence of a dissipative process whose effectiveness increased as $1/\rho$—and not on the particular dissipative process. In particular I showed that the inclusion of Newtonian cooling with a rate coefficient proportional to $1/\rho$ (a completely unrealistic process) led to Yanowitch's main results with much simpler mathematics [**16**]. It is, I feel, worth going over this problem not only for its results, but also because it shows with considerable clarity the relative advantages of analytical and numerical approaches.

V.2.a. **Tides (or internal gravity waves) in an atmosphere with Newtonian cooling inversely proportional to density.** The inclusion of Newtonian cooling changes Equation (II. 18) to the following

(4) $$\frac{dT}{dt} = \frac{gH}{\rho_0}\frac{1}{c_v}\frac{d\rho}{dt} - a(z)\delta T,$$

where $a(z)$ = Newtonian cooling rate coefficient, and where we have omitted thermal excitation. We will simply assume the tide is excited at some lower boundary.

As already mentioned, our latitude structure functions are still determined by Laplace's Tidal Equation, but our vertical structure equation becomes (assuming an isothermal basic state)

(5) $$\frac{d^2y_n}{dx^2} + \left\{\frac{\kappa H}{h_n}\left(1 + \frac{a}{i\sigma\gamma}\right)^{-1} - \frac{1}{4}\right\}y_n = 0.$$

We are assuming $a\alpha(1/\rho_0)$ and since $\rho_0 = \rho_0(0)e^{-x}$, we may write

(6) $$\chi \equiv (a/\sigma\gamma) = \varepsilon e^x$$

where we wish ε to be $\ll 1$ in order that dissipation be unimportant locally for $x \sim O(1)$. Also, we will restrict ourselves to vertically propagating waves for which $\kappa H/h_n > \frac{1}{4}$.

If we let $x \to \infty$ (5) approaches

(7) $$d^2y_n/dx^2 - y_n/4 = 0$$

and our solutions approach

(8) $$y_n \sim A_1e^{-x/2} + B_1e^{x/2}.$$

Requiring boundedness as $x \to \infty$ implies $B_1 = 0$. For x small, (5) assumes its adiabatic form and has solutions

$$(9) \qquad y_n \sim A_2 \exp(i\lambda x) + B_2 \exp(-i\lambda x)$$

where $\lambda = (\kappa H/h_n - \frac{1}{4})^{1/2}$. The radiation condition would imply $B_2 = 0$. Whether $B_1 = 0$ implies that $B_2 = 0$ will be seen in what follows.

In order to solve (5) exactly, the following changes of variables prove useful:

$$\mu = i\chi,$$
$$\varphi = \mu^{-i\lambda} y,$$
$$a = i\lambda + \tfrac{1}{2},$$
$$b = i\lambda - \tfrac{1}{2},$$
$$c = 2a.$$

(5) becomes

$$(10) \qquad \mu(1 - \mu)\varphi'' + [c - (1 + a + b)\mu]\varphi' - ab\varphi = 0$$

which is simply the hypergeometric equation. The primes refer to differentiation with respect to μ. Since we know our solution for large values of x (and hence μ)—Equation (8) with $B_1 = 0$—it will be convenient to focus on large values of μ which is facilitated by the following further change of variables:

$$\Theta = (-\mu)^a \varphi,$$
$$\xi = 1/\mu,$$
$$a' = a = i\lambda + \tfrac{1}{2},$$
$$b' = a + 1 - c = -i\lambda + \tfrac{1}{2},$$
$$c' = a + 1 - b = 2.$$

(10) becomes

$$(11) \qquad \xi(1 - \xi)\Theta'' + [c' - (1 + a' + b')\xi]\Theta' - a'b'\Theta = 0$$

where primes now refer to ξ-derivatives. (11) is also the hypergeometric equation. One of its solutions is

$$(12) \qquad \Theta_1 = F(a', b', c'; \xi),$$

where F is the hypergeometric function. Since $c' = 2$ the second solution is

$$(13) \qquad \Theta_2 = \Theta_1 \ln \xi - \frac{1}{(a' - 1)(b' - 1)} \frac{1}{\xi} - \sum_{k=1}^{\infty} B_k \xi^{\mu},$$

where

$$B_k = \frac{(a')_k(b')_k}{(c')_k k!}\left[\frac{1}{a'} + \frac{1}{a'+1} + \cdots + \frac{1}{a'+k+1} + \frac{1}{b'} + \cdots + \frac{1}{b'+k+1} - 1 - \frac{1}{2} - \cdots - \frac{1}{k} - \frac{1}{2} - \frac{1}{3} - \cdots - \frac{1}{1-k}\right],$$

and

$$(v)_k = v(v+1)(v+2)\cdots(v+k-1) \quad \text{for any } v.$$

Thus, the solution of (10) for $|\mu| > 1$ is

$$\varphi = A(-\mu)^{-a}(\Theta_1 + B\Theta_2). \tag{14}$$

Now, it may be shown that Θ_1 corresponds to the solution we want as $x \to \infty$, while Θ_2 leads to the unbounded solution. Hence, $B = 0$ and

$$\varphi = A(-\mu)^{-a}F(a', b', c'; 1/\mu) \tag{15}$$

for $|\mu| > 1$. The analytic continuation of (15) for $|\mu| < 1$ may be taken directly from Erdelyi, et al. [**17**]:

$$\varphi = A\left\{\frac{\Gamma(1-c)\Gamma(a+1-b)}{\Gamma(1-b)\Gamma(a+1-c)}u_1 - \frac{\Gamma(c)\Gamma(1-c)\Gamma(a+1-b)}{\Gamma(2-c)\Gamma(c-b)\Gamma(a)}\exp[i\pi(c-1)]u_5\right\} \tag{16}$$

where

$$u_1 = F(a, b, c; \mu),$$

$$u_5 = (\mu)^{1-c}F(a+1-c, b+1-c, 2-c; \mu)$$

and $\Gamma(v)$ is the gamma function of v. Now,

$$u_1 \to 1 \tag{17}$$

and

$$u_5 \to \mu^{1-c} \tag{18}$$

as $|\mu| \to 0$. If we let

$$y_1 \equiv \mu^{i\lambda}u_1 \tag{19}$$

and

$$y_2 \equiv \mu^{i\lambda}u_5 \tag{20}$$

then

$$y_1 \xrightarrow[|\mu|\to 0]{} \mu^{i\lambda} = (i\varepsilon)^{i\lambda} e^{i\lambda x} \tag{21}$$

and

$$y_2 \xrightarrow[|\mu|\to 0]{} \mu^{i\lambda}\mu^{1-c} = (i\varepsilon)^{-i\lambda} e^{-i\lambda x}. \tag{22}$$

Thus (21) and (22) are seen to correspond to the two terms in (9), and the ratio B_2/A_2 is given by

$$\frac{B_2}{A_2} = -\frac{\Gamma(c)\Gamma(1-b)\Gamma(a+1-c)}{\Gamma(2-c)\Gamma(c-b)\Gamma(a)}(i\varepsilon)^{-2i\lambda}\exp(i\pi(c-1)). \tag{23}$$

Our asymptotics assumed $|\arg \mu| < \pi$; therefore

$$i\varepsilon = \varepsilon e^{\pi i/2}.$$

Also, we shall define

$$\varepsilon = e^{-x_0}.$$

Thus

$$(i\varepsilon)^{-2i\lambda} = \exp(2i\lambda x_0)\exp(\pi\lambda)$$

and

$$\exp[i\pi(c-1)] = \exp(-2\pi\lambda).$$

We may now use the fact that $\Gamma(v^*) = \Gamma^*(v)$ (where * refers to complex conjugate) and the definitions of a, b and c to show

$$\Gamma(c)/\Gamma(2-c) = \exp(iv_1), \qquad \Gamma(1-b)/\Gamma(c-b) = \exp(iv_2)$$

and

$$\Gamma(a+1-c)/\Gamma(a) = \exp(iv_3)$$

where v_1, v_2 and v_3 are real constants. Finally we obtain

$$B_2/A_2 = -\exp[i(v_1 + v_2 + v_3 + 2\lambda x_0)]\exp(-\pi\lambda). \tag{24}$$

Now

$$|B_2/A_2| = \exp(-\pi\lambda) \tag{25}$$

or

$$|B_2/A_2| = \exp(-2\pi^2 H/L) \tag{25a}$$

where

$$L = \text{adiabatic vertical wavelength.}$$

(25) is exactly the result obtained by Yanowitch; it also describes numerical results obtained for realistic physical systems.

We now wish to inquire as to the form of our solutions. We have at least two choices. The first is simply to evaluate our analytic solution as given by (15) and (16). This turns out to be a substantially more cumbersome task than the numerical integration of (5) using the method given in Part III. In Figures 1 and 2 we show the results of such numerical integrations (β in these figures corresponds to λ above). They may be summarized as follows:

(i) For $0 < \lambda \gtrsim 0.3$ there is an amplitude growth as $e^{x/2}$ up to $x \sim x_0$; however, there is also significant reflection from this level. Above $x \sim x_0$ amplitudes are constant and there is no phase variation.

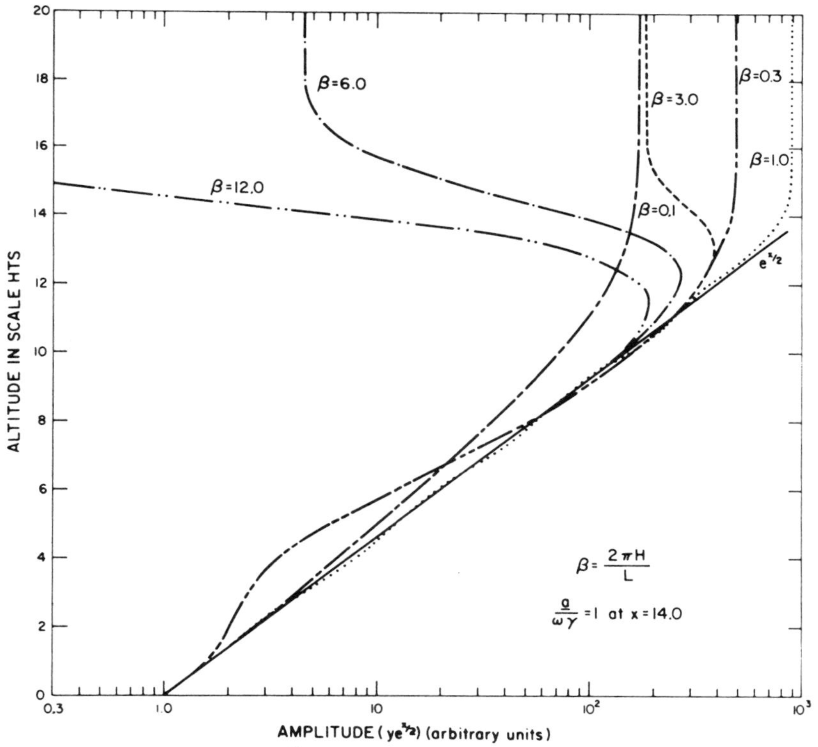

FIGURE 1. The amplitude of $ye^{x/2}$ as a function of height (in scale heights) for various values of β. $\beta = 2\pi H/L$ where L is the vertical wavelength of a wave in the absence of Newtonian cooling, and $H = RT_0/gM$ = scale height. Newtonian cooling is comparable with other processes at $x = 14$ where $a/\omega\gamma = 1$. Also shown is the variation of $ye^{x/2}$ with height in the absence of Newtonian cooling, namely $e^{x/2}$.

Reproduced by permission of the National Research Council of Canada from the Canadian Journal of Physics, **46**, pp. 1842–1843 (1968).

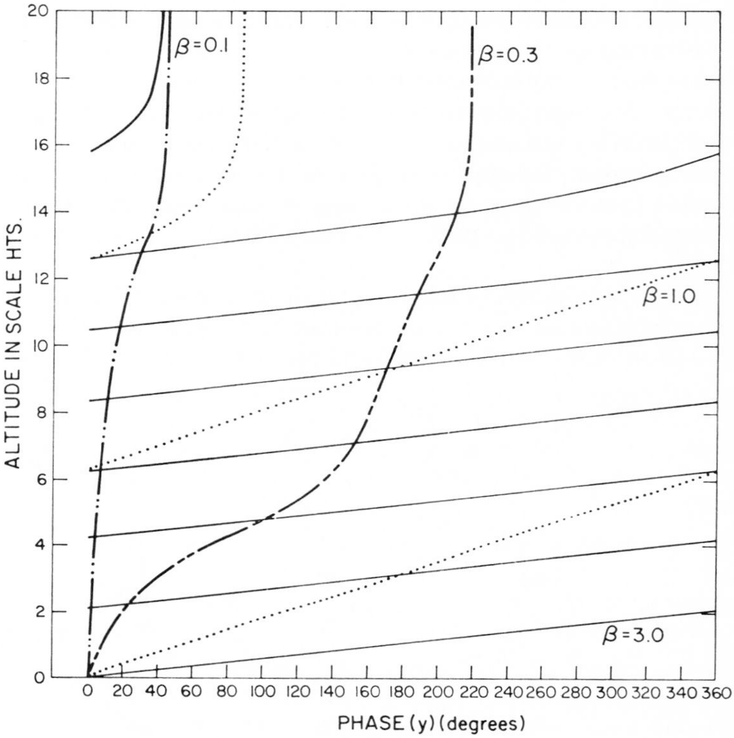

FIGURE 2. The phase of y as a function of height (in scale heights) for various values of β. See caption for Figure 1 for further details.

Reproduced by permission of the National Research Council of Canada from the Canadian Journal of Physics, **46**, *pp.* 1842–1843 (1968).

(ii) For $0.3 \lesssim \lambda \lesssim 2.0$ amplitudes grow as $e^{x/2}$ up to $x \sim x_0$. There is negligible reflection, and above $x \sim x_0$ amplitudes and phases are constant.

(iii) For $\lambda > 2.0$ amplitudes grow as $e^{x/2}$ until some height $x < x_0$ which decreases as λ increases. Above this height amplitudes decrease sharply with altitude eventually asymptoting to a constant at some height $x > x_0$. Phase variations with height continues to $x \sim x_0$.

The above description is in agreement with our earlier qualitative discussion, Yanowitch's results, and the results of a recent numerical study of internal gravity waves in compressible atmospheres with realistic distributions of basic temperature, viscosity and thermal conductivity.[2]

[2] The numerical method is the analog of the method given in Part III for a system of four coupled second order equations [**18**], [**19**].

In the last mentioned study the variation of temperature meant that λ varied with height. It was the value of λ in the height range where dissipative time scales become comparable with the wave period that determined wave behavior. Results of this study for modes with the same frequencies and inviscid vertical structures as the main tidal modes are shown in Figures 3 and 4. A standard atmosphere with an 800°K exosphere was assumed. The results, to the extent that they are applicable to real tides, are of considerable practical importance. We see that energy excited in the

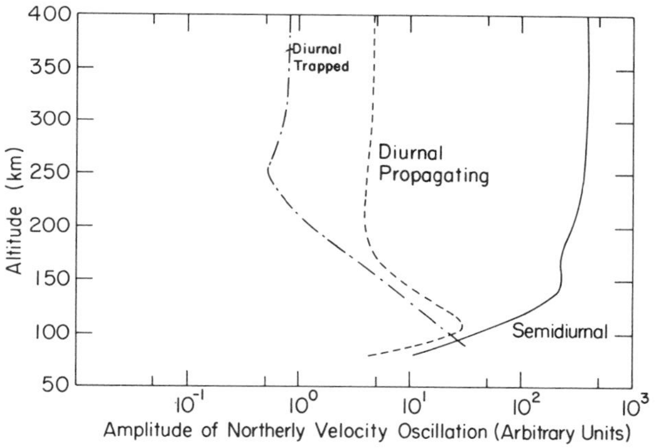

FIGURE 3. Theoretical altitude distribution of the amplitude of u for the main semidiurnal (——), the main propagating diurnal (– – –), and the main trapped diurnal (— — — –) modes on Earth. Molecular viscosity and thermal conductivity have been taken into account.

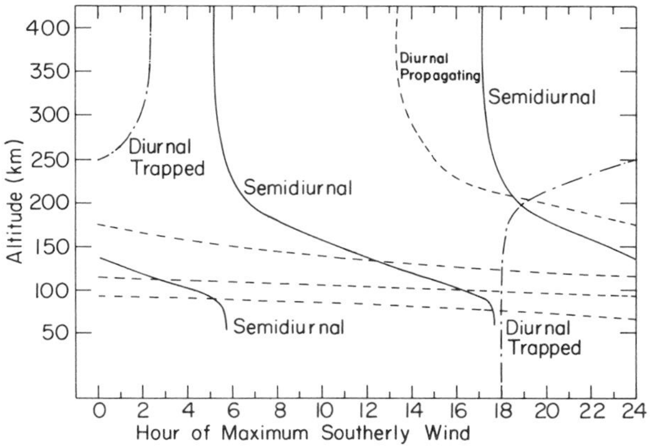

FIGURE 4. Same as Figure 3, except for the phases of the various modes.

diurnal trapped mode below 90 km will decay by 2 orders of magnitude before asymptoting to a constant in the thermosphere. The diurnal propagating mode continues to grow till about 108 km but decays sharply above this altitude. For these reasons as well as for some to be presented in the next section we conclude that the diurnal oscillation of the thermosphere must be excited in situ and not from below. The semidiurnal mode (for which $\lambda \sim O(1)$), on the other hand, shows $e^{x/2}$ growth till above 130 km and no amplitude decay above this height. This suggests that the thermosphere will have a strong semidiurnal oscillation ($\delta T \sim$ 200°K at the equator) excited by ozone and water vapor absorption. This has not been observed; however, it may be shown that satellite drag techniques strongly smooth semidiurnal oscillations.

Parenthetically, it should be remarked that both infrared cooling and turbulence seem unlikely to reduce amplitudes by more than 30% in the region below the heights where molecular processes become important.

V.3. **Nonlinear effects.** Had we approached this approximation by means of amplitude expansions for inviscid, adiabatic solutions we would have found that successive terms would have grown as $e^{x/2}$, e^{x}, $e^{3x/2}$, etc.—clearly an untenable situation. This difficulty can, in principle, be eliminated by expanding in terms of viscous, conducting solutions. The difficulty of this approach should be evident from the discussion of §V.2. A qualitative examination of the nature of likely nonlinear effects serves, moreover, to suggest that an amplitude expansion may often be fundamentally inappropriate.

As internal gravity waves, tides are basically transverse oscillations for which advective (i.e. $\boldsymbol{V} \cdot \nabla$) terms are very small. Thus the most important nonlinear terms will result from ρ/ρ_0 approaching (or exceeding) unity. For the main semidiurnal mode this quantity approaches ~ 0.5 before viscosity and conductivity put an end to amplitude growth. For the diurnal propagating mode, another amplitude dependent effect proves more important; namely the instability of the wave. It appears possible for still linearizable wave fields to have shears and/or temperature gradients which are unstable to perturbations that grow rapidly compared to the wave frequency. This seems likely to be true for most Richardson Number instabilities; i.e., if we write $T = T_0(z) + \delta T$ then

$$Ri = \frac{(g/T)(\partial T/\partial z + g/C_p)}{(\partial u'/\partial z)^2 + (\partial v'/\partial z)^2}$$

and when $Ri < \frac{1}{4}$ we have instability [**20**], [**13**]. This occurs for the propagating diurnal mode at 88 km over the equator and must continue until

108 km above which height dissipative processes cause amplitude decay. Amplitude expansions would be useless for this case.

V.4. **Orographic effects.** Orographic effects do not primarily affect migrating tides. Instead they generate additional oscillations of tidal periods which do not travel with the sun. As we saw in Part I such oscillations are important for tidal winds below 30 km. However, in these lectures we will restrict ourselves to migrating tides. The interested listener will find some further discussion of this matter in Lindzen [**21**] and in Chapman and Lindzen [**22**].

ACKNOWLEDGEMENT. The preparation of the above lectures was supported by Grant GA-1622 of the National Science Foundation.

REFERENCES, PART V

1. B. Haurwitz, *Atmospheric oscillations and meridional temperature gradient*, Beitrage Phys. Atmos. **30** (1957), 46–54.

2. W. C. Chiu, *On the oscillations of the atmosphere*, Arch. Meteorol. Geophys. Bioklimatol. Ser. A **5** (1952), 280–303. MR **14**, 1034.

3. R. Sawada, *The effect of zonal winds on the atmospheric lunar tide*, Arch. Meteorol. Geophys. Bioklimatol. Ser. A **15** (1966), 129–167.

4. L. A. Dikii, *Allowance for mean wind in calculating the frequencies of free atmospheric oscillations*, Izv. Atmos. Ocean. Phys., English ed., **4** (1967), 583–584.

5. J. R. Booker and F. P. Bretherton, *The critical layer for internal gravity waves in a shear flow*, J. Fluid Mech. **27** (1967), 513–539.

6. W. Jones, *Propagation of internal gravity waves in fluids with shear flow and rotation*, J. Fluid Mech. **30** (1967), 439–448.

7. R. S. Lindzen, *Internal equatorial planetary-scale waves in shear flow*, J. Atmospheric Sci. **27** (1970), 394–407.

8. J. R. Holton, *The influence of mean wind shear on the propagation of Kelvin waves*, Tellus **22** (1970), 186–193.

9. S. T. Butler and K. A. Small, *The excitation of atmospheric oscillations*, Proc. Roy. Soc. Ser. A **274** (1963), 91–121.

10. R. S. Lindzen, *Thermally driven diurnal tide in the atmosphere*, Quart. J. Roy. Meteorol. Soc. **93** (1967), 18–42.

11. R. S. Lindzen and D. J. McKenzie, *Tidal theory with Newtonian cooling*, Pure Appl. Geophys. **66** (1967), 90–96.

12. R. E. Dickinson and M. A. Geller, *A generalization of "tidal theory with Newtonian cooling"*, J. Atmospheric Sci. **25** (1968), 932–933.

13. R. S. Lindzen, *The application of classical atmospheric tidal theory*, Proc. Roy. Soc. Ser. A **303** (1968), 299–316.

14. M. L. V. Pitteway and C. O. Hines, *The viscous damping of atmospheric gravity waves*, Canad. J. Phys. **41** (1963), 1935.

15. M. Yanowitch, *Effect of viscosity on gravity waves and the upper boundary condition*, J. Fluid Mech. **29** (1967), 209–231.

16. R. S. Lindzen, *Vertically propagating waves in an atmosphere with Newtonian cooling inversely proportional to density*, Canad. J. Phys. **46** (1968), 1835–1840.

17. A. Erdelyi, M. Magnus, F. Oberhettinger and F. G. Tricomi, *Higher transcendental functions*, McGraw-Hill, New York. MR **15**, 419, **16**, 586.

18. R. S. Lindzen and H.-L. Kuo, *A reliable method for the numerical integration of a large class of ordinary and partial differential equations*, Mon. Wea. Rev. **97** (1969), 732–734.

19. R. S. Lindzen, *Internal gravity waves in atmospheres with realistic dissipation and temperature; Part* I, *Mathematical development and propagation of waves into the thermosphere*, Geophys. Fluid Dynamics **1** (1970), 303–355.

20. R. R. Hodges, Jr., *Generation of turbulence in the upper atmosphere by internal gravity waves*, J. Geophys. Res. **72** (1967), 3455–3458.

21. R. S. Lindzen, *The application and applicability of terrestrial atmospheric tidal theory to Venus and Mars*, J. Atmospheric Sci. **27** (1970), 536–549.

22. See [**28**] of References for Part I.

UNIVERSITY OF CHICAGO

Author Index

Roman numbers refer to pages on which a reference is made to an author or a work of an author.

Italic numbers refer to pages on which a complete reference to a work by the author is given.

Boldface numbers indicate the first page of the articles in the book.

Subject Index